NUMERICAL MATHEMATICS AND SCIENTIFIC COMPUTATION

Series Editors

NUMERICAL MATHEMATICS AND SCIENTIFIC COMPUTATION

Books in the series

Monographs marked with an asterix (*) appeared in the series 'Monographs in Numerical Analysis' which has been folded into, and is continued by, the current series.

* P. Dierckx: *Curve and surface fittings with splines*
* J. H. Wilkinson: *The algebraic eigenvalue problem*
* I. Duff, A. Erisman, and J. Reid: *Direct methods for sparse matrices*
* M. J. Baines: *Moving finite elements*
* J. D. Pryce: *Numerical solution of Sturm–Liouville problems*

K. Burrage: *Parallel and sequential methods for ordinary differential equations*
Y. Censor and S. A. Zenios: *Parallel optimization: theory, algorithms and applications.*
M. Ainsworth, J. Levesley, W. Light, and M. Marletta: *Wavelets, multilevel methods and elliptic PDEs*
W. Freeden, T. Gervens, and M. Schreiner: *Constructive approximation on the sphere: theory and applications to geomathematics*
Ch. Schwab: *p- and hp-finite element methods: theory and applications to solid and fluid mechanics*
J. W. Jerome: *Modelling and computation for applications in mathematics, science, and engineering*
Alfio Quarteroni and Alberto Valli: *Domain decomposition methods for partial differential equations*
G. E. Karniadakis and S. J. Sherwin: *Spectral/hp element methods for CFD*
I. Babuška and T. Strouboulis: *The finite element method and its reliability*
B. Mohammadi and O. Pironneau: *Applied shape optimization for fluids*
S. Succi: *The Lattice Boltzmann Equation for fluid dynamics and beyond*
P. Monk: *Finite element methods for Maxwell's equations*
A. Bellen & M. Zennaro: *Numerical methods for delay differential equations*
J. Modersitzki: *Numerical methods for image registration*
M. Feistauer, J. Felcman, and I. Straškraba: *Mathematical and computational methods for compressible flow*
W. Gautschi: *Orthogonal polynomials: computation and approximation*
M. K. Ng: *Iterative methods for Toeplitz systems*
Michael Metcalf, John Reid, and Malcolm Cohen: *Fortran 95/2003 explained*
George Em Karniadakis and Spencer Sherwin: *Spectral/hp element methods for CFD, second edition*
Dario A. Bini, Guy Latouche, and Beatrice Meini: *Numerical methods for structured Markov chains*
Howard Elman, David Silvester, and Andy Wathen: *Finite elements and fast iterative solvers: with applications in incompressible fluid dynamics*
Moody Chu and Gene Golub: *Inverse eigenvalue problems: Theory and applications*
Jean-Frédéric Gerbeau, Claude Le Bris, and Tony Lelièvre: *Mathematical methods for the magnetohydrodynamics of liquid metals*

Mathematical methods for the Magnetohydrodynamics of Liquid Metals

Jean-Frédéric Gerbeau
Institut National de Recherche en Informatique et en Automatique
FRANCE

Claude Le Bris and Tony Lelièvre
Ecole Nationale des Ponts et Chaussées
FRANCE

OXFORD
UNIVERSITY PRESS

OXFORD
UNIVERSITY PRESS

Great Clarendon Street, Oxford OX2 6DP

Oxford University Press is a department of the University of Oxford.
It furthers the University's objective of excellence in research, scholarship,
and education by publishing worldwide in

Oxford New York

Auckland Cape Town Dar es Salaam Hong Kong Karachi
Kuala Lumpur Madrid Melbourne Mexico City Nairobi
New Delhi Shanghai Taipei Toronto

With offices in

Argentina Austria Brazil Chile Czech Republic France Greece
Guatemala Hungary Italy Japan Poland Portugal Singapore
South Korea Switzerland Thailand Turkey Ukraine Vietnam

Oxford is a registered trade mark of Oxford University Press
in the UK and in certain other countries

Published in the United States
by Oxford University Press Inc., New York

First published 2006

British Library Cataloguing in Publication Data
Data available

Library of Congress Cataloging in Publication Data
Data available

Typeset by SPI Publisher Services, Pondicherry, India
Printed in Great Britain
on acid-free paper by
Biddles Ltd., King's Lynn, Norfolk
ISBN 0-19-856665-4 978-0-19-856665-6

1 3 5 7 9 10 8 6 4 2

PREFACE

This book focuses on mathematical and numerical techniques for the simulation of magnetohydrodynamics phenomena. The emphasis is laid on the magnetohydrodynamics of liquid metals, and on a prototypical industrial application. Starting from a good understanding of the physics at play, the approach is mathematical in nature, based on the rigorous analysis of the equations at hand, and a solid numerical analysis of the algorithmic strategies used in the simulations. At each stage of the exposition, examples of numerical simulations are provided, first on academic test cases to illustrate the approach, next on benchmarks well documented in the professional literature. The guideline followed throughout the book is the study of a particular setting of the MHD equations that models a real industrial case. This specific industrial case is the main topic addressed in the final chapter of the book.

In a nutshell, one could probably say that this book starts from the balance equations of continuum mechanics, passes through the notion of weak convergence in Hilbert spaces, and goes all the way to the simulation of a process of metal industry.

Chapter 1 presents how some standard MHD equations can be derived from the general conservation equations for fluid mechanics coupled with the Maxwell equations modeling the electromagnetic phenomena. A hierarchy of models is examined, from the most general one (a full time-dependent system consisting of the incompressible Navier–Stokes equations with a Lorentz body force calculated from the Maxwell equations and Ohm's law) to the most simplified one. Depending on the physical context, one model or the other may be meaningful. Several of these models will be evaluated from the standpoint of the mathematical analysis in the rest of the book. The most sophisticated model raises unsolved questions of existence and uniqueness, mainly related with the hyperbolic nature of the Maxwell equations, but some simplified models can be fully analyzed.

The matter of Chapter 2 is the modeling of one-fluid problems. The crucial point under consideration, both mathematically and numerically, is the understanding of the coupling between hydrodynamics phenomena and electromagnetic phenomena. From the mathematical viewpoint, the coupling induces a nonlinearity, additional with respect to the nonlinearities already present in the hydrodynamics. A series of difficult, thus interesting, problems follows. With a reasonable amount of theoretical effort, these problems can be dealt with. For instance, it can be shown that a system coupling the time-dependent incompressible Navier–Stokes equations with a simplified form of the Maxwell equations (the so-called low-frequency approximation) is well-posed when the electromagnetic equation is taken to be time-dependent, *i.e.* in parabolic form. In contrast, the same model is likely to be ill-posed when the electromagnetic equation is taken to be time-independent, *i.e.* in elliptic form, while the hydrodynamic equations are still in a time-dependent form.

At the numerical level, the difficulties due to nonlinearities translate into several numerical challenges. Dedicated finite elements methods, with *ad hoc* mixed formulations or stabilization techniques must be designed. This is the purpose of Chapter 3.

Chapters 4 and 5 are the central chapters of the book. They deal with multi-fluid MHD problems, on the theoretical side and on the numerical side, respectively. In addition to the coupling between hydrodynamics and electromagnetics examined in Chapter 2, the high nonlinearity that needs to be addressed is geometrical in nature, and is due to the presence of one (or many) free interface(s) separating the fluids, assumed non-miscible. Already at the mathematical level, the analysis is substantially more intricate, and a long list of simple, but open, problems can be drawn up. Numerically, one has also to resort to up-to-date techniques for the simulation of moving interfaces. In particular, we present a numerical method based on the Arbitrary Lagrangian Eulerian formulation, and lay special emphasis on the stability of the time-advancing schemes.

In Chapters 2 and 4 where theoretical issues are investigated, we have chosen to put the emphasis on the notion of global-in-time weak solutions. In contrast, we only briefly study the strong solutions and their properties. Our choice is motivated by the fact that, in the state of art of mathematical knowledge, strong solutions are only known to exist for small times and sufficiently regular data. For most practical situations, the time of existence of a strong solution cannot be explicitly evaluated in terms of the data. The situation cannot therefore be considered satisfactory. It is not ready to be discussed in such a textbook.

The only exception we make regards the long-time behavior of the solutions of our equations, because that seems to us an important issue in view of the practical purposes we have. Then, we assume the solutions are sufficiently regular and concentrate on their limit as time goes to infinity.

The closing chapter of the book, Chapter 6, is entirely devoted to one industrial application, the simulation of the industrial production of aluminum in electrolytic cells. The simulation of this specific problem has indeed motivated the whole scientific strategy described in the first five chapters. It serves as an illustration of the efficiency of the approach presented throughout this book. We have of course idealized the real industrial situation for the clarity of exposition. A schematic description of the problem is as follows. An electric current of huge intensity runs downwards through two horizontal layers of incompressible non-miscible conducting fluids. The system can be modeled by the equations of Chapters 4 and 5, namely a system coupling the two-fluid incompressible equations with the parabolic form of the Maxwell equations in each fluid. Owing to the magnetohydrodynamic coupling, the interface between the fluids moves, and, in view of the very high intensity of the electric current, the system is very sensitive to instabilities. The industrial challenge is to model, understand and, further, control these instabilities. Numerical simulation of nonlinear systems can help to reach such a goal. Other techniques (such as a stability analysis for the linearized system) are also employed, and are overviewed for comparison.

To conclude this introduction, let us mention that the present book is based upon two families of references in the literature.

For the basic ingredients of Chapter 2, regarding the one-fluid problem, our exposition is mostly based upon the famous textbooks by J.L. Lions [153, 154], G. Duvaut and J-L. Lions [73], R. Temam [227, 229], along with the reference article M. Sermange and R. Temam [208]. All these references are landmark references which are remarkably lively today for the problems under consideration in this book. Our presentation in the theoretical sections of Chapter 2 is only a rearrangement, biased toward our specific purpose here, of the material contained there.

The finite element discretization presented in Chapter 3 mainly relies on the theory of problems with Lagrange multipliers developed in the early 1970s by I. Babuska [11] and F. Brezzi [31]. Our presentation of the general considerations is inspired by the textbooks by V. Girault and P.A. Raviart [106] and A. Ern and J.-L. Guermond [75]. Results related to the discretization of the MHD equations are adapted from various articles, in particular M.D. Gunzburger *et al.* [118] and D. Schötzau [205].

The early sections of Chapter 4 recall the theory for non-homogeneous fluids, essentially developed in the monograph of P-L. Lions [155]. We also make use of the related work of A. Nouri and F. Poupaud [182].

The second set of references is a series of articles published in the last decade, and authored or co-authored by the three of us [87–95, 98–101, 140, 141], along with two works in collaboration with M. Bercovier and B. Desjardins [66, 96]. The material contained in those references has been complemented, updated and adapted, in order to best fit with our pedagogic endeavor here and the specific industrial application targeted. We will here and there rely heavily on the articles cited above, but we have done our best to give hereunder a self-contained and hopefully accessible presentation.

JFG, CLB, TL,
Rocquencourt and Marne-la-Vallée, 2006.

ACKNOWLEDGEMENTS

We wish to express our deep gratitude to those who assisted us in this endeavor. The help of P.-L. Lions (Collège de France) was, of course, priceless. J.-P. Boujot pointed out the practical relevance of the subject, and initiated our modelization work in the early 1990s. M. Bercovier (Hebrew University of Jerusalem) constantly provided us with precious advice on discretization methods and implementation issues. We also benefited very much from the interaction with the board and the engineers at Alcan-Pechiney (Laboratoire de Recherche des Fabrications and Centre de Recherche de Voreppe): P. Homsi, C. Vanvoren and O. Martin, J. Colin de Verdière, J.-M. Gaillard, N. Ligonesche, M. Le Hervet, Th. Tomasino. We are grateful to a number of colleagues, who went over the manuscript and suggested several improvements: M. Bercovier (Hebrew University of Jerusalem), X. Blanc (Paris VI), A. Ern (ENPC), M. Flueck (EPFL), L. Formaggia (Politecnico di Milano), B. Maury (Paris XI), A. Orriols (ENPC), M. Picasso (EPFL), J. Rappaz (EPFL), M.V. Romerio (EPFL), Th. Tomasino (Alcan). The financial support of Alcan-Pechiney is gratefully acknowledged.

CONTENTS

NOTATION

Throughout this book, scalar-valued fields, such as the pressure p, are denoted by roman letters. On the other hand, vector-valued fields such as the velocity $\boldsymbol{u}$ and tensor-valued fields such as the stress tensor $\boldsymbol{\tau}$ are denoted by bold-face letters.

The space variable is generically denoted by $\mathbf{x}$, and varies in a bounded domain Ω, a subset of $\mathbb{R}^d$, $d = 2$ or 3. The boundary of Ω is denoted by $\partial\Omega$ and is supposed to be at least Lipschitz continuous. The outward normal to Ω is denoted by $\boldsymbol{n}$. The time variable t varies in an interval $[0, T]$, with final time T that may possibly be $+\infty$.

The differential operators gradient, divergence and curl are, respectively, denoted by $\boldsymbol{\nabla}$, div and **curl**. For example we write,

$$\operatorname{div} \boldsymbol{u} = \sum_{i=1}^{d} \frac{\partial u_i}{\partial x_i}, \qquad \boldsymbol{\nabla} p = \left(\frac{\partial p}{\partial x_i} \right)_{i=1..d},$$

and, for $d = 3$,

$$\mathbf{curl}\, \boldsymbol{B} = \left(\frac{\partial B_3}{\partial x_2} - \frac{\partial B_2}{\partial x_3}, \frac{\partial B_1}{\partial x_3} - \frac{\partial B_3}{\partial x_1}, \frac{\partial B_2}{\partial x_1} - \frac{\partial B_1}{\partial x_2} \right).$$

An arbitrary body force term is denoted by $\boldsymbol{f}$.

Throughout the book, when there is a risk of ambiguity, we shall overline the parameters that are assumed constant. For instance, when the density ρ, which in general varies in the computational domain, is assumed constant, we denote it by $\overline{\rho}$.

The transpose of a matrix A is denoted by A^T.

For any domain Ω, meas(Ω) denotes the Lebesgue measure of the domain. We denote by $\mathcal{D}(\Omega)$ the space of smooth functions with compact support in Ω.

In addition, the following Lebesgue and Sobolev type spaces will be used:

$L^2_0(\Omega) = \{p \in L^2(\Omega), \int_\Omega p = 0\}$,

$\mathbb{L}^2(\Omega) = (L^2(\Omega))^d$,
$\mathbb{H}^1(\Omega) = (H^1(\Omega))^d$,
$\mathbb{H}^1_0(\Omega) = \{\boldsymbol{u} \in \mathbb{H}^1(\Omega), \boldsymbol{u} = 0 \text{ on } \partial\Omega\}$,

$\mathbb{H}^1_n(\Omega) = \{\boldsymbol{u} \in \mathbb{H}^1(\Omega), \boldsymbol{u}\cdot\boldsymbol{n} = 0 \text{ on } \partial\Omega\}$,
$\mathbb{H}^1_\tau(\Omega) = \{\boldsymbol{u} \in \mathbb{H}^1(\Omega), \boldsymbol{u}\times\boldsymbol{n} = 0 \text{ on } \partial\Omega\}$,
$H(\mathbf{curl}\,,\Omega) = \{\boldsymbol{v} \in \mathbb{L}^2(\Omega), \mathbf{curl}\,\boldsymbol{v} \in \mathbb{L}^2(\Omega)\}$,
$H(\mathrm{div}\,,\Omega) = \{\boldsymbol{v} \in \mathbb{L}^2(\Omega), \mathrm{div}\,\boldsymbol{v} \in L^2(\Omega)\}$,
$H_0(\mathbf{curl}\,,\Omega) = \{\boldsymbol{v} \in H(\mathbf{curl}\,,\Omega), \boldsymbol{v}\times\boldsymbol{n} = 0 \text{ on } \partial\Omega\}$,
$H_0(\mathrm{div}\,,\Omega) = \{\boldsymbol{v} \in H(\mathrm{div}\,,\Omega), \boldsymbol{v}\cdot\boldsymbol{n} = 0 \text{ on } \partial\Omega\}$,
$H(\mathbf{curl}\,^0,\Omega) = \{\boldsymbol{v} \in H(\mathbf{curl}\,,\Omega), \mathbf{curl}\,\boldsymbol{v} = 0\}$,
$H(\mathrm{div}\,^0,\Omega) = \{\boldsymbol{v} \in H(\mathrm{div}\,,\Omega), \mathrm{div}\,\boldsymbol{v} = 0\}$,
$H_0(\mathbf{curl}\,^0,\Omega) = H(\mathbf{curl}\,^0,\Omega) \cap H_0(\mathbf{curl}\,,\Omega)$,
$H_0(\mathrm{div}\,^0,\Omega) = H(\mathrm{div}\,^0,\Omega) \cap H_0(\mathrm{div}\,,\Omega)$.

1

THE MAGNETOHYDRODYNAMICS EQUATIONS

This first chapter presents the basic modeling of magnetohydrodynamics we will need in the sequel. The equations of continuum fluid mechanics are first established, then the equations of classical electromagnetism. The two sets of equations are next coupled, thereby forming the magnetohydrodynamics (abbreviated henceforth as MHD) equations. We proceed from the most general setting to the most specific one. A hierarchy of models is thus introduced. In this hierarchy, we particularly concentrate on the setting that will serve as a testbed for the whole book, namely the multifluid incompressible MHD equations. But more general equations such as the full time-dependent system consisting of the incompressible Navier–Stokes equations with a Lorentz body force calculated from the Maxwell equations and Ohm's law, are also mentionned. Likewise, simplifications of our prototype system are examined. Depending on the physical context one model or the other may be meaningful. A section presenting the non-dimensionalized form of the equations concludes the chapter. This form will be that used for the numerical analysis and the numerical simulations.

1.1 The general fluid equations

1.1.1 The conservation equations

The usual description for fluids follows from continuum mechanics. There is a huge number of textbooks presenting the fundamental equations of fluid mechanics (see, *e.g.* [139]). In addition, one may read the sections devoted to modeling in the following mathematically oriented books: R. Dautray and J.-L. Lions [45, Chapter 1], P.-L. Lions [155, Chapter 1], O Pironneau [186], P.M. Gresho and R.L. Sani [110, 111].

The governing equations are first the equation of *conservation of mass*

$$\frac{\partial \rho}{\partial t} + \operatorname{div}(\rho\, \boldsymbol{u}) = 0, \tag{1.1}$$

and second the equation for the *conservation of momentum*

$$\frac{\partial(\rho\, \boldsymbol{u})}{\partial t} + \operatorname{div}(\rho\, \boldsymbol{u} \otimes \boldsymbol{u}) - \operatorname{div} \boldsymbol{\tau} = \boldsymbol{f}, \tag{1.2}$$

where ρ denotes the *density* of the fluid, $\boldsymbol{u}$ its *velocity*, $\boldsymbol{\tau}$ the *stress tensor*, and $\boldsymbol{f}$ the density of body forces applied to the fluid. A useful decomposition of the latter is

$$\boldsymbol{f} = \rho \boldsymbol{f}_m + \boldsymbol{f}_v, \tag{1.3}$$

where $\boldsymbol{f}_m$ and $\boldsymbol{f}_v$ are the body forces per unit mass and forces per unit volume applied to the fluid, respectively. Anticipating the sequel of the present book (see (1.34) in Section 1.3), it is useful in order to fix ideas to mention that a standard example of such forces $\boldsymbol{f}_m$ and $\boldsymbol{f}_v$ is the gravity force $\boldsymbol{f}_m = \boldsymbol{g}$ and the Lorentz force

$$\boldsymbol{f}_v = \frac{1}{\overline{\mu}}\,\mathbf{curl}\,\boldsymbol{B} \times \boldsymbol{B},$$

where $\boldsymbol{B}$ is the magnetic induction and $\overline{\mu}$ the magnetic permeability.

The stress tensor can be written

$$\boldsymbol{\tau} = -p\,\mathrm{Id} + \boldsymbol{T} \tag{1.4}$$

where p is the (hydrodynamic)[1] *pressure* of the fluid and $\boldsymbol{T}$ the *viscous stress tensor*. Generally, the latter is related to the velocity field by a so-called *constitutive relation*, or *closure relation*,

$$\boldsymbol{T} = \boldsymbol{T}(\boldsymbol{u}, p, ...) \tag{1.5}$$

which can be a pointwise relation, but also a more complicated relation such as, *e.g.* a partial differential equation, or an integral equation.

In addition to the above two equations of conservation of mass and momentum (1.1)–(1.2), there is first an equation for the conservation of the energy of the fluid, and second, for the closure of the full system, an equation of state relating p, ρ and the temperature T. For simplicity, and because it is the physically relevant case for the main application we have in mind in the present book, we will consider the temperature as fixed and constant throughout the domain occupied by the fluid and for all times. Then no conservation law other than the above two are needed. Let us however mention that there are some cases, even in the context of the application we target, where temperature effects do play an important role. We will neglect such effects here and will not treat further the temperature-dependent case. An exception is the setting treated in Section 2.2.6.2.

Using the fact that the laws of mechanics need to be invariant under a change of Galilean referential system, assuming that the physical properties of the fluid under consideration are isotropic, and more importantly assuming that $\boldsymbol{T}$ is a linear function of $\boldsymbol{u}$, it can be shown that necessarily, $\boldsymbol{T}$ can be written as

$$\boldsymbol{T} = \lambda\,(\mathrm{div}\,\boldsymbol{u})\,\mathrm{Id} + 2\eta \boldsymbol{D}(\boldsymbol{u}), \tag{1.6}$$

where λ and η are two real scalar coefficients called the *Lamé coefficients*[2], and

$$\boldsymbol{D}(\boldsymbol{u}) = \frac{1}{2}\,(\boldsymbol{\nabla u} + \boldsymbol{\nabla u}^T) \tag{1.7}$$

[1]We mention the adjective *hydrodynamic* because in the context of MHD some other pressure, of magnetic nature, appears.

[2]The standard notation for the Lamé coefficients is λ and μ. However, in the present context of MHD, the notation μ is commonly used for the magnetic permeability (see (1.31) below), so we denote the second Lamé coefficient by η.

is the *rate of deformation tensor*, also called the *shear rate tensor*. When the relation (1.6) holds (*i.e.* essentially when the linearity of the relation between $\boldsymbol{T}$ and $\boldsymbol{u}$ is assumed), the fluid is said to be a *Newtonian fluid*. On the other hand, when the closure relation (1.5) cannot be simplified into (1.6), the fluid is said to be *non-Newtonian*. An outstanding modeling issue is then to derive a correct closure relation (1.5). Fortunately, the fluids we will treat in the sequel can be considered as Newtonian with a good level of accuracy. We therefore make this assumption henceforth.

The coefficients λ and η generally depend on the density ρ and on the temperature. Since we have considered the temperature as fixed, they only depend on the density.

The kinetic theory of monatomic gas tells that the two coefficients are related by

$$\lambda = -\frac{2}{3}\eta, \tag{1.8}$$

and in fact this relation holds true in practice for most fluids. The only coefficient describing the physics of the fluid is then

$$\eta = \eta(\rho) \tag{1.9}$$

called the *viscosity* of the fluid. In practice, it is non-negative. If this coefficient is zero, then the fluid is said to be *inviscid*, and the equations that follow are called the *Euler equations*. In the present book, we will not treat the case of inviscid fluids, and concentrate on the case $\eta > 0$ (see however Chapter 4 for the modeling of free-surface problems). The case $\eta > 0$ is physically relevant for the application we have in mind[3]. Then the fluid is said to be *viscous*. The equations of conservation of momentum for a Newtonian viscous fluid are called the *Navier–Stokes equations*. In particular, the nonlinear term $\operatorname{div}(\rho\boldsymbol{u}\otimes\boldsymbol{u})$ (which indeed simplifies to $\overline{\rho}\,\boldsymbol{u}.\boldsymbol{\nabla}\boldsymbol{u}$ when ρ is a constant) is called the *Navier term*, after the French engineer Navier. When this term can be neglected, the equation of conservation of momentum is called the *generalized Stokes equation*, or more concisely the *Stokes equation*[4].

Considering the application we will address in the following, namely some industrial applications involving liquid metals, it is standard to assume the fluid is *incompressible*, *i.e.*

$$\operatorname{div}\boldsymbol{u} = 0. \tag{1.10}$$

[3]That is indeed fortunate since the mathematical analysis of the Euler equation is even more difficult than that of the Navier–Stokes equation, and the numerical simulation is also more challenging.

[4]There is an ambiguity as the Stokes equation, or more precisely the Stokes *problem* is also the terminology used for the stationary system (2.32) introduced in Chapter 2.

Then the fluid is an *incompressible viscous Newtonian* fluid, and is modeled by the following system, called the *incompressible (density-dependent) Navier–Stokes equations*:

$$\begin{cases} \dfrac{\partial}{\partial t}(\rho \boldsymbol{u}) + \operatorname{div}(\rho \boldsymbol{u} \otimes \boldsymbol{u}) - \operatorname{div}(\eta(\rho)(\boldsymbol{\nabla u} + \boldsymbol{\nabla u}^T)) = -\boldsymbol{\nabla} p + \boldsymbol{f}, \\ \operatorname{div} \boldsymbol{u} = 0, \\ \dfrac{\partial \rho}{\partial t} + \operatorname{div}(\rho \boldsymbol{u}) = 0. \end{cases} \tag{1.11}$$

To write the equation of conservation of momentum, we could equally use:

$$\operatorname{div}(\rho \boldsymbol{u} \otimes \boldsymbol{u}) = \boldsymbol{u} \operatorname{div}(\rho \boldsymbol{u}) + \rho\,(\boldsymbol{u}.\boldsymbol{\nabla})\boldsymbol{u}.$$

In addition to the incompressibility assumption, a further simplification is to consider a *homogeneous fluid*, for which the density ρ is a positive constant[5]:

$$\rho = \overline{\rho} = \text{constant}. \tag{1.12}$$

This implies that the viscosity η is also a constant, with value $\overline{\eta} > 0$. Then the above system simplifies into the famous *incompressible homogeneous Navier–Stokes equations*:

$$\begin{cases} \overline{\rho}\dfrac{\partial \boldsymbol{u}}{\partial t} + \overline{\rho}\boldsymbol{u}.\boldsymbol{\nabla u} - \overline{\eta}\,\Delta \boldsymbol{u} = -\boldsymbol{\nabla} p + \boldsymbol{f}, \\ \operatorname{div} \boldsymbol{u} = 0. \end{cases} \tag{1.13}$$

It is worth mentionning that, often, and especially in the mathematical literature, the word "homogeneous" is omitted, and one simply speaks of the *incompressible Navier–Stokes equations* to refer to (1.13).

The case of particular interest for us in the present book is not as simple as the homogeneous case. It is the case when the fluid under consideration indeed consists of several (say q) immiscible fluids, each of them with a given homogeneous density $\overline{\rho}_i$ and viscosity $\eta(\overline{\rho}_i) = \overline{\eta}_i$, $i = 1, \ldots, q$. Then the system considered is the *multifluid incompressible Navier–Stokes equations*:

$$\begin{cases} \dfrac{\partial}{\partial t}(\rho \boldsymbol{u}) + \operatorname{div}(\rho \boldsymbol{u} \otimes \boldsymbol{u}) - \operatorname{div}(\eta(\rho)\,(\boldsymbol{\nabla u} + \boldsymbol{\nabla u}^T)) = -\boldsymbol{\nabla} p + \boldsymbol{f}, \\ \operatorname{div} \boldsymbol{u} = 0, \\ \dfrac{\partial \rho}{\partial t} + \operatorname{div}(\rho \boldsymbol{u}) = 0, \\ \rho = \overline{\rho}_i, \quad \text{on each fluid} \quad \text{and} \quad \eta(\rho_i) = \overline{\eta}_i > 0. \end{cases} \tag{1.14}$$

[5]We recall that throughout the book, when there is a risk of ambiguity, we overline the parameters that are assumed constant.

A particular occurence of the above multifluid setting is that when one of the fluid is the vacuum. To fix the idea, let us consider the case of $q = 2$ "fluids", with $\overline{\rho}_1 > 0$ and $\overline{\rho}_2 \equiv 0$. Then the system under consideration is indeed a model for the evolution of one incompressible homogeneous fluid with a free surface (namely the surface separating the fluid from the vacuum). Therefore, in the mathematical analysis and the numerical simulations addressed in the sequel, in particular in Chapters 4 and 5, it should be borne in mind that the multifluid hydrodynamic case indeed embodies the free-surface case, if one of the densities is allowed to vanish.

However, we will not be able to treat system (1.14) with this degree of generality when we couple it to the electromagnetism equations, that is to say in the MHD context. The reason for this is a difficulty related to the setting of the MHD system when a vacuum is present. We shall come back to this in Section 1.3.

We will therefore only treat in the sequel system (1.14) with positive $\overline{\rho}_i$, thereby excluding the case where one of the fluids in the system is the vacuum. The case of one (or many) fluid(s) enclosed in a free surface and surrounded by a vacuum is consequently beyond the scope of our analysis, but the case of a free *interface* (by this we mean, *e.g.* two fluids separated by a free interface and filling a whole fixed domain with rigid outer boundaries) is included. This will be a major issue of interest in the sequel.

Remark 1.1.1 [Surface tension and turbulence effects] In the above discussion on hydrodynamic modeling, two questions have been left aside.

First, surface tension effects may be accounted for in the multifluid equations (1.14). We will see this point in more detail in Chapters 4 and 5.

Second, for very large Reynolds number (see Section 1.6 for the precise definition of this non-dimensional number) the direct numerical simulation of the hydrodynamics equations (1.11) is out of reach. The common practice is then to resort to turbulence models (see M. Lesieur [149, 148] or H. Tennekes and J.L. Lumley [230] for a mechanical introduction and B. Mohammadi and O. Pironneau [168] for a mathematical presentation). In the specific context of MHD flows, the modeling of turbulence effects is mostly an unsolved issue, still subject to debate. We refer to P.A. Davidson [47, 48] and R. Moreau [172] for an introduction. The use of the usual $k-\epsilon$ model does not seem to be adequate (see [175]). As explained in [174], MHD turbulence is for example well understood in the case of a uniform DC magnetic field. But in many other situations such as non-uniform DC fields[6] or AC fields at mean frequencies, the effect of the magnetic field on turbulence is not yet understood (see [242] for some numerical experiments). ◇

So far we have only concentrated on the equations, and we now need to address the question of boundary and initial conditions.

[6] As is the case in the industrial application we intend to deal with, namely the modeling of aluminum electrolysis cells.

1.1.2 Boundary and initial conditions

First, we need to make precise on which domain the above equations are posed. Many situations are generically addressed:

(i) a bounded domain Ω included in $\mathbb{R}^d$, $d = 2, 3$;
(ii) a bounded periodic domain Ω included in $\mathbb{R}^d$, $d = 2, 3$, say $\Omega = [0, 1]^d$, with periodic boundary condition (which is also expressed mathematically as the torus Π^d);
(iii) the exterior of a bounded domain: Ω^c included in $\mathbb{R}^d$, $d = 2, 3$;
(iv) the half-space $\mathbb{R}^d_+$, $d = 2, 3$;
(v) the whole space $\mathbb{R}^d$, $d = 2, 3$.

More specific applications can also require us to consider other cases.

In the mathematical literature, cases (i), (ii) and (v) are the most commonly addressed. In view of the application we target, we shall mainly concentrate on the case (i). For some specific issues, we shall mention here and there whether cases (ii) and (v) can (or cannot) be treated analogously.

For theoretical purposes, we will assume that the bounded domain Ω is enclosed by a sufficiently regular boundary $\partial\Omega$. Mathematically, this will allow for some standard theoretical results to hold. We will see this in Chapter 2. Correspondingly, the normal outward unit vector $\boldsymbol{n}$ is unambiguously defined.

Having said that, the question of boundary conditions is now simple[7]. We will consider the setting of *Dirichlet boundary conditions* on the velocity

$$\boldsymbol{u} = \boldsymbol{u}_{bc}, \quad \text{on the boundary} \quad \partial\Omega, \tag{1.15}$$

and, often, *homogeneous Dirichlet boundary conditions, i.e.*

$$\boldsymbol{u} = 0, \quad \text{on the boundary} \quad \partial\Omega. \tag{1.16}$$

The latter condition is to be seen as the superposition of the no-slip boundary condition

$$\boldsymbol{u} \times \boldsymbol{n} = 0 \tag{1.17}$$

which is a condition implied by the *viscous* nature of the fluid, and the no-penetration boundary condition

$$\boldsymbol{u} \cdot \boldsymbol{n} = 0, \tag{1.18}$$

which translates the fact that the system we consider is hydrodynamically closed (no fluid enters nor escapes through the boundary).

[7]There is a noticeable exception: when the motion of a free surface is to be accounted for, there are delicate questions about the boundary condition to impose at the intersection of this surface and rigid boundaries of the fluid domain. We shall come back to this in Chapter 4.

A noteworthy consequence of condition (1.16) is that the equation of conservation of mass (1.1) which can also be written

$$\frac{\partial \rho}{\partial t} + \boldsymbol{u} \cdot \boldsymbol{\nabla} \rho = 0,$$

does not require any boundary condition on the density ρ to be well-posed[8].

Let us now turn to the initial conditions the equations need to be supplied with.

There are *a priori* three unknowns fields in equations (1.11) and (1.14): the velocity $\boldsymbol{u}$, the pressure p, and the density ρ. In (1.13), there are only two of them, $\boldsymbol{u}$ and p. However, there is no time evolution on the pressure p in the three systems (1.11), (1.13) and (1.14). Let us mention incidentally that this implies that the evolution problem is not mathematically a *Cauchy problem* (see (2.10) for a definition). Somewhat vaguely stated, the pressure field instantaneously adjusts itself in order for the fluid velocity to satisfy the incompressibility constraint (1.10). We will come back to the difficulties caused by this fact in Chapter 2, Section 2.1.1.2, and in Chapter 3.

Therefore only an initial condition on the velocity and on the density needs to be provided (and in the case of (1.13) only the initial velocity is needed). For (1.11) and (1.14), it is clear that the initial conditions should be set as

$$\rho \boldsymbol{u}(t=0,\cdot) \ \text{ given,} \tag{1.19}$$

$$\rho(t=0,\cdot) \ \text{ given,} \tag{1.20}$$

for the time derivative operator acts on $\rho \boldsymbol{u}$ and ρ in both systems. There is nevertheless a slight difficulty in the initial condition, again related to the fact that the density ρ may vanish somewhere in the domain. Intuitively, on the set where $\rho = 0$, or more generally where ρ is very small, poor information on $\boldsymbol{u}$ itself is carried by (1.19). We shall come back to this difficulty in Chapter 4.

On the other hand, the initial condition for (1.13) is evidently

$$\boldsymbol{u}(t=0,\cdot) \ \text{ given.} \tag{1.21}$$

1.1.3 Steady-state equations

Before turning to the equations of electromagnetism, we mention that all the above equations (1.11), (1.13), and (1.14) have a stationary variant, respectively the *steady-state incompressible Navier–Stokes equations*:

$$\begin{cases} \operatorname{div}(\rho \boldsymbol{u} \otimes \boldsymbol{u}) - \operatorname{div}(\eta(\rho)(\boldsymbol{\nabla} \boldsymbol{u} + \boldsymbol{\nabla} \boldsymbol{u}^T)) = -\boldsymbol{\nabla} p + \boldsymbol{f}, \\ \operatorname{div} \boldsymbol{u} = 0, \\ \boldsymbol{u} \cdot \boldsymbol{\nabla} \rho = 0, \end{cases} \tag{1.22}$$

[8] At least formally, and we shall see this is indeed the case mathematically in Chapter 4.

the *steady-state incompressible homogeneous Navier–Stokes equations*:

$$\begin{cases} \overline{\rho}\boldsymbol{u}.\boldsymbol{\nabla}\boldsymbol{u} - \overline{\eta}\,\Delta\boldsymbol{u} = -\boldsymbol{\nabla}p + \boldsymbol{f}, \\ \operatorname{div}\boldsymbol{u} = 0. \end{cases} \tag{1.23}$$

and the *steady-state multifluid incompressible Navier–Stokes equations*:

$$\begin{cases} \operatorname{div}(\rho\boldsymbol{u}\otimes\boldsymbol{u}) - \operatorname{div}(\eta(\rho)\,(\boldsymbol{\nabla}\boldsymbol{u} + \boldsymbol{\nabla}\boldsymbol{u}^T)) = -\boldsymbol{\nabla}p + \boldsymbol{f}, \\ \operatorname{div}\boldsymbol{u} = 0, \\ \boldsymbol{u}\cdot\boldsymbol{\nabla}\rho = 0, \\ \rho = \overline{\rho}_i, \quad \text{on each subdomain}\,\Omega_i \quad \text{and} \quad \eta(\rho_i) = \overline{\eta}_i, \end{cases} \tag{1.24}$$

with the same condition of positivity on $\eta(\rho)$ as in the time-dependent case.

Each variant is supplied with boundary conditions on the velocity, namely (1.16). For such stationary systems, there is of course no need for an initial condition, but in order for the non-homogeneous problem (1.22) (and thus in particular (1.22)) to be well-posed, at least an additional constraint on the density ρ must be imposed. For instance, for (1.24), the mass of each fluid $\int_{\Omega_i}\rho_i$ needs to be fixed, which amounts to prescribing the measures of the subsets Ω_i in (1.24).

1.2 The electromagnetic description

Classical electromagnetism is described by the *Maxwell equations.* The fundamental equations for this theory can be read in many textbooks, such as R. Dautray and J.-L. Lions [45, Volume 1] or L. Landau and E. Lifchitz [139]. We reproduce the basic ingredients here for consistency and for the convenience of the reader.

We recall that the *Maxwell equations* are the *Maxwell–Ampère equation*

$$-\frac{\partial\boldsymbol{D}}{\partial t} + \mathbf{curl}\,\boldsymbol{H} = \boldsymbol{j}, \tag{1.25}$$

the *Maxwell–Coulomb equation*

$$\operatorname{div}\boldsymbol{D} = \rho_c, \tag{1.26}$$

the *Maxwell–Faraday equation*

$$\frac{\partial\boldsymbol{B}}{\partial t} + \mathbf{curl}\,\boldsymbol{E} = 0, \tag{1.27}$$

and the *Maxwell–Gauss equation*

$$\operatorname{div}\boldsymbol{B} = 0. \tag{1.28}$$

In the above equations, the three-dimensional vector fields $\boldsymbol{D}$, $\boldsymbol{B}$, $\boldsymbol{E}$, $\boldsymbol{H}$, denote the *electric induction* and *magnetic induction*, and the *electric field* and *magnetic field*, respectively[9]. On the other hand, the three-dimensional vector field $\boldsymbol{j}$ denotes the *current density*, and the scalar field ρ_c denotes the *charge density*[10].

The above four equations are not independent, in the following sense. First, if equation (1.28) is assumed to hold at initial time, then it is a simple matter, taking the divergence of (1.27), to show that (1.28) will necessarily hold for all time $t > 0$. Second, using the *conservation of electric charge*:

$$\frac{\partial \rho_c}{\partial t} + \operatorname{div} \boldsymbol{j} = 0, \tag{1.29}$$

it is simple to prove that if equation (1.26) holds at initial time, then, taking the divergence of (1.25), equation (1.26) will necessarily hold for all time $t > 0$.

It is important to note that the above equations hold generically in the *whole, three-dimensional,* space $\mathbb{R}^3$. This is indeed a two-fold difficulty. First, the fact that these equations are posed in the *whole* space has important consequences on the mathematical analysis and the numerical simulation. We shall see below, when studying the MHD of fluids enclosed in a bounded domain, that the main approach that restricts the Maxwell equations to the same bounded domain raises the difficult questions of boundary conditions for the Maxwell equations. See the next section for a first discussion. Second, the fact that the Maxwell equations are three-dimensional in nature is also both a serious mathematical difficulty (because the functional analysis is then significantly more complicated than for the two-dimensional situation) and a difficulty for the numerical simulation (which is much more expensive in three dimensions).

There are some settings where the Maxwell equations may be simplified into two-dimensional equations, owing to some specific geometrical properties of the system under consideration. Apart from those rare situations, two-dimensional simulations of the Maxwell equations have poor physical relevance (at least for the applications we have in mind). They provide no particular insight into the three-dimensional case. In that respect (like in many others), the Maxwell equations are very different in nature from the fluid mechanics equations, where many two-dimensional, very enlightening, situations can be considered.

We now need to make specific the physical nature of the medium where the electromagnetic fields propagate. This information is indeed encoded in the relations that link $\boldsymbol{D}$, $\boldsymbol{E}$, $\boldsymbol{H}$ and $\boldsymbol{B}$. The latter will be one step needed to make a closed system of equations from (1.25)–(1.27).

When the ambient medium is the vacuum, such a relation is

$$\left\{ \begin{aligned} \boldsymbol{D} &= \varepsilon_0 \, \boldsymbol{E}, \\ \boldsymbol{H} &= \frac{1}{\mu_0} \, \boldsymbol{B}, \end{aligned} \right. \tag{1.30}$$

[9]Note that a slight abuse of language consists in commonly calling $\boldsymbol{B}$ the magnetic field, instead of $\boldsymbol{H}$. The reason is that the two fields are often proportional, see (1.31) and below.

[10]The charge density is commonly denoted by ρ, but here we denote it by ρ_c in order to avoid confusion with the density ρ of the fluid considered in the MHD setting.

where ε_0 and μ_0, respectively, denote the *electric permittivity* and the *magnetic permeability* of the vacuum. They satisfy $\varepsilon_0\mu_0 = \frac{1}{c^2}$, with c denoting the *speed of light*.

On the other hand, inside an electrically conducting medium, (1.30) hold under the form

$$\begin{cases} \boldsymbol{D} = \varepsilon\,\boldsymbol{E}, \\ \boldsymbol{H} = \mu^{-1}\,\boldsymbol{B}, \end{cases} \tag{1.31}$$

for some ε and μ, respectively, the *electric permittivity* and the *magnetic permeability* of the medium. These parameters may depend on $\boldsymbol{E}$ and $\boldsymbol{B}$, respectively, and are in general tensor-valued. Equation (1.31) is a type of "constitutive" law. In contrast to (1.5), it is electromagnetic in nature. In the simple isotropic homogeneous case, both parameters ε and μ are scalar and constant and the medium is called a *perfect medium*. It is convenient to measure ε and μ using as a reference their values in vacuum, and thus to write

$$\begin{cases} \varepsilon = \varepsilon_r\,\varepsilon_0, \\ \mu = \mu_r\,\mu_0, \end{cases} \tag{1.32}$$

where ε_r, μ_r are the *permittivity relative to vacuum* and the *permeability relative to vacuum*, or *relative permittivity* and *relative permeability*.

Collecting (1.25), (1.26), (1.27), (1.28), together with (1.31), (1.32), the following general system of Maxwell equations in a continuum (perfect dielectric) medium is obtained:

$$\begin{cases} -\dfrac{\partial(\varepsilon\,\mathbf{E})}{\partial t} + \mathbf{curl}\left(\dfrac{1}{\mu}\,\boldsymbol{B}\right) = \mathbf{j}, \\ \operatorname{div}\left(\varepsilon\,\boldsymbol{E}\right) = \rho_c, \\ \dfrac{\partial \boldsymbol{B}}{\partial t} + \mathbf{curl}\,\mathbf{E} = 0, \\ \operatorname{div}\boldsymbol{B} = 0. \end{cases} \tag{1.33}$$

System (1.33) is not sufficient to determine the fields $\boldsymbol{E}$ and $\boldsymbol{B}$. There is a two-fold reason for this.

First, (1.33) is not closed, because it allows us to identify $\boldsymbol{E}$ and $\boldsymbol{B}$ only when $\boldsymbol{j}$ is known, ρ_c being an auxiliary variable that we can see as defined by (1.26). In order to close the system, we need yet another equation belonging to the family of "constitutive laws," like (1.29), that is a characteristic feature of the medium under consideration. This law is known as *Ohm's law* and relates $\boldsymbol{j}$ to $\boldsymbol{E}$ and $\boldsymbol{B}$. We shall see this in (1.36) below.

Second, even with the additional equation provided by Ohm's law, system (1.33) needs to be supplied with initial conditions on the fields $\boldsymbol{B}$ and $\boldsymbol{E}$, and, more importantly, boundary conditions. As mentioned above, the latter might be necessary when the equations are restricted to a bounded domain.

For both the question of Ohm's law, and the question of boundary conditions, we prefer to postpone the discussion until the next section. Indeed, both issues are very much problem-dependent. We rather present them in our specific MHD context.

1.3 The MHD coupling

Magnetohydrodynamics (MHD) is the study of the interaction of electromagnetic fields and conducting fluids. The modeling consists of a coupling between the equations of continuum fluid mechanics and the Maxwell equations of electromagnetism, respectively seen in Sections 1.1 and 1.2. We now present the basic ingredients for this coupling, in the various, more or less simplified, forms it may take. We will not treat all the possible situations, and, for the situation we will treat, we will not enter all details. For a comprehensive description, we refer the reader to the numerous monographs devoted to the subject: for example, P.A. Davidson [47], R. Moreau [172], P.A. Davidson and A. Thess [51], W.F. Hughes and F.J. Young [130].

For simplicity, we first consider the case of *one* homogeneous fluid (modeled hydrodynamically by system (1.13)), and in Section 1.3.3 adapt the modeling to cover the case of a non-homogeneous fluid or several fluids, the latter being our main concern here.

Before getting to the heart of the matter, let us briefly draw up the list of the tasks we have to complete:

- we need to define the body force term $\boldsymbol{f}$ inserted in the right-hand side of equation (1.2);
- we need to account for the physical nature of the medium under consideration (here a conducting fluid), which implies
  - * to decide whether the permeability μ and the permittivity ε can be considered as scalar and constant;
  - * to write Ohm's law, relating $\boldsymbol{j}$ to $\boldsymbol{E}$ and $\boldsymbol{B}$;
- we need to make specific the boundary conditions for the Maxwell equations.

1.3.1 The general MHD system

In order to model MHD phenomena, we need a description that couples the hydrodynamics equation, say (1.11), and the Maxwell equations (1.33).

First, the body force term in (1.2) needs to be made precise:

$$\boldsymbol{f} = \boldsymbol{j} \times \boldsymbol{B} + \boldsymbol{f}_{\text{ext}}. \tag{1.34}$$

The first term in the right-hand side is the *Lorentz force*, a consequence of the electric current $\boldsymbol{j}$ running within the magnetic field $\boldsymbol{B}$. It is a force that influences the motion, along the velocity field $\boldsymbol{u}$, of the particles of the conducting fluid. The second term is due to possible external forces. A typical case for such forces is that of the gravity forces

$$\boldsymbol{f}_{\text{ext}} = \overline{\rho}\,\mathbf{g}. \tag{1.35}$$

Second, we recall that in order to be a mathematically closed system, the Maxwell system (1.33) needs to be complemented by *Ohm's law*. In the MHD context, Ohm's law most often reads in the form

$$\boldsymbol{j} = \sigma\big(\boldsymbol{E} + \boldsymbol{u} \times \boldsymbol{B}\big) \tag{1.36}$$

where σ denotes the *electric conductivity* of the fluid. The second term of (1.36) explicitly accounts for the deviation of the lines of electric current by the hydrodynamics flow. In some oversimplified situations, it can be neglected, leading to an Ohm's law in the more usual form $\mathbf{j} = \sigma \boldsymbol{E}$, which is typically the form valid for solid media. On the other hand, for fluids the term $\boldsymbol{u} \times \boldsymbol{B}$ often contains crucial information, and thus *cannot* be neglected.

For other more general situations in MHD, it may happen that Ohm's law is more complicated than (1.36). For example it might even be a time-dependent differential equation. We will not treat such complicated cases, and adopt the form (1.36) henceforth.

Regarding the physical parameters of the fluid, namely the permittivity ε, the permeability μ, and the electric conductivity σ, we choose to consider all of them as fixed scalar constants, homogeneous throughout the fluid. Let us anticipate the multifluid case, which will be specifically addressed in Section 1.3.3, and mention that for the applications we treat in this book, ε and μ can be considered as constant over the whole domain even if many different fluids are occupying the domain. In fact, these two parameters do not significantly vary for different fluids in most of the MHD applications. On the other hand, the heterogeneities of σ need to be accounted for. As shown on Table 6.1 in Chapter 6, they are large and indeed play a crucial role.

Our system of equations for one-fluid MHD modeling thus reads

$$\left\{\begin{array}{l} \left\{\begin{array}{l} \bar{\rho}\,\dfrac{\partial \boldsymbol{u}}{\partial t} + \bar{\rho}\,\boldsymbol{u}\cdot\nabla\boldsymbol{u} - \bar{\eta}\,\Delta\boldsymbol{u} + \nabla p = \boldsymbol{j}\times\boldsymbol{B} + \bar{\rho}\,\mathbf{g}, \\ \operatorname{div}\boldsymbol{u} = 0, \end{array}\right. \\ \\ \left\{\begin{array}{l} -\bar{\varepsilon}\,\dfrac{\partial \boldsymbol{E}}{\partial t} + \dfrac{1}{\bar{\mu}}\,\mathbf{curl}\,\boldsymbol{B} = \mathbf{j}, \\ \operatorname{div}\boldsymbol{E} = \dfrac{1}{\varepsilon}\,\rho_c, \\ \dfrac{\partial \boldsymbol{B}}{\partial t} + \mathbf{curl}\,\mathbf{E} = 0, \\ \operatorname{div}\boldsymbol{B} = 0, \\ \boldsymbol{j} = \bar{\sigma}\big(\boldsymbol{E} + \boldsymbol{u}\times\boldsymbol{B}\big). \end{array}\right. \end{array}\right. \tag{1.37}$$

We now turn to our third task, regarding the boundary conditions on the fields $\boldsymbol{E}$ and $\boldsymbol{B}$.

Apart from the constitutive laws (1.31) and Ohm's law (1.36), the specificity of the Maxwell equations for conducting fluids, as opposed to the same equations written, *e.g.* in the vacuum, resides in the possible need for supplying the system with *ad hoc* boundary conditions. Indeed, as mentioned above, the Maxwell equations are valid in the whole physical space $\mathbb{R}^3$. On the other hand, as the goal here is to simulate a MHD fluid that most often occupies only a bounded domain Ω in $\mathbb{R}^3$, there is the need to adequately define the simulation domain.

A first option is to set the Maxwell equations in the whole space, while solving the hydrodynamics equation on the domain Ω occupied by the fluid. Regarding only the Maxwell equations (1.33), this seems to be the method of choice. But then an extension of Ohm's law (1.36) is needed outside the fluid domain, unless the current $\boldsymbol{j}$ is assumed known there. Notice indeed that $\boldsymbol{u}$ appears in (1.36). In addition to this, the fact that the physical confinement device for the fluid is then embedded in the domain where the Maxwell equations are posed may be the source of various difficulties. Such a device is often delicate to model and treat. Therefore, alternative tracks may be followed.

A second option is to restrict the Maxwell equation to a bounded domain. In turn, this option divides into two: taking as the domain for the Maxwell equations the domain Ω occupied by the fluid, or choosing a domain larger than Ω. We cannot discuss this choice with full generality. In either situation, boundary conditions are needed. We only consider the former in the sequel.

A standard choice for the boundary conditions for (1.33) is the following

$$\begin{cases} \boldsymbol{E} \times \boldsymbol{n} = \boldsymbol{k}, \\ \boldsymbol{B} \cdot \boldsymbol{n} = q, \end{cases} \tag{1.38}$$

where $\boldsymbol{k}$ and q, respectively, are given vector and scalar functions on the boundary. We require of course that $\boldsymbol{k} \cdot \boldsymbol{n} = 0$. For a perfectly conducting boundary, $\boldsymbol{E} \times \boldsymbol{n} = \boldsymbol{k} = 0$ and $\boldsymbol{B} \cdot \boldsymbol{n} = q = 0$.

Yet another option is (see for example [118] in the framework of the steady form of the equations that will be mentioned below in (1.44)):

$$\begin{cases} \boldsymbol{B} \times \boldsymbol{n} = \boldsymbol{q}, \\ \boldsymbol{E} \cdot \boldsymbol{n} = k, \end{cases} \tag{1.39}$$

with $\boldsymbol{q} \cdot \boldsymbol{n} = 0$.

We emphasize that designing accurate boundary conditions, *i.e.* evaluations of $(\boldsymbol{k}, q)$ or $(\boldsymbol{q}, k)$, is not easy, especially because accurate experimental measures of magnetic quantities are often delicate to obtain. This is often the case in industrial environments, see Chapter 6.

1.3.2 A commonly used simplified MHD coupling

For the MHD applications that are the focus of the present book, system (1.37) may be simplified. A commonly used assumption is to neglect the first term $\partial(\varepsilon\, \boldsymbol{E})/\partial t$, often called the *displacement current*, in the Maxwell–Ampère equation (1.25). See the physical argument advocated to justify this in [47, p. 31] or [179, p. 38–40]. Then system (1.37) can be reorganized as follows. The field $\boldsymbol{j}$ is expressed from $\boldsymbol{B}$ with (1.25), next $\boldsymbol{E}$ is expressed from $\boldsymbol{j}$, $\boldsymbol{u}$ and $\boldsymbol{B}$ using (1.36), and inserted in (1.27). On the other hand, (1.26) and (1.36) are left aside. We thus obtain the following system with the triple of unknown fields $(\boldsymbol{u}, p, \boldsymbol{B})$:

$$\begin{cases} \begin{cases} \overline{\rho}\, \dfrac{\partial \boldsymbol{u}}{\partial t} + \overline{\rho}\, \boldsymbol{u} \cdot \nabla\, \boldsymbol{u} - \overline{\eta}\, \Delta \boldsymbol{u} + \nabla p = \dfrac{1}{\overline{\mu}}\, \mathbf{curl}\, \boldsymbol{B} \times \boldsymbol{B} + \overline{\rho}\, \mathbf{g}, \\[2ex] \qquad\qquad\qquad\qquad \operatorname{div} \boldsymbol{u} = 0, \end{cases} \\[4ex] \begin{cases} \dfrac{\partial \boldsymbol{B}}{\partial t} + \dfrac{1}{\overline{\mu}}\, \mathbf{curl} \left(\dfrac{1}{\overline{\sigma}}\, \mathbf{curl}\, \boldsymbol{B} \right) = \mathbf{curl}\, (\boldsymbol{u} \times \boldsymbol{B}), \\ \qquad\qquad\qquad \operatorname{div} \boldsymbol{B} = 0. \end{cases} \end{cases} \tag{1.40}$$

Correspondingly, the initial conditions are now only on the pair $(\mathbf{u}, \boldsymbol{B})$. We have already discussed the boundary condition on $\boldsymbol{u}$, that is usually a Dirichlet homogeneous boundary condition (1.16). Inserting it into Ohm's law (1.36), using $\boldsymbol{j} = \mathbf{curl}\,\frac{1}{\mu}\boldsymbol{B}$, we obtain from (1.38) the following boundary conditions on $\boldsymbol{B}$:

$$\begin{cases} \mathbf{curl}\left(\dfrac{1}{\mu}\boldsymbol{B}\right) \times \boldsymbol{n} = \tilde{\boldsymbol{k}}, \\ \boldsymbol{B} \cdot \boldsymbol{n} = q, \end{cases} \tag{1.41}$$

for some $\tilde{\boldsymbol{k}}$ depending on the parameters of the problems and the boundary condition set in (1.38).

1.3.3 *The density-dependent case*

In this section, we modify the previous MHD setting to account for the *multifluid* (or, more generally, *density-dependent*) case. It is straightforward to follow the derivations of the previous section, now keeping track of the fact that the density is not homogeneous throughout the domain. In particular, it may consist of many different, piecewise constant, densities. Correspondingly, the viscosity $\eta = \eta(\rho)$ and the electric conductivity $\sigma = \sigma(\rho)$ also vary, while the magnetic permeability μ is kept constant at the value $\overline{\mu}$, for reasons that have been discussed above. We thus consider the system:

$$\begin{cases} \begin{cases} \dfrac{\partial}{\partial t}(\rho\boldsymbol{u}) + \operatorname{div}(\rho\boldsymbol{u}\otimes\boldsymbol{u}) - \operatorname{div}\left(\eta(\rho)\left(\boldsymbol{\nabla}\boldsymbol{u} + \boldsymbol{\nabla}\boldsymbol{u}^T\right)\right) = -\boldsymbol{\nabla}p + \rho\boldsymbol{g} \\ \qquad\qquad + \dfrac{1}{\overline{\mu}}\mathbf{curl}\,\boldsymbol{B}\times\boldsymbol{B}, \\ \operatorname{div}\boldsymbol{u} = 0, \end{cases} \\ \dfrac{\partial\rho}{\partial t} + \operatorname{div}(\rho\boldsymbol{u}) = 0, \\ \begin{cases} \dfrac{\partial\boldsymbol{B}}{\partial t} + \dfrac{1}{\overline{\mu}}\mathbf{curl}\left(\dfrac{1}{\sigma(\rho)}\mathbf{curl}\,\boldsymbol{B}\right) = \mathbf{curl}\,(\boldsymbol{u}\times\boldsymbol{B}), \\ \operatorname{div}\boldsymbol{B} = 0. \end{cases} \end{cases} \tag{1.42}$$

1.4 Other MHD models

Other simplifications of system (1.37) can be adopted, making use of some steady-state approximations.

In particular, it is often considered that electromagnetic phenomena have characteristic times that are so short in comparison with the characteristic time of hydrodynamics phenomena that the stationary form of the Maxwell equations may be coupled to the time-dependent hydrodynamics equations. This gives rise to the following system:

$$\left\{\begin{array}{l}\left\{\begin{array}{l}\bar{\rho}\,\dfrac{\partial \boldsymbol{u}}{\partial t}+\bar{\rho}\,\boldsymbol{u}\cdot\nabla\boldsymbol{u}-\bar{\eta}\,\Delta\boldsymbol{u}+\nabla p=\dfrac{1}{\mu}\,\mathbf{curl}\,\boldsymbol{B}\times\boldsymbol{B}+\bar{\rho}\,\mathbf{g},\\[2ex] \qquad\qquad\qquad \operatorname{div}\boldsymbol{u}=0,\end{array}\right.\\[4ex] \left\{\begin{array}{l}\dfrac{1}{\mu}\,\mathbf{curl}\left(\dfrac{1}{\bar{\sigma}}\,\mathbf{curl}\,\boldsymbol{B}\right)=\mathbf{curl}\,(\boldsymbol{u}\times\boldsymbol{B}),\\ \qquad \operatorname{div}\boldsymbol{B}=0.\end{array}\right.\end{array}\right. \tag{1.43}$$

This system will be studied mathematically in Chapter 2, and we will question there its well-posedness. A variant consists in setting the right-hand side of the magnetic equation to zero, so that $\dfrac{1}{\mu}\,\mathbf{curl}\,\big(\dfrac{1}{\bar{\sigma}}\,\mathbf{curl}\,\boldsymbol{B}\big)=0$ replaces the magnetic equation above in (1.43). These are the one-fluid equations related to the multifluid setting considered in the linearized studies that will be examined in Chapter 6.

A further approximation is to consider that all the phenomena are steady, and thus consider the system

$$\left\{\begin{array}{l}\left\{\begin{array}{l}\bar{\rho}\,\boldsymbol{u}\cdot\nabla\boldsymbol{u}-\eta\,\Delta\boldsymbol{u}+\nabla p=\dfrac{1}{\mu}\,\mathbf{curl}\,\boldsymbol{B}\times\boldsymbol{B}+\bar{\rho}\,\mathbf{g},\\[2ex] \qquad\qquad\qquad \operatorname{div}\boldsymbol{u}=0,\end{array}\right.\\[4ex] \left\{\begin{array}{l}\dfrac{1}{\mu}\,\mathbf{curl}\left(\dfrac{1}{\sigma}\,\mathbf{curl}\,\boldsymbol{B}\right)=\mathbf{curl}\,(\boldsymbol{u}\times\boldsymbol{B}),\\ \qquad \operatorname{div}\boldsymbol{B}=0.\end{array}\right.\end{array}\right. \tag{1.44}$$

The mathematical analysis of this system and its finite element discretization will be presented in detail in Chapters 2 and 3.

1.5 The MHD system considered in the sequel

We have overviewed several variants of the MHD problem. For the industrial application we mainly target, namely the simulation of reduction cells for the production of aluminum, the MHD system we use consists of the two-fluid incompressible Navier–Stokes equations coupled with the parabolic form of the Maxwell equations through the body force term $\boldsymbol{f}=\mathbf{curl}\,\boldsymbol{B}\times\boldsymbol{B}$. This gives rise to system (1.42).

The equations (1.42) are posed on a simply connected regular domain Ω. They are supplied with the following boundary conditions on $\partial\Omega$

$$\boldsymbol{u}|_{\partial\Omega}=0, \tag{1.45}$$

$$\left\{\begin{array}{l}\mathbf{curl}\left(\dfrac{1}{\mu}\,\boldsymbol{B}\right)\times\boldsymbol{n}=\tilde{\boldsymbol{k}},\\ \qquad\qquad \boldsymbol{B}\cdot\boldsymbol{n}=q.\end{array}\right. \tag{1.46}$$

Most often, in mathematical studies, we shall take $\tilde{\mathbf{k}} = 0$ and $q = 0$ (see Chapter 3, Section 3.8 for the necessary adaptations otherwise).

Regarding the initial data, we shall assume

$$\begin{cases} \boldsymbol{u}(t=0,\mathbf{x}) = \boldsymbol{u}_0, \\ \boldsymbol{B}(t=0,\mathbf{x}) = \boldsymbol{B}_0, \end{cases} \tag{1.47}$$

and also, and this will be the key feature of our system, the following initial condition of a *piecewise constant* density[11]:

$$\rho(t=0,\mathbf{x}) = \begin{cases} \overline{\rho}_1 > 0, \quad \text{constant on}\, \Omega_1, \\ \overline{\rho}_2 > 0, \quad \text{constant on}\, \Omega_2, \\ \text{with} \quad \overline{\Omega}_1 \cup \overline{\Omega}_2 = \overline{\Omega}, \quad \text{meas}(\Omega_i) > 0, \ \ i = 1,2. \end{cases} \tag{1.48}$$

This property means that we consider two fluids with possibly different, but constant, density (and consequently viscosity and conductivity).

So the problem under consideration can be written

$$\begin{cases} \begin{cases} \begin{cases} \dfrac{\partial}{\partial t}(\rho \boldsymbol{u}) + \operatorname{div}(\rho \boldsymbol{u} \otimes \boldsymbol{u}) - \operatorname{div}(\eta(\rho)(\boldsymbol{\nabla} \boldsymbol{u} + \boldsymbol{\nabla} \boldsymbol{u}^T)) = -\boldsymbol{\nabla} p + \rho \boldsymbol{g} \\ \qquad\qquad + \dfrac{1}{\mu} \mathbf{curl}\, \boldsymbol{B} \times \boldsymbol{B}, \\ \operatorname{div} \boldsymbol{u} = 0, \end{cases} \\ \dfrac{\partial \rho}{\partial t} + \operatorname{div}(\rho \boldsymbol{u}) = 0, \\ \begin{cases} \dfrac{\partial \boldsymbol{B}}{\partial t} + \dfrac{1}{\mu} \mathbf{curl} \left(\dfrac{1}{\sigma(\rho)} \mathbf{curl}\, \boldsymbol{B} \right) = \mathbf{curl}\,(\boldsymbol{u} \times \boldsymbol{B}), \\ \operatorname{div} \boldsymbol{B} = 0, \end{cases} \end{cases} \\ \begin{cases} \boldsymbol{u}|_{\partial\Omega} = 0, \\ \boldsymbol{B} \cdot \boldsymbol{n} = 0, \\ \mathbf{curl}\, \boldsymbol{B} \times \boldsymbol{n} = 0, \end{cases} \\ \begin{cases} \boldsymbol{u}(t=0,.) = \boldsymbol{u}_0, \\ \boldsymbol{B}(t=0,.) = \boldsymbol{B}_0, \\ \rho(t=0,.) = \begin{cases} \overline{\rho}_1 > 0, \quad \text{constant on}\, \Omega_1, \\ \overline{\rho}_2 > 0, \quad \text{constant on}\, \Omega_2, \\ \text{with} \quad \overline{\Omega}_1 \cup \overline{\Omega}_2 = \overline{\Omega}, \quad \text{meas}(\Omega_i) > 0, \ i = 1,2. \end{cases} \end{cases} \end{cases} \tag{1.49}$$

[11] This property will propagate forward in time.

The mathematical study of this system, together with its numerical simulation, is the main topic of the present book. Some other models in the list of those introduced in this chapter will however be briefly addressed, not in all details.

We find it useful to now emphasize the various properties assumed in order for the fluid under consideration to satisfy the equations (1.49):

- the fluid[12] is isotropic and Newtonian (equations (1.6), (1.7));
- when there are several of them, the fluids are immiscible (thus the propagation in time of the values in (1.48);
- the flow is incompressible (condition (1.10));
- the temperature is fixed and constant;
- the density is either constant throughout the domain for system (1.13), or piecewise constant, the latter meaning we are treating several non-miscible homogenous fluids[13];
- the displacement current $\partial(\varepsilon\,\mathbf{E})/\partial t$ is neglected,
- Ohm's law is taken in the form (1.36);
- the magnetic permeability is assumed constant;
- we neglect in (1.34) the electric force on the fluid since the charge density within the conductor is zero in our framework (see [47] pp. 29–30 or [179] pp. 40–41).

Although the above system (1.49) is the focus of the present book, we must mention that all our arguments readily apply, possibly through easy modifications, to the following situations:

- more than two fluids are considered;
- other types of boundary conditions for the magnetic field (see Section 3.8), namely

$$\boldsymbol{B} \times \boldsymbol{n} = \boldsymbol{q}; \tag{1.50}$$

- other types of initial conditions, in particular corresponding to the case of more than two fluids, or more generally to the case

$$\rho(t = 0, \mathbf{x}) = \rho_0(\mathbf{x}) \geq \underline{\rho}_0 \text{ constant} > 0. \tag{1.51}$$

Finally, there are some situations we are unable to treat with all the mathematical rigor we would like. Among others, we can quote:

- the case when surface tension effects need to be accounted for;
- the case when the parabolic approximation for the evolution of the magnetic field is no valid longer, and one has to treat the "original" Maxwell system (1.33).

[12]In the present list, the denomination "the fluid" of course refers to "the set of fluids" in the case of several fluids modeled by system (1.14).

[13]This property is *assumed* on the initial condition and shown to be propagated forward in time, when the mathematical setting is convenient.

1.6 Non-dimensionalized equations

For the twofold purpose of numerical analysis and implementation, the above equations have to be expressed in non-dimensionalized form.

To this end, we introduce a characteristic velocity U, a characteristic length L, and a characteristic magnetic field B. The characteristic time and pressure are then, respectively, L/U and $\rho_1 U^2$. We define the following non-dimensional parameters (where $i = 1, 2$ denotes the fluid):

$$
\begin{aligned}
Re_i &= \frac{\rho_i U L}{\eta(\rho_i)}, && \text{(Reynolds numbers)},\\
Rm_i &= \mu\,\sigma(\rho_i)\, LU, && \text{(magnetic Reynolds numbers)},\\
S &= \frac{B^2}{\mu\,\rho_1\, U^2}, && \text{(coupling parameter)},\\
Fr &= \frac{U^2}{gL}, && \text{(Froude number)},\\
M &= \frac{\rho_2}{\rho_1}, && \text{(density ratio)}.
\end{aligned}
$$

The coupling parameter S can also be expressed as a function of the more widely used Hartmann number

$$
Ha_i = BL\left(\frac{\sigma(\rho_i)}{\eta(\rho_i)}\right)^{1/2} \quad \text{and } S = \frac{(Ha_1)^2}{Re_1 Rm_1} = M\frac{(Ha_2)^2}{Re_2 Rm_2}.
$$

By rewriting the MHD system (1.49) in terms of the new variables $\tilde{x} = x/L$, $\tilde{t} = t/T$, $\tilde{\boldsymbol{u}} = \boldsymbol{u}/U$, $\tilde{p} = p/P$, $\tilde{\boldsymbol{B}} = \boldsymbol{B}/B$ $\tilde{\rho} = \rho/\rho_1$, $\tilde{\boldsymbol{g}} = \boldsymbol{g}/g$, one obtains (omitting the *tilde* for the sake of clarity):

$$
\left\{
\begin{array}{l}
\text{in } \Omega_{1,t}, \left\{
\begin{array}{l}
\dfrac{\partial \boldsymbol{u}}{\partial t} + \boldsymbol{u}\cdot\boldsymbol{\nabla}\boldsymbol{u} + \boldsymbol{\nabla}p - \dfrac{2}{Re_1}\operatorname{div}(\boldsymbol{D}(\boldsymbol{u})) = \dfrac{1}{Fr}\boldsymbol{g} + S\,\mathbf{curl}\,\boldsymbol{B}\times\boldsymbol{B},\\
\dfrac{\partial \boldsymbol{B}}{\partial t} + \dfrac{1}{Rm_1}\mathbf{curl}\,(\mathbf{curl}\,\boldsymbol{B}) = \mathbf{curl}\,(\boldsymbol{u}\times\boldsymbol{B}),
\end{array}\right.\\[2ex]
\text{in } \Omega_{2,t}, \left\{
\begin{array}{l}
M\dfrac{\partial \boldsymbol{u}}{\partial t} + M\boldsymbol{u}\cdot\boldsymbol{\nabla}\boldsymbol{u} + \boldsymbol{\nabla}p - \dfrac{2M}{Re_2}\operatorname{div}(\boldsymbol{D}(\boldsymbol{u})) = \dfrac{M}{Fr}\boldsymbol{g} + S\,\mathbf{curl}\,\boldsymbol{B}\times\boldsymbol{B},\\
\dfrac{\partial \boldsymbol{B}}{\partial t} + \dfrac{1}{Rm_2}\mathbf{curl}\,(\mathbf{curl}\,\boldsymbol{B}) = \mathbf{curl}\,(\boldsymbol{u}\times\boldsymbol{B}),
\end{array}\right.\\
\operatorname{div}\boldsymbol{u} = 0,\\
\operatorname{div}\boldsymbol{B} = 0,\\
\dfrac{\partial \rho}{\partial t} + \operatorname{div}(\rho\boldsymbol{u}) = 0.
\end{array}
\right.
\tag{1.52}
$$

This system is complemented with adequate boundary and initial conditions on $\boldsymbol{u}$ and $\boldsymbol{B}$, as in (1.49):

$$
\text{on } \partial\Omega, \left\{
\begin{array}{r}
\boldsymbol{u} = 0,\\
\boldsymbol{B}\cdot\boldsymbol{n} = 0,\\
\mathbf{curl}\,\boldsymbol{B}\times\boldsymbol{n} = 0,
\end{array}\right.
\tag{1.53}
$$

$$\begin{cases} \boldsymbol{u}(t=0,.) = \boldsymbol{u}_0, \\ \boldsymbol{B}(t=0,.) = \boldsymbol{B}_0, \\ \rho(t=0,.) = \begin{cases} 1, & \text{constant on}\,\Omega_{1,0}, \\ M, & \text{constant on}\,\Omega_{2,0}, \\ \text{with} & \overline{\Omega}_{1,0} \cup \overline{\Omega}_{2,0} = \overline{\Omega}, \quad \text{meas}(\Omega_{i,0}) > 0,\ i = 1,2. \end{cases} \end{cases} \tag{1.54}$$

The domains $\Omega_{i,t}$ in (1.52) are defined by: $\Omega_{1,t} = \{\mathbf{x} \in \Omega, \rho(t,\mathbf{x}) = 1\}$ and $\Omega_{2,t} = \{\mathbf{x} \in \Omega, \rho(t,\mathbf{x}) = M\}$.

Moreover, since we wrote the MHD equations separately on $\Omega_{1,t}$ and $\Omega_{2,t}$ in (1.52), we need to add the following transmission conditions at the interface $\Sigma_t = \partial\Omega_{1,t} \cap \partial\Omega_{2,t}$ to obtain a system equivalent to (1.49): on Σ_t,

$$\begin{cases} \left(-p\mathrm{Id} + \dfrac{2}{Re_1}\boldsymbol{D}(\boldsymbol{u})\right)\Big|_{\Omega_1} \cdot \boldsymbol{n} = \left(-p\mathrm{Id} + \dfrac{2}{Re_2}\boldsymbol{D}(\boldsymbol{u})\right)\Big|_{\Omega_2} \cdot \boldsymbol{n}, \\ \left(\dfrac{1}{Rm_1}\mathbf{curl}\,(\boldsymbol{B}) - \boldsymbol{u} \times \boldsymbol{B}\right)\Big|_{\Omega_1} \times \boldsymbol{n} = \left(\dfrac{1}{Rm_2}\mathbf{curl}\,(\boldsymbol{B}) - \boldsymbol{u} \times \boldsymbol{B}\right)\Big|_{\Omega_2} \times \boldsymbol{n}, \end{cases} \tag{1.55}$$

where $\boldsymbol{n}$ denotes the normal to Σ_t, directed from $\Omega_{1,t}$ to $\Omega_{2,t}$. The first condition is the continuity of the normal stress at the interface[14] while the second condition expresses the continuity of the tangential component of the electric field. Notice that we also implicitely require that $\boldsymbol{u}$ and $\boldsymbol{B}$ are continuous at the interface.

[14] This is the condition which will be modified to take into account surface tension effects (see Section 5.1.4 for details).

2

MATHEMATICAL ANALYSIS OF ONE-FLUID PROBLEMS

The focus of this book is the nonlinearity of the MHD system (1.49) which we reproduce here for the convenience of the reader

$$
\begin{cases}
\begin{cases}
\begin{cases}
\dfrac{\partial}{\partial t}(\rho \boldsymbol{u}) + \operatorname{div}(\rho \boldsymbol{u} \otimes \boldsymbol{u}) - \operatorname{div}(\eta(\rho)(\boldsymbol{\nabla} \boldsymbol{u} + \boldsymbol{\nabla} \boldsymbol{u}^T)) = -\boldsymbol{\nabla} p + \rho \boldsymbol{g} \\
\qquad\qquad\qquad\qquad\qquad\qquad\qquad\qquad + \dfrac{1}{\overline{\mu}} \mathbf{curl}\, \boldsymbol{B} \times \boldsymbol{B}, \\
\qquad\qquad\qquad\qquad\qquad\qquad \operatorname{div} \boldsymbol{u} = 0,
\end{cases} \\
\dfrac{\partial \rho}{\partial t} + \operatorname{div}(\rho \boldsymbol{u}) = 0, \\
\begin{cases}
\dfrac{\partial \boldsymbol{B}}{\partial t} + \dfrac{1}{\overline{\mu}} \mathbf{curl}\,(\dfrac{1}{\sigma(\rho)} \mathbf{curl}\, \boldsymbol{B}) = \mathbf{curl}\,(\boldsymbol{u} \times \boldsymbol{B}), \\
\qquad\qquad \operatorname{div} \boldsymbol{B} = 0,
\end{cases}
\end{cases} \\
\rho(t = 0, \mathbf{x}) = \begin{cases}
\overline{\rho}_1 > 0, & \text{constant on } \Omega_1, \\
\overline{\rho}_2 > 0, & \text{constant on } \Omega_2, \\
\text{with } \ \overline{\Omega}_1 \cup \overline{\Omega}_2 = \overline{\Omega}, & \operatorname{meas}(\Omega_i) > 0, \ i = 1, 2.
\end{cases}
\end{cases}
$$

We omit for brevity the initial conditions on $\boldsymbol{u}$, $\boldsymbol{B}$, and the boundary conditions. Instead of *the nonlinearity* of the above system, we should more appropriately write *the nonlinearities*, since there are schematically three sources of nonlinearity:

(i) the Navier term $\boldsymbol{u} \cdot \nabla \boldsymbol{u}$ (or $\operatorname{div}(\rho \boldsymbol{u} \otimes \boldsymbol{u})$) in the left-hand side of the Navier–Stokes equation, which is a nonlinearity of purely hydrodynamic nature;

(ii) the coupling between the major two equations through the right-hand side terms $\mathbf{curl}\, \boldsymbol{B} \times \boldsymbol{B}$, $\mathbf{curl}\,(\boldsymbol{u} \times \boldsymbol{B})$, respectively; and

(iii) the fact that several fluids with different densities are involved, or equivalently that the density ρ is not homogeneous, which amounts to saying that the nonlinear term $\operatorname{div}(\rho \boldsymbol{u})$ in the equation of conservation of mass is not trivial.

The present chapter along with Chapter 3 address the difficulties arising from the first two sources of nonlinearity, leaving for Chapter 4 and Chapter 5 the additional difficulties caused by the third nonlinearity in the system, in fact the nastiest one. We thus restrict ourselves here to the case of *one* homogeneous fluid, first modeled purely hydrodynamically by the incompressible homogeneous Navier–Stokes equation (1.13), and next in the setting of the MHD equations (1.40). In order to lighten the notation, we assume throughout the chapter that the value of the constant density $\overline{\rho}$, of the constant magnetic permeability $\overline{\mu}$, and of the constant electric conductivity $\overline{\sigma}$ (the latter except in Sections 2.2.5 and 2.2.6) are all set to one. This may be done without loss of generality. We could equally assume $\overline{\eta} = 1$ but we will not do so, and explicitly keep $\overline{\eta}$ in our equations, in order to emphasize the crucial role played by the viscous (mathematically regularizing) term $\overline{\eta}\,\Delta \boldsymbol{u}$. The analysis of the Euler equation ($\overline{\eta} = 0$) would be completely different, several orders of magnitude more difficult. Some issues are even unsolved to date. Our equations (1.13) and (1.40) thus respectively read:

$$\begin{cases} \dfrac{\partial \boldsymbol{u}}{\partial t} + \boldsymbol{u} \cdot \boldsymbol{\nabla} \boldsymbol{u} - \overline{\eta}\,\Delta \boldsymbol{u} = -\boldsymbol{\nabla} p + \boldsymbol{f}, \\ \\ \operatorname{div} \boldsymbol{u} = 0, \end{cases}$$

and

$$\begin{cases} \begin{cases} \dfrac{\partial \boldsymbol{u}}{\partial t} + \boldsymbol{u} \cdot \nabla \boldsymbol{u} - \overline{\eta}\,\Delta \boldsymbol{u} + \nabla p = \mathbf{curl}\,\boldsymbol{B} \times \boldsymbol{B} + \boldsymbol{g}, \\ \\ \operatorname{div} \boldsymbol{u} = 0, \end{cases} \\ \\ \begin{cases} \dfrac{\partial \boldsymbol{B}}{\partial t} + \mathbf{curl}\,(\mathbf{curl}\,\boldsymbol{B}) = \mathbf{curl}\,(\boldsymbol{u} \times \boldsymbol{B}), \\ \operatorname{div} \boldsymbol{B} = 0. \end{cases} \end{cases}$$

We focus here on the mathematical theory. The numerical approximation is dealt with in Chapter 3. The material contained in the present chapter is now standard. In particular, the mathematical analysis of the (purely hydrodynamic) homogeneous incompressible Navier–Stokes equations (1.13) is well known (although of course still in development for its most advanced aspects). Following the seminal work of Leray [145–147], a tremendous number of works have been devoted to the subject. Among a long list of authors who have devoted a whole monograph or a significant part of a textbook to the subject, we would like to cite J.-L. Lions in [153, 154], R. Temam in the two monographs [227] and [229], and P.-L. Lions in [155].

The mathematical analysis of the one-fluid MHD equations is also a standard topic. However, it has not been explored to the extent of the purely hydrodynamic case, mostly because it is related to some specific applications, and therefore not so generic. Two standard mathematical references are the book by G. Duvaut and J.-L. Lions [73], and the article by M. Sermange and R. Temam [208].

In such a context, there is no need to reproduce in detail the complete mathematical analysis performed to date, covering all possible contexts (depending on the dimension of the ambient space, the domain where the equations are posed, the boundary conditions, the regularity of the data, *etc.*) and all possible settings (weak solutions, strong solutions, mild solutions[15]). Should we do this, the present book would be a pale mirror of existing textbooks of outstanding quality and completeness. However, for the convenience of the reader, for consistency, and also in order to prepare the ground for the analysis of the non-homogeneous case, we find it useful to still devote the present chapter and the following to the one-fluid case. The rule of the game we have fixed for this purpose is the following. After presenting in Section 2.1.1 some elementary mathematical ingredients that will be useful throughout the book, we treat in (almost) all details two examples of one-fluid systems: the two-dimensional purely hydrodynamic case, and the three-dimensional MHD case. These two detailed examples will be the matter of Sections 2.1.2 and 2.2, respectively. This allows us, for all the other problems addressed in the rest of the book, to only concentrate on the key issues and on the bottom lines for the mathematical arguments, with a special emphasis on the nonlinear phenomena. In Chapter 3, the same rule will be applied when describing the discretization strategy and outlining the various issues raised by the numerical analysis.

In addition to this, and as announced in the introduction of the book, we will mainly concentrate on manipulating *weak* solutions. We shall mention *strong* solutions only incidentally, and, above all, for comparison. The reason why we concentrate on weak solutions is that, except for very specific situations (some of them are studied in the present chapter such as the two-dimensional setting, or the small-force setting), strong solutions are not known to exist for all times for practical cases of interest.

Our overview of the mathematical analysis will hopefully provide the reader not familiar with such issues with a flavor of the mathematical challenges, difficulties and techniques. On the other hand, the expert in such theoretical issues may easily skip some parts of Section 2.1, in particular 2.1.2, or even the whole section itself. He may directly proceed either to Section 2.2, or to Chapter 3 for the discretization issues, or even to Chapters 4 and 5 for the non-homogeneous case. Should the need arise, the reader will only have to come back to the present chapter to see the most basic definitions used throughout the book.

Without betraying our main purpose, which is the study of (1.49), we will incidentally allow ourselves some incursions in the mathematical study of some other settings. This is the topic of Section 2.2.3, where we briefly discuss the mathematical difficulties arising from (1.37), of Section 2.2.4 where we consider steady-state problems such as (1.44), and also Sections 2.2.5 and 2.2.6. In the

[15]The reader not familiar with this terminology will be introduced to it in the course of the chapter; let us only say for the time being that weak solutions are solutions of the equations in a sense broader than the usual sense, strong solutions are sufficiently regular solutions so that the equation has its usual meaning, and the notion of mild solution is intermediate between the above two.

latter, we shall see that (1.43), although well-posed mathematically in some circumstances (*i.e.* it admits a unique solution), is also likely to be ill-posed in some others.

Let us mention that, again for avoiding redundancy with the existing literature, we shall not provide all the details of the proofs either, when we address the much more advanced questions related to the non-homogeneous case in Chapter 4. As we get to such more advanced issues, the text will become less elementary, of course. We will concentrate on the crucial issues and on the major mathematical difficulties to overcome in such an intricate situation. All the unnecessary technical details are omitted on purpose. They can be read in the various articles this book relies upon.

To conclude this introductory section, we would like to stress the following convention. Throughout this chapter, we will assume for simplicity that the domain Ω on which the equations are posed has a boundary as regular as needed for all our results of regularity of the solutions to hold. Most of the time, we will not make precise the regularity of the boundary in all the statements of the results below. We are aware that this is an abuse, that the regularity of the boundary is an important issue and that some results may be invalid, or at least some substantial difficulties may arise, when the boundary is not sufficiently regular. Such technicalities will be very much detailed in the next chapter, because they have important consequences on the numerics. However, following our line of simplicity for the mathematical exposition, we will omit them in the present chapter.

2.1 Mathematical results on the incompressible homogeneous Navier–Stokes equations

The present section is devoted to the mathematical analysis of the purely hydrodynamic incompressible homogeneous Navier–Stokes equations, namely

$$\begin{cases} \dfrac{\partial \boldsymbol{u}}{\partial t} + \boldsymbol{u} \cdot \boldsymbol{\nabla} \boldsymbol{u} - \overline{\eta} \Delta \boldsymbol{u} = -\boldsymbol{\nabla} p + \boldsymbol{f}, \\ \operatorname{div} \boldsymbol{u} = 0. \end{cases} \tag{2.1}$$

As already mentioned this is a well-explored subject, which we shall not treat in the finest details.

To start with, we lay in Section 2.1.1 some mathematical basic results. Then, in Section 2.1.2, we consider in some details the two-dimensional case, on a bounded domain, with homogeneous Dirichlet boundary conditions. This case is known to be the simplest one, and to some extent it is the only case when complete results are available. In addition to this completeness, the fact is that the proofs are rather easy. Many delicate questions related to functional analysis can be avoided. The proofs can be considered as prototypical for the proofs in dimension 3, even if of course the arguments become significantly more complicated in the latter case, or even do not allow us to conclude. By focusing on this "easy" case, we will be able to highlight (a) the major questions that are of mathematical interest, (b) the main theoretical difficulties encountered, and (c) how they are overcome by now standard techniques.

Next, we turn (more rapidly) in Section 2.1.3 to the three-dimensional case. Section 2.1.4 overviews some specific questions, such as the case of other geometrical settings.

2.1.1 Some basics

The present section is primarily aimed at beginners in the field of the mathematical analysis of the models of fluid mechanics, or, more generally, of time-dependent nonlinear partial differential equations. The notions introduced are basic, the properties mentioned are simple. The inequalities and the embeddings of functional spaces we mention are not sharp. However, we cannot start ex nihilo. We are obliged, for the sake of brevity, to consider some notions as known. The prerequisite is some basic notions of functional analysis, such as the definition and the main properties of Lebesgue spaces L^p, $1 \leq p \leq +\infty$, and of Sobolev spaces H^k.

Additionally, we consider as known the elementary properties of (infinite-dimensional) Banach and Hilbert spaces, the definition, and the basic properties, of the weak topologies of these spaces. Those are much less elementary notions. The most commonly used Sobolev embeddings will also be considered known. For all these notions, we refer, *e.g.* to the textbooks by H. Brézis [30], E.H. Lieb and M. Loss [152], in addition to the more focused monographs [153, 154, 227, 229, 155] already cited above. Let us, however, mention that whenever it does not significantly affect the length of our exposition, we shall recall some of these basic notions. This will be the case in Subsection 2.1.1.1, where we recall the fundamental notion of weak topology, first in the specific case of the L^2 space.

Experts can see the present section as a warm-up for the sequel, or may simply skip it.

2.1.1.1 Basic mathematics tools To start with, we recall the definition of the weak topology of the space $L^2(0,1)$.

Definition 2.1 *A sequence u_n of functions in $L^2(0,1)$ is said to weakly converge to the function $u \in L^2(0,1)$ when for any $v \in L^2(0,1)$,*

$$\lim_{n \longrightarrow +\infty} \int_0^1 u_n\, v = \int_0^1 u\, v. \tag{2.2}$$

A simple example is

$$u_n(x) = \sin(2\pi\, n\, x), \tag{2.3}$$

which weakly converges to the function $u \equiv 0$ on $(0,1)$. It is indeed easy to verify that (2.2) holds by integration by parts when v is C^1, and then argue by density of the C^1 functions in L^2. On the other hand, the sequence (2.3) *does not* converge to zero for the L^2 topology (often called the *strong topology*, to emphasize the difference from the weak one). Indeed, should it be the case we would have

$$\lim_{n \longrightarrow 0} \int_0^1 |u_n - u|^2 = 0 \tag{2.4}$$

thus the L^2 norm of u_n would converge to zero, but

$$\int_0^1 |u_n|^2 = \int_0^1 \sin^2(2\pi nx) = \int_0^1 \left(\frac{1}{2} - \frac{1}{2}\cos(4\pi nx)\right) = \frac{1}{2}.$$

However, it is true that *strong convergence implies weak convergence*, by a simple application of the Cauchy–Schwarz inequality. Weak convergence is thus a weaker mathematical notion than strong convergence, and the two do not coincide.

Likewise, we may define the weak topology on any Hilbert space with scalar product $(\cdot,\cdot)$:

$$u_n \text{ weakly converges to } u \text{ in } V \text{ if } (u_n, v) \overset{n \longrightarrow +\infty}{\longrightarrow} (u, v) \;\; \forall v \in V. \tag{2.5}$$

This allows us to define the weak topology again of any L^2 space, but also, *e.g.* that of $H^1(0,1)$:

$$u_n \text{ weakly converges to } u \text{ in } H^1(0,1) \text{ when}$$
$$\int_0^1 (u_n' v' + u_n v) \overset{n \longrightarrow +\infty}{\longrightarrow} \int_0^1 (u'v' + uv) \;\; \forall v \in H^1(0,1),$$

or that of $H^1(\Omega)$, for $\Omega \subset \mathbb{R}^d$, by a straightforward adaptation of the above definition.

Another generalization is the notion of weak topology for the L^p spaces, $1 \leq p \leq +\infty$, which are Banach, but (for $p \neq 2$) not Hilbert, spaces.

Definition 2.2 *Let Ω be an open domain in $\mathbb{R}^d$, $d \geq 1$. For $1 \leq p < +\infty$, the sequence $u_n \in L^p(\Omega)$ is said to weakly converge to $u \in L^p(\Omega)$ when for all $v \in L^q(\Omega)$, $\frac{1}{p} + \frac{1}{q} = 1$,*

$$\lim_{n \longrightarrow +\infty} \int_\Omega u_n\, v = \int_\Omega u\, v. \tag{2.6}$$

In the special case $p = \infty$, the notion needs to be adapted (owing to the fact that the dual space of L^∞ is not an L^p space): the sequence $u_n \in L^\infty(\Omega)$ is said to weakly-$\star$ converge to $u \in L^\infty(\Omega)$ when for all functions $v \in L^1(\Omega)$, (2.6) holds. There are some properties of the weak topology that are not valid for the weak-$\star$ topology. We will mention them in the sequel when needed (see, e.g. Proposition 2.4 and Remark 2.1.2 below).

The notation

$$u_n \overset{n \to \infty}{\rightharpoonup} u \tag{2.7}$$

is used for the weak (or the weak-$\star$) convergence, in order to emphasize the difference with the strong convergence usually denoted by

$$u_n \overset{n \to \infty}{\longrightarrow} u. \tag{2.8}$$

Remark 2.1.1 The definition of the weak topology and the weak-$\star$ topology of Banach spaces is the obvious generalization of the above definition. A sequence u_n in E is said to weakly converge in the Banach space E when

$$\lim_{n \longrightarrow +\infty} \langle u_n, v \rangle = \langle u, v \rangle$$

for all $v \in E'$ (the dual space of E, *i.e.* the space of *continuous* linear forms on E). On the other hand, a sequence $v_n \in E'$ is said to weakly-$\star$ converge if

$$\lim_{n \longrightarrow +\infty} \langle u, v_n \rangle = \langle u, v \rangle$$

for all $u \in E$. In the above formula, $\langle \cdot, \cdot \rangle$ of course denotes the duality between E and E'. $\diamondsuit$

We now investigate the basic properties of the weak topology regarding products of sequences. It is a standard fact that when u_n (strongly) converges to u in L^p and v_n (strongly) converges to v in L^q, $1 \le p \le +\infty$, $\frac{1}{p} + \frac{1}{q} = 1$, then $u_n v_n$ (strongly) converges to uv in L^1. This is a straightforward consequence of the *Hölder inequality*

$$\|uv\|_{L^1(\Omega)} \le \|u\|_{L^p(\Omega)} \|v\|_{L^q(\Omega)}, \tag{2.9}$$

for u in $L^p(\Omega)$, v in $L^q(\Omega)$, $\frac{1}{p} + \frac{1}{q} = 1$.

The following proposition addresses the same question when weak convergence comes into play.

Proposition 2.3 *Let $1 \le p \le +\infty$ and q such that $\frac{1}{p} + \frac{1}{q} = 1$.*

(i) If u_n strongly converges to u in $L^p(\Omega)$, and v_n weakly converges to v in $L^q(\Omega)$, then $u_n v_n$ weakly converges to uv in $L^1(\Omega)$.
(ii) If u_n weakly converges (respectively, weakly-$\star$ converges if $p = +\infty$) to u in $L^p(\Omega)$, and v_n weakly converges to v in $L^q(\Omega)$, then nothing can be said on the sequence $u_n v_n$.

As for (ii), we emphasize that for the sequence $u_n v_n$ to converge (even weakly, or even in the sense of distributions), more needs to be known about either of the sequence u_n or v_n. This may come from Proposition 2.5 below.

The property of the weak topology that is the very motivation for the introduction of this somewhat delicate notion, is:

Proposition 2.4 *Any bounded sequence u_n in $L^p(\Omega)$, $1 < p < +\infty$ (respectively $p = \infty$) is weakly convergent (respectively weakly-$\star$ convergent) up to an extraction.*

Remark 2.1.2 Note that the case $p = 1$ is not addressed by the above proposition (the result is then false). Note also that the case $p = \infty$ is specific. $\diamondsuit$

Indeed, Proposition 2.4 *creates* an object, namely the weak limit of the extraction. The mathematical analysis may, in a second step, provide additional information on this object. Typically, in the present book, the sequence is a sequence of approximate solutions of the equations under study (in a sense to be made precise later on), and the weak limit is the tentative solution of the equations.

A key property is the following: if, in addition to the bounds on the sequence under consideration, we dispose of bounds on the sequence of derivatives then we not only have weak convergence, but indeed strong convergence. This is formalized, *e.g.* by the celebrated

Proposition 2.5 [Rellich's theorem] *Assume that* Ω *is a bounded regular domain in* $\mathbb{R}^d$. *Then a weakly convergent sequence in* $H^1(\Omega)$ *is, up to an extraction, strongly convergent in* $L^2(\Omega)$.

Remark 2.1.3 Note that the boundedness of the domain plays a crucial role, and that, *e.g.* the result does not hold when $\Omega = \mathbb{R}^d$. ◊

The above theorem is *one* instance of the so-called *compact Sobolev embeddings* that state the same type of result, replacing H^1 by H^k, or even $W^{k,p}$, and L^2 by L^q, respectively, for appropriate choices of k, p, q. We refer to the bibliography for the definition of these spaces, along with more details on their basic properties and the related Sobolev embeddings. We will use henceforth such embeddings without further justification.

2.1.1.2 A formal study Without loss of generality we may assume $\overline{\rho} = 1$ and consider

$$\text{(NS)} \quad \begin{cases} \dfrac{\partial \boldsymbol{u}}{\partial t} + \boldsymbol{u} \cdot \boldsymbol{\nabla} \boldsymbol{u} - \overline{\eta}\, \Delta \boldsymbol{u} = -\boldsymbol{\nabla} p + \boldsymbol{f}, \\ \\ \operatorname{div} \boldsymbol{u} = 0, \end{cases}$$

supplied with the initial condition

$$\boldsymbol{u}(t = 0, \cdot) = \boldsymbol{u}_0,$$

and the boundary condition

$$\boldsymbol{u}|_{\partial\Omega} = 0.$$

The first mathematical observation is that, as already mentioned in Chapter 1, this system is not a Cauchy problem from the mathematical viewpoint, *i.e.* it is not an evolution system of the form

$$\frac{dY}{dt} = F(t, Y). \tag{2.10}$$

Indeed, the vector of unknowns Y is $\begin{pmatrix} \boldsymbol{u} \\ p \end{pmatrix}$ and there is no time evolution equation on the unknown field p. Even when forgetting this difficulty, another observation is that the divergence-free constraint $\operatorname{div} \boldsymbol{u} = 0$ needs to be fulfilled for all time. Formally, the system under consideration would rather look like a differential algebraic system

$$\begin{cases} \dfrac{dY}{dt} = F(t, Y), \\ G(Y) = 0. \end{cases}$$

Actually, the above two difficulties are two sides of one sole mathematical fact. Somewhat vaguely stated, the pressure p is indeed only an auxiliary unknown, playing the role of the Lagrange multiplier of the incompressibility constraint. In the mathematical analysis, these two difficulties will be solved simultaneously, indeed by transforming the system into a Cauchy problem. The latter will be performed by setting the equation on an adequate functional space (namely the space V defined in (2.19) below). We shall see in Chapter 3 that the numerical approach will be somewhat different.

The second mathematical observation concerns the nonlinear character of the system, owing to the presence of the Navier term $\boldsymbol{u} \cdot \boldsymbol{\nabla u}$. It is intuitive that nonlinear equations are more difficult to study than linear ones, and this is indeed the case here.

To set the stage, let us momentarily suppress the above two difficulties, by considering the much simpler *heat equation*

$$\frac{\partial y}{\partial t} - \Delta y = 0, \tag{2.11}$$

supplied with homogeneous Dirichlet boundary conditions. We would like to prove there exists a solution to this equation, for some given initial condition $y(t = 0) = z$. To this end, many methods exist. Let us describe the following one. Equation (2.11) is a partial differential equation, and may therefore be seen as an ordinary differential set on the unknown function $y(t, \cdot)$, valued in an *infinite* dimensional space (of functions of the variable $\mathbf{x}$). One way to simplify the proof of the existence of a solution to (2.11) is to introduce an orthonormal basis set of $L^2(\Omega)$, denoted by $\{w_i\}_{i \in \mathbb{N}}$, to search for y in the form

$$y(t, \mathbf{x}) = \sum_{i \in \mathbf{N}} y_i(t)\, w_i(\mathbf{x}), \tag{2.12}$$

and next to project (2.11) on this basis set. We correspondingly project the initial condition on the same basis set. We then see that (2.11) is formally equivalent to

$$\frac{dy_i(t)}{dt} - \sum_{k \in \mathbf{N}} y_k(t)(\Delta w_k, w_i) = 0, \quad \forall i \in \mathbb{N}. \tag{2.13}$$

In this manner, the problem of proving the existence of a solution to a time-dependent *partial* differential equation amounts to solving a Cauchy problem for an *infinite system* of *ordinary* differential equations. Despite the very formal nature of the above lines, we shall see in the sequel that it illustrates a standard strategy:

(Fact 1) { Sooner or later, the proof of existence relies upon the Cauchy–Lipschitz theorem for a system of ordinary differential equations. The latter is usually obtained by projecting the equation on an adequate basis set.

In fact, let us try to formalize the above argument. When the decomposition (2.12) is truncated at order n, the set of equations (2.13) indeed defines, instead of y_i, the i-th component y_i^n of the approximation

$$y^n = \sum_{i=0}^{n} y_i^n(t) w_i$$

of $y(t,x)$. Equation (2.13) indeed holds for all indices $1 \le i \le n$, *i.e.*

$$\frac{dy_i^n(t)}{dt} - \sum_{k=0}^{n} y_k^n(t)(\Delta w_k, w_i) = 0, \quad \forall i \le n, \tag{2.14}$$

which is a *finite* system of ODEs. In other words, y^n formally solves (2.11) with $y^n(t=0,\cdot) = z^n = \sum_{i=1}^{n}(z, w_i)w_i$ as an initial condition. Once the existence of y^n is obtained from the Cauchy–Lipschitz theorem, we let n go to infinity so that y^n approaches in the limit a solution to (2.11), denoted by y. For this limit process to be valid, we need y^n to satisfy adequate bounds. This is clearly necessary, but this is also sufficient since we know from basic results of functional analysis that bounded sequences converge up to an extraction, when the topology is adequate (recall Proposition 2.4). In a linear equation such as (2.11) the weak convergence is sufficient to pass to the limit. It successively shows that y^n weakly converges to y and that y is a solution.

Proving the existence thus amounts to proving some bounds on the y^n, that are called *a priori* bounds. To understand why such bounds hold, it suffices to formally multiply (2.11) by y (the tentative solution) which yields

$$\frac{1}{2}\frac{d}{dt}\int_\Omega |y|^2 + \int_\Omega |\nabla y|^2 = 0,$$

thus

$$\frac{1}{2}\int_\Omega |y(T,\cdot)|^2 + \int_0^T \int_\Omega |\nabla y|^2 = \frac{1}{2}\int_\Omega |y(t=0,\cdot)|^2. \tag{2.15}$$

This is a formal estimate on the tentative solution y in the functional space $L^\infty(0,T;L^2(\Omega)) \cap L^2(0,T;H_0^1(\Omega))$. It holds provided $y(t=0,\cdot) \in L^2(\Omega)$. The same estimation technique, with some more difficult technical details though, allows us to *rigorously* prove that the sequence y^n is bounded in the same functional space[16]. This allows us to pass to the limit for the weak topology in (2.14),

[16] We will come back to this in Section 2.1.2.1.

and eventually obtain the existence of a solution to (2.11). We summarize this in

(Fact 2) $\left\{ \begin{array}{l} \text{In the proof of existence, } \textit{a priori} \text{ estimates are needed} \\ \text{in order to set the equations in the correct functional space,} \\ \text{and pass to the limit in the sequence approximating the solution.} \end{array} \right.$

We have considered so far the linear case, assimilating (NS) to (2.11). This is not sufficient since our equation (NS) is nonlinear.

For (NS), the *a priori* estimate analogous to (2.15) then reads

$$\frac{1}{2}\int_\Omega |\boldsymbol{u}(T,\cdot)|^2 + \overline{\eta}\int_0^T\int_\Omega |\boldsymbol{\nabla u}|^2 = \frac{1}{2}\int_\Omega |\boldsymbol{u}(t=0,\cdot)|^2 + \int_0^T\int_\Omega \boldsymbol{f}\cdot\boldsymbol{u}. \qquad (2.16)$$

Here we have used both the divergence-free constraint and the homogeneous Dirichlet boundary condition to eliminate the Navier term and the pressure term, as well as all boundary terms originating from the integration by parts (such calculations will be made clear in the next section). Assume that the initial condition $\boldsymbol{u}_0$ has components in $L^2(\Omega)$ and that the body force term $\boldsymbol{f}$ may be controlled in an appropriate space (we will make all this precise below and momentarily assume that the right-hand side of (2.16) is bounded). Then (2.16) shows first that the right functional spaces where the components of $\boldsymbol{u}$ should be considered are at least $L^2(0,T;H^1(\Omega))\cap L^\infty(0,T;L^2(\Omega))$, and, second, that the sequence of approximate solutions is (formally) bounded in this space. It is, however, well known that this estimate is not sufficient to pass to the limit in the nonlinear term $\boldsymbol{u}\cdot\boldsymbol{\nabla u}$. This is exemplified by Proposition 2.3, part (ii). Indeed, some compactness argument (in the vein of Proposition 2.5) is necessary for passing to the limit in the nonlinear terms. Weak convergence does not suffice. We thus observe

(Fact 3) $\left\{ \begin{array}{l} \text{For nonlinear terms, additional } \textit{a priori} \text{ estimates} \\ \text{and compactness results are needed to pass to the limit.} \end{array} \right.$

We appreciate how vague the previous argument is. The rest of the section will now make it rigorous by considering a specific case. The above guidelines will be explicitly illustrated.

2.1.1.3 Functional setting of the equations In what follows, for the sake of generality, d denotes the dimension of the ambient space. Momentarily for the present section, $d = 2$. Soon and for the rest of the book, we will consider $d = 3$.

For $m \geq 0$, we denote by $H^m(\Omega)$ the Sobolev space

$$H^m(\Omega) = \{u \in L^2(\Omega); D^\gamma u \in L^2(\Omega), \forall\gamma, |\gamma| \leq m\} \qquad (2.17)$$

where $\gamma = (\gamma_i)_{1\leq i\leq d}$ is a multi-index and $|\gamma| = \sum_{i=1}^{d} \gamma_i$. The norm on $H^m(\Omega)$ is:

$$||u||_{H^m(\Omega)} = \left(\sum_{|\gamma|=0}^{m} ||D^\gamma u||^2_{L^2(\Omega)}\right)^{1/2}.$$

The space $H_0^1(\Omega)$ is the subspace of $H^1(\Omega)$ consisting of the functions vanishing on $\partial\Omega$.

For any space X, we shall denote $(X)^d$ by $\mathbb{X}$. For example, $(L^2(\Omega))^d$ is denoted by $\mathbb{L}^2(\Omega)$, $(H^m(\Omega))^d$ by $\mathbb{H}^m(\Omega)$.

Let $T > 0$ and let X be a Banach space. The space $L^p(0,T;X)$, $1 \le p \le \infty$, is the space of classes of L^p functions from $(0,T)$ into X. We recall that this is a Banach space for the norm

$$\left(\int_0^T ||u(t)||_X^p \, dt\right)^{1/p} \quad \text{if } 1 \le p < \infty, \qquad \operatorname*{ess\,sup}_{t\in[0,T]} ||u(t)||_X \text{ if } p = \infty.$$

We denote by $\mathcal{D}(\Omega)$ (resp. $\mathcal{D}(\overline{\Omega})$) the space of real functions infinitely differentiable with compact support in Ω (resp. $\overline{\Omega}$).

We now introduce the spaces

$$\mathcal{V} = \{\boldsymbol{v} \in (\mathcal{D}(\Omega))^d, \operatorname{div} \boldsymbol{v} = 0\}, \tag{2.18}$$

$$V = \{\boldsymbol{v} \in \mathbb{H}_0^1(\Omega), \operatorname{div} \boldsymbol{v} = 0\}, \tag{2.19}$$

$$H = \{\boldsymbol{v} \in \mathbb{L}^2(\Omega), \operatorname{div} \boldsymbol{v} = 0, \boldsymbol{v}.\boldsymbol{n}|_{\partial\Omega} = 0\}. \tag{2.20}$$

The space V is the closure of $\mathcal{V}$ in $\mathbb{H}_0^1(\Omega)$. On the other hand, the space H is defined as the closure of $\mathcal{V}$ in $\mathbb{L}^2(\Omega)$. Throughout the book, this space will be denoted by $H_0(\operatorname{div}^0, \Omega)$. We however prefer to employ the notation H for the introductory material discussed in the present chapter.

Let us briefly explain why the boundary condition $\boldsymbol{v}.\boldsymbol{n}|_{\partial\Omega} = 0$ makes sense in (2.20). For $\boldsymbol{v} \in \mathbb{L}^2(\Omega)$, divergence-free, we may define $\boldsymbol{v}.\boldsymbol{n}|_{\partial\Omega}$ by

$$\langle \boldsymbol{v}.\boldsymbol{n}|_{\partial\Omega}, \varphi\rangle = \int_\Omega \boldsymbol{v} \cdot \boldsymbol{\nabla}\varphi,$$

for all $\varphi \in H^1(\Omega)$. This gives a meaning to $\boldsymbol{v}.\boldsymbol{n}|_{\partial\Omega}$ in the dual space of the space of all traces of functions of $H^1(\Omega)$. The latter is a non-integer negative Sobolev space, denoted by $H^{-1/2}(\partial\Omega)$ (see Section 2.2.5.1 below). In fact, the above integration by parts also shows that we may define by

$$\langle \boldsymbol{v}.\boldsymbol{n}|_{\partial\Omega}, \varphi\rangle = \int_\Omega \boldsymbol{v} \cdot \boldsymbol{\nabla}\varphi + \int_\Omega \varphi \operatorname{div} \boldsymbol{v},$$

the function $\boldsymbol{v}.\boldsymbol{n}|_{\partial\Omega}$ (as a function in $H^{-1/2}(\partial\Omega)$) for all $\boldsymbol{v} \in L^2(\Omega)$ with $\operatorname{div} \boldsymbol{v} \in L^2(\Omega)$.

Considering the above spaces allows us to encode the incompressiblity constraint $\operatorname{div} \boldsymbol{u} = 0$ in the variational space itself, so that we only have then to treat the main equation (first line of (NS)). In fact, proceeding in this manner also eliminates the pressure and therefore allows us to transform (NS) into a Cauchy problem. This strategy significantly simplifies the analytical study.

For the numerical approximation, encoding the divergence-free constraint in the variational space (namely the finite element space) is also a possible strategy. It is definitely not the easiest and the most efficient one: the incompressiblity condition is usually treated in an explicit manner, *i.e.* as a constraint and not as a restriction of the variational space. A new version of the mathematical analysis of the equations has to be performed. This will be detailed in Chapter 3.

For $\boldsymbol{v} \in V$ we denote

$$||\boldsymbol{v}||_V = \left(\int_\Omega |\boldsymbol{\nabla v}|^2\, dx\right)^{1/2}. \tag{2.21}$$

It can be established that $||.||_V$ defines a norm which is equivalent to that induced by $\mathbb{H}^1(\Omega)$ on V.

Let us first detail the derivation of the *a priori* estimate established in a somewhat vague sense in the previous section. We momentarily assume that the force $\boldsymbol{f}$ and the solution $\boldsymbol{u}$ we manipulate are sufficiently regular, so that all the computations below make sense. Multiplying

$$\frac{\partial \boldsymbol{u}}{\partial t} + \boldsymbol{u}\cdot\boldsymbol{\nabla u} - \overline{\eta}\Delta\boldsymbol{u} = -\boldsymbol{\nabla} p + \boldsymbol{f}$$

by $\boldsymbol{u}$ and integrating over the domain Ω, we obtain

$$\int_\Omega \frac{\partial \boldsymbol{u}}{\partial t}\cdot\boldsymbol{u} + \int_\Omega (\boldsymbol{u}\cdot\boldsymbol{\nabla u})\cdot\boldsymbol{u} - \overline{\eta}\int_\Omega \Delta\boldsymbol{u}\cdot\boldsymbol{u} = -\int_\Omega \boldsymbol{\nabla} p\cdot\boldsymbol{u} + \int_\Omega \boldsymbol{f}\cdot\boldsymbol{u}.$$

We have

$$\frac{1}{2}\frac{d}{dt}\int_\Omega |\boldsymbol{u}|^2 = \int_\Omega \frac{\partial \boldsymbol{u}}{\partial t}\cdot\boldsymbol{u}, \tag{2.22}$$

and, by integration by parts (or, more appropriately, by the application of the so-called Green formula), respectively

$$\int_\Omega (\boldsymbol{u}\cdot\boldsymbol{\nabla u})\cdot\boldsymbol{u} = \int_\Omega \boldsymbol{u}\cdot\boldsymbol{\nabla}\frac{|\boldsymbol{u}|^2}{2} = -\int_\Omega \frac{|\boldsymbol{u}|^2}{2}\operatorname{div}\boldsymbol{u} + \int_{\partial\Omega}(\boldsymbol{u}\cdot\boldsymbol{n})\frac{|\boldsymbol{u}|^2}{2}, \tag{2.23}$$

$$-\int_\Omega \Delta\boldsymbol{u}\cdot\boldsymbol{u} = \int_\Omega |\boldsymbol{\nabla u}|^2 - \int_{\partial\Omega}(\boldsymbol{\nabla u}\cdot\boldsymbol{n})\cdot\boldsymbol{u}, \tag{2.24}$$

and

$$\int_\Omega \boldsymbol{\nabla} p\cdot\boldsymbol{u} = -\int_\Omega p\operatorname{div}\boldsymbol{u} + \int_{\partial\Omega} p\,\boldsymbol{u}\cdot\boldsymbol{n}. \tag{2.25}$$

All boundary terms in (2.23), (2.24) and (2.25) vanish because of the homogeneous boundary condition set on $\boldsymbol{u}$. In addition, because of the divergence-free constraint on $\boldsymbol{u}$, the first terms of the right-hand side of (2.23) and (2.25) also vanish. We thus obtain

$$\frac{1}{2}\frac{d}{dt}\int_\Omega |\boldsymbol{u}|^2 + \overline{\eta}\int_\Omega |\boldsymbol{\nabla u}|^2 = \int_\Omega \boldsymbol{f}\cdot\boldsymbol{u}.$$

This equation, which has already been formally established above, is sometimes called the *first energy equality*, as it expresses the balance of energy in the system:

the rate of decay of the kinetic energy (first term) is a balance between the viscous dissipation (second term) and the energy provided to the system by the forces (right-hand side). In the previous sentence, the terms "decay" and "energy provided" must of course be understood *algebraically*, since they may equally be "growth" or "energy dissipated," depending on the signs of the quantities involved.

Integrating in time from $t = 0$ to $t = T$, we obtain

$$\frac{1}{2}\int_\Omega |\boldsymbol{u}(T,\cdot)|^2 + \overline{\eta}\int_0^T\int_\Omega |\boldsymbol{\nabla u}|^2 = \frac{1}{2}\int_\Omega |\boldsymbol{u}(t=0,\cdot)|^2 + \int_0^T\int_\Omega \boldsymbol{f}\cdot\boldsymbol{u}, \quad (2.26)$$

(identical to (2.16)) and consequently

$$\frac{1}{2}\int_\Omega |\boldsymbol{u}(T,\cdot)|^2 + \frac{1}{2}\overline{\eta}\int_0^T\int_\Omega |\boldsymbol{\nabla u}|^2 \leq \frac{1}{2}\int_\Omega |\boldsymbol{u}(t=0,\cdot)|^2 + C\int_0^T \|\boldsymbol{f}\|_{V'}^2, \quad (2.27)$$

using the Young inequality:

$$\int_\Omega \boldsymbol{f}\cdot\boldsymbol{u} \leq \frac{1}{2}\overline{\eta}\,\|\boldsymbol{u}\|_V^2 + C\,\|\boldsymbol{f}\|_{V'}^2.$$

This equation is often called the *first energy inequality*, or *first a priori estimate*. As outlined in the previous section, but now using the notation introduced above and accounting for the divergence-free constraint and the boundary condition, it suggests that the right functional space for the solution $\boldsymbol{u}$ is

$$L^2(0,T;V)\cap L^\infty(0,T;H),$$

for an initial condition $\boldsymbol{u}_0 \in H$. Regarding the body force term $\boldsymbol{f}$, we see that it is sufficient to assume $\boldsymbol{f} \in L^2(0,T;V')$, which requires us to slightly modify the above lines replacing the integral

$$\int_\Omega \boldsymbol{f}\cdot\boldsymbol{v}.$$

by

$$\langle \boldsymbol{f}\,,\,\boldsymbol{v}\rangle_{V',V}.$$

In addition, the above computation also suggests what sense is to be given to the equation itself. Indeed, multiplying formally the equation by a test function $\boldsymbol{v}$, also tentatively belonging to the space V (thus constant in time), and performing the same computation yields the following *weak formulation* of the equation

$$\frac{d}{dt}\int_\Omega \boldsymbol{u}\cdot\boldsymbol{v} + \int_\Omega (\boldsymbol{u}\cdot\boldsymbol{\nabla})\boldsymbol{u}\cdot\boldsymbol{v} + \overline{\eta}\int_\Omega \boldsymbol{\nabla u} : \boldsymbol{\nabla v} = \langle \boldsymbol{f}\,,\,\boldsymbol{v}\rangle_{V',V}. \quad (2.28)$$

At this stage, we are thus in a position to introduce the following

Definition 2.6 *Assume $\boldsymbol{f} \in L^2(0,T;V')$ and $\boldsymbol{u}_0 \in H$. The velocity field $\boldsymbol{u}$ is said to be a weak solution of (NS) when*

$$\boldsymbol{u} \in L^2(0,T;V) \cap L^\infty(0,T;H) \tag{2.29}$$

satisfies

$$\left\langle \frac{\partial \boldsymbol{u}}{\partial t}, \boldsymbol{v} \right\rangle_{V',V} + \int_\Omega (\boldsymbol{u} \cdot \nabla)\boldsymbol{u} \cdot \boldsymbol{v} + \overline{\eta} \int_\Omega \nabla \boldsymbol{u} : \nabla \boldsymbol{v} = \langle \boldsymbol{f}, \boldsymbol{v} \rangle_{V',V}, \tag{2.30}$$

in the sense of distributions in time, for all $\boldsymbol{v} \in V$, and

$$\boldsymbol{u}(t=0,\cdot) = \boldsymbol{u}_0. \tag{2.31}$$

A number of remarks are in order.

Remark 2.1.4 The terminology "weak solution" originates from the fact that the term $-\Delta \boldsymbol{u}$ in (NS) is not defined almost everywhere, and has to be understood through an integration by parts, see (2.30). In contrast, a "strong" solution is sufficiently regular, so that $-\Delta \boldsymbol{u}$ has a meaning almost everywhere in space and time (see Proposition 2.15). ◇

Remark 2.1.5 From a rigorous viewpoint, the above statement may seem incorrect, at least for three reasons. First, in order for $\left\langle \frac{\partial \boldsymbol{u}}{\partial t}, \boldsymbol{v} \right\rangle$ to make sense in (2.30), the function $\partial \boldsymbol{u}/\partial t$ needs to be in V' for all time. It is not clear from the formulation that this property holds. Likewise, condition (2.31) *a priori* makes no sense for a function $\boldsymbol{u}$ that is only L^2 in time. More regularity with respect to the time variable should be known, in order for (2.31) to make sense. Actually, the two observations are related and we shall indeed see that sufficient regularity comes out of the conditions (2.29) (2.30) so that both issues can be settled (see Lemma 2.7). In addition to this, for the term $\int_\Omega (\boldsymbol{u} \cdot \nabla)\boldsymbol{u} \cdot \boldsymbol{v}$ to make sense, we need that, for all times, $\boldsymbol{u} \cdot \boldsymbol{v} \in L^2(\Omega)$ when both $\boldsymbol{u}$ and $\boldsymbol{v}$ belong to V. We shall see with (2.40) that it is indeed the case in dimension 2. We shall see a similar argument for dimension 3. ◇

Remark 2.1.6 The above formulation indeed amounts to integrating, first in space and next in time, the formal equation (NS) against test functions that are products of the form $\psi(t)\boldsymbol{v}(\mathbf{x})$. In Chapter 4, when dealing with non-homogeneous equations, we shall see another weak formulation of the equations, which of course applies to the present setting. It consists in taking test functions $\boldsymbol{v} = \boldsymbol{v}(t,x)$ that also vary in time. In the present context, these functions would be chosen in $L^2(0,T;V) \cap L^\infty(0,T;H)$. The two formulations can be shown to be equivalent, using density arguments. ◇

Remark 2.1.7 Actually owing to the integrability in time of $\boldsymbol{u}$ (and $\partial \boldsymbol{u}/\partial t$, see Remark 2.1.5), equation (2.30) holds not only in the distributional sense but also L^1, and thus *almost everywhere*, in time. ◇

Remark 2.1.8 Surprisingly, the pressure p has disappeared from the notion of solution. The pressure may indeed be recovered *a posteriori*, as follows. If $\boldsymbol{w} \in V'$ satisfies $\langle \boldsymbol{w}, \boldsymbol{v} \rangle = 0$ for all $\boldsymbol{v} \in V$, then there exists some distribution p such that $\boldsymbol{w} = \boldsymbol{\nabla} p$. Applying this argument to

$$\boldsymbol{w} = \frac{\partial}{\partial t}\boldsymbol{u} - \overline{\eta}\Delta \boldsymbol{u} + \boldsymbol{u} \cdot \boldsymbol{\nabla} \boldsymbol{u} - \boldsymbol{f}$$

for almost all time (see Remark 2.1.7), we obtain the existence of a pressure p so that (1.40) holds. We refer to Chapter 3 for another discussion on this question (see in particular Remark 3.1.2). ♢

Remark 2.1.9 As mentioned above, the elimination of the pressure from the notion of solution has the advantage of transforming the problem into a Cauchy problem. ♢

2.1.1.4 Some preliminary tools for the Navier–Stokes setting We now need to lay some groundwork before turning to the proof of the existence of a solution to our equation, in the sense of Definition 2.6.

The Stokes problem A preliminary task is to consider the so-called *Stokes problem*

$$\begin{cases} -\Delta \boldsymbol{u} + \boldsymbol{\nabla} p = \boldsymbol{f}, \\ \operatorname{div} \boldsymbol{u} = 0, \\ \boldsymbol{u}|_{\partial\Omega} = 0. \end{cases} \tag{2.32}$$

We indeed need to understand how to deal with a formulation set on spaces of divergence-free functions, such as V. It is a standard fact (which we will recall in Section 2.2.4) that, for all $\boldsymbol{f} \in \mathbb{L}^2(\Omega)$, and even for all $\boldsymbol{f} \in \mathbb{H}^{-1}(\Omega)$, there exists a unique solution $\boldsymbol{u} \in V$ of this equation. It is indeed the unique solution to the following weak formulation of (2.32):

$$\boldsymbol{u} \in V \qquad \text{such that} \qquad \int_\Omega \boldsymbol{\nabla} \boldsymbol{u} : \boldsymbol{\nabla} \boldsymbol{v} = \langle \boldsymbol{f}, \boldsymbol{v} \rangle, \qquad \forall \boldsymbol{v} \in V, \tag{2.33}$$

where the bracket on the right-hand side denotes the duality between V and V'. The latter stems from the duality between H_0^1 functions (the components of $\boldsymbol{v}$) and H^{-1} functions (those of $\boldsymbol{f}$). The pressure is then recovered in a second step, as explained in Remark 2.1.8. From this, it is evident (see, *e.g.* R. Temam [229]) how to construct a linear operator Λ, continuous from $\mathbb{L}^2$ to V, that associates to $\boldsymbol{f} \in \mathbb{L}^2$ the solution

$$\Lambda \boldsymbol{f} := \boldsymbol{u} \tag{2.34}$$

of the Stokes problem. The continuity of Λ holds since the consideration of $\boldsymbol{v} = \boldsymbol{u}$ and $\boldsymbol{f} \in \mathbb{L}^2$ in (2.33) yields the estimate

$$\int_\Omega |\boldsymbol{\nabla} \boldsymbol{u}|^2 \leq \int_\Omega |\boldsymbol{f}|^2.$$

From the compact embedding of V in $\mathbb{L}^2$ (due to the Rellich theorem recalled in Proposition 2.5), it follows that Λ is compact. As it is also self-adjoint, the sequence of its eigenfunctions $\boldsymbol{w}_j$ (by eigenvalues λ_j^{-1}) may be considered. They are the (normalized in $\mathbb{L}^2$) solutions to

$$\begin{cases} -\Delta \boldsymbol{w}_j + \boldsymbol{\nabla} p_j = \lambda_j \boldsymbol{w}_j, \\ \operatorname{div} \boldsymbol{w}_j = 0, \\ \boldsymbol{w}_j|_{\partial\Omega} = 0, \end{cases} \tag{2.35}$$

or in other words the non-zero functions of V that are solutions to

$$\int_\Omega \boldsymbol{\nabla} \boldsymbol{w}_j : \boldsymbol{\nabla} \boldsymbol{v} = \lambda_j \int_\Omega \boldsymbol{w}_j \cdot \boldsymbol{v}, \qquad \forall \boldsymbol{v} \in V. \tag{2.36}$$

The functions $\boldsymbol{w}_j$ are countably infinite in number. They are associated to a sequence λ_i, increasing to infinity. To see the positivity, set $\boldsymbol{v} = \boldsymbol{w}_j$ in (2.36). These functions form an orthonormal family in H that indeed generates H. These eigenfunctions will play a crucial role in the existence proof below. They provide an appropriate finite-dimensional approximation of the solution which allows *a priori* estimates to hold.

In particular, an interesting property of the functions $\boldsymbol{w}_j$ is the following. Define

$$P_n \boldsymbol{v} = \sum_{j=1}^n \left(\int_\Omega \boldsymbol{w}_j \cdot \boldsymbol{v} \right) \boldsymbol{w}_j, \tag{2.37}$$

the projector of the space H on $\mathrm{Span}(\boldsymbol{w}_1, \ldots, \boldsymbol{w}_n)$. Then, as a projector in H, P_n obviously has norm less than one. The point is that, as an operator on V, it *also* has norm less than one. Indeed, because of (2.36), the $\frac{1}{\sqrt{\lambda_j}} \boldsymbol{w}_j$ form an orthonormal family of V. Thus for any $\boldsymbol{v} \in V$,

$$P_n \boldsymbol{v} = \sum_{j=1}^n \left(\int_\Omega \boldsymbol{w}_j \cdot \boldsymbol{v} \right) \boldsymbol{w}_j = \sum_{j=1}^n \left(\int_\Omega \frac{1}{\sqrt{\lambda_j}} \boldsymbol{\nabla} \boldsymbol{w}_j : \boldsymbol{\nabla} \boldsymbol{v} \right) \frac{1}{\sqrt{\lambda_j}} \boldsymbol{w}_j,$$

and our claim follows taking the norm in V. This will be useful below.

We shall need another mathematical object constructed from the operator Λ. As for any fixed $\boldsymbol{u} \in V$, the linear map $\boldsymbol{v} \longrightarrow \int_\Omega \boldsymbol{\nabla} \boldsymbol{u} : \boldsymbol{\nabla} \boldsymbol{v}$ is continuous on V, we may represent it, using the Riesz theorem, by a linear application $\boldsymbol{v} \longrightarrow \langle \boldsymbol{w}, \boldsymbol{v} \rangle$, the bracket denoting the duality bracket between V and its dual space V'. Of course, $\boldsymbol{w}$ depends linearly on $\boldsymbol{u}$, and may be denoted by $\boldsymbol{w} = A\boldsymbol{u}$, and doing so we define the operator A, from V to V', such that

$$\int_\Omega \boldsymbol{\nabla} \boldsymbol{u} : \boldsymbol{\nabla} \boldsymbol{v} = \langle A\boldsymbol{u}, \boldsymbol{v} \rangle, \qquad \boldsymbol{u} \in V, \qquad \forall \boldsymbol{v} \in V. \tag{2.38}$$

Coming back to (2.33), we see that A is the inverse of the operator Λ.

Somewhat vaguely stated, the operator A is a generalization of the Laplacian operator $-\Delta$. For illustration, take two functions $\boldsymbol{f}_1$ and $\boldsymbol{f}_2$, both in $\mathbb{L}^2$ that only differ from a gradient field $\boldsymbol{f}_1 - \boldsymbol{f}_2 = \boldsymbol{\nabla} p$. Then the solution $\boldsymbol{u}$ of the two associated Stokes problem with right-hand sides $\boldsymbol{f}_1$ and $\boldsymbol{f}_2$ is the same function $\boldsymbol{u}$, that is indeed in $V \cap \mathbb{H}^2(\Omega)$. While $\Delta \boldsymbol{u}$ picks one particular $\boldsymbol{f}$ (namely $\boldsymbol{f} = -\Delta \boldsymbol{u}$, corresponding to $p \equiv C^e$), there is the need to adequately describe *all* the right-hand sides $\boldsymbol{f}$ that would give $\boldsymbol{u}$ as a solution of the Stokes problem. The purpose of the operator A is to define the element of the dual space (V') that conveniently describes this set.

Often, with a slight abuse of language and notation, $A\boldsymbol{u}$ is simply denoted by $-\Delta \boldsymbol{u}$.

Remark 2.1.10 The reader may know that the eigenfunctions of the Laplacian operator play a crucial role in the analysis of the heat equation. This is of course a simpler occurrence of the present analysis. ◊

Remark 2.1.11 Anticipating the MHD system studied in Section 2.2, we wish to mention that some analogous considerations will be made there on the eigenvalues of the steady-state magnetic equation

$$\begin{cases} \mathbf{curl}\,\mathbf{curl}\,\boldsymbol{B} = \boldsymbol{f}, \text{ in } \Omega, \\ \operatorname{div} \boldsymbol{B} = 0, \text{ in } \Omega, \\ \boldsymbol{B} \cdot \boldsymbol{n} = 0, \text{ on } \partial\Omega, \\ \mathbf{curl}\,\boldsymbol{B} \times \boldsymbol{n} = 0, \text{ on } \partial\Omega. \end{cases}$$

◊

Dependance upon time We shall also need the following two results regarding the time variable.

Lemma 2.7 [227, Chapter 3, Lemma 1.2] *Recall V and H are defined by (2.19) and (2.20). A function*

$$\boldsymbol{v} \in L^2(0,T;V) \quad \text{such that} \quad \frac{\partial \boldsymbol{v}}{\partial t} \in L^2(0,T;V')$$

is equal to a continuous function

$$\boldsymbol{v} \in C^0(0,T;H),$$

up to a possible modification on a set of times of measure zero.

The following lemma is a compactness lemma. It can be seen as a time-dependent version of the Rellich theorem. It states the compactness of a set of time-dependent functions when there exists both bounds on the derivatives in time and bounds on the derivatives in space.

Lemma 2.8 [153, Theorem 5.1, p. 58] *Let B be a Banach space, and B_0 and B_1 be two reflexive Banach spaces. Assume $B_0 \subset B$ with compact injection, $B \subset B_1$ with continuous injection. Fix $T < +\infty$, $1 < p_0 < +\infty$, $1 < p_1 < +\infty$. Then the injection*

$$\left\{ v \in L^{p_0}(0,T;B_0) \quad \text{s.t.} \quad \frac{\partial v}{\partial t} \in L^{p_1}(0,T;B_1) \right\} \subset L^{p_0}(0,T;B) \tag{2.39}$$

is compact.

2.1.2 *The illustrative example of the two-dimensional case*

We now make the discussion specific to the two-dimensional case. To start with, and for consistency, we state some continuity and compactness lemmas without proving them. They are all "basic" tools of functional analysis, that will play a crucial role in the sequel. We refer to the bibliography for their proofs.

First, we concentrate on the space variable only.

Lemma 2.9 [153, Lemma 6.2, p. 70] **[Restricted to dimension 2]** *For Ω bounded in $\mathbb{R}^2$, there exists a constant $c(\Omega)$ such that, for all $v \in H_0^1(\Omega)$,*

$$\|v\|_{L^4(\Omega)} \leq c(\Omega) \, \|v\|_{H_0^1(\Omega)}^{1/2} \, \|v\|_{L^2(\Omega)}^{1/2}. \tag{2.40}$$

Lemma 2.10 [229, p. 11] **[Restricted to dimension 2]** *For Ω bounded in $\mathbb{R}^2$, there exists a constant $c(\Omega)$ such that, for all $\boldsymbol{v} \in \mathbb{H}^2(\Omega)$,*

$$\|\boldsymbol{v}\|_{\mathbb{L}^\infty(\Omega)} \leq \begin{cases} c(\Omega) \, \|\boldsymbol{v}\|_{\mathbb{L}^2(\Omega)}^{1/2} \, \|\boldsymbol{v}\|_{\mathbb{H}^2(\Omega)}^{1/2}, \\ \\ c(\Omega) \, \|\boldsymbol{v}\|_{\mathbb{H}^1(\Omega)}^{3/4} \, \|\boldsymbol{v}\|_{\mathbb{H}^2(\Omega)}^{1/4}, \end{cases} \tag{2.41}$$

and, for all $\boldsymbol{v} \in H$ such that $\Delta \boldsymbol{v} \in H$,

$$\|\boldsymbol{v}\|_{\mathbb{L}^\infty(\Omega)} \leq \begin{cases} c(\Omega) \, \|\boldsymbol{v}\|_{\mathbb{L}^2(\Omega)}^{1/2} \, \|A\boldsymbol{v}\|_{\mathbb{L}^2(\Omega)}^{1/2}, \\ \\ c(\Omega) \, \|\boldsymbol{\nabla v}\|_{\mathbb{L}^2(\Omega)}^{3/4} \, \|A\boldsymbol{v}\|_{\mathbb{L}^2(\Omega)}^{1/4}. \end{cases} \tag{2.42}$$

Remark 2.1.12 Some other lemmas in the same spirit will be stated for the three-dimensional case in Section 2.1.3. ♢

2.1.2.1 Existence of weak solutions We are now in a position to prove

Proposition 2.11 [Two-dimensional case] *There exists a solution (in the sense of Definition 2.6) to problem (NS).*

The proof of Proposition 2.11 falls in three steps, which basically follow the pattern introduced in Section 2.1.1.2 (and the three "Facts" emphasized there).

Step 1: Construction of a finite-dimensional approximation To start with, we define an approximate solution at order $n \geq 1$, namely $\boldsymbol{u}_n$, as follows. Recall we denote by $\boldsymbol{w}_j$ the eigenfunctions of the Stokes problem (2.35). First, we introduce $\boldsymbol{u}_{0n} \in \mathrm{Span}\,(\boldsymbol{w}_1, \ldots, \boldsymbol{w}_n)$ that converges in H to $\boldsymbol{u}_0$ (the function defining the initial condition (2.31)) as n goes to infinity. Then, for all t, we define

$$\boldsymbol{u}_n(t, \mathbf{x}) = \sum_{j=1}^{n} g_{jn}(t) \boldsymbol{w}_j(\mathbf{x}), \tag{2.43}$$

in Span $(\boldsymbol{w}_1, \ldots, \boldsymbol{w}_n)$, such that the time-dependent functions $g_{jn}(t)$ solve, for all $1 \leq j \leq n$,

$$\left\langle \frac{\partial \boldsymbol{u}_n}{\partial t}, \boldsymbol{w}_j \right\rangle_{V',V} + \int_\Omega (\boldsymbol{u}_n \cdot \boldsymbol{\nabla})\boldsymbol{u}_n \cdot \boldsymbol{w}_j + \overline{\eta} \int_\Omega \boldsymbol{\nabla} \boldsymbol{u}_n \cdot \boldsymbol{\nabla} \boldsymbol{w}_j = \langle \boldsymbol{f}, \boldsymbol{w}_j \rangle_{V',V}, \tag{2.44}$$

supplied with the initial condition

$$\boldsymbol{u}_n(0, \cdot) = \boldsymbol{u}_{0n}. \tag{2.45}$$

Equation (2.44), along with the initial condition (2.45), forms a system of n ordinary differential equations on the n unknown functions g_{jn}. Using Cauchy–Lipschitz theory, it is straightforward to see that this system is well-posed. It thus defines a unique solution $\boldsymbol{u}_n$, on some time interval $[0, T_n]$ with final time T_n *a priori* depending on n (in fact the sequel of the argument shows that we may take $T_n = T$).

Step 2: A priori estimates on the finite-dimensional approximation The second step now consists in rigorously establishing the *a priori* estimate (2.16), using $\boldsymbol{u}_n$ instead of $\boldsymbol{u}$. The third step will show that (2.16) indeed holds true for our solution $\boldsymbol{u}$. For this purpose, we mimic the formal argument made above. It consisted in multiplying the equation by $\boldsymbol{u}$ and integrating both in space and in time. We therefore multiply the j-th line of (2.44) by $g_{jn}(t)$, then sum over j running from 1 to n. Using the same algebraic properties as above for the formal argument, we obtain

$$\begin{aligned} \frac{1}{2} \frac{d}{dt} \int_\Omega |\boldsymbol{u}_n|^2 + \overline{\eta} \int_\Omega |\boldsymbol{\nabla} \boldsymbol{u}_n|^2 &= \langle \boldsymbol{f}, \boldsymbol{u}_n \rangle_{V',V} \\ &\leq \frac{1}{2} \overline{\eta} \int_\Omega |\boldsymbol{\nabla} \boldsymbol{u}_n|^2 + C \, \|\boldsymbol{f}\|_{V'}^2 \end{aligned}$$

by the Young inequality and using the fact that the norm of $\boldsymbol{u}_n$ in V is $\|\boldsymbol{u}_n\|_V^2 = \int_\Omega |\boldsymbol{\nabla} \boldsymbol{u}_n|^2$. Integrating now in time, we obtain

$$\frac{1}{2} \int_\Omega |\boldsymbol{u}_n(t, \cdot)|^2 + \frac{1}{2} \overline{\eta} \int_0^t \int_\Omega |\boldsymbol{\nabla} \boldsymbol{u}_n(s, \cdot)|^2 \leq \int_\Omega |\boldsymbol{u}_{0n}|^2 + C \int_0^t \|\boldsymbol{f}\|_{V'}^2. \tag{2.46}$$

Since the right-hand side is uniformly bounded with respect to n, this shows that we may take $T_n = T$ and that

$$\boldsymbol{u}_n \quad \text{is bounded in} \quad L^2(0, T; V) \cap L^\infty(0, T; H). \tag{2.47}$$

Another *a priori* estimate that will turn out to be crucial for the rest of our proof, is now obtained as follows. We first remark that, for all $\boldsymbol{v} \in V$,

$$\begin{aligned} \left| \int_\Omega (\boldsymbol{u}_n \cdot \boldsymbol{\nabla}) \boldsymbol{u}_n \cdot \boldsymbol{v} \right| &= \left| - \int_\Omega (\boldsymbol{u}_n \cdot \boldsymbol{\nabla}) \boldsymbol{v} \cdot \boldsymbol{u}_n \right| \\ &\leq \|\boldsymbol{u}_n\|_{\mathbb{L}^4(\Omega)}^2 \, \|\boldsymbol{v}\|_V \\ &\leq C \, \|\boldsymbol{u}_n\|_H \, \|\boldsymbol{u}_n\|_V \, \|\boldsymbol{v}\|_V, \end{aligned} \tag{2.48}$$

thus $(\boldsymbol{u}_n \cdot \boldsymbol{\nabla})\boldsymbol{u}_n$ may be represented by an element, denoted by $g(\boldsymbol{u}_n)$ of V' (as was the case for $\Delta \boldsymbol{u}$ that is represented by $A\boldsymbol{u}$, see (2.38)). In addition, owing to the bound in $L^2(0,T;V) \cap L^\infty(0,T;H)$ already known on $\boldsymbol{u}_n$, which is uniform with respect to n, we know from (2.48) that $g(\boldsymbol{u}_n)$ is bounded in $L^2(0,T;V')$. This will be useful below. Using $g(\boldsymbol{u}_n)$, we next see that (2.44) implies

$$\frac{\partial \boldsymbol{u}_n}{\partial t} = -P_n(g(\boldsymbol{u}_n)) - \overline{\eta}\, P_n\, A\boldsymbol{u}_n + P_n\, \boldsymbol{f}, \tag{2.49}$$

where P_n is the orthogonal projector in H onto Span $(\boldsymbol{w}_1, \ldots, \boldsymbol{w}_n)$ defined in (2.37). We now make use of the fact mentioned above: in addition to being of norm less than one as an operator on H, P_n is also of norm less than one when considered as an operator on V. The latter holds because of the special choice of the functions $\boldsymbol{w}_j$.

From this, we deduce that the adjoint operator of P_n, that is P_n itself since P_n is obviously self-adjoint, is an operator of norm less than one when considered on V'. Therefore, as we know that $g(\boldsymbol{u}_n)$, $A\boldsymbol{u}_n$ and $\boldsymbol{f}$ are all bounded in $L^2(0,T;V')$, we deduce from (2.49) that

$$\frac{\partial \boldsymbol{u}_n}{\partial t} \quad \text{is bounded in} \quad L^2(0,T;V'). \tag{2.50}$$

Step 3: Passage to the limit We are now in a position to pass to the limit in the system of ordinary differential equations (2.44), which, we recall, has size growing with n.

From the bounds (2.50) and (2.47), we respectively deduce that, up to an extraction we do not make explicit, we may assume that

$$\boldsymbol{u}_n \rightharpoonup \boldsymbol{u} \,\text{weakly in}\, L^2(0,T;V)\,, \text{weakly} - \star \,\text{in}\; L^\infty(0,T;H)$$

$$\text{and} \quad \frac{\partial \boldsymbol{u}_n}{\partial t} \rightharpoonup \frac{\partial \boldsymbol{u}}{\partial t} \text{weakly in}\, L^2(0,T;V') \tag{2.51}$$

as n goes to infinity. This information suffices to pass to the limit in the three linear terms of (2.44), for $\boldsymbol{w}_j$ fixed and n going to infinity.

In addition, combining the $L^2(0,T;V)$ bound in (2.47) with (2.50) and using Lemma 2.8 (with $B_0 = V$, $B = H$, $B_1 = V'$[17], $p_0 = p_1 = 2$), we know that

$$\boldsymbol{u}_n \longrightarrow \boldsymbol{u} \,\text{strongly in}\, L^2(0,T;H) \,\text{and thus also a.e. in time and space.} \tag{2.52}$$

This allows for passing to the limit in the nonlinear term of (2.44). Indeed, we have

$$\int_\Omega (\boldsymbol{u}_n \cdot \boldsymbol{\nabla})\boldsymbol{u}_n \cdot \boldsymbol{w}_j = -\int_\Omega (\boldsymbol{u}_n \cdot \boldsymbol{\nabla})\boldsymbol{w}_j \cdot \boldsymbol{u}_n.$$

In view of the bound (2.47) on $\boldsymbol{u}_n$, we know from Lemma 2.9 that $\boldsymbol{u}_n$ is bounded in $L^4(0,T;\mathbb{L}^4(\Omega))$ and thus that all components of the tensor product $\boldsymbol{u}_n \otimes \boldsymbol{u}_n$

[17] Note that the assumption about the compactness of the injection of B_0 into B is indeed true because of Proposition 2.5.

are bounded in $L^2(0,T;L^2(\Omega))$. Therefore they may be assumed to converge weakly in that space. But, owing to Proposition 2.3, part (i), we know that in the sense of distributions at least, they also converge to the corresponding components of $\boldsymbol{u}\otimes\boldsymbol{u}$, because $\boldsymbol{u}_n$ strongly converge in $L^2(0,T,\mathbb{L}^2(\Omega))$. Thus we have weak convergence in $L^2(0,T,L^2(\Omega))$ to the correct limit, *i.e.*

$$\int_\Omega (\boldsymbol{u}_n\cdot\boldsymbol{\nabla})\boldsymbol{u}_n\cdot\boldsymbol{w}_j \quad \rightharpoonup \quad \int_\Omega (\boldsymbol{u}\cdot\boldsymbol{\nabla})\boldsymbol{u}\cdot\boldsymbol{w}_j \quad \text{weakly in } L^2(0,T).$$

Collecting the above information, we now may pass to the limit in (2.44) and obtain that

$$\left\langle \frac{\partial \boldsymbol{u}}{\partial t}\,,\,\boldsymbol{w}_j\right\rangle_{V',V} + \int_\Omega (\boldsymbol{u}\cdot\boldsymbol{\nabla})\boldsymbol{u}\cdot\boldsymbol{w}_j + \overline{\eta}\int_\Omega \boldsymbol{\nabla u}:\boldsymbol{\nabla w}_j = \langle \boldsymbol{f}\,,\,\boldsymbol{w}_j\rangle_{V',V} \tag{2.53}$$

holds in the distributional sense, for all $\boldsymbol{w}_j$. In fact, as we *know* that $\partial\boldsymbol{u}/\partial t\in L^2(0,T;V')$, (2.53) holds for almost all time. Since the latter functions asymptotically span V, this concludes the proof of (2.30). It remains now to deal with the initial condition. We fix $\boldsymbol{w}_j$ and first multiply (2.44) by a function ψ, C^1 in time, that we assume to vanish at time T, but not necessarily at time 0 (otherwise we could never recover the initial condition). Then we integrate in time:

$$\begin{aligned}\int_0^T \psi(t)\left(\left\langle \frac{\partial \boldsymbol{u}_n}{\partial t}\,,\,\boldsymbol{w}_j\right\rangle_{V',V} + \int_\Omega (\boldsymbol{u}_n\cdot\boldsymbol{\nabla})\boldsymbol{u}_n\cdot\boldsymbol{w}_j + \overline{\eta}\int_\Omega \boldsymbol{\nabla u}_n:\boldsymbol{\nabla w}_j\right)dt \\ = \int_0^T \psi(t)\,\langle \boldsymbol{f},\ \boldsymbol{w}_j\rangle_{V',V}\,dt.\end{aligned} \tag{2.54}$$

We may write the first term as follows

$$\int_0^T \psi(t)\left\langle \frac{\partial \boldsymbol{u}_n}{\partial t}\,,\,\boldsymbol{w}_j\right\rangle_{V',V} dt = -(\boldsymbol{u}_n,\boldsymbol{w}_j)\psi(0) - \int_0^T \psi'(t)\,\langle \boldsymbol{u}_n\,,\,\boldsymbol{w}_j\rangle_{V',V}\,dt$$

and next pass to the limit as n goes to infinity. We thus obtain

$$\begin{aligned}&-(\boldsymbol{u}_0,\boldsymbol{w}_j)\psi(0) - \int_0^T \psi'(t)\,\langle \boldsymbol{u}\,,\,\boldsymbol{w}_j\rangle_{V',V}\,dt \\ &+\int_0^T \psi(t)\left(\int_\Omega (\boldsymbol{u}\cdot\boldsymbol{\nabla})\boldsymbol{u}\cdot\boldsymbol{w}_j + \overline{\eta}\int_\Omega \boldsymbol{\nabla u}:\boldsymbol{\nabla w}_j\right)dt = \int_0^T \psi(t)\,\langle \boldsymbol{f}\,,\,\boldsymbol{w}_j\rangle_{V',V}\,dt,\end{aligned}$$

where we can remark that

$$-\int_0^T \psi'(t)\,\langle \boldsymbol{u}\,,\,\boldsymbol{w}_j\rangle_{V',V}\,dt = \int_0^T \psi(t)\left\langle \frac{\partial \boldsymbol{u}}{\partial t}\,,\,\boldsymbol{w}_j\right\rangle_{V',V} dt + (\boldsymbol{u}(0),\boldsymbol{w}_j)\psi(0).$$

Therefore,

$$
\begin{aligned}
&(-\boldsymbol{u}_0 + \boldsymbol{u}(0), \boldsymbol{w}_j)\psi(0) \\
&= \int_0^T \psi(t) \left(\left\langle \frac{\partial \boldsymbol{u}}{\partial t}, \boldsymbol{w}_j \right\rangle_{V',V} \right. \\
&\qquad \left. + \int_\Omega (\boldsymbol{u} \cdot \boldsymbol{\nabla})\boldsymbol{u} \cdot \boldsymbol{w}_j + \overline{\eta} \int_\Omega \boldsymbol{\nabla u} : \boldsymbol{\nabla w}_j - \langle \boldsymbol{f}, \boldsymbol{w}_j \rangle_{V',V} \right) dt.
\end{aligned}
$$

But as (2.53) holds almost everywhere in time, this simplifies to

$$(-\boldsymbol{u}_0 + \boldsymbol{u}(0), \boldsymbol{w}_j)\, \psi(0) = 0,$$

which holds for all j, and all values $\psi(0)$. This of course establishes the initial condition.

This concludes the proof of the existence stated in Proposition 2.11. Note that a straightforward corollary of the proof (passing to the limit in (2.46)), is that the tentative *a priori* estimate (2.27) indeed holds rigorously for the weak solution we have constructed (or for *any* weak solution constructed in this manner).

Remark 2.1.13 The methodology employed above to pass to the limit is quite general: one passes to the limit in the equation first in the sense of distributions (at least) (*i.e.* using test functions that eliminate the initial value), next get more information of the regularity of the solution, and finally recover the initial condition by an adequate integration by parts using test functions that do not eliminate the initial value any more. This also applies to boundary value problems. ◊

Remark 2.1.14 We would like to emphasize the role played by our specific choice of basis functions $\boldsymbol{w}_j$. It allows for the bound (2.50) to hold, which in turn implies the strong convergence (2.52), and allows for passing to the limit in the nonlinear term. Otherwise stated, if the equation was linear, such a specific choice would not be necessary. We shall see an example of this in Chapter 4, Section 4.2.2.1. Actually, the bound (2.50) is also used above for proving $\partial \boldsymbol{u}/\partial t \in L^2(0,T;V')$ and thus obtaining (2.53) almost everywhere in time, and recovering the initial condition (2.31). But the latter may be shown directly using the equation. ◊

2.1.2.2 Uniqueness of weak solutions Regarding the uniqueness, we have:

Proposition 2.12 [Two-dimensional case] *The solution whose existence is stated in Proposition 2.11 is unique.*

The proof of this proposition uses the following

Proposition 2.13 [Two-dimensional case] *A function* $\boldsymbol{u} \in L^2(0,T;V) \cap L^\infty(0,T;H)$ *satisfying (2.30) (in the distributional sense in time) is such that*

$$\frac{\partial \boldsymbol{u}}{\partial t} \in L^2(0,T;V'). \tag{2.55}$$

It follows then from Lemma 2.7 that $\boldsymbol{u} \in C^0(0,T;H)$*, and thus the initial condition (2.31) makes sense for solutions satisfying (2.30).*

Remark 2.1.15 Note of course that the point is to show (2.55) for *any* solution. The particular solution obtained in the previous section by the Galerkin method does satisfy this, by construction. ◇

To prove (2.55), it is sufficient to show that

$$\boldsymbol{u} \cdot \boldsymbol{\nabla} \boldsymbol{u} \in L^2(0,T;V').$$

For this purpose, we note that, successively using an integration by parts, the Hölder inequality and Lemma 2.9,

$$\left| \int_\Omega \boldsymbol{u} \cdot \boldsymbol{\nabla} \boldsymbol{u} \cdot \boldsymbol{v} \right| = \left| - \int_\Omega \boldsymbol{u} \cdot \boldsymbol{\nabla} \boldsymbol{v} \cdot \boldsymbol{u} \right| \leq \|\boldsymbol{u}\|^2_{\mathbb{L}^4(\Omega)} \|\boldsymbol{v}\|_{\mathbb{H}^1(\Omega)}$$
$$\leq C \, \|\boldsymbol{u}\|_{\mathbb{L}^2(\Omega)} \|\boldsymbol{u}\|_{\mathbb{H}^1(\Omega)} \|\boldsymbol{v}\|_{\mathbb{H}^1(\Omega)} \, .$$

Since $\boldsymbol{u} \in L^2(0,T;V) \cap L^\infty(0,T;H)$, $\boldsymbol{u} \cdot \boldsymbol{\nabla} \boldsymbol{u} \in L^2(0,T;V')$ follows. This concludes the proof of Proposition 2.13.

We next sketch the proof of Proposition 2.12. Let us now denote by $\boldsymbol{w} = \boldsymbol{u}_1 - \boldsymbol{u}_2$ the difference of two solutions. We deduce from (2.30) that $\boldsymbol{w}$ satisfies

$$\left\langle \frac{\partial \boldsymbol{w}}{\partial t}, \boldsymbol{v} \right\rangle_{V',V} + \int_\Omega (\boldsymbol{w} \cdot \boldsymbol{\nabla}) \boldsymbol{u}_1 \cdot \boldsymbol{v} + \int_\Omega (\boldsymbol{u}_2 \cdot \boldsymbol{\nabla}) \boldsymbol{w} \cdot \boldsymbol{v} + \overline{\eta} \int_\Omega \boldsymbol{\nabla} \boldsymbol{w} : \boldsymbol{\nabla} \boldsymbol{v} = 0,$$

in the sense of distributions in time, for all $\boldsymbol{v} \in V$, and with zero as initial condition. In fact, using Proposition 2.13, both $\boldsymbol{u}_1$ and $\boldsymbol{u}_2$ satisfy (2.55), thus $\partial \boldsymbol{w}/\partial t \in L^2(0,T;V')$, and the above equation also holds almost everywhere in time. Choosing $\boldsymbol{v} = \boldsymbol{w}(t)$ as a test function, we integrate in time, from $t = 0$ to some arbitrary T. To this end, we only need to remark (see [227]) that, since $\boldsymbol{w} \in L^2(0,T,V)$ and $\partial \boldsymbol{w}/\partial t \in L^2(0,T;V')$, we have

$$\frac{1}{2} \frac{d}{dt} \int_\Omega |\boldsymbol{w}|^2 = \left\langle \frac{\partial \boldsymbol{w}}{\partial t}, \boldsymbol{w} \right\rangle_{V',V}$$

almost everywhere in time, and, on the other hand, that a simple integration by parts yields

$$\int_\Omega (\boldsymbol{u}_2 \cdot \boldsymbol{\nabla}) \boldsymbol{w} \cdot \boldsymbol{w} = 0.$$

We thus obtain:

$$\frac{1}{2} \|\boldsymbol{w}(T,\cdot)\|^2_{\mathbb{L}^2(\Omega)} + \overline{\eta} \int_0^T \int_\Omega |\boldsymbol{\nabla} \boldsymbol{w}|^2 = - \int_0^T \int_\Omega (\boldsymbol{w} \cdot \boldsymbol{\nabla}) \boldsymbol{u}_1 \cdot \boldsymbol{w}. \qquad (2.56)$$

Next, we successively use the Hölder inequality, Lemma 2.9, and the Young inequality to estimate from above the right-hand side:

$$\begin{aligned}
\left|\int_0^T \int_\Omega (\boldsymbol{w}\cdot\boldsymbol{\nabla})\boldsymbol{u}_1\cdot\boldsymbol{w}\right| &\leq \int_0^T \|\boldsymbol{w}(t,\cdot)\|^2_{\mathbb{L}^4(\Omega)}\,\|\boldsymbol{\nabla}\boldsymbol{u}_1(t,\cdot)\|_{\mathbb{L}^2(\Omega)} \\
&\leq C\int_0^T \|\boldsymbol{w}(t,\cdot)\|_{\mathbb{H}^1(\Omega)}\,\|\boldsymbol{w}(t,\cdot)\|_{\mathbb{L}^2(\Omega)}\,\|\boldsymbol{\nabla}\boldsymbol{u}_1(t,\cdot)\|_{\mathbb{L}^2(\Omega)} \\
&\leq \overline{\eta}\int_0^T \|\boldsymbol{w}(t,\cdot)\|^2_{\mathbb{H}^1(\Omega)} \\
&\quad + C\int_0^T \|\boldsymbol{w}(t,\cdot)\|^2_{\mathbb{L}^2(\Omega)}\,\|\boldsymbol{\nabla}\boldsymbol{u}_1(t,\cdot)\|^2_{\mathbb{L}^2(\Omega)},
\end{aligned}$$

where C is some irrelevant constant. We now insert this in (2.56) and use

$$f(t) := \|\boldsymbol{\nabla}\boldsymbol{u}_1(t,\cdot)\|^2_{\mathbb{L}^2(\Omega)} \quad \in L^1(0,T).$$

We readily obtain for all $T \geq 0$,

$$\frac{1}{2}\,\|\boldsymbol{w}(T,\cdot)\|^2_{\mathbb{L}^2(\Omega)} \leq C^{te}\int_0^T f(t)\,\|\boldsymbol{w}(t,\cdot)\|^2_{\mathbb{L}^2(\Omega)}.$$

A simple application of the Gronwall lemma shows that $\boldsymbol{w} \equiv 0$ for all times. This concludes the proof of Proposition 2.12.

It is *very important* to note that we have established a *global-in-time* existence and uniqueness result of weak solutions. Indeed, given some T such that the force term $\boldsymbol{f}$ belongs to $L^2(0,T,V')$, we have obtained the solution $\boldsymbol{u}$ without any restriction on the size of T. The same observation applies to the regularity results we turn to in the next section. We shall see in Section 2.1.3 that the situation is drastically different when the ambient dimension is 3. Then it is not clear whether uniqueness and regularity hold regardless of the size of T. On the other hand, the *existence* of a weak solution will then remain *global.*

2.1.2.3 Further regularity It is a natural question to ask whether the solution $\boldsymbol{u}$ inherits additional regularity of the data, namely the force $\boldsymbol{f}$ and the initial condition $\boldsymbol{u}_0$.

Based on our experience of the previous proof, we know that the bottom line of the proof of existence relies upon some *a priori* estimates. We have indeed seen, and it is important to emphasize this from the methodological viewpoint, that

(Fact 4) { Once formal *a priori* estimates are established on the tentative solution, we rigorously prove such estimates for some finite-dimensional approximation of the solution (which may require a special choice of the basis set), and pass to the limit.

Remark 2.1.16 Actually, rather than a finite-dimensional approximation of the solution, other approximations, such as, *e.g.* solutions of regularized versions of the equation, may be employed. We shall see this in Chapter 4. ◇

Therefore, for the additional regularity we wish to prove here, we are going to follow this line and mainly concentrate on the formal *a priori* estimates. At the very end of the present section, we will briefly explain how to make them rigorous.

In order to shorten the exposition, we shall only consider two cases of additional regularity of the data, and only outline the proof for these two cases. The reason why we consider these two cases is that they show a prototypical technique for obtaining more regularity. Indeed, as seen above, the usual *a priori* estimate (*i.e.* the first energy equality (2.16), or the first estimate (2.27)) is essentially obtained by multiplying the equation by $\boldsymbol{u}$ and integrating. On the other hand, further regularity is obtained by the so-called *second energy inequality* or *second a priori estimate*, which is an *a priori* estimate obtained by multiplying the equation by *derivatives* of $\boldsymbol{u}$, or by differentiating the equation itself. The following two propositions, along with the outline of their proof, illustrate the technique.

Proposition 2.14 *Assume that in addition to* $\boldsymbol{f} \in L^2(0,T;V')$, *we have* $\partial \boldsymbol{f}/\partial t \in L^2(0,T;V')$ *and* $\boldsymbol{f}(t=0,\cdot) \in H$. *Assume also that* $\boldsymbol{u}_0 \in V \cap \mathbb{H}^2(\Omega)$. *Then the solution provided by Propositions 2.11 and 2.12 satisfies*

$$\frac{\partial \boldsymbol{u}}{\partial t} \in L^2(0,T;V) \cap L^\infty(0,T;H). \tag{2.57}$$

Proposition 2.15 *Assume that* $\boldsymbol{f} \in L^2(0,T;H)$ *(and not only* $\boldsymbol{f} \in L^2(0,T;V')$*) and that* $\boldsymbol{u}_0 \in V$. *Then the solution provided by Propositions 2.11 and 2.12 satisfies*

$$\boldsymbol{u} \in L^\infty(0,T;V) \cap L^2(0,T;\mathbb{H}^2(\Omega)). \tag{2.58}$$

It is called a strong solution to (NS).

To justify both propositions above, we simply establish some formal *a priori* estimate. The details of the rigorous derivation are left to the reader.

In the first case, the additional information regards the derivative in time of the force. This suggests we formally differentiate the equation with respect to time

$$\frac{\partial^2 \boldsymbol{u}}{\partial t^2} + \frac{\partial \boldsymbol{u}}{\partial t} \cdot \boldsymbol{\nabla} \boldsymbol{u} + \boldsymbol{u} \cdot \boldsymbol{\nabla} \frac{\partial \boldsymbol{u}}{\partial t} - \overline{\eta} \Delta \frac{\partial \boldsymbol{u}}{\partial t} = -\boldsymbol{\nabla} \frac{\partial p}{\partial t} + \frac{\partial \boldsymbol{f}}{\partial t},$$

next multiply it by $\partial \boldsymbol{u}/\partial t$, and integrate in space:

$$\frac{1}{2}\frac{d}{dt}\int_\Omega \left|\frac{\partial \boldsymbol{u}}{\partial t}\right|^2 + \int_\Omega \frac{\partial \boldsymbol{u}}{\partial t} \cdot \boldsymbol{\nabla}\boldsymbol{u} \cdot \frac{\partial \boldsymbol{u}}{\partial t} + \overline{\eta}\int_\Omega \left|\boldsymbol{\nabla}\frac{\partial \boldsymbol{u}}{\partial t}\right|^2 = \int_\Omega \frac{\partial \boldsymbol{f}}{\partial t} \cdot \frac{\partial \boldsymbol{u}}{\partial t}.$$

Using

$$\begin{aligned}\left|\int_\Omega \frac{\partial \boldsymbol{u}}{\partial t}\cdot\boldsymbol{\nabla}\boldsymbol{u}\cdot\frac{\partial \boldsymbol{u}}{\partial t}\right| &= \left|-\int_\Omega \frac{\partial \boldsymbol{u}}{\partial t}\cdot\boldsymbol{\nabla}\frac{\partial \boldsymbol{u}}{\partial t}\cdot\boldsymbol{u}\right|\\
&\le \left\|\boldsymbol{\nabla}\frac{\partial \boldsymbol{u}}{\partial t}\right\|_{\mathbb{L}^2(\Omega)}\left\|\frac{\partial \boldsymbol{u}}{\partial t}\right\|_{\mathbb{L}^4(\Omega)}\|\boldsymbol{u}\|_{\mathbb{L}^4(\Omega)},\\
&\le C\left\|\boldsymbol{\nabla}\frac{\partial \boldsymbol{u}}{\partial t}\right\|^{3/2}_{\mathbb{L}^2(\Omega)}\left\|\frac{\partial \boldsymbol{u}}{\partial t}\right\|^{1/2}_{\mathbb{L}^2(\Omega)}\|\boldsymbol{u}\|_{\mathbb{L}^4(\Omega)},\\
&\le \frac{1}{4}\overline{\eta}\left\|\boldsymbol{\nabla}\frac{\partial \boldsymbol{u}}{\partial t}\right\|^{2}_{\mathbb{L}^2(\Omega)}+C\left\|\frac{\partial \boldsymbol{u}}{\partial t}\right\|^{2}_{\mathbb{L}^2(\Omega)}\|\boldsymbol{u}\|^4_{\mathbb{L}^4(\Omega)}\end{aligned}$$

$$\int_\Omega \frac{\partial \boldsymbol{f}}{\partial t}\cdot\frac{\partial \boldsymbol{u}}{\partial t}\le \frac{1}{4}\overline{\eta}\left\|\boldsymbol{\nabla}\frac{\partial \boldsymbol{u}}{\partial t}\right\|^{2}_{\mathbb{L}^2(\Omega)}+C\left\|\frac{\partial \boldsymbol{f}}{\partial t}\right\|^{2}_{V'},$$

and other elementary manipulations, we obtain

$$\frac{1}{2}\frac{d}{dt}\int_\Omega\left|\frac{\partial \boldsymbol{u}}{\partial t}\right|^2+\frac{1}{2}\overline{\eta}\int_\Omega\left|\boldsymbol{\nabla}\frac{\partial \boldsymbol{u}}{\partial t}\right|^2\le C\left\|\frac{\partial \boldsymbol{f}}{\partial t}\right\|^{2}_{V'}+C\left\|\frac{\partial \boldsymbol{u}}{\partial t}\right\|^{2}_{\mathbb{L}^2(\Omega)}\|\boldsymbol{u}\|^4_{\mathbb{L}^4(\Omega)}.$$

Integrating in time we obtain:

$$\begin{aligned}&\frac{1}{2}\int_\Omega\left|\frac{\partial \boldsymbol{u}}{\partial t}(t,\cdot)\right|^2+\frac{1}{2}\overline{\eta}\int_0^t\int_\Omega\left|\boldsymbol{\nabla}\frac{\partial \boldsymbol{u}}{\partial t}\right|^2\\
&\quad\le \frac{1}{2}\int_\Omega\left|\frac{\partial \boldsymbol{u}}{\partial t}(0,\cdot)\right|^2+C\int_0^t\left\|\frac{\partial \boldsymbol{f}}{\partial t}\right\|^{2}_{V'}+C\int_0^t\left\|\frac{\partial \boldsymbol{u}}{\partial t}\right\|^{2}_{\mathbb{L}^2(\Omega)}\|\boldsymbol{u}\|^4_{\mathbb{L}^4(\Omega)}.\end{aligned}$$

Now, we know that $\boldsymbol{u}\in L^4(0,T;\mathbb{L}^4)$, which was already true without the additional assumptions on the data. Thus the above inequality can easily be treated by the Gronwall lemma. This concludes the formal proof of Proposition 2.14. Note that we make use of the regularity of the initial condition in the following formal way: using the equation formally, we have

$$\frac{\partial \boldsymbol{u}}{\partial t}(0,\cdot)=-\boldsymbol{u}_0\cdot\boldsymbol{\nabla}\boldsymbol{u}_0+\overline{\eta}\Delta\boldsymbol{u}_0+\boldsymbol{f}(0,\cdot),$$

up to a gradient term that we omit. Next each term of the right-hand side is in $L^2(\Omega)$, and we conclude.

Remark 2.1.17 As emphasized above, it is important at this stage to note that we did not impose any condition of smallness on the final time T. So if the initial condition is regular and the forces remain regular for all times, the solution is also regular for all times. This fact is specific to dimension 2, and it is not known whether it also holds in dimension 3 (see below). ♢

Let us now turn to the, again formal, proof of Proposition 2.15. Here, the assumption on $\boldsymbol{f}$ suggests that we may multiply the equation by the highest possible derivative of $\boldsymbol{u}$, namely $-\Delta \boldsymbol{u}$:

$$\frac{1}{2}\frac{d}{dt}\int_\Omega |\nabla \boldsymbol{u}|^2 - \int_\Omega \boldsymbol{u}\cdot\nabla\boldsymbol{u}\cdot\Delta\boldsymbol{u} + \overline{\eta}\int_\Omega |\Delta\boldsymbol{u}|^2 = -\int_\Omega \boldsymbol{f}\cdot\Delta\boldsymbol{u}.$$

The right-hand side is easy to deal with using the Young inequality:

$$\left|\int_\Omega \boldsymbol{f}\cdot\Delta\boldsymbol{u}\right| \le \frac{1}{4}\overline{\eta}\int_\Omega |\Delta\boldsymbol{u}|^2 + C\int_\Omega |\boldsymbol{f}|^2.$$

The transport term is treated as follows:

$$\begin{aligned}\left|\int_\Omega \boldsymbol{u}\cdot\nabla\boldsymbol{u}\cdot\Delta\boldsymbol{u}\right| &\le \|\boldsymbol{u}\|_{\mathbb{L}^\infty(\Omega)}\|\nabla\boldsymbol{u}\|_{\mathbb{L}^2(\Omega)}\|\Delta\boldsymbol{u}\|_{\mathbb{L}^2(\Omega)},\\ &\le C\|\boldsymbol{u}\|^{1/2}_{\mathbb{L}^2(\Omega)}\|\nabla\boldsymbol{u}\|_{\mathbb{L}^2(\Omega)}\|\Delta\boldsymbol{u}\|^{3/2}_{\mathbb{L}^2(\Omega)},\\ &\le \frac{1}{4}\overline{\eta}\|\Delta\boldsymbol{u}\|^2_{\mathbb{L}^2(\Omega)} + C\|\boldsymbol{u}\|^2_{\mathbb{L}^2(\Omega)}\|\nabla\boldsymbol{u}\|^4_{\mathbb{L}^2(\Omega)},\end{aligned}$$

using as above Lemma 2.10 and the Young inequality. The right-hand side can be further estimated from above by

$$\frac{1}{4}\overline{\eta}\|\Delta\boldsymbol{u}\|^2_{\mathbb{L}^2(\Omega)} + C\|\nabla\boldsymbol{u}\|^4_{\mathbb{L}^2(\Omega)}$$

for another irrelevant constant C, as we already know that $\boldsymbol{u}\in L^\infty(0,T;H)$. Collecting all the informations, we obtain the inequality

$$\frac{1}{2}\frac{d}{dt}\int_\Omega |\nabla\boldsymbol{u}|^2 + \frac{1}{4}\overline{\eta}\int_\Omega|\Delta\boldsymbol{u}|^2 \le C\|\nabla\boldsymbol{u}\|^4_{\mathbb{L}^2(\Omega)} + C\int_\Omega |\boldsymbol{f}|^2.$$

We now insert the information $\|\nabla\boldsymbol{u}\|_{\mathbb{L}^2(\Omega)}\in L^2(0,T)$ to rewrite the inequality as

$$\frac{1}{2}\frac{d}{dt}\int_\Omega |\nabla\boldsymbol{u}|^2 + \frac{1}{4}\overline{\eta}\int_\Omega|\Delta\boldsymbol{u}|^2 \le C\,f(t)\,\|\nabla\boldsymbol{u}\|^2_{\mathbb{L}^2(\Omega)} + C\int_\Omega |\boldsymbol{f}|^2,$$

for a function $f(t)$ that belongs to $L^1(0,T)$. It follows from an application of the Gronwall lemma that the regularity stated in Proposition 2.14 indeed holds (formally).

To make the above proof rigorous requires us in particular to mimic the formal proof at the finite-dimensional level, and thus to be able to use $\boldsymbol{v} = -\Delta\boldsymbol{u}_n$ as a test function in the approximation of equation (2.44). This is performed by noting that, owing to our special choice of basis functions, we have

$$\int_\Omega \nabla\boldsymbol{w}_j : \nabla\boldsymbol{v} = \langle A\boldsymbol{w}_j, \boldsymbol{v}\rangle_{V',V} = \lambda_j\int_\Omega \boldsymbol{w}_j\cdot\boldsymbol{v},$$

for all $\boldsymbol{v}\in V$. Multiplying equation (2.44) by $\lambda_j\boldsymbol{w}_j$ therefore amounts to formally using $-\Delta\boldsymbol{w}_j$ (or, more appropriately stated, $A\boldsymbol{w}_j$) as test function instead of $\boldsymbol{w}_j$.

2.1.3 The three-dimensional hydrodynamic case

Here we turn to the three-dimensional case. We will only indicate where and how the proofs of the previous section needs to be modified, when they can be extended to cover the three-dimensional case. We will also indicate which questions, solved in the two-dimensional setting, are open in the three-dimensional one.

2.1.3.1 Existence We have the following proposition, analogous to Proposition 2.11.

Proposition 2.16 [Three-dimensional case] *There exists a solution (in the sense of Definition 2.6) to problem (NS).*

The proof of Proposition 2.16 follows the same pattern as that of Proposition 2.11. We will therefore only indicate the slight modifications. All the arguments regarding the linear terms are still valid, and the modifications all concern the specific treatment of the nonlinear term $\boldsymbol{u} \cdot \boldsymbol{\nabla u}$ (and, to a much smaller extent, that of the time derivative, thus in particular that of the initial condition).

Step 1 mimics that of the proof of Proposition 2.11. The same holds for Step 2, and we obtain (2.47). However, as for the additional *a priori* estimate (2.50), we have to slightly change the setting. In the vein of (2.19), we introduce the functional space

$$V_2 = \left\{ \boldsymbol{v} \in \mathbb{H}^2(\Omega), \operatorname{div} \boldsymbol{v} = 0 \, \boldsymbol{v}\Big|_{\partial\Omega} = 0, \frac{\partial \boldsymbol{v}}{\partial n}\Big|_{\partial\Omega} = 0 \right\}, \tag{2.59}$$

which is the closure of $\mathcal{V}$ (defined by (2.18)) in $\mathbb{H}^2(\Omega)$. We are then going to show that, instead of (2.50), the property

$$\frac{\partial \boldsymbol{u}_n}{\partial t} \quad \text{is bounded in} \quad L^2(0,T;V_2') \tag{2.60}$$

holds. For this purpose, we adapt the argument made above on (2.49). From the Hölder inequality, we know that

$$\|\boldsymbol{u}\|_{\mathbb{L}^3(\Omega)} \leq \|\boldsymbol{u}\|_{\mathbb{L}^2(\Omega)}^{1/2} \|\boldsymbol{u}\|_{\mathbb{L}^6(\Omega)}^{1/2}. \tag{2.61}$$

Using the Sobolev embedding $H^1(\Omega) \subset L^6(\Omega)$, which holds since Ω is now a regular bounded domain in $\mathbb{R}^3$ (recall that in $\mathbb{R}^2$, we have $H^1(\Omega) \subset L^p(\Omega)$, for all $2 \leq p < +\infty$), and

$$\|\boldsymbol{u}\|_{\mathbb{L}^3(\Omega)} \leq \|\boldsymbol{u}\|_{\mathbb{L}^2(\Omega)}^{1/2} \|\boldsymbol{u}\|_{\mathbb{H}^1(\Omega)}^{1/2}. \tag{2.62}$$

This allows us to bound the nonlinear term $g(\boldsymbol{u}_n)$ as follows:

$$\begin{aligned} |\langle g(\boldsymbol{u}_n), \boldsymbol{v}\rangle| = \left| \int_\Omega (\boldsymbol{u}_n \cdot \boldsymbol{\nabla}) \boldsymbol{u}_n \, \boldsymbol{v} \right| &= \left| - \int_\Omega \boldsymbol{u}_n \cdot \boldsymbol{\nabla} \boldsymbol{v} \, \boldsymbol{u}_n \right| \\ &\leq \|\boldsymbol{u}_n\|_{\mathbb{L}^3(\Omega)}^2 \|\boldsymbol{\nabla v}\|_{\mathbb{L}^3(\Omega)} \\ &\leq C \, \|\boldsymbol{u}_n\|_{\mathbb{L}^2(\Omega)} \|\boldsymbol{u}_n\|_{\mathbb{H}^1(\Omega)} \|\boldsymbol{v}\|_{\mathbb{H}^2(\Omega)} \end{aligned}$$

where the latter majoration holds because $H^1 \subset L^3$ and thus $\|\boldsymbol{\nabla v}\|_{\mathbb{L}^3(\Omega)} \leq C\, \|\boldsymbol{v}\|_{\mathbb{H}^2(\Omega)}$. It follows that

$$\|g(\boldsymbol{u}_n(t,\cdot))\|_{V_2'} \leq C\, \|\boldsymbol{u}_n\|_H\, \|\boldsymbol{u}_n\|_V\,,$$

and thus, integrating in time,

$$\|g(\boldsymbol{u}_n)\|_{L^2(0,T;V_2')} \leq C\, \|\boldsymbol{u}_n\|_{L^\infty(0,T;H)}\, \|\boldsymbol{u}_n\|_{L^2(0,T;V)}\,.$$

Accounting for (2.47), we have $g(\boldsymbol{u}_n)$ bounded in $L^2(0,T;V_2')$, which we insert in (2.49). We next obtain (2.60).

As for Step 3, the treatment of the linear terms is left unchanged, but we again need an adaptation for the nonlinear term. Using (2.62), and the bound (2.47), we obtain

$$\boldsymbol{u}_n \quad \text{is bounded in} \quad L^4(0,T;\mathbb{L}^3(\Omega)), \tag{2.63}$$

from which we infer that all the components of $\boldsymbol{u}_n \otimes \boldsymbol{u}_n$ are bounded in $L^2(0,T;L^{3/2}(\Omega))$. We pass to the weak limit in this latter space (instead of $L^2(0,T;\mathbb{L}^2(\Omega))$ as above), and obtain the weak convergence of the nonlinear term

$$\int_\Omega (\boldsymbol{u}_n\cdot\boldsymbol{\nabla})\boldsymbol{u}_n\cdot\boldsymbol{w}_j \quad \rightharpoonup \quad \int_\Omega (\boldsymbol{u}\cdot\boldsymbol{\nabla})\boldsymbol{u}\cdot\boldsymbol{w}_j \quad \text{weakly in } L^2(0,T).$$

Next, we obtain a limit equation analogous to (2.53), *i.e.*

$$\left\langle \frac{\partial \boldsymbol{u}}{\partial t}\,,\,\boldsymbol{w}_j\right\rangle_{V_2',V_2} + \int_\Omega (\boldsymbol{u}\cdot\boldsymbol{\nabla})\boldsymbol{u}\cdot\boldsymbol{w}_j + \overline{\eta}\int_\Omega \boldsymbol{\nabla u} : \boldsymbol{\nabla w}_j = \langle \boldsymbol{f}\,,\,\boldsymbol{w}_j\rangle, \tag{2.64}$$

(in the distributional sense in time, and for all j) the only difference lying in the formulation of the first term

$$\left\langle \frac{\partial \boldsymbol{u}}{\partial t}\,,\,\boldsymbol{w}_j\right\rangle_{V_2',V_2}$$

which now makes sense *via* the duality (V_2',V_2) instead of (V',V). This holds true because $\boldsymbol{w}_j$ is not only in V, but indeed in V_2.

Remark 2.1.18 Actually, the "natural" functional space to perform the above proof is not the space V_2 defined by (2.59), but the closure of $\mathcal{V}$ in the fractional Sobolev space $\mathbb{H}^{3/2}(\Omega)$. The reader has been spared such unnecessary technicalities here. ◇

Remark 2.1.19 Related to the above remark on the choice of the natural functional space is the issue of invariance of the norm under scaling. In dimension 2, it can be noticed that the $\mathbb{H}^1(\Omega)$ seminorm $\int |\nabla \boldsymbol{u}|^2$ satisfies the invariance relation: $\int |\nabla(\boldsymbol{u}(\lambda \boldsymbol{x}))|^2 d\boldsymbol{x} = \int |\nabla \boldsymbol{u}|^2$, for all λ. This is related to the fact that $\mathbb{H}^1(\Omega)$ is the natural functional space for the arguments in two dimensions. In dimension 3, the seminorm invariant under scaling is that of $\mathbb{H}^{3/2}(\Omega)$. This confirms the relevance of the functional space $\mathbb{H}^{3/2}(\Omega)$, as stressed in the above remark. The importance of the invariance under scaling will be again mentioned in Remark 2.1.25 below, with another viewpoint though. ◊

Remark 2.1.20 It is worth noticing that since we only obtain

$$\frac{\partial \boldsymbol{u}}{\partial t} \in L^2(0,T;V_2')$$

(in the vein of (2.60)), and not

$$\frac{\partial \boldsymbol{u}}{\partial t} \in L^2(0,T;V'),$$

we cannot apply Lemma 2.7. In fact, we do not have here $\boldsymbol{u} \in C^0(0,T;H)$: $\boldsymbol{u}$ is continuous with values in a fractional negative Sobolev space $\mathbb{H}^{-1/4}(\Omega)$. This is the optimal meaning which may be given to the initial condition. As for the above remark, we have preferred to avoid such technicalities. Actually, we also have the so-called *weak continuity* $\boldsymbol{u} \in C^0(0,T;H_w)$. See Chapter 4, Section 4.2.1 for the definition of this notion. ◊

The above shows that the three-dimensional case is not that different from the two-dimensional case, as far as existence of a global-in-time weak solution is concerned. We only had to slightly modify the functional inequalities in order to obtain the right bounds in the right functional spaces. But nothing dramatic. In fact, this is for uniqueness, and related to this, for regularity, that the situation is radically different.

2.1.3.2 Uniqueness Let us formally explain which difficulties arise and where. A glance at the proof of the uniqueness of weak solutions in two dimensions reveals that the bottom line of the proof of Proposition 2.12 is to use $\boldsymbol{w} = \boldsymbol{u}_1 - \boldsymbol{u}_2$ as a test function in the equation satisfied by $\boldsymbol{w} = \boldsymbol{u}_1 - \boldsymbol{u}_2$ itself. This requires the term $\left\langle \frac{\partial \boldsymbol{w}}{\partial t}, \boldsymbol{w} \right\rangle$ to have a meaning. Unfortunately, this is not the case in general, since we only know that the solutions (and consequently $\boldsymbol{w}$) belong to $L^2(0,T;V)$ with derivative in time in $L^2(0,T;V_2')$ and not in $L^2(0,T;V')$.

The other side of the same difficulty is to consider the nonlinear term. We say this is "the other side" because of course the regularity of $\partial \boldsymbol{u}/\partial t$ is bootstrapped from the regularity of the nonlinear term $g(\boldsymbol{u})$ as demonstrated above on two occasions. If we try to obtain uniqueness, we see that we are led to consider terms such as $(g(\boldsymbol{u}_1) - g(\boldsymbol{u}_2), \boldsymbol{u}_1 - \boldsymbol{u}_2)$ and here again, we cannot conclude.

Of course, whatever the viewpoint, if a better regularity of the solution is assumed, then uniqueness holds. Such a regularity is typically obtained in two settings:

- for small times (*i.e.* T sufficiently small) and regular data (*i.e.* initial condition $\boldsymbol{u}_0$ and forces $\boldsymbol{f}$);
- or for arbitrarily large times provided the data are sufficiently small in regular norms.

Remark 2.1.21 Of course, another option is to omit the nonlinear term, and to treat the *linear* time-dependent Stokes equation. Existence, uniqueness, and regularity of solutions, all global-in-time, may then be established. ◇

In both the above settings, it is possible to show that the impact of the nonlinear term can be controlled. We refer to the literature for many results in the above two directions. We only wish to briefly formally explain how the *technical* obstruction arises. Let us give ourselves two solutions $\boldsymbol{u}_1$ and $\boldsymbol{u}_2$. Then, we may easily establish the following formal estimate:

$$\frac{1}{2}\frac{d}{dt}\int_\Omega (\boldsymbol{u}_1-\boldsymbol{u}_2)^2+\int_\Omega |\boldsymbol{\nabla}(\boldsymbol{u}_1-\boldsymbol{u}_2)|^2 = -\int_\Omega (\boldsymbol{u}_1\cdot\boldsymbol{\nabla}\boldsymbol{u}_1-\boldsymbol{u}_2\cdot\boldsymbol{\nabla}\boldsymbol{u}_2).(\boldsymbol{u}_1-\boldsymbol{u}_2),$$

$$= -\int_\Omega (\boldsymbol{u}_1-\boldsymbol{u}_2)\cdot\boldsymbol{\nabla}\boldsymbol{u}_1\cdot(\boldsymbol{u}_1-\boldsymbol{u}_2). \quad (2.65)$$

In the context of weak solutions, the best estimate known on $\boldsymbol{u}_1$ is

$$\boldsymbol{u}_1 \in L^\infty(0,T;\mathbb{L}^2(\Omega)) \cap L^2(0,T;\mathbb{H}^1(\Omega)).$$

Therefore, the best possible majoration for the right-hand side of (2.65) yields

$$\frac{1}{2}\frac{d}{dt}\int_\Omega (\boldsymbol{u}_1-\boldsymbol{u}_2)^2+\int_\Omega |\boldsymbol{\nabla}(\boldsymbol{u}_1-\boldsymbol{u}_2)|^2 \leq \|\boldsymbol{u}_1\|_{L^2(0,T;\mathbb{H}^1(\Omega))}\ \|\boldsymbol{u}_1-\boldsymbol{u}_2\|^2_{L^4(0,T;\mathbb{L}^4(\Omega))}. \quad (2.66)$$

Now the norm $L^4(0,T;\mathbb{L}^4(\Omega))$ *cannot* be majorized using only the norm in $L^\infty(0,T;\mathbb{L}^2(\Omega))\cap L^2(0,T;\mathbb{H}^1(\Omega))$, which is the norm appearing in the left-hand side of (2.66). In fact, and we become somehow technical here, interpolation inequalities show that, at best, the $L^\infty(0,T;\mathbb{L}^2(\Omega))\cap L^2(0,T;\mathbb{H}^1(\Omega))$ norm controls $L^4(0,T;\mathbb{L}^3(\Omega))\cap L^{8/3}(0,T;\mathbb{L}^4(\Omega))$.

At this point, it thus appears necessary to get some additional information on the regularity of both $\boldsymbol{u}_1$ and $\boldsymbol{u}_2$. To this end, we argue on, say, $\boldsymbol{u}_1$, formally multiply the equation by $A\boldsymbol{u}_1$ and integrate. For clarity, we even assume that the force field $\boldsymbol{f}$ is zero. This yields:

$$\frac{1}{2}\frac{d}{dt}\int_\Omega |\boldsymbol{\nabla}\boldsymbol{u}_1|^2+\int_\Omega |A\,\boldsymbol{u}_1|^2 = \int_\Omega (\boldsymbol{u}_1\cdot\boldsymbol{\nabla}\boldsymbol{u}_1)\cdot A\,\boldsymbol{u}_1. \quad (2.67)$$

It is then possible to show the following inequality (in dimension 3)

$$\left|\int_\Omega \boldsymbol{u}\cdot\boldsymbol{\nabla}\boldsymbol{v}\cdot\boldsymbol{w}\right| \leq C\ \|\boldsymbol{u}\|_V\ \|\boldsymbol{v}\|_V^{1/2}\ \|A\,\boldsymbol{v}\|_H^{1/2}\ \|\boldsymbol{w}\|_H \quad (2.68)$$

for all $\boldsymbol{u}\in V$, $\boldsymbol{v}\in H$ such that $A\boldsymbol{v}\in H$, $\boldsymbol{w}\in H$. This inequality can be read in [229]. Using it in the right-hand side of (2.67), we obtain

$$\frac{1}{2}\frac{d}{dt}\int_\Omega |\boldsymbol{\nabla}\boldsymbol{u}_1|^2+\int_\Omega |A\,\boldsymbol{u}_1|^2 \leq C\ \|\boldsymbol{u}_1\|_V^{3/2}\ \|A\,\boldsymbol{u}_1\|_H^{3/2}$$

$$\leq \frac{1}{2}\int_\Omega |A\,\boldsymbol{u}_1|^2 + C\ \|\boldsymbol{u}_1\|_V^6, \quad (2.69)$$

using the Young inequality. It is straightforward to see that the above inequality implies an inequality of the form

$$\dot{X}(t) \leq X^3(t), \tag{2.70}$$

where, up to some irrelevant multiplicative constant, $X(t) = \int_\Omega |\boldsymbol{\nabla} \boldsymbol{u}_1|^2$. By a simple argument on ordinary differential equations, (2.70) yields a bound on X *for small times and not for all times*:

$$X^2(t) \leq \frac{X^2(0)}{1 - 2t\, X^2(0)}.$$

For T sufficiently small (as a function of the data), it follows that $\boldsymbol{u}_1 \in L^\infty(0,T;\mathbb{H}^1(\Omega)) \cap L^2(0,T;\mathbb{H}^2(\Omega))$ (in other words, $\boldsymbol{u}_1$ is a *strong* solution), and this can be inserted in (2.65), to get an estimate better than (2.66) and conclude that uniqueness holds.

We have outlined the proof of the following proposition (to be compared with Proposition 2.15).

Proposition 2.17 *Assume (with possibly $T = +\infty$) $\boldsymbol{f} \in L^2(0,T;H)$ (and not only $\boldsymbol{f} \in L^2(0,T;V')$) and $\boldsymbol{u}_0 \in V$. Then there exists some time $T^\star \leq T$ (possibly $T^\star = +\infty$), depending on the data, such that on $[0,T^\star]$ the solution provided by Proposition 2.16 satisfies*

$$\boldsymbol{u} \in L^\infty(0,T^\star;V) \cap L^2(0,T^\star;\mathbb{H}^2(\Omega)), \tag{2.71}$$

and is unique in the class of weak solutions.

Remark 2.1.22 It is enlightening to revisit here the two-dimensional setting. The inequality (2.68) is then replaced by (see again [229]):

$$\left| \int_\Omega \boldsymbol{u} \cdot \boldsymbol{\nabla} \boldsymbol{v} \cdot \boldsymbol{w} \right| \leq C\, \|\boldsymbol{u}\|_H^{1/2}\, \|\boldsymbol{u}\|_V^{1/2}\, \|\boldsymbol{v}\|_V^{1/2}\, \|A\,\boldsymbol{v}\|_H^{1/2}\, \|\boldsymbol{w}\|_H \tag{2.72}$$

which, using the bound $\boldsymbol{u}_1 \in L^\infty(0,T;H)$, allows us to prove

$$\frac{1}{2}\frac{d}{dt}\int_\Omega |\boldsymbol{\nabla} \boldsymbol{u}_1|^2 + \int_\Omega |A\,\boldsymbol{u}_1|^2 \leq \frac{1}{2}\int_\Omega |A\,\boldsymbol{u}_1|^2 + C\, \|\boldsymbol{u}_1\|_V^4,$$

instead of (2.69). We thus have

$$\frac{1}{2}\frac{d}{dt}\int_\Omega |\boldsymbol{\nabla} \boldsymbol{u}_1|^2 \leq C\, \|\boldsymbol{u}_1\|_V^2\, \|\boldsymbol{u}_1\|_V^2,$$

where $\|\boldsymbol{u}_1\|_V^2$ is already known to be a $L^1(0,T)$ function. The above inequality is thus of the form

$$\dot{Y} \leq f(t)\, Y(t) \tag{2.73}$$

where $f \in L^1(0,T)$, rather than of the form (2.70). A simple Gronwall type argument thus allows us to conclude that Y is bounded by a function of T, however large T is. ◊

Remark 2.1.23 Likewise, slightly modifying the above argument, it is possible to prove that for a sufficiently small initial condition (and also for small forces indeed), the solution is regular for all times. ◇

Remark 2.1.24 It is a surprising thing, somehow related to (Fact 1) in Section 2.1.1.2, that the argument reduces to an estimate on the solution of an ordinary differential equation, (2.70). It is more or less always the case in such mathematical studies. The comparison of (2.70) and (2.73) illustrates the technical difference between dimensions 3 and 2. ◇

In its present state, the theory of solutions of the (three-dimensional) Navier–Stokes equation is still unsatisfactory. A crucial point is indeed that it is not known whether the above-mentioned difficulties in proving regularity and/or uniqueness of weak solution are technical in nature, or simply the manifestation of a mathematical obstruction.

It is not known whether weak solutions are unique (or what additional "natural" property would make them unique). It is not known either whether they stay regular for all times if the data are regular, or if they indeed show singularities. In spite of tremendous efforts by many outstanding mathematicians, progress is slow in this field. Some related questions have been somewhat clarified, but the above questions basically remain unsolved since the work of J. Leray in the 1930s. They are of course of primary interest from the standpoint of mathematical analysis, but also for numerical simulations in fluid mechanics.

In the absence of a definite understanding of the regularity of weak solutions, the efforts concentrate on alternative questions. For instance, should a weak solution become singular at some particular instants, the sets of these particular times and the type of pathologies (*i.e.* the type of blow-up or discontinuities of the various norms of $\boldsymbol{u}(t,\cdot)$) that may occur at such times are studied. It is known that a weak solution is a succession in time of strong (thus unique) solutions, defined respectively between two consecutive "accident" times t_n and t_{n+1}. The set of such times has been shown to be small in $\mathbb{R}_+$ (in the sense of their Hausdorff measure). Likewise, the set of initial conditions such that the solution is not globally-in-time regular and unique has been shown to be small in some sense.

2.1.4 Related issues

As briefly mentioned above, our arguments have been conducted in the setting of equations posed on a bounded regular domain Ω, with homogeneous boundary conditions. The proof applies *mutatis mutandis* to three other settings:

- non-homogeneous Dirichlet boundary conditions, or more generally other types of boundary conditions, still in bounded domains;
- periodic boundary conditions;
- equations posed over the whole space $\mathbb{R}^d$, or a half space, *etc.*

As mentioned above, a possible alternative to the notion of *weak* solutions is the notion of *mild solutions*. Mild solutions are proved to exist by the following strategy: (a) rewriting of the equation the form of a fixed-point equation; and (b) application of a fixed-point procedure (Banach fixed-point theorem).

We argue on the equation set on the whole space for simplicity. It may be rephrased as follows:

$$\boldsymbol{u}(t,\cdot) = S(t)\,\boldsymbol{u}_0(\cdot) - \int_0^t S(t-s)\,\mathbb{P}\,\mathrm{div}\,(\boldsymbol{u}\otimes\boldsymbol{u})\,ds + \int_0^t S(t-s)\,\mathbb{P}\,\boldsymbol{f}\,ds \quad (2.74)$$

where $\mathbb{P}$ is the projector on divergence-free fields[18]. The operator $S(t)$ is the kernel of the heat equation, *i.e.* the operator $S(t) = \exp(t\Delta)$ such that $v(t,\cdot) = S(t)\,v_0$ is the solution v to

$$\begin{cases} \dfrac{\partial v}{\partial t} - \Delta v = 0, \\ v(t=0,\cdot) = v_0. \end{cases} \quad (2.75)$$

Specifically, in dimension 3, it is defined by

$$S(t)\psi(\cdot) = \left(\frac{1}{4\pi t}\right)^{3/2} \exp\left(-\frac{|\cdot|^2}{4t}\right) \star \psi(\cdot),$$

where $\star$ denotes the convolution.

The existence of $\boldsymbol{u}$ is obtained in functional spaces such as $C(0,T;\mathbb{L}^3(\mathbb{R}^3))$.

A major advantage of the mild formulation is that uniqueness may be proved, say in the space $C(0,T;\mathbb{L}^3(\mathbb{R}^3))$. Continuous dependence on the data also holds true[19].

Remark 2.1.25 The fact that the space $\mathbb{L}^d(\mathbb{R}^d)$ plays a specific role for the Navier–Stokes equations in dimension d is related to some invariance properties of the equation under rescaling of solutions (both in time and space): if $\boldsymbol{u}$ is a solution with initial condition $\boldsymbol{u}_0$, then, for any λ, $\lambda\,\boldsymbol{u}(\lambda^2 t, \lambda\,\mathbf{x})$ is a solution with initial condition $\boldsymbol{u}_0(\mathbf{x}) = \lambda\,\boldsymbol{u}_0(\lambda\,\mathbf{x})$. The latter rescaling precisely leaves the $\mathbb{L}^d$ norm of $\boldsymbol{u}_0$ invariant. This is related to the notion of *self-similar solutions*. ◊

Two important comments are in order.

The physical interpretation of mild solutions is uneasy. In contrast, weak solutions belong to the natural "energy" spaces, for which the integrals making sense are $\int_{\mathbf{R}^3} |\boldsymbol{u}|^2$ and $\int_{\mathbf{R}^3} |\boldsymbol{\nabla}\boldsymbol{u}|^2$. The first quantity corresponding to the kinetic energy of the fluid, and the second one to the rate of viscous dissipation.

Existence, regularity and uniqueness of mild solutions come at a price. Rather restrictive assumptions on the initial condition need to be imposed. Basically, the initial velocity should be small in appropriate norms accounting for oscillations. Quite sophisticated functional spaces such as the *Besov spaces* are used

[18] Note that again, like in the weak formulation, the pressure has been eliminated in the mild formulation. It will be recovered by the standard arguments already seen above.

[19] The latter fact is known to be false for weak solutions that do not depend continuously on the data; see the bibliography.

to formalize this. In such spaces, the norm is small if the function is small and oscillates sufficiently. The existence of mild solutions for initial condition small in Besov norms is related to the fact that one can pass to the limit in nonlinear terms if there are enough oscillations. In practice, such assumptions on the initial condition may be difficult to check.

For more details on the theory of mild solutions, we refer to the seminal work by T. Kato and his collaborators (see in particular [83]), and next to the numerous works by Y. Meyer, M. Cannone, F. Planchon, *e.g.* see in particular [34, 35] (both in French).

Another setting important to address is the stationary setting, *i.e.* the question of finding $(\boldsymbol{u}(\mathbf{x}), p(\mathbf{x}))$ a solution to the system

$$\begin{cases} \boldsymbol{u}\cdot\nabla\,\boldsymbol{u} - \Delta\boldsymbol{u} + \nabla p = \boldsymbol{f}, \\ \operatorname{div}\boldsymbol{u} = 0. \end{cases} \tag{2.76}$$

We postpone the study of this system until Section 2.2.4 where we deal in fact with the stationary version of the MHD system, namely system (1.44) from Chapter 1.

2.2 Mathematical results on the one-fluid MHD equations

The focus of this section is system (1.40):

$$\begin{cases} \begin{cases} \dfrac{\partial \boldsymbol{u}}{\partial t} + \boldsymbol{u}\cdot\nabla\,\boldsymbol{u} - \overline{\eta}\,\Delta\boldsymbol{u} + \nabla p = \mathbf{curl}\,\boldsymbol{B}\times\boldsymbol{B} + \boldsymbol{g}, \\ \operatorname{div}\boldsymbol{u} = 0. \end{cases} \\ \begin{cases} \dfrac{\partial \boldsymbol{B}}{\partial t} + \mathbf{curl}\,\mathbf{curl}\,\boldsymbol{B} = \mathbf{curl}\,\big(\boldsymbol{u}\times\boldsymbol{B}\big), \\ \operatorname{div}\boldsymbol{B} = 0, \end{cases} \end{cases} \tag{2.77}$$

where we have set $\overline{\rho} = 1$, as in the previous section, and also here $\overline{\mu} = 1$, $\overline{\sigma} = 1$. This is performed without loss of generality. We keep $\overline{\eta}$ explicit in order to emphasize the role played by the viscous term.

The above system is supplied with the following homogeneous boundary conditions on $\partial\Omega$:

$$\begin{aligned} \boldsymbol{u} &= 0, \\ \boldsymbol{B}\cdot\boldsymbol{n} &= 0, \\ \mathbf{curl}\,\boldsymbol{B}\times\boldsymbol{n} &= 0. \end{aligned} \tag{2.78}$$

Considering such conditions is sufficient to understand the basic mathematical features of system (1.40). Non-homogeneous conditions may also be considered. They are treated by lifting, and one is eventually left with a system analogous to the homogeneous case above, with a different source term $\boldsymbol{g}$ though, and other additional terms whose treatment does not raise major difficulties.

The reason why we concentrate on the above system is that this is the relevant mathematical system for the applications we target. We will show that this system is well posed mathematically. Two alternative, somewhat natural, systems could be considered. We will see below that the well-posedness of both the more general system (1.37) and the simpler system (1.43) is unclear mathematically. This will be the purpose of Sections 2.2.3 and 2.2.5, respectively.

We begin our study with an overview of the mathematical literature on MHD systems in Section 2.2.1, and then proceed with the mathematical analysis in Section 2.2.2.

2.2.1 A brief overview of the literature

Several works have already been devoted to the study of MHD systems for *one* fluid with constant density. We now offer a brief overview on those we are aware of.

Existence and uniqueness results are established by G. Duvaut and J.-L. Lions in [73] for the case of the time-dependent MHD equations (without displacement current) posed on a simply connected bounded domain in the framework of Bingham fluids. These results are complemented by M. Sermange and R. Temam in [208] for classical Newtonian fluids. These authors show that the classical properties of the Navier–Stokes equations can be extended to the MHD system. More precisely, they prove in the two-dimensional case the existence and the uniqueness of a global weak solution, which, in addition, is a strong solution for regular data. When the space dimension is three, they prove that a global weak solution exists and that for more regular data, a strong solution exists and is unique for small times. Also, they study the large-time behavior and the Haussdorf dimension of a functional invariant set. Some of these results are also presented by R. Temam in [228] and by J.-M. Ghidaglia in [104].

The works [73, 208], applied to the three-dimensional case, will be the major material used in Section 2.2.2.

Other related studies include the following. The case of multiply connected bounded sets is studied by J.-M. Domingez de la Rasilla in [69] for the stationary equation and by K. Kerieff in the time-dependent problem [132]. The work [69] also provides a numerical analysis of the finite element method in this context. So does the work by M.D. Gunzburger, A.J. Meir, and J.S. Peterson [118]. E. Sanchez-Palancia has treated in [202, 203] an MHD problem in an exterior domain both in the stationary and the time-dependent cases (without displacement currents). J. Rappaz and R. Touzani have studied the MHD equations in a particular two-dimensional non-connected domain relevant for some industrial applications such as electromagnetic casting. They establish existence results in [194] (summarized in [193]) and provide a numerical analysis of the problem in [195]. Let us also mention an interesting alternative viewpoint which consists in considering the electrical current rather than the magnetic field as the main electromagnetic unknown (see A.J. Meir, P.G. Schmidt [161–163]). This setting will be examined in Section 2.2.6.3 below.

2.2.2 *Mathematical analysis*

2.2.2.1 Functional spaces In order to treat the additional unknown field $\boldsymbol{B}$, we introduce the spaces

$$\mathcal{W} = \{\boldsymbol{C} \in (\mathcal{D}(\overline{\Omega}))^3, \operatorname{div} \boldsymbol{C} = 0, \boldsymbol{C} \cdot \boldsymbol{n}|_{\partial\Omega} = 0\}, \tag{2.79}$$

$$W = \{\boldsymbol{C} \in \mathbb{H}^1(\Omega), \operatorname{div} \boldsymbol{C} = 0, \boldsymbol{C} \cdot \boldsymbol{n}|_{\partial\Omega} = 0\}. \tag{2.80}$$

The space W is the closure of $\mathcal{W}$ in $\mathbb{H}^1(\Omega)$. For $\boldsymbol{C} \in W$ we denote

$$||\boldsymbol{C}||_W = \left(\int_\Omega |\mathbf{curl}\, \boldsymbol{C}|^2 \, d\boldsymbol{x} \right)^{1/2}. \tag{2.81}$$

One can establish that $||.||_W$ defines a norm on W which is equivalent to that induced by $\mathbb{H}^1(\Omega)$ on W (cf. G. Duvaut and J.-L. Lions [73]). The fact that Ω is simply connected (and regular) is essential for this point.

We also recall from (2.20) the notation

$$H = \{\boldsymbol{v} \in \mathbb{L}^2(\Omega), \operatorname{div} \boldsymbol{v} = 0, \boldsymbol{v}.\boldsymbol{n}|_{\partial\Omega} = 0\},$$

which is a shorthand notation for $H_0(div^0, \Omega)$.

2.2.2.2 The magnetic problem For the hydrodynamics problem, we have introduced above the Stokes problem (2.32), the operators Λ and A, respectively defined by (2.34) and (2.38), and the eigenelements $(\boldsymbol{w}_j, \lambda_j)$ for (2.35). In the same fashion, we now introduce the magnetic problem

$$\begin{cases} \mathbf{curl}\,\mathbf{curl}\, \boldsymbol{B} = \boldsymbol{f}, \\ \operatorname{div} \boldsymbol{B} = 0, \\ \boldsymbol{B} \cdot \boldsymbol{n} = 0, \\ \mathbf{curl}\, \boldsymbol{B} \times \boldsymbol{n} = 0. \end{cases} \tag{2.82}$$

A vector-valued function $\boldsymbol{f}$ being given in H, the above system admits[20] a unique solution $\boldsymbol{B} \in \mathbb{H}^1(\Omega)$, that indeed belongs to W (and also to $\mathbb{H}^2(\Omega)$). We denote

$$\boldsymbol{B} = \tilde{\Lambda} \boldsymbol{f} \tag{2.83}$$

thereby defining the linear operator $\tilde{\Lambda}$ from H to W. Equivalently, we may define the operator $\tilde{A}$ by

$$\int_\Omega \mathbf{curl}\, \boldsymbol{B} \cdot \mathbf{curl}\, \boldsymbol{C} = \langle \tilde{A}\boldsymbol{B}, \boldsymbol{C} \rangle, \tag{2.84}$$

for all $\boldsymbol{C} \in W$, the bracket in the right-hand side denoting the duality between W and its dual space W'. The operator $\tilde{A}$ is the inverse of the operator $\tilde{\Lambda}$, which means that for $\boldsymbol{f} \in H$ given, it is equivalent to say that $\boldsymbol{B} \in W$ satisfies $\tilde{A}\boldsymbol{B} = \boldsymbol{f}$ or that $\boldsymbol{B} \in \mathbb{H}^1(\Omega)$ solves (2.82).

[20] This is a standard result that we will recall in Section 2.2.4.

Formally, the key ingredient for the proof of this latter fact is the following one: saying that $\boldsymbol{B}$ solves (2.82) implies that

$$\int_\Omega \mathbf{curl}\,\boldsymbol{B}\cdot\mathbf{curl}\,\boldsymbol{C} - \int_{\partial\Omega} \boldsymbol{C}\cdot(\mathbf{curl}\,\boldsymbol{B}\times\boldsymbol{n}) = \int_\Omega \boldsymbol{f}\cdot\boldsymbol{C},$$

thus

$$\int_\Omega \mathbf{curl}\,\boldsymbol{B}\cdot\mathbf{curl}\,\boldsymbol{C} = \int_\Omega \boldsymbol{f}\cdot\boldsymbol{C}, \tag{2.85}$$

which is exactly $\tilde{A}\boldsymbol{B} = \boldsymbol{f}$. Conversely, $\tilde{A}\boldsymbol{B} = \boldsymbol{f}$ implies that (2.85) holds for all divergence-free field $\boldsymbol{C}$ such that $\boldsymbol{C}\cdot\boldsymbol{n} = 0$. Then it suffices to perform the *Helmholtz decomposition* of a general $\boldsymbol{C}\in\mathbb{H}^1(\Omega)$ such that $\boldsymbol{C}\cdot\boldsymbol{n} = 0$ as a sum $\boldsymbol{C} = \tilde{\boldsymbol{C}} + \boldsymbol{\nabla}\Phi$ where $\tilde{\boldsymbol{C}}$ is divergence-free and $\Delta\Phi = \operatorname{div}\boldsymbol{C}$, $\partial\Phi/\partial n = \boldsymbol{C}\cdot\boldsymbol{n}$, to infer (2.82) from (2.85). See Section 2.2.4.1 for the details of this argument.

For the same reasons as above, the operator $\tilde{\Lambda}$ admits an infinite sequence of eigenfunctions, henceforth denoted by $\tilde{\boldsymbol{w}}_j$, associated with eigenvalues $\tilde{\lambda}_j^{-1}$. They form a basis of the space W and solve:

$$\begin{cases} \mathbf{curl}\,\mathbf{curl}\,\tilde{\boldsymbol{w}}_j = \tilde{\lambda}_j\tilde{\boldsymbol{w}}_j, \\ \operatorname{div}\tilde{\boldsymbol{w}}_j = 0, \\ \tilde{\boldsymbol{w}}_j\cdot\boldsymbol{n}|_{\partial\Omega} = 0, \\ \mathbf{curl}\,\tilde{\boldsymbol{w}}_j\times\boldsymbol{n}|_{\partial\Omega} = 0. \end{cases} \tag{2.86}$$

The magnetostatic problem (2.82) will be examined in Section 2.2.4 below, where some of the arguments only outlined above will be detailed. We also refer to [208] and the references therein.

2.2.2.3 A priori estimates, setting and main result From system (1.40)–(2.78), it is immediate to obtain, multiplying the first equation by $\boldsymbol{u}$, the second by $\boldsymbol{B}$ and integrating over Ω:

$$\frac{1}{2}\frac{d}{dt}\int_\Omega |\boldsymbol{u}|^2 + \overline{\eta}\int_\Omega |\boldsymbol{\nabla}\boldsymbol{u}|^2 = \int_\Omega (\mathbf{curl}\,\boldsymbol{B}\times\boldsymbol{B})\cdot\boldsymbol{u} + \int_\Omega \boldsymbol{g}\cdot\boldsymbol{u}$$

$$\frac{1}{2}\frac{d}{dt}\int_\Omega |\boldsymbol{B}|^2 + \int_\Omega |\mathbf{curl}\,\boldsymbol{B}|^2 = \int_\Omega \mathbf{curl}\,(\boldsymbol{u}\times\boldsymbol{B})\cdot\boldsymbol{B}.$$

Owing to the cancellation of the two terms in the right-hand side, we deduce the following *a priori* estimate

$$\frac{1}{2}\frac{d}{dt}\int_\Omega (|\boldsymbol{u}|^2 + |\boldsymbol{B}|^2) + \overline{\eta}\int_\Omega |\boldsymbol{\nabla}\boldsymbol{u}|^2 + \int_\Omega |\mathbf{curl}\,\boldsymbol{B}|^2 = \int_\Omega \boldsymbol{g}\cdot\boldsymbol{u}. \tag{2.87}$$

This suggests that the natural notion of weak solution is the following:

Definition 2.18 *Let $\boldsymbol{g} \in L^2(0,T;V')$, $\boldsymbol{u}_0 \in H$, $\boldsymbol{B}_0 \in H$. The pair $(\boldsymbol{u}, \boldsymbol{B})$ is said to be a weak solution to (2.77)–(2.78), if*

$$\boldsymbol{u} \in L^\infty(0,T;H) \cap L^2(0,T;V), \quad \boldsymbol{B} \in L^\infty(0,T;H) \cap L^2(0,T;W), \tag{2.88}$$

satisfy, for all $\boldsymbol{v} \in V$ and all $\boldsymbol{C} \in W$,

$$\begin{cases} \left\langle \dfrac{\partial \boldsymbol{u}}{\partial t}, \boldsymbol{v} \right\rangle_{V',V} + \displaystyle\int_\Omega (\boldsymbol{u}\cdot\boldsymbol{\nabla})\boldsymbol{u}\cdot\boldsymbol{v} + \overline{\eta}\int_\Omega \boldsymbol{\nabla}\boldsymbol{u}\cdot\boldsymbol{\nabla}\boldsymbol{v} = \int_\Omega (\mathbf{curl}\,\boldsymbol{B}\times\boldsymbol{B})\cdot\boldsymbol{v} \\ \qquad\qquad\qquad\qquad\qquad\qquad\qquad\qquad + \langle \boldsymbol{g}, \boldsymbol{v}\rangle_{V',V}, \\ \left\langle \dfrac{\partial \boldsymbol{B}}{\partial t}, \boldsymbol{C} \right\rangle_{W',W} + \displaystyle\int_\Omega \mathbf{curl}\,\boldsymbol{B}\cdot\mathbf{curl}\,\boldsymbol{C} = \int_\Omega \mathbf{curl}\,(\boldsymbol{u}\times\boldsymbol{B})\cdot\boldsymbol{C}, \end{cases} \tag{2.89}$$

supplied with the initial condition

$$\begin{cases} \boldsymbol{u}(t=0,\cdot) = \boldsymbol{u}_0, \\ \boldsymbol{B}(t=0,\cdot) = \boldsymbol{B}_0. \end{cases} \tag{2.90}$$

Proposition 2.19 *There exists a solution (in the sense of Definition 2.18) to problem (2.77)–(2.78).*

The proof of Proposition 2.19 mimics those of Propositions 2.11 and 2.16. We use the eigenvectors $\boldsymbol{w}_j$ of the Stokes problem, and the eigenvectors $\tilde{\boldsymbol{w}}_j$ of the magnetostatic equation, respectively, defined by (2.35) and (2.86). We build with these vectors a basis of the space $V \times W$, and thus write the following finite-dimensional formulation:

$$\begin{cases} \left\langle \dfrac{\partial \boldsymbol{u}_n}{\partial t}, \boldsymbol{w}_j \right\rangle_{V',V} + \displaystyle\int_\Omega (\boldsymbol{u}_n\cdot\boldsymbol{\nabla})\boldsymbol{u}_n\cdot\boldsymbol{w}_j + \overline{\eta}\int_\Omega \boldsymbol{\nabla}\boldsymbol{u}_n : \boldsymbol{\nabla}\boldsymbol{w}_j \\ \qquad = \displaystyle\int_\Omega (\mathbf{curl}\,\boldsymbol{B}_n\times\boldsymbol{B}_n)\cdot\boldsymbol{w}_j + (\boldsymbol{g}, \boldsymbol{w}_j), \\ \left\langle \dfrac{\partial \boldsymbol{B}_n}{\partial t}, \tilde{\boldsymbol{w}}_j \right\rangle_{W',W} + \displaystyle\int_\Omega \mathbf{curl}\,\boldsymbol{B}_n\cdot\mathbf{curl}\,\tilde{\boldsymbol{w}}_j \\ \qquad = \displaystyle\int_\Omega \mathbf{curl}\,(\boldsymbol{u}_n\times\boldsymbol{B}_n)\cdot\tilde{\boldsymbol{w}}_j, \end{cases} \tag{2.91}$$

for all $1 \le j \le n$. System (2.91) is supplied with the initial conditions

$$\begin{cases} \boldsymbol{u}_n(0,\cdot) = \boldsymbol{u}_{0n}, \\ \boldsymbol{B}_n(0,\cdot) = \boldsymbol{B}_{0n}, \end{cases} \tag{2.92}$$

where $\boldsymbol{u}_{0n} \in \mathrm{Span}\,(\boldsymbol{w}_1, ..., \boldsymbol{w}_n)$ converges in H to $\boldsymbol{u}_0$, and $\boldsymbol{B}_{0n} \in \mathrm{Span}\,(\tilde{\boldsymbol{w}}_1, ..., \tilde{\boldsymbol{w}}_n)$ converges in H to $\boldsymbol{B}_0$.

Next, exactly as in the proofs of Propositions 2.11 and 2.16, we successively obtain that

$$\begin{aligned} &\boldsymbol{u}_n \quad \text{is bounded in} \quad L^2(0,T;V) \cap L^\infty(0,T;H), \\ &\boldsymbol{B}_n \quad \text{is bounded in} \quad L^2(0,T;W) \cap L^\infty(0,T;H), \end{aligned} \tag{2.93}$$

and that

$$\begin{aligned} &\frac{\partial \boldsymbol{u}_n}{\partial t} \quad \text{is bounded in} \quad L^2(0,T;V'), \\ &\frac{\partial \boldsymbol{B}_n}{\partial t} \quad \text{is bounded in} \quad L^2(0,T;W'). \end{aligned} \tag{2.94}$$

The latter owes to the specific choice of our basis functions $\boldsymbol{w}_j$ and $\tilde{\boldsymbol{w}}_j$. Up to an extraction, we may assume that

$$\begin{aligned} &\boldsymbol{u}_n \rightharpoonup \boldsymbol{u} \,\text{weakly in}\, L^2(0,T;V)\,,\text{weakly} - \star\,\text{in}\; L^\infty(0,T;H), \\ &\quad \text{and} \quad \frac{\partial \boldsymbol{u}_n}{\partial t} \rightharpoonup \frac{\partial \boldsymbol{u}}{\partial t} \quad \text{weakly in} \quad L^2(0,T;V'), \\ &\boldsymbol{B}_n \rightharpoonup \boldsymbol{B} \quad \text{weakly in}\, L^2(0,T;W)\,,\text{weakly} - \star\,\text{in}\; L^\infty(0,T;H), \\ &\quad \text{and} \quad \frac{\partial \boldsymbol{B}_n}{\partial t} \rightharpoonup \frac{\partial \boldsymbol{B}}{\partial t} \quad \text{weakly in} \quad L^2(0,T;W'), \end{aligned} \tag{2.95}$$

as n goes to infinity. Using the same compactness Lemma 2.8, we also obtain

$$\boldsymbol{u}_n \longrightarrow \boldsymbol{u} \quad \text{and} \quad \boldsymbol{B}_n \longrightarrow \boldsymbol{B} \quad \text{strongly in}\, L^2(0,T;H) \quad \text{and} \quad L^2(0,T;H)$$

$$\text{and thus also a.e. in time and space.} \tag{2.96}$$

As above, such convergence on $\boldsymbol{u}_n$ is sufficient to pass to the limit in the nonlinear term $\boldsymbol{u}_n \cdot \boldsymbol{\nabla} \boldsymbol{u}_n$. The only additional task in order to conclude here is to show that the convergences allow us to pass to the limit in both the nonlinear terms appearing in the right-hand sides of (2.91), namely

$$\mathbf{curl}\, \boldsymbol{B}_n \times \boldsymbol{B}_n \quad \text{and} \quad \mathbf{curl}\,(\boldsymbol{u}_n \times \boldsymbol{B}_n).$$

The argument mimics that for the Navier term. The strong convergences (2.96) are combined with the weak convergences (2.95) to obtain the following convergences

$$\begin{aligned} &\mathbf{curl}\, \boldsymbol{B}_n \times \boldsymbol{B}_n \rightharpoonup \mathbf{curl}\, \boldsymbol{B} \times \boldsymbol{B} \,\text{weakly in}\, L^1((0,T) \times \Omega), \\ &\boldsymbol{u}_n \times \boldsymbol{B}_n \longrightarrow \boldsymbol{u} \times \boldsymbol{B} \,\text{strongly in}\, L^1((0,T) \times \Omega). \end{aligned}$$

This allows us to conclude the existence of a solution to (1.40) (in the sense of Definition 2.18). Then the boundary condition $\mathbf{curl}\, \boldsymbol{B} \times \boldsymbol{n} = 0$ and the initial conditions (2.90) are recovered by standard arguments.

Remark 2.2.1 It is worth mentioning that the weak solution we have constructed indeed satisfies the energy inequality

$$\frac{1}{2}\int_\Omega \left(|\boldsymbol{u}|^2 + |\boldsymbol{B}|^2\right)(T) + \overline{\eta}\int_0^T\int_\Omega |\boldsymbol{\nabla u}|^2 + \int_0^T\int_\Omega |\mathbf{curl}\,\boldsymbol{B}|^2$$

$$\leq \int_0^T \langle \boldsymbol{g}\,,\,\boldsymbol{u}\rangle + \frac{1}{2}\int_\Omega \left(|\boldsymbol{u}_0|^2 + |\boldsymbol{B}_0|^2\right).$$

This can be proved simply by integrating in time (2.87) and passing to the limit. ♢

2.2.2.4 Related results The state of the art of the mathematical knowledge on the one-fluid MHD equation (1.40) very much agrees with that of the purely hydrodynamics Navier–Stokes equation (2.1). We have proved the existence of a (global-in-time) weak solution above. In three dimensions, which is definitely the physically relevant case for MHD, the uniqueness of such a weak solution, without further regularity assumptions, is not known. The regulariy, and consequently the uniqueness, may however be proved for small times under appropriate regularity conditions on the data.

The reason for the above similarity primarily lies in the fact that the magnetic equation in (1.40), being *linear* in $\boldsymbol{B}$, does not cause any additional difficulty. The nonlinear term $\mathbf{curl}\,(\boldsymbol{u}\times\boldsymbol{B})$ in the right-hand side is formally like the Navier term. On the other hand, the magnetic equation does not provide any regularizing effect on the velocity $\boldsymbol{u}$ either. Thus it does not simplify the situation in comparison to the purely hydrodynamics case.

We shall see in Chapter 4 that this similarity survives in the case of a non-homogeneous density ρ. Our analysis of the two-fluid MHD system (namely system (1.49) of Chapter 1) draws much of its inspiration from the treatment of the purely hydrodynamic case.

To conclude this section, let us quote without proof the following proposition, extracted from [208]. It provides some prototypical regularity/uniqueness results for the MHD system (1.40). More details may be found in the bibliography.

Proposition 2.20 [208, Theorems 3.1 and 3.2]

(i) Assume $\boldsymbol{g}\in L^2(0,T;V')$, $\boldsymbol{u}_0\in H$, $\boldsymbol{B}_0\in H$. Assume also that a solution to (1.40), in the sense of Definition 2.18, satisfies

$$\boldsymbol{u}\in L^4(0,T;V) \qquad \boldsymbol{B}\in L^4(0,T;W).$$

Then it is unique in this class.

(ii) Assume $\boldsymbol{g}\in L^\infty(0,T;H)$, $\boldsymbol{u}_0\in V$, $\boldsymbol{B}_0\in W$. Then there exists $T^\star$ (possibly $T^\star=+\infty$ if $T=+\infty$), depending on the data, such that on $[0,T^\star]$ there exists a unique weak solution, that is in addition a strong solution, i.e.

$$\boldsymbol{u}\in L^2(0,T^\star;\mathbb{H}^2(\Omega))\cap L^\infty(0,T^\star;V), \quad \boldsymbol{B}\in L^2(0,T^\star;\mathbb{H}^2(\Omega))\cap L^\infty(0,T^\star;W).$$

2.2.3 Back to the hyperbolic system

We are now in a position to address the mathematical issues raised by system (1.37), *i.e.* (again setting $\overline{\rho} = 1$, $\overline{\mu} = 1$, and, here, $\overline{\varepsilon} = 1$, for clarity of exposition)

$$\left\{\begin{array}{l} \left\{\begin{array}{l} \dfrac{\partial \boldsymbol{u}}{\partial t} + \boldsymbol{u} \cdot \nabla \boldsymbol{u} - \overline{\eta}\, \Delta \boldsymbol{u} + \nabla p = \boldsymbol{j} \times \boldsymbol{B} + \boldsymbol{g}, \\ \qquad\qquad\qquad\qquad \operatorname{div} \boldsymbol{u} = 0. \end{array}\right. \\ \left\{\begin{array}{l} -\dfrac{\partial \boldsymbol{E}}{\partial t} + \mathbf{curl}\, \boldsymbol{B} = \boldsymbol{j}, \\ \qquad \operatorname{div} \boldsymbol{E} = \rho_c, \\ \dfrac{\partial \boldsymbol{B}}{\partial t} + \mathbf{curl}\, \boldsymbol{E} = 0, \\ \qquad \operatorname{div} \boldsymbol{B} = 0, \\ \qquad\qquad \boldsymbol{j} = \boldsymbol{E} + \boldsymbol{u} \times \boldsymbol{B}. \end{array}\right. \end{array}\right.$$

In the light of the previous section, the main difference between systems (1.37) and (1.40) may now be explained in mathematical terms.

Due to the presence of the Maxwell equations in their general form, which is hyperbolic, the above system is indeed *very* difficult mathematically. The existence of a solution, even in a weak sense, is unclear.

With a view to appreciating this, let us recall from the previous section that the first step in the proof of the existence of solutions to such a system of equations is to establish the *a priori* energy estimate. It is a simple manipulation to show that, formally, a solution satisfies

$$\frac{1}{2}\frac{d}{dt}\int_\Omega |\boldsymbol{u}|^2 + \overline{\eta}\int_\Omega |\nabla \boldsymbol{u}|^2 = \int_\Omega (\boldsymbol{j} \times \boldsymbol{B}) \cdot \boldsymbol{u}, \tag{2.97}$$

and

$$\frac{1}{2}\frac{d}{dt}\int_\Omega |\boldsymbol{E}|^2 + \frac{1}{2}\frac{d}{dt}\int_\Omega |\boldsymbol{B}|^2 = -\int_\Omega \boldsymbol{j} \cdot \boldsymbol{E}. \tag{2.98}$$

Next, the right-hand side of (2.98) can be modified, accounting for Ohm's law:

$$\frac{1}{2}\frac{d}{dt}\int_\Omega |\boldsymbol{E}|^2 + \frac{1}{2}\frac{d}{dt}\int_\Omega |\boldsymbol{B}|^2 = -\int_\Omega |\boldsymbol{j}|^2 - \int_\Omega (\boldsymbol{j} \times \boldsymbol{B}) \cdot \boldsymbol{u}. \tag{2.99}$$

Summing up (2.97) and (2.99) yields the energy estimate:

$$\frac{1}{2}\frac{d}{dt}\int_\Omega (|\boldsymbol{u}|^2 + |\boldsymbol{E}|^2 + |\boldsymbol{B}|^2) + \int_\Omega |\boldsymbol{j}|^2 + \overline{\eta}\int_\Omega |\nabla \boldsymbol{u}|^2 = 0. \tag{2.100}$$

Notice that, in the above manipulations, we set the external forces and all boundary conditions to zero, for simplicity. Estimate (2.100) clearly indicates that we dispose of $L^\infty(0,T;L^2(\Omega))$ bounds on the vector fields $\boldsymbol{E}$ and $\boldsymbol{B}$ together with a $L^2((0,T) \times \Omega)$ bound on the current $\boldsymbol{j}$, and with the (classical) $L^\infty(0,T;\mathbb{L}^2(\Omega)) \cap L^2(0,T;\mathbb{H}^1(\Omega))$ bounds on the velocity $\boldsymbol{u}$. No additional bound seems to be available that would allow us to pass to the limit in the nonlinear term $\boldsymbol{j} \times \boldsymbol{B}$ of the right-hand side of the Navier–Stokes equation. It is unclear how to derive further energy estimates on system (1.37) that provide more *a priori* regularity on the fields $\boldsymbol{E}$, $\boldsymbol{B}$ and $\boldsymbol{j}$, and allow us to pass to the limit in approximations of the system. To the best of our knowledge, system (1.37) presents an unsolved mathematical difficulty.

2.2.4 Stationary problems

We would now like to address some questions related to the stationary MHD system (1.44) in the case of only *one* homogeneous fluid:

$$\left\{\begin{array}{l}\left\{\begin{array}{r}\boldsymbol{u}\cdot\nabla\boldsymbol{u}-\Delta\boldsymbol{u}+\nabla p=\mathbf{curl}\,\boldsymbol{B}\times\boldsymbol{B}+\boldsymbol{f},\\ \operatorname{div}\boldsymbol{u}=0.\end{array}\right.\\ \\ \left\{\begin{array}{r}\mathbf{curl}\,\mathbf{curl}\,\boldsymbol{B}-\mathbf{curl}\left(\boldsymbol{u}\times\boldsymbol{B}\right)=\boldsymbol{g},\\ \operatorname{div}\boldsymbol{B}=0.\end{array}\right.\end{array}\right. \tag{2.101}$$

All constants have been fixed to one for simplicity. Notice that, in contrast to (1.44), we have (a) restored the notation $\boldsymbol{f}$ for the body force term in the hydrodynamic equation, and (b) added a source term $\boldsymbol{g}$ in the magnetostatic equation. The latter may come from an appropriate lifting of the boundary conditions, and should not be confused with the gravity forces, of course. We shall see that the stationary problems are very different in mathematical nature from the time-dependent problems. The techniques employed to establish existence of solutions may significantly differ.

The comprehensive analysis of system (2.101) will be performed in Chapter 3, Section 3.3. The analysis will then be specifically oriented for numerical purposes. To prepare the ground, we investigate in the present section some preliminary questions related to system (2.101). Those will be useful to set the stage for the analysis conducted in Chapter 3.

We will argue in the specific case of dimension 3. The two-dimensional setting is much simpler[21]. However, as our focus is the MHD setting, the physical relevance of the dimension 2 setting is unclear and thus we prefer to restrict ourselves to the three-dimensional case.

In fact, in order to deal with system (2.101), it is useful to first recall some results concerning two *linear* problems: the Stokes problem (2.32)

$$\left\{\begin{array}{r}-\Delta\boldsymbol{u}+\boldsymbol{\nabla}p=\boldsymbol{f},\\ \operatorname{div}\boldsymbol{u}=0,\\ \boldsymbol{u}|_{\partial\Omega}=0,\end{array}\right.$$

[21]The difference between dimension 2 and dimension 3 is however less striking for the stationary setting than in the time-dependent context.

and the magnetostatic problem (2.82)

$$\begin{cases} \mathbf{curl}\,\mathbf{curl}\,\boldsymbol{B} = \boldsymbol{f}, \\ \operatorname{div}\boldsymbol{B} = 0, \\ \boldsymbol{B}\cdot\boldsymbol{n}|_{\partial\Omega} = 0, \\ \mathbf{curl}\,\boldsymbol{B}\times\boldsymbol{n}|_{\partial\Omega} = 0. \end{cases}$$

Both problems have been encountered so far in the text, but we have considered their basic properties as known. It is now time to (briefly) detail them. In a second step we shall turn to the stationary Navier–Stokes problem (1.23). As announced above, the study of the stationary MHD system (2.101) will then be postponed until Chapter 3. We emphasize that the Stokes problem (2.32) will be again addressed in Chapter 3, Sections 3.1.1–3.1.4, when preparing for discretization issues. Likewise, the magnetostatic problem (2.82) will also be mentioned in Chapter 3.

2.2.4.1 The linear case To start with, we briefly study system (2.32)

$$\begin{cases} -\Delta\boldsymbol{u} + \boldsymbol{\nabla} p = \boldsymbol{f}, \\ \operatorname{div}\boldsymbol{u} = 0, \\ \boldsymbol{u}|_{\partial\Omega} = 0. \end{cases}$$

One possible way to prove the existence of a solution to this system is to see it as the Euler–Lagrange equation (or the optimality system) corresponding to a minimization of some energy. More precisely, we introduce for $\boldsymbol{f} \in \mathbb{L}^2(\Omega)$ fixed, the energy functional

$$J(\boldsymbol{v}) = \frac{1}{2}\int_\Omega |\boldsymbol{\nabla}\boldsymbol{v}|^2 - \int_\Omega \boldsymbol{f}\cdot\boldsymbol{v} \tag{2.102}$$

and the minimization problem

$$\inf\left\{J(\boldsymbol{v});\quad \boldsymbol{v}\in V\right\}, \tag{2.103}$$

where, we recall, the space V is defined by (2.19).

Then we remark that the functional J is continuous on V and, more importantly, that its quadratic part is coercive, *i.e.* there exists some positive constant $C_0 > 0$ such that

$$\frac{1}{2}\int_\Omega |\boldsymbol{\nabla}\boldsymbol{v}|^2 \geq C_0\,\|\boldsymbol{v}\|_V^2\,. \tag{2.104}$$

This property follows from a simple application of the Poincaré inequality. Next, we know that such a strongly convex functional admits a unique minimizer on a Hilbert space (in fact in any (non-empty) closed convex subset of a Hilbert space). In addition, since it is differentiable, the unique minimizer of J is also the unique solution to the Euler–Lagrange equation of the problem. We henceforth denote it by $\boldsymbol{u}$. Now, saying that $\boldsymbol{u}$ solves the Euler–Lagrange equation of (2.103) is exactly saying that $\boldsymbol{u} \in V$ satisfies

$$\int_\Omega \nabla \boldsymbol{u} : \nabla \boldsymbol{v} = \int_\Omega \boldsymbol{f} \cdot \boldsymbol{v} \tag{2.105}$$

for all $\boldsymbol{v} \in V$. It follows that for $\boldsymbol{f} \in \mathbb{L}^2(\Omega)$, (2.105) has a unique solution $\boldsymbol{u} \in V$. The pressure p now needs to be recovered. To this end, let us introduce the form

$$\mathcal{L}(\boldsymbol{v}) = \int_\Omega \nabla \boldsymbol{u} : \nabla \boldsymbol{v} - \int_\Omega \boldsymbol{f} \cdot \boldsymbol{v}. \tag{2.106}$$

It is a linear continuous form on $\mathbb{H}_0^1(\Omega)$. A theoretical result then tells us that such a form vanishes on the subspace V if and only if there exists a function $p \in L^2(\Omega)$ such that, for all $\boldsymbol{v} \in \mathbb{H}_0^1(\Omega)$,

$$\mathcal{L}(\boldsymbol{v}) = \int_\Omega p \operatorname{div} \boldsymbol{v}. \tag{2.107}$$

In addition, the function p is unique, up to an additive constant. This theoretical result is actually a variant, for L^2 integrable functions, of the *De Rham theorem*. In fact, this variant may in turn be established as a particular case of a more general theorem, related to the closed range theorem. This will be detailed in Chapter 3, Section 3.1.3. See in particular Remark 3.1.2.

At this stage, we have therefore established the existence and uniqueness of $\boldsymbol{u} \in V$ satisfying, for any $\boldsymbol{v} \in \mathbb{H}_0^1(\Omega)$,

$$\int_\Omega \nabla \boldsymbol{u} : \nabla \boldsymbol{v} - \int_\Omega p \operatorname{div} \boldsymbol{v} = \int_\Omega \boldsymbol{f} \cdot \boldsymbol{v}. \tag{2.108}$$

Choosing $\boldsymbol{v} \in \mathcal{D}(\Omega)$, we first recover the equation

$$-\Delta \boldsymbol{u} + \nabla p = \boldsymbol{f} \tag{2.109}$$

in the sense of distributions, and, owing to the regularity of $\boldsymbol{u}$, p and $\boldsymbol{f}$, in the sense of $\mathbb{H}^{-1}(\Omega)$. On the other hand, we have both the constraint $\operatorname{div} \boldsymbol{u} = 0$ in $L^2(\Omega)$ (thus almost everywhere), and the boundary condition $\boldsymbol{u} = 0$ on $\partial\Omega$, since we know that $\boldsymbol{u} \in V$.

We also remark that it is a simple extension of the above arguments to treat the case of a force $\boldsymbol{f} \in \mathbb{H}^{-1}(\Omega)$. We have therefore proved.

Theorem 2.21 *Let Ω be a bounded connected domain with a Lipschitz continuous boundary. Let $\boldsymbol{f} \in \mathbb{H}^{-1}(\Omega)$. Then there exists a unique solution $\boldsymbol{u} \in \mathbb{H}_0^1(\Omega)$ (or equivalently $\boldsymbol{u} \in V$), $p \in L^2(\Omega)$ (the latter unique up to an additive constant) to system (2.32). The notion of solution is understood in the sense of (2.105) or equivalently (2.108).*

Remark 2.2.2 An alternative proof of existence and uniqueness for the Stokes problem is provided by the Lax–Milgram theorem. We have rather presented the energetic approach here, that takes advantage of the symmetry of the problem under consideration. In contrast, the Lax–Milgram theorem will be used below, for the linearized problem (2.119), which is not symmetric and thus cannot be treated *via* energy considerations. ♢

When the force $\boldsymbol{f}$ is in $\mathbb{L}^2(\Omega)$, some additional regularity on the solution $\boldsymbol{u}$, p to the Stokes problem, may be proved. We will however not do so here and only quote (without proof) the following result (see, *e.g.* Theorem I.5.4 in [106]). Many variants of this result exist in the literature, providing various regularity results on the solution, either in Sobolev or Schauder spaces.

Theorem 2.22 *Assume that $\partial\Omega$ is of class $\mathcal{C}^2$. There exists a constant $C(\Omega)$ depending only on the domain Ω such that, for any $\boldsymbol{f} \in \mathbb{L}^2(\Omega)$, the solution to the Stokes problem with $\boldsymbol{f}$ as right-hand side, provided by Theorem 2.21, satisfies $\boldsymbol{u} \in \mathbb{H}^2(\Omega)$, $p \in \mathbb{H}^1(\Omega)$ and*

$$\|\boldsymbol{u}\|_{\mathbb{H}^2(\Omega)} + \left\| p - \int_\Omega p \right\|_{\mathbb{H}^1(\Omega)} \leq C(\Omega)\ \|\boldsymbol{f}\|_{\mathbb{L}^2(\Omega)}. \tag{2.110}$$

Of course, in such circumstances the solution $\boldsymbol{u}$ is called strong, since (2.109) makes sense almost everywhere.

We now turn to the magnetostatic problem (2.82)

$$\begin{cases} \mathbf{curl}\,\mathbf{curl}\,\boldsymbol{B} = \boldsymbol{f}, \\ \operatorname{div}\boldsymbol{B} = 0, \\ \boldsymbol{B}\cdot\boldsymbol{n}|_{\partial\Omega} = 0, \\ \mathbf{curl}\,\boldsymbol{B}\times\boldsymbol{n}|_{\partial\Omega} = 0. \end{cases}$$

The same variational approach as above allows us to also study this problem (and in fact Remark 2.2.2 equally applies). First of all, we remark that we may only hope to solve the problem when the right-hand side $\boldsymbol{f}$ satisfies $\operatorname{div}\boldsymbol{f} = 0$ (since $\operatorname{div}\mathbf{curl}\,\mathbf{curl}\,\boldsymbol{B} = 0$) and $\boldsymbol{f}\cdot\boldsymbol{n}|_{\partial\Omega} = 0$ (since $\boldsymbol{B}\cdot\boldsymbol{n}|_{\partial\Omega} = 0$ and $\mathbf{curl}\,\boldsymbol{B}\times\boldsymbol{n}|_{\partial\Omega} = 0$ together imply and $\mathbf{curl}\,\mathbf{curl}\,\boldsymbol{B}\cdot\boldsymbol{n}|_{\partial\Omega} = 0$). Therefore we henceforth assume that $\operatorname{div}\boldsymbol{f} = 0$, and $\boldsymbol{f}\cdot\boldsymbol{n} = 0$. We also impose $\boldsymbol{f} \in \mathbb{L}^2(\Omega)$: in contrast to the study of the Stokes problem (2.32), we cannot simply accomodate here a force $\boldsymbol{f} \in \mathbb{H}^{-1}(\Omega)$. The reason is that, at least formally, $\boldsymbol{f} \in \mathbb{H}^{-1}(\Omega)$ corresponds to $\boldsymbol{B} \in \mathbb{H}^1(\Omega)$, and no more regularity than that. Then the boundary condition $\mathbf{curl}\,\boldsymbol{B}\times\boldsymbol{n}$ only has a very weak sense (indeed in negative Sobolev spaces). We introduce the functional

$$K(\boldsymbol{C}) = \frac{1}{2}\int_\Omega |\mathbf{curl}\,\boldsymbol{C}|^2 - \int_\Omega \boldsymbol{f}\cdot\boldsymbol{C} \tag{2.111}$$

and the minimization problem

$$\inf\{K(\boldsymbol{C});\quad \boldsymbol{C}\in W\}, \tag{2.112}$$

for W defined by (2.80).

Then we remark as above that the functional K is continuous, convex, with a coercive quadratic part. It follows that there exists a unique minimizer of (2.112) that is the unique function $\boldsymbol{B} \in W$ satisfying, for all $\boldsymbol{C} \in W$,

$$\int_\Omega \mathbf{curl}\,\boldsymbol{B} \cdot \mathbf{curl}\,\boldsymbol{C} = \int_\Omega \boldsymbol{f} \cdot \boldsymbol{C} \tag{2.113}$$

for all $\boldsymbol{C} \in W$. We now explain why (2.113), that is the Euler–Lagrange equation of the minimization problem (2.112), is indeed a weak formulation of (2.82).

For this purpose, as outlined in Section 2.2.2 above, we now enlarge the set of test functions $\boldsymbol{C}$. We again perform the Helmholtz decomposition of an arbitrary function $\boldsymbol{C} \in \mathbb{H}^1(\Omega)$ (here satisfying $\boldsymbol{C}\cdot\boldsymbol{n} = 0$), writing it as a sum $\boldsymbol{C} = \tilde{\boldsymbol{C}} + \boldsymbol{\nabla}\Phi$ where $\tilde{\boldsymbol{C}}$ is divergence-free, $\tilde{\boldsymbol{C}} \cdot \boldsymbol{n} = 0$, and Φ is the solution to

$$\begin{cases} \Delta\Phi = \operatorname{div}\boldsymbol{C}, \\ \dfrac{\partial\Phi}{\partial n}|_{\partial\Omega} = \boldsymbol{C}\cdot\boldsymbol{n} = 0. \end{cases}$$

In fact we may prove, by a standard elliptic regularity argument on the Laplace equation, that $\Phi \in H^2(\Omega)$, thus $\tilde{\boldsymbol{C}} \in \mathbb{H}^1(\Omega)$, thus $\tilde{\boldsymbol{C}} \in W$. Inserting $\tilde{\boldsymbol{C}}$ as a test function in (2.113), we obtain that for all $\boldsymbol{C} \in \mathbb{H}^1(\Omega)$ with $\boldsymbol{C}\cdot\boldsymbol{n} = 0$ we have

$$\int_\Omega \mathbf{curl}\,\boldsymbol{B} \cdot \mathbf{curl}\,\boldsymbol{C} = \int_\Omega \boldsymbol{f} \cdot \boldsymbol{C}. \tag{2.114}$$

Formula (2.114) is inferred from (2.113) using on the one hand $\mathbf{curl}\,\boldsymbol{\nabla}\Phi = 0$ and on the other hand $\operatorname{div}\boldsymbol{f} = 0$ and $\boldsymbol{f}\cdot\boldsymbol{n} = 0$ thus

$$\int_\Omega \boldsymbol{f}\cdot\boldsymbol{\nabla}\Phi = \int_{\partial\Omega} \boldsymbol{f}\cdot\boldsymbol{n}\,\Phi = 0.$$

We next choose $\boldsymbol{C} \in \mathcal{D}(\Omega)$ in (2.114) and recover

$$\mathbf{curl}\,\mathbf{curl}\,\boldsymbol{B} = \boldsymbol{f}, \tag{2.115}$$

in the sense of distributions. In addition, a regularity result (see Theorem 2.24 below) allows us to deduce from (2.114) that, when $\boldsymbol{f} \in \mathbb{L}^2(\Omega)$, we have $\boldsymbol{B} \in \mathbb{H}^2(\Omega)$, and (2.115) holds in $\mathbb{L}^2(\Omega)$. The following integration by parts

$$\int_\Omega \mathbf{curl}\,\boldsymbol{B}\cdot\mathbf{curl}\,\boldsymbol{C} = \int_\Omega \mathbf{curl}\,\mathbf{curl}\,\boldsymbol{B}\cdot\boldsymbol{C} + \int_{\partial\Omega} \boldsymbol{C}\cdot(\mathbf{curl}\,\boldsymbol{B}\times\boldsymbol{n})$$

then makes sense, and we deduce from (2.114) and (2.115) that

$$\mathbf{curl}\,\boldsymbol{B}\times\boldsymbol{n} = 0 \tag{2.116}$$

on the boundary $\partial\Omega$. Note that the meaning of (2.116) is precisely provided by

$$\int_{\partial\Omega} \boldsymbol{C}\cdot(\mathbf{curl}\,\boldsymbol{B}\times\boldsymbol{n}) = 0$$

for all $\boldsymbol{C} \in \mathbb{H}^1(\Omega)$ with $\boldsymbol{C}\cdot\boldsymbol{n} = 0$. On the other hand, we have both the constraint $\operatorname{div}\boldsymbol{B} = 0$ in $L^2(\Omega)$ (thus almost everywhere), and the boundary condition $\boldsymbol{B} \cdot \boldsymbol{n} = 0$ on $\partial\Omega$, since we know that $\boldsymbol{B} \in W$. We have thus recovered the whole system (2.82).

We have proved

Theorem 2.23 *Let Ω be a bounded connected regular domain. Let $\boldsymbol{f} \in \mathbb{L}^2(\Omega)$ with $\operatorname{div} \boldsymbol{f} = 0$ and $\boldsymbol{f} \cdot \boldsymbol{n} = 0$ (i.e. $\boldsymbol{f} \in H$). Then there exists a unique solution $\boldsymbol{B} \in W$ to system (2.82), in the sense of (2.113) or (2.114).*

In addition, we have mentioned $\mathbb{H}^2(\Omega)$ regularity on $\boldsymbol{B}$, which indeed comes from the following precise result.

Theorem 2.24 [86, Theorem 3.2.2] and [208] *Assume again that $\boldsymbol{f} \in \mathbb{L}^2(\Omega)$. Then the solution $\boldsymbol{B}$ provided by the previous theorem satisfies $\boldsymbol{B} \in \mathbb{H}^2(\Omega)$, all equations have a strong sense, and we have*

$$\|\boldsymbol{B}\|_{\mathbb{H}^2(\Omega)} \leq C(\Omega)\ \|\boldsymbol{f}\|_{\mathbb{L}^2(\Omega)} \tag{2.117}$$

for some constant $C(\Omega)$ that only depends on the domain Ω and not on $\boldsymbol{f}$.

We now turn to the nonlinear setting, first purely hydrodynamic, and next MHD.

2.2.4.2 The purely hydrodynamic case The system we now consider is the steady-state incompressible (homogeneous) Navier–Stokes equation, *i.e.*

$$\begin{cases} \boldsymbol{u} \cdot \nabla \boldsymbol{u} - \Delta \boldsymbol{u} + \nabla p = \boldsymbol{f}, \\ \operatorname{div} \boldsymbol{u} = 0, \\ \boldsymbol{u}|_{\partial\Omega} = 0. \end{cases} \tag{2.118}$$

We give ourselves a function $\boldsymbol{f} \in \mathbb{H}^{-1}(\Omega)$ and ask the question of the existence and uniqueness of the solution $(\boldsymbol{u}, p)$, in a sense to be made precise, to the above system. As above, the uniqueness of the pressure p is understood up to an additive constant.

There are several ways to address this question (see Remark 2.2.3 below). Here we proceed by a fixed point argument.

We will first prove that problem (2.118) admits a solution when $\boldsymbol{f} \in \mathbb{L}^2(\Omega)$; next generalize the existence to cover the case $\boldsymbol{f} \in \mathbb{H}^{-1}(\Omega)$. Uniqueness under an appropriate assumption is finally proved.

Let us fix $\boldsymbol{u}_0 \in V$, and define the *linearized* problem

$$\begin{cases} \boldsymbol{u}_0 \cdot \nabla \boldsymbol{u} - \Delta \boldsymbol{u} + \nabla p = \boldsymbol{f}, \\ \operatorname{div} \boldsymbol{u} = 0, \\ \boldsymbol{u}|_{\partial\Omega} = 0. \end{cases} \tag{2.119}$$

The existence and uniqueness of the solution to (2.119) follows from the application of the Lax–Milgram theorem. Indeed, the bilinear form

$$\mathcal{A}(\boldsymbol{u},\boldsymbol{v}) = \int_\Omega \boldsymbol{\nabla}\boldsymbol{u} : \boldsymbol{\nabla}\boldsymbol{v} + \int_\Omega (\boldsymbol{u}_0 \cdot \nabla)\,\boldsymbol{u}\cdot\boldsymbol{v} \tag{2.120}$$

is continuous on V, and coercive since

$$\mathcal{A}(\boldsymbol{u},\boldsymbol{u}) = \int_\Omega |\boldsymbol{\nabla}\boldsymbol{u}|^2,$$

and we may apply the Poincaré inequality. As the right-hand side of (2.119) defines a continuous map on V, namely

$$\mathcal{L}(\boldsymbol{v}) = \langle \boldsymbol{f}, \boldsymbol{v}\rangle_{\mathbb{H}^{-1},\mathbb{H}^1_0}, \tag{2.121}$$

we may claim that the problem

$$\text{Find}\,\boldsymbol{u}\in V \quad \text{s.t.} \quad \mathcal{A}(\boldsymbol{u},\boldsymbol{v}) = \mathcal{L}(\boldsymbol{v}) \quad \text{for all}\,\boldsymbol{v}\in V \tag{2.122}$$

admits a unique solution. This readily shows the existence and uniqueness of the (weak) solution $\boldsymbol{u}$ to (2.119). We denote this solution $\boldsymbol{u}$ by $\Phi(\boldsymbol{u}_0)$, thereby defining a map Φ from V to V. We are now going to investigate the properties of this map Φ.

First, using (2.122) with $\boldsymbol{v} = \boldsymbol{u}$, we see that

$$c_0 \left\|\boldsymbol{u}\right\|_V^2 \le \mathcal{A}(\boldsymbol{u},\boldsymbol{u}) = \mathcal{L}(\boldsymbol{u}) \le \left\|\boldsymbol{f}\right\|_{\mathbb{H}^{-1}(\Omega)} \left\|\boldsymbol{u}\right\|_V,$$

where c_0 is the coercivity constant of $\mathcal{A}$, thus that

$$\left\|\Phi(\boldsymbol{u}_0)\right\|_V \le C \left\|\boldsymbol{f}\right\|_{\mathbb{H}^{-1}(\Omega)}, \tag{2.123}$$

which shows that Φ maps any $\boldsymbol{u}_0$ on a bounded set in V. Here and below C denotes various constants depending only on the domain.

Second, considering $\Phi(\boldsymbol{u}_1)$ and $\Phi(\boldsymbol{u}_2)$ for $\boldsymbol{u}_1$ and $\boldsymbol{u}_2$ in V, we have, denoting $\boldsymbol{w} = \Phi(\boldsymbol{u}_1) - \Phi(\boldsymbol{u}_2)$:

$$-\Delta\boldsymbol{w} + \boldsymbol{u}_1\cdot\boldsymbol{\nabla}\boldsymbol{w} = -(\boldsymbol{u}_1 - \boldsymbol{u}_2)\cdot\boldsymbol{\nabla}(\Phi(\boldsymbol{u}_2)) - \boldsymbol{\nabla}\Pi,$$

where $\Pi = p_1 - p_2$. We thus have

$$\begin{aligned}\left\|\boldsymbol{\nabla}\boldsymbol{w}\right\|^2_{\mathbb{L}^2(\Omega)} &\le \left\|\boldsymbol{u}_1 - \boldsymbol{u}_2\right\|_{\mathbb{L}^3(\Omega)} \left\|\boldsymbol{\nabla}(\Phi(\boldsymbol{u}_2))\right\|_{\mathbb{L}^2(\Omega)} \left\|\boldsymbol{w}\right\|_{\mathbb{L}^6(\Omega)} \\ &\le C \left\|\boldsymbol{u}_1 - \boldsymbol{u}_2\right\|_{\mathbb{H}^1(\Omega)} \left\|\boldsymbol{\nabla}(\Phi(\boldsymbol{u}_2))\right\|_{\mathbb{L}^2(\Omega)} \left\|\boldsymbol{w}\right\|_{\mathbb{H}^1(\Omega)}. \end{aligned} \tag{2.124}$$

In view of (2.123), this shows

$$\left\|\boldsymbol{w}\right\|_{\mathbb{H}^1(\Omega)} \le C \left\|\boldsymbol{u}_1 - \boldsymbol{u}_2\right\|_{\mathbb{H}^1(\Omega)} \left\|\boldsymbol{f}\right\|_{\mathbb{H}^{-1}(\Omega)} \tag{2.125}$$

thus the (in fact Lipschitz) continuity of the map Φ from V to V.

Third, writing (2.119) in the form

$$-\Delta \boldsymbol{u} = \boldsymbol{f} - \boldsymbol{u}_0 \cdot \boldsymbol{\nabla} \boldsymbol{u} - \boldsymbol{\nabla} p,$$

we see that $\boldsymbol{u}$ is also the solution to the linear Stokes problem (2.32) with $\boldsymbol{f} - \boldsymbol{u}_0 \cdot \boldsymbol{\nabla} \boldsymbol{u}$ as right-hand side. Using the regularity result quoted in Theorem 2.22, we have that

$$\|\boldsymbol{u}\|_{\mathbb{H}^2(\Omega)} \leq C\,(\|\boldsymbol{f}\|_{\mathbb{L}^2(\Omega)} + \|\boldsymbol{u}_0 \cdot \boldsymbol{\nabla} \boldsymbol{u}\|_{\mathbb{L}^2(\Omega)}),$$

where it should be noted that the $\mathbb{L}^2$ regularity of the force $\boldsymbol{f}$ is used. We next estimate the last term of the right-hand side:

$$\begin{aligned}
\|\boldsymbol{u}_0 \cdot \boldsymbol{\nabla} \boldsymbol{u}\|_{\mathbb{L}^2(\Omega)} &\leq \|\boldsymbol{u}_0\|_{\mathbb{L}^6(\Omega)}\ \|\boldsymbol{\nabla} \boldsymbol{u}\|_{\mathbb{L}^3(\Omega)} \\
&\leq C\ \|\boldsymbol{u}_0\|_{\mathbb{H}^1(\Omega)}\ \|\boldsymbol{\nabla} \boldsymbol{u}\|^{1/2}_{\mathbb{L}^2(\Omega)}\ \|\boldsymbol{\nabla} \boldsymbol{u}\|^{1/2}_{\mathbb{L}^6(\Omega)} \\
&\leq C\ \|\boldsymbol{u}_0\|_{\mathbb{H}^1(\Omega)}\ \|\boldsymbol{u}\|^{1/2}_{\mathbb{H}^1(\Omega)}\ \|\boldsymbol{u}\|^{1/2}_{\mathbb{H}^2(\Omega)} \\
&\leq C\ \|\boldsymbol{u}_0\|_{\mathbb{H}^1(\Omega)}\ \|\boldsymbol{f}\|^{1/2}_{\mathbb{L}^2(\Omega)}\ \|\boldsymbol{u}\|^{1/2}_{\mathbb{H}^2(\Omega)} \qquad (2.126)
\end{aligned}$$

and thus obtain:

$$\|\boldsymbol{u}\|_{\mathbb{H}^2(\Omega)} \leq C\,\big(\|\boldsymbol{f}\|_{\mathbb{L}^2(\Omega)} + \|\boldsymbol{u}_0\|_{\mathbb{H}^1(\Omega)}\ \|\boldsymbol{f}\|^{1/2}_{\mathbb{L}^2(\Omega)}\ \|\boldsymbol{u}\|^{1/2}_{\mathbb{H}^2(\Omega)}\big), \qquad (2.127)$$

whence

$$\|\Phi(\boldsymbol{u}_0)\|_{\mathbb{H}^2(\Omega)} \leq C\ \|\boldsymbol{f}\|_{\mathbb{L}^2(\Omega)}\ \big(1 + \|\boldsymbol{u}_0\|^2_{\mathbb{H}^1(\Omega)}\big). \qquad (2.128)$$

As $\mathbb{H}^2(\Omega)$ is compactly embedded in $\mathbb{H}^1(\Omega)$, this shows that the map Φ is compact from V to V.

We are now in a position to apply the Schauder fixed point theorem: owing to (2.123), we may fix a (convex) ball $\mathcal{B}$ in V such that Φ maps $\mathcal{B}$ in $\mathcal{B}$. As Φ is compact, it thus has a fixed point in $\mathcal{B}$, which is of course a solution to (2.118). In fact, the fixed point solution is automatically a strong solution since we have remarked that Φ maps $\mathbb{H}^1(\Omega)$ in $\mathbb{H}^2(\Omega)$. The pressure p is recovered in the usual way.

We now extend the above result to a force $\boldsymbol{f} \in \mathbb{H}^{-1}(\Omega)$. For this purpose, we argue by density, considering a sequence of forces $\boldsymbol{f}_n \in \mathbb{L}^2(\Omega)$ that converges to $\boldsymbol{f}$ for the $\mathbb{H}^{-1}(\Omega)$ topology. We denote by $\boldsymbol{u}_n$ the solution to (2.118) with right-hand side $\boldsymbol{f}_n$ provided by the above argument. Clearly, precisely because of the arguments developed above, we know that $\boldsymbol{u}_n$ is bounded in $\mathbb{H}^1(\Omega)$: note indeed that we only make use of the $\mathbb{H}^{-1}(\Omega)$ norm of the right-hand side in (2.123). It follows that we may assume that $\boldsymbol{u}_n$ converges to some $\boldsymbol{u} \in V$, weakly in $\mathbb{H}^1(\Omega)$, and, owing to the Sobolev embedding theorems, strongly in $\mathbb{L}^p(\Omega)$ for all $1 \leq p < 6$. Using the weak formulation of (2.118):

$$\int_\Omega \boldsymbol{\nabla} \boldsymbol{u}_n : \boldsymbol{\nabla} \boldsymbol{v} + \int_\Omega (\boldsymbol{u}_n \cdot \nabla)\, \boldsymbol{u}_n \cdot \boldsymbol{v} = \langle \boldsymbol{f}_n, \boldsymbol{v} \rangle_{\mathbb{H}^{-1}(\Omega),\mathbb{H}^1_0(\Omega)},$$

for all $\boldsymbol{v} \in V$, we now pass to the limit in each term. The only non-trivial term is the nonlinear term, for which we remark that

$$\int_\Omega (\boldsymbol{u}_n \cdot \nabla)\,\boldsymbol{u}_n \cdot \boldsymbol{v} = -\int_\Omega (\boldsymbol{u}_n \cdot \nabla)\,\boldsymbol{v} \cdot \boldsymbol{u}_n$$

where in the right-hand side $\boldsymbol{u}_n$ strongly converges to $\boldsymbol{u}$ in $\mathbb{L}^4(\Omega)$. This readily shows, using an argument already detailed in Section 2.1.2.1, that we obtain

$$\int_\Omega \boldsymbol{\nabla u} : \boldsymbol{\nabla v} + \int_\Omega (\boldsymbol{u} \cdot \nabla)\,\boldsymbol{u} \cdot \boldsymbol{v} = \langle \boldsymbol{f}, \boldsymbol{v} \rangle_{\mathbb{H}^{-1}(\Omega),\mathbb{H}^1_0(\Omega)},$$

for all $\boldsymbol{v} \in V$. We have thus obtained a weak solution to (2.118). Of course, we cannot hope to have a strong solution since we only have $\boldsymbol{f} \in \mathbb{H}^{-1}(\Omega)$, and not $\boldsymbol{f} \in \mathbb{L}^2(\Omega)$. We have proved:

Proposition 2.25 *Assume $\boldsymbol{f} \in \mathbb{H}^{-1}(\Omega)$. Then there exists a weak solution $\boldsymbol{u} \in V$ to problem (2.118).*

Remark 2.2.3 An alternative proof of the existence of a solution can be performed with the approach used for the time-dependent case. The approach consists of three steps: Step 1: project equation (2.118) on a finite-dimensional basis set (Galerkin approach); Step 2: prove that the nonlinear finite-dimensional problem has a solution; Step 3: pass to the limit as the dimension goes to infinity. In the present stationary setting, the second step is typically performed using the Brouwer fixed point theorem (in contrast with the time-dependent setting where the Cauchy–Lipschitz theorem for ordinary differential equations is used). The third step makes use of estimates analogous to those used in the proof we followed. Actually, this alternative approach is to some extent consistent with (or equivalent to) the above proof: indeed, the usual proof of the Schauder fixed point theorem is based on the Brouwer fixed point theorem. ♢

Let us now turn to the uniqueness of our solution. In full generality, it is not known whether the solution provided by Proposition 2.25 is unique, even if we assume it is a strong solution *i.e.* we take $\boldsymbol{f} \in \mathbb{L}^2(\Omega)$. The context of *small data*, however, allows for proving uniqueness:

Proposition 2.26 *Assume that $\boldsymbol{f}$ is sufficiently small in $\mathbb{H}^{-1}(\Omega)$ norm. Then the solution is unique.*

At this stage of our exposition, the proof is rather straightforward. Considering two solutions $\boldsymbol{u}_1$ and $\boldsymbol{u}_2$, we form the difference $\boldsymbol{w} = \boldsymbol{u}_1 - \boldsymbol{u}_2$. We then remark

$$\|\nabla \boldsymbol{w}\|^2_{\mathbb{L}^2(\Omega)} \leq \|\boldsymbol{w}\|^2_{\mathbb{H}^1(\Omega)} \|\nabla \boldsymbol{u}_2\|_{\mathbb{L}^2(\Omega)} \leq C \|\boldsymbol{w}\|^2_{\mathbb{H}^1(\Omega)} \|\boldsymbol{f}\|_{\mathbb{H}^{-1}(\Omega)},$$

successively using (2.124) and (2.123). Taking $\|\boldsymbol{f}\|_{\mathbb{H}^{-1}(\Omega)}$ sufficiently small, we obtain $\boldsymbol{w} = 0$, and thus conclude that uniqueness holds.

Remark 2.2.4 In fact, in the statement of the proposition (and it may be clarified in the above proof by making all the constants explicit), the term *"small"* means *small in comparison to the viscosity.* The latter is fixed here to one. ◇

2.2.4.3 The MHD case We now briefly turn to the steady-state MHD system

$$\left\{\begin{array}{l}\left\{\begin{array}{r}\boldsymbol{u}\cdot\nabla\,\boldsymbol{u} - \Delta\boldsymbol{u} + \nabla p = \mathbf{curl}\,\boldsymbol{B}\times\boldsymbol{B} + \boldsymbol{f},\\ \operatorname{div}\boldsymbol{u} = 0,\\ \boldsymbol{u}|_{\partial\Omega} = 0,\end{array}\right.\\ \left\{\begin{array}{r}\mathbf{curl}\,\mathbf{curl}\,\boldsymbol{B} - \mathbf{curl}\left(\boldsymbol{u}\times\boldsymbol{B}\right) = \boldsymbol{g},\\ \operatorname{div}\boldsymbol{B} = 0,\\ \boldsymbol{B}\cdot\boldsymbol{n}|_{\partial\Omega} = 0,\\ \mathbf{curl}\,\boldsymbol{B}\times\boldsymbol{n}|_{\partial\Omega} = 0.\end{array}\right.\end{array}\right. \tag{2.129}$$

Actually, this system may be proved to admit a weak solution using the general machinery that will be introduced in Chapter 3. There, it is the system (3.34)–(3.35) and it will be treated using in particular Theorem 3.15. The result is stated in Theorem 3.22. In the present chapter, we wish to only mention that a self-consistent proof can be performed. It treats system (2.129) specifically, without using an abstract framework, and, in the same manner as we have treated (2.118) above, it uses a fixed point procedure based upon a linearization.

For this purpose, a pair $(\boldsymbol{u}_0, B_0) \in V \times W$ is fixed and the following linear system is introduced

$$\left\{\begin{array}{l}\left\{\begin{array}{r}\boldsymbol{u}_0\cdot\nabla\,\boldsymbol{u} - \Delta\boldsymbol{u} + \nabla p = \mathbf{curl}\,\boldsymbol{B}\times\boldsymbol{B}_0 + \boldsymbol{f},\\ \operatorname{div}\boldsymbol{u} = 0\\ \boldsymbol{u}|_{\partial\Omega} = 0,\end{array}\right.\\ \left\{\begin{array}{r}\mathbf{curl}\,\mathbf{curl}\,\boldsymbol{B} - \mathbf{curl}\left(\boldsymbol{u}\times\boldsymbol{B}_0\right) = \boldsymbol{g},\\ \operatorname{div}\boldsymbol{B} = 0,\\ \boldsymbol{B}\cdot\boldsymbol{n}|_{\partial\Omega} = 0,\\ \mathbf{curl}\,\boldsymbol{B}\times\boldsymbol{n}|_{\partial\Omega} = 0.\end{array}\right.\end{array}\right. \tag{2.130}$$

The existence of a (unique) solution to system (2.130) is proved by an application of the Lax–Milgram theorem. Next the Schauder fixed point theorem is applied, and a weak solution to (2.129) is demonstrated to exist.

We wish to indicate that system (2.130) will be approximated numerically in Chapter 3, Section 3.5 (it is identical to system (3.71)–(3.77)).

Let us conclude this section by mentioning that the stationary MHD equations have been treated in detail by M.D. Gunzburger, A.J. Meir and J.S. Peterson in [118]. They prove the existence of a solution and its uniqueness in particular cases. Non-homogeneous boundary conditions for $\boldsymbol{u}$ and $\boldsymbol{B}$ are used in this work and the authors propose two types of boundary conditions for the electromagnetic field. We will extensively return to this setting in Section 3.3 of Chapter 3.

2.2.5 A hybrid problem

The purpose of this section is the mathematical study of system (1.43) that we reproduce here in the following form

$$\frac{\partial \boldsymbol{u}}{\partial t} + \boldsymbol{u} \cdot \boldsymbol{\nabla} \boldsymbol{u} - \overline{\eta} \Delta\, \boldsymbol{u} = \boldsymbol{f} - \boldsymbol{\nabla} p + \mathbf{curl}\, \boldsymbol{b} \times \boldsymbol{b} \qquad \text{in } \Omega, \tag{2.131}$$

$$\operatorname{div} \boldsymbol{u} = 0 \qquad \text{in } \Omega, \tag{2.132}$$

$$\frac{1}{\overline{\sigma}} \mathbf{curl}\, (\mathbf{curl}\, \boldsymbol{b}) = \mathbf{curl}\, (\boldsymbol{u} \times \boldsymbol{b}) \qquad \text{in } \Omega, \tag{2.133}$$

$$\operatorname{div} \boldsymbol{b} = 0 \qquad \text{in } \Omega, \tag{2.134}$$

and that is supplied with the following initial and boundary conditions:

$$\boldsymbol{u} = 0 \qquad \text{on } \partial\Omega, \tag{2.135}$$

$$\boldsymbol{b} \cdot \boldsymbol{n} = q \qquad \text{on } \partial\Omega, \tag{2.136}$$

$$\mathbf{curl}\, \boldsymbol{b} \times \boldsymbol{n} = \boldsymbol{k} \times \boldsymbol{n} \qquad \text{on } \partial\Omega, \tag{2.137}$$

$$\boldsymbol{u}|_{t=0} = \boldsymbol{u}_0 \qquad \text{in } \Omega. \tag{2.138}$$

To write the above system from (1.43), we have:

- used the fact that, without loss of generality, the value of the constant parameters $\overline{\rho}$ and $\overline{\mu}$ may be set to unity;
- momentarily denoted by $\boldsymbol{b}$ the magnetic field $\boldsymbol{B}$ (see the decomposition (2.144) below);
- more importantly, replaced the gravity field $\boldsymbol{g}$ by a general force field $\boldsymbol{f}$, which will be used in the sequel to illustrate some particular properties of this system;
- kept explicit the parameter $\overline{\sigma}$ (considered to be one so far in the chapter), because we shall need some particular assumption on the size of this parameter below.

The hydrodynamics equation is the same for (1.43) and (1.40): it is a time-dependent equation. In contrast, the magnetic equation is taken to be stationary in (1.43). This is why we call (1.43) a *hybrid* system. More precisely, instead of a parabolic equation on $\boldsymbol{B}$, we have, at least formally, an *elliptic* equation. The difficulty is that the ellipticity of the equation for $\boldsymbol{B}$ depends on the velocity field $\boldsymbol{u}$. When the latter becomes too large, ellipticity may be lost.

Under adequate assumptions upon the physical data, we can prove that a strong solution exists and is unique at least on a time interval $[0, T^*]$ for some time T^* depending on the data (see Theorem 2.31).

On the other hand, as soon as the magnetic operator is no longer invertible – which may occur if the velocity gets large – we can construct two distinct solutions to the system.

This latter observation shows that, despite the fact that it seems very appealing as it constitutes an apparent simplification of system (1.40), the model considered here should be used only with great care in numerical simulations.

The results overviewed here have been announced in J.F. Gerbeau and C. Le Bris [91], and detailed in J.F. Gerbeau and C. Le Bris [95].

2.2.5.1 Mathematical setting and preliminary results We begin with a presentation of some functional spaces, and some preliminary existence and regularity results regarding the magnetic equation.

Functional spaces In addition to the spaces introduced so far, the following trace spaces and norms will also be needed:

$$H^{1/2}(\partial\Omega) = \{v|_{\partial\Omega}, v \in H^1(\Omega)\}, \qquad ||q||_{H^{1/2}(\partial\Omega)} = \inf_{w\in H^1(\Omega), w|_{\partial\Omega}=q} ||w||_{H^1(\Omega)},$$

$$\mathbb{H}^{1/2}(\partial\Omega) = \{\boldsymbol{v}|_{\partial\Omega}, v_i \in H^{1/2}(\partial\Omega), i = 1, .., 3\},$$
$$||\boldsymbol{g}||_{\mathbb{H}^{1/2}(\partial\Omega)} = \inf_{\boldsymbol{w}\in\mathbb{H}^1(\Omega), \boldsymbol{w}|_{\partial\Omega}=\boldsymbol{g}} ||\boldsymbol{w}||_{\mathbb{H}^1(\Omega)},$$

$$\mathbb{H}^{-1/2}(\partial\Omega) = (\mathbb{H}^{1/2}(\partial\Omega))', \qquad ||\boldsymbol{k}||_{\mathbb{H}^{-1/2}(\partial\Omega)} = \sup_{\boldsymbol{g}\in\mathbb{H}^{1/2}(\partial\Omega), \boldsymbol{g}\neq 0} \frac{\langle \boldsymbol{k}, \boldsymbol{g}\rangle}{||\boldsymbol{g}||_{\mathbb{H}^{1/2}(\partial\Omega)}}.$$

Regularity of the data We shall suppose in the sequel that

$$\boldsymbol{u}_0 \in \mathbb{H}^1_0(\Omega) \cap \mathbb{H}^2(\Omega), \text{ with } \operatorname{div} \boldsymbol{u}_0 = 0, \tag{2.139}$$

$$q \in \mathcal{C}(0, T; H^{3/2}(\partial\Omega)), \text{ with } \int_{\partial\Omega} q{=}0, \tag{2.140}$$

$$\boldsymbol{k} \in \mathcal{C}(0, T; \mathbb{H}^{1/2}(\partial\Omega)), \tag{2.141}$$

$$\boldsymbol{f} \in L^\infty(0, T; \mathbb{L}^2(\Omega)). \tag{2.142}$$

From a physical viewpoint, it is natural to assume that $\boldsymbol{k}$ is the trace on $\partial\Omega$ of the gradient of the electrical potential:

$$\boldsymbol{k} = \overline{\sigma}\boldsymbol{\nabla}\phi|_{\partial\Omega}. \tag{2.143}$$

First of all, we notice that we can split the magnetic field $b(t) \in \mathbb{H}^1(\Omega)$ satisfying (2.136) and (2.134) into the sum of a function $\boldsymbol{B}^d(t)$ that satisfies (2.136) and a function $\boldsymbol{B}(t) \in W$. Indeed, we have:

Proposition 2.27 *Let $q \in \mathcal{C}(0,T;H^{k-1/2}(\Omega))$ for $k=1$ or $k=2$. Then there exist $\boldsymbol{B}^d \in \mathcal{C}(0,T;\mathbb{H}^k(\Omega))$ and a constant c such that*

$$\boldsymbol{B}^d \cdot \boldsymbol{n} = q \quad \text{on } [0,T]\times\partial\Omega \text{ and} \qquad ||\boldsymbol{B}^d||_{\mathcal{C}(0,T;\mathbb{H}^k(\Omega))} \le c||q||_{\mathcal{C}(0,T;H^{k-1/2}(\Omega))}.$$

Moreover, we can impose that

$$div\, \boldsymbol{B}^d(t) = 0 \qquad \text{and} \qquad \mathbf{curl}\, \boldsymbol{B}^d(t) = 0 \qquad \text{for } t \in [0,T].$$

The proof of this proposition follows by defining $\boldsymbol{B}^d(t) = \boldsymbol{\nabla}\theta(t)$ where $\theta(t)$ is a solution of the Neumann problem

$$\begin{cases} -\Delta\,\theta = 0 & \text{in } \Omega \\ \dfrac{\partial\theta}{\partial n} = q(t) & \text{on } \partial\Omega. \end{cases}$$

Let us now set

$$\boldsymbol{B}(t) = \boldsymbol{b}(t) - \boldsymbol{B}^d(t). \tag{2.144}$$

We replace the original problem (2.131)–(2.138) with the following one:

$$\frac{\partial \boldsymbol{u}}{\partial t} + \boldsymbol{u}\cdot\boldsymbol{\nabla}\boldsymbol{u} - \overline{\eta}\Delta\,\boldsymbol{u} = \boldsymbol{f} - \boldsymbol{\nabla}p + \mathbf{curl}\,\boldsymbol{B}\times\boldsymbol{B} + \mathbf{curl}\,\boldsymbol{B}\times\boldsymbol{B}^d \text{ in } \Omega \tag{2.145}$$

$$\operatorname{div}\boldsymbol{u} = 0 \qquad \text{in } \Omega, \tag{2.146}$$

$$\frac{1}{\overline{\sigma}}\,\mathbf{curl}\,(\mathbf{curl}\,\boldsymbol{B}) = \mathbf{curl}\,(\boldsymbol{u}\times\boldsymbol{B}) + \mathbf{curl}\,(\boldsymbol{u}\times\boldsymbol{B}^d) \qquad \text{in } \Omega, \tag{2.147}$$

$$\operatorname{div}\boldsymbol{B} = 0 \qquad \text{in } \Omega, \tag{2.148}$$

with the following initial and boundary conditions:

$$\boldsymbol{u} = 0 \qquad \text{on } \partial\Omega, \tag{2.149}$$

$$\boldsymbol{B}\cdot\boldsymbol{n} = 0 \qquad \text{on } \partial\Omega, \tag{2.150}$$

$$\mathbf{curl}\,\boldsymbol{B}\times\boldsymbol{n} = \boldsymbol{k}\times\boldsymbol{n} \qquad \text{on } \partial\Omega, \tag{2.151}$$

$$\boldsymbol{u}|_{t=0} = \boldsymbol{u}_0 \qquad \text{in } \Omega, \tag{2.152}$$

For a time T and some constant M, we now fix some $\boldsymbol{v} \in \mathbb{K}_M$ where $\mathbb{K}_M$ is the (non-empty) convex set $\mathbb{K}_M$ defined by

$$\begin{aligned} \mathbb{K}_M = \{\boldsymbol{v} \in L^2(0,T;V),\ \sup\nolimits_{t\in[0,T]} ||\boldsymbol{v}(t)||_V \le M,& \\ ||\boldsymbol{v}||_{L^2(0,T;\mathbb{H}^2(\Omega))} \le M,& \\ ||\tfrac{\partial \boldsymbol{v}}{\partial t}||_{L^2(0,T;\mathbb{L}^2(\Omega))} \le M\}.& \end{aligned}$$

Then, we consider the following problem, which is the magnetic part of the above problem. Find $B \in \mathcal{C}(0,T;W)$ such that

$$\frac{1}{\bar{\sigma}}\,\mathbf{curl}\,(\mathbf{curl}\,\boldsymbol{B}) = \mathbf{curl}\,(\boldsymbol{v} \times \boldsymbol{B}) + \mathbf{curl}\,(\boldsymbol{v} \times \boldsymbol{B}^d) \qquad \text{in } \Omega, \tag{2.153}$$

$$\operatorname{div} \boldsymbol{B} = 0 \qquad \text{in } \Omega, \tag{2.154}$$

satisfying the following boundary conditions

$$\boldsymbol{B} \cdot \boldsymbol{n} = 0, \tag{2.155}$$

$$\mathbf{curl}\,\boldsymbol{B} \times \boldsymbol{n} = \boldsymbol{k} \times \boldsymbol{n} \qquad \text{on } \partial\Omega. \tag{2.156}$$

Regarding this problem, it is easy to prove the following.

Proposition 2.28 *For $\boldsymbol{v} \in \mathbb{K}_M$ and M sufficiently small, problem (2.153)–(2.156) has a unique solution $\boldsymbol{B} \in \mathcal{C}(0,T;W)$. In addition, we have the following estimate:*

$$\sup_{t\in[0,T]} ||\boldsymbol{B}(t)||_W \leq \frac{\alpha_1 + \beta_1 ||\boldsymbol{v}||_{L^\infty(0,T;V)}}{1 - \gamma_1 ||\boldsymbol{v}||_{L^\infty(0,T;V)}}, \tag{2.157}$$

where α_1, β_1 and γ_1 are some constants that may be made explicit in terms of $\boldsymbol{k}, \boldsymbol{B}^d$ and $\bar{\sigma}$.

In the next section, the vector field $\boldsymbol{B}$ defined above will appear on the right-hand side of the Navier–Stokes equation in the Lorentz force $\mathbf{curl}\,\boldsymbol{B} \times \boldsymbol{B}$. An estimate on $\boldsymbol{u}$ is needed, in the norm $L^\infty(0,T;\mathbb{H}^1(\Omega))$, in order to prove the coercivity of problem (2.153)–(2.156). Such a control on $\boldsymbol{u}$ is typically obtained with strong solutions of the Navier–Stokes equations. To define strong solutions, the force term in the Navier–Stokes equations has to belong to $L^\infty(0,T;\mathbb{L}^2(\Omega))$ (see R. Temam [227], Section 2.1.4 above, and *e.g.* Proposition 2.15 in the two-dimensional case). To this end, the estimate on $\boldsymbol{B}$ in $L^\infty(0,T;W)$ is not sufficient. An additional estimate on $\boldsymbol{B}$ is needed. This is the following.

Proposition 2.29 *Assume M sufficiently small. Then the solution of problem (2.153)–(2.156) provided by Proposition 2.28 satisfies*

$$||\boldsymbol{B}||_{L^\infty(0,T;\mathbb{W}^{1,3}(\Omega))} \leq \alpha_2 + \gamma_2 ||\boldsymbol{v}||_{L^\infty(0,T;V)} \frac{\alpha_1 + \beta_1 ||\boldsymbol{v}||_{L^\infty(0,T;V)}}{1 - \gamma_1 ||\boldsymbol{v}||_{L^\infty(0,T;V)}} + \beta_2 ||\boldsymbol{v}||_{L^\infty(0,T;V)} \tag{2.158}$$

where α_2, β_2 and γ_2 are some constants that may be made explicit.

Proposition 2.29 is in fact a consequence (for $m = 0, p = 3/2$) of the following Proposition 2.30. The latter is a straightforward extension (in the non-homogeneous case) of B. Saramito [204, Proposition 2.1] (see also Lemma 2.1 and Remark 2.3 in this reference).

Proposition 2.30 *Let m be a non-negative integer and $1 < p < \infty$. Let $\boldsymbol{g} \in \mathbb{W}^{m,p}(\Omega)$, with $div\, \boldsymbol{g} = 0$ and $\boldsymbol{g} \cdot \boldsymbol{n} = 0$ on $\partial\Omega$, $\boldsymbol{k} \in \mathbb{W}^{m+1-1/p,p}(\partial\Omega)$, $q \in \mathbb{W}^{m+2-1/p,p}(\partial\Omega)$.*

Then, there exists a unique $\boldsymbol{B} \in \mathbb{W}^{m+2,p}(\Omega)$ such that

$$\begin{cases} \mathbf{curl}\,(\mathbf{curl}\,\boldsymbol{B}) = \boldsymbol{g} & \text{in } \Omega, \\ div\, \boldsymbol{B} = 0 & \text{in } \Omega, \\ \boldsymbol{B} \cdot \boldsymbol{n} = q & \text{on } \partial\Omega, \\ \mathbf{curl}\,\boldsymbol{B} \times \boldsymbol{n} = \boldsymbol{k} \times \boldsymbol{n} & \text{on } \partial\Omega, \end{cases}$$

and $||\boldsymbol{B}||_{\mathbb{W}^{m+2,p}(\Omega)} \leq c_2(||\boldsymbol{g}||_{\mathbb{W}^{m,p}(\Omega)} + ||\boldsymbol{k}||_{\mathbb{W}^{m+1-1/p,p}(\partial\Omega)} + ||q||_{W^{m+2-1/p,p}(\partial\Omega)})$.

2.2.5.2 Well-posedness of the system for small data We are now in a position to prove the well-posedness of our problem when the data are sufficiently small. We denote $\alpha_0 = ||f||_{L^\infty(0,T;\mathbb{L}^2(\Omega))}$ and we recall we have defined in the previous sections some constants α_i, β_i and γ_i. Let us fix $M > 0$. For c denoting various constants which actually do not depend on the data, we define

$$\Theta(M) = \alpha_0 + c\left(\alpha_2 + \gamma_2 M \frac{\alpha_1 + \beta_1 M}{1 - \gamma_1 M} + \beta_2 M\right)\left(1 + \frac{\alpha_1 + \beta_1 M}{1 - \gamma_1 M}\right),$$

$$\mu_1(M)^2 = 4 \max\left(||\boldsymbol{u}_0||_V^2, \frac{2}{c\overline{\eta}^2}\Theta(M)^2\right), \tag{2.159}$$

$$\mu_2(M)^2 = \frac{c}{\overline{\eta}}\left(||\boldsymbol{u}_0||_V^2 + \frac{2T}{\overline{\eta}}\Theta(M)^2 + \mu_1(M)^3\right), \tag{2.160}$$

$$\mu_3(M) = \Theta(M) + c\mu_2(M) + c\mu_1(M)\mu_2(M). \tag{2.161}$$

Theorem 2.31 *Assume the physical data $\boldsymbol{u}_0$, $1/\overline{\eta}$, $\overline{\sigma}$, $\boldsymbol{f}$, q, $\boldsymbol{k}$, are sufficiently small in the sense they allow for the following to hold:*

$$\textit{There exists } 0 < M < 1/\gamma_1 \textit{ such that } \mu_i(M) \leq M, i = 1, 2, 3 \tag{2.162}$$

(which indeed may be proved to be possible). Then, there exists a time $T^* > 0$ such that the MHD problem (2.131)–(2.138) has a unique solution on $[0, T^*]$. This solution satisfies $\boldsymbol{u} \in L^2(0, T^*; \mathbb{H}^2(\Omega)) \cap L^\infty(0, T^*; \mathbb{H}_0^1(\Omega))$ and $\boldsymbol{b} \in \mathcal{C}(0, T^*; \mathbb{H}^1(\Omega)) \cap L^\infty(0, T^*; \mathbb{H}^2(\Omega))$.

The proof of Theorem 2.31 falls in three steps. The existence part is a consequence of the Schauder fixed point theorem. The regularity essentially comes from Proposition 2.30. The uniqueness is, as often, based on a Gronwall type argument. The smallness of the data is a key ingredient.

2.2.5.3 Remark on the non-uniqueness It has been proved in the previous section that the MHD problem (2.145)–(2.152) has a unique solution for small data, at least on an interval $[0, T^*]$, $T^* > 0$. The idea of the proof is to ensure the coercivity of equation (2.147) by controlling the $\mathbb{H}^1(\Omega)$ norm of $\boldsymbol{u}$ on $[0, T^*]$.

We exhibit in this section an example of non-uniqueness in the case when the operator $T_{\boldsymbol{u}} : \boldsymbol{B} \to \mathbf{curl}\,(\mathbf{curl}\,\boldsymbol{B}) - \mathbf{curl}\,(\boldsymbol{u} \times \boldsymbol{B})$ is not invertible.

We henceforth assume for simplicity that $k = 0$, $q = 0$, thus we deal with the homogeneous boundary conditions on $\partial\Omega$:

$$\begin{cases} \boldsymbol{u} = 0, \\ \boldsymbol{B} \cdot \boldsymbol{n} = 0, \\ \mathbf{curl}\,\boldsymbol{B} \times \boldsymbol{n} = 0. \end{cases}$$

Let us assume that for some t_0 and some $\tilde{\boldsymbol{u}} = \tilde{\boldsymbol{u}}(t_0, x)$ the operator $T_{\tilde{\boldsymbol{u}}} : \boldsymbol{B} \to \mathbf{curl}\,(\mathbf{curl}\,\boldsymbol{B}) - \mathbf{curl}\,(\tilde{\boldsymbol{u}} \times \boldsymbol{B})$ is not invertible. Then, there exists a divergence-free field $\tilde{\boldsymbol{B}} \not\equiv 0$ satisfying the above boundary conditions and

$$\mathbf{curl}\,(\mathbf{curl}\,\tilde{\boldsymbol{B}}) = \mathbf{curl}\,(\tilde{\boldsymbol{u}} \times \tilde{\boldsymbol{B}}).$$

Note that such a $\tilde{\boldsymbol{u}}$ is necessarilly "large enough," otherwise, $T_{\tilde{\boldsymbol{u}}}$ would be coercive. If we consider the force $\tilde{\boldsymbol{f}} = \tilde{\boldsymbol{u}} \cdot \boldsymbol{\nabla} \tilde{\boldsymbol{u}} - \overline{\eta} \Delta\, \tilde{\boldsymbol{u}} - \mathbf{curl}\,\tilde{\boldsymbol{B}} \times \tilde{\boldsymbol{B}}$, then $(\tilde{\boldsymbol{u}}, \tilde{\boldsymbol{B}})$ is a (stationary) solution to

$$\begin{cases} \dfrac{\partial \boldsymbol{u}}{\partial t} + \boldsymbol{u} \cdot \boldsymbol{\nabla} \boldsymbol{u} - \overline{\eta} \Delta\, \boldsymbol{u} + \boldsymbol{\nabla} p = \tilde{\boldsymbol{f}} + \mathbf{curl}\,\boldsymbol{B} \times \boldsymbol{B}, \\ \operatorname{div} \boldsymbol{u} = 0, \\ \mathbf{curl}\,(\mathbf{curl}\,\boldsymbol{B}) = \mathbf{curl}\,(\boldsymbol{u} \times \boldsymbol{B}), \\ \operatorname{div} \boldsymbol{B} = 0. \end{cases} \tag{2.163}$$

Next, we define $\boldsymbol{u}'$ as the solution of

$$\begin{cases} \dfrac{\partial \boldsymbol{u}}{\partial t} + \boldsymbol{u} \cdot \boldsymbol{\nabla} \boldsymbol{u} - \overline{\eta} \Delta\, \boldsymbol{u} + \boldsymbol{\nabla} p = \tilde{\boldsymbol{f}}, \\ \operatorname{div} \boldsymbol{u} = 0. \end{cases}$$

with the "initial" condition $\boldsymbol{u}'(t_0, .) = \tilde{\boldsymbol{u}}(t_0, .)$.

We finally observe that $(\tilde{\boldsymbol{u}}, \tilde{\boldsymbol{B}})$ and $(\boldsymbol{u}', 0)$ are different (since $\tilde{\boldsymbol{B}} \not\equiv 0$) while they both satisfy (2.163) on $[t_0, +\infty)$. Thus, we have two different solutions of the MHD problem with homogeneous boundary conditions.

Of course, in order to be complete, the present analysis would require either to prove that for large enough initial velocities (or more generally, large enough data), the operator $\boldsymbol{B} \to \mathbf{curl}\,(\mathbf{curl}\,\boldsymbol{B}) - \mathbf{curl}\,(\boldsymbol{u}(t, \cdot) \times \boldsymbol{B})$ does indeed become non-invertible in finite time. Or at least, it would be enlightening to prove that for some given very particular velocity field $\boldsymbol{u}$ (then taken as an initial condition for the system), it is indeed not invertible. Unfortunately, we are not able to prove either of the two facts. It remains that the above analysis shows the model should be very carefuly used in numerical simulations since it *is likely* to be ill-posed as soon as the velocity becomes too large.

2.2.6 *Other MHD models and formulations*

In this book we mainly focus on the formulation (1.42) and on its stationary version. The purpose of this section is to present other formulations of the same problem, or slightly different models.

2.2.6.1 The undisturbed magnetic field model In some cases, the magnetic field is completely determined by the environment of the computational domain and is almost undisturbed by the flow. Consequently $\boldsymbol{B}$ is assumed to be known within a good degree of accuracy. In the time-independent setting, the electric field is given by

$$\boldsymbol{E} = -\boldsymbol{\nabla}\phi,$$

where ϕ denotes the electric potential. Ohm's law therefore reads (see (1.36) in Chapter 1),

$$\boldsymbol{j} = \sigma(-\boldsymbol{\nabla}\phi + \boldsymbol{u} \times \boldsymbol{B}),$$

and the Lorentz force is then given by

$$\boldsymbol{j} \times \boldsymbol{B} = \sigma(-\boldsymbol{\nabla}\phi + \boldsymbol{u} \times \boldsymbol{B}) \times \boldsymbol{B}.$$

Since moreover $\operatorname{div} \boldsymbol{j} = 0$, the stationary MHD equations can be written with $(\boldsymbol{u}, p, \phi)$ as unknowns:

$$\begin{cases} -\eta\Delta\,\boldsymbol{u} + \rho\boldsymbol{u}\cdot\boldsymbol{\nabla}\boldsymbol{u} + \boldsymbol{\nabla}p + \sigma\boldsymbol{\nabla}\phi \times \boldsymbol{B} - \sigma(\boldsymbol{u} \times \boldsymbol{B}) \times \boldsymbol{B} = 0, \\ \hfill \operatorname{div}\boldsymbol{u} = 0, \\ \hfill -\operatorname{div}(\sigma\boldsymbol{\nabla}\phi) + \operatorname{div}(\boldsymbol{u} \times \boldsymbol{B}) = 0, \end{cases}$$

where we recall that $\boldsymbol{B}$ is assumed to be known. Of course this system has to be supplied with boundary conditions. Note that it is usually easier to prescribe the boundary conditions on ϕ than on $\boldsymbol{B}$. This model is studied by J.S. Peterson in [184]. In particular existence and (conditional) uniqueness of weak solutions are established and approximation results are presented. Examples of applications proposed by Peterson include electromagnetic pumps and the flow of liquid lithium for fusion reaction cooling blankets.

2.2.6.2 Coupling the MHD and heat equations Thermal effects often play a significant role in industrial MHD flows (aluminum electrolysis, continuous metal casting, *etc.*). In [164], A.J. Meir considers the MHD equations coupled with the heat equation within the Boussinesq approximation. We recall that the Boussinesq approximation consists in assuming that the variations in density ρ are negligible except in the body force $\rho\boldsymbol{g}$ due to buoyancy ($\boldsymbol{g}$ being the acceleration of gravity). In this body force, the density is assumed to depend on the temperature T in the following simplified way:

$$\rho = \rho_r\left[1 - \beta(T - T_r)\right],$$

where ρ_r and T_r are a reference density and temperature, respectively, and β is the so-called thermal expansion coefficient.

The equations governing the unknowns $(\boldsymbol{u}, \boldsymbol{B}, p, T)$ then read:

$$\begin{cases} -\eta\Delta\,\boldsymbol{u} + \rho_r\boldsymbol{u}\cdot\boldsymbol{\nabla}\boldsymbol{u} + \boldsymbol{\nabla}p = \rho_r\boldsymbol{g}\,[1-\beta(T-T_r)] + \dfrac{1}{\mu}\mathbf{curl}\,\boldsymbol{B}\times\boldsymbol{B}, \\ \operatorname{div}\boldsymbol{u} = 0, \\ \mathbf{curl}\left(\dfrac{1}{\mu\sigma}\mathbf{curl}\,\boldsymbol{B}\right) = \mathbf{curl}\,(\boldsymbol{u}\times\boldsymbol{B}), \\ -\kappa\Delta\,T + \rho_r c_p\boldsymbol{u}\cdot\boldsymbol{\nabla}T = \dfrac{1}{\mu^2\sigma}|\mathbf{curl}\,\boldsymbol{B}|^2 + 2\eta|\boldsymbol{D}(\boldsymbol{u})|^2 + \psi, \end{cases}$$

where ψ denotes a heat source, κ is the heat conductivity, and c_p is the specific heat at constant pressure. The other parameters have been defined earlier in this book. The terms $|\mathbf{curl}\,\boldsymbol{B}|^2/(\mu^2\sigma)$ and $2\eta|\boldsymbol{D}(\boldsymbol{u})|^2$ model the Joule heating and the viscous heating, respectively.

A simplified version of this system, where the viscous and Joule heating are neglected, is studied in [164] (existence, uniqueness, error estimates of a finite element approximation). Note that this simplification is motivated by technical reasons (owing to the mathematical difficulty related to the nonlinearity of these terms). However, in many practical applications, the Joule effect is in fact the main source of heat.

2.2.6.3 A velocity–electric current formulation for unbounded domains Throughout this book, the fluid and magnetic equations are supposed to be defined on the same bounded domain. We thus are concerned with *boundary value problems* with boundary conditions prescribed at the outer surface. This is a common choice in most studies on the MHD equations. Nevertheless it is well known that the magnetic field extends through the whole space and may satisfy different equations within and outside the fluid. Starting from this observation, A.J. Meir and G. Schmidt have proposed a formulation that we now present without entering into any technical detail. The basic idea has been introduced in [161] in the case of an exterior domain $\mathbb{R}^3\backslash\overline{\Omega}$ filled with a homogeneous insulating dielectric (no external current). The first results have then been extended in [162] in those cases when the magnetic permeability μ takes different values within and outside the fluid. Well-posedness is proved for small data. Finally the cases with external electric current are treated in [163] and a solution procedure applied to an academic test case is proposed.

The magnetic field $\boldsymbol{B}$ is decomposed as

$$\boldsymbol{B} = \boldsymbol{B}_0 + \mathcal{B}(\boldsymbol{j}), \qquad \text{in } \mathbb{R}^3, \tag{2.164}$$

where $\boldsymbol{B}_0$ is assumed to be known and $\mathcal{B}(\boldsymbol{j})$ is created by the unknown current $\boldsymbol{j}$ in the fluid. The field $\boldsymbol{B}_0$ can be generated by electric currents flowing in circuits *outside* the fluid domain (and computed for example with the Biot–Savart law).

We search for the velocity $\boldsymbol{u}$, the pressure p, the electric current density $\boldsymbol{j}$, the electric potential ϕ and the magnetic field $\boldsymbol{B}$ solution of the following problem inside Ω:

$$\begin{cases} -\eta\Delta\,\boldsymbol{u} + \rho\,\boldsymbol{u}\cdot\boldsymbol{\nabla}\boldsymbol{u} + \boldsymbol{\nabla}p - \boldsymbol{j}\times\boldsymbol{B} = 0, \\ \operatorname{div}\boldsymbol{u} = 0, \\ \dfrac{1}{\sigma}\boldsymbol{j} + \boldsymbol{\nabla}\phi - \boldsymbol{u}\times\boldsymbol{B} = 0, \\ \operatorname{div}\boldsymbol{j} = 0. \end{cases} \tag{2.165}$$

Usually, the electric current $\boldsymbol{j}$ and ϕ are eliminated from these equations to derive the MHD equations in $(\boldsymbol{u}, p, \boldsymbol{B})$. Here, we rather eliminate the magnetic field $\boldsymbol{B}$ using the Biot–Savart law:

$$\mathcal{B}(\boldsymbol{j})(\boldsymbol{x}) = -\frac{\mu}{4\pi}\int_\Omega \frac{\boldsymbol{x}-\boldsymbol{y}}{|\boldsymbol{x}-\boldsymbol{y}|^3}\times\boldsymbol{j}(\boldsymbol{y})\,d\boldsymbol{y}.$$

Using the decomposition (2.164), system (2.165) can then be rewritten with the unknowns $(\boldsymbol{u}, p, \boldsymbol{j}, \phi)$ in Ω:

$$\begin{cases} -\eta\Delta\,\boldsymbol{u} + \rho\,\boldsymbol{u}\cdot\boldsymbol{\nabla}\boldsymbol{u} + \boldsymbol{\nabla}p - \boldsymbol{j}\times\mathcal{B}(\boldsymbol{j}) - \boldsymbol{j}\times\boldsymbol{B}_0 = 0, \\ \operatorname{div}\boldsymbol{u} = 0, \\ \dfrac{1}{\sigma}\boldsymbol{j} + \boldsymbol{\nabla}\phi - \boldsymbol{u}\times\mathcal{B}(\boldsymbol{j}) - \boldsymbol{u}\times\boldsymbol{B}_0 = 0, \\ \operatorname{div}\boldsymbol{j} = 0. \end{cases} \tag{2.166}$$

This system has to be completed by boundary conditions on $\boldsymbol{u}$ and $\boldsymbol{j}\cdot\boldsymbol{n}$ on $\partial\Omega$. The electric potential ϕ plays for the constraint $\operatorname{div}\boldsymbol{j} = 0$ the same role as the pressure p for the incompressibility constraint (see [161], p. 95). System (2.166) is set on the bounded space Ω whereas relation (2.164) gives $\boldsymbol{B}$ in the whole space $\mathbb{R}^3$. Note that this approach is restricted to the stationary equations and the discretization of this system is more involved than discretization of the traditional MHD equations.

3

NUMERICAL APPROXIMATION OF ONE-FLUID PROBLEMS

The purpose of this chapter is to present variational formulations of the MHD equations, their discretization with the finite element method and some algorithms to solve the resulting nonlinear problems.

For clarity of exposition and pedagogic purposes, we first consider the Stokes problem and we give a concise presentation of the general theory in an abstract framework. We next briefly address some stabilized finite element techniques. The material presented in Section 3.1 and 3.2 is well documented: we refer the reader, for instance, to the textbooks by V. Girault and P.A. Raviart [106], A. Quarteroni and A. Valli [192] and A. Ern and J.-L. Guermond [75].

The application of the general results to the MHD equations is presented in Section 3.3 where two variational formulations are considered. Mixed finite element discretizations of both formulations are addressed in Section 3.4 and a stabilized discretization of the linearized equations is proposed in Section 3.5. In Section 3.6, we present some solution strategies for the nonlinear MHD equations, and we conclude the chapter in Section 3.7 with numerical illustrations.

In this chapter, Ω denotes a domain of $\mathbb{R}^3$ which is at least simply connected Lipschitz and $\boldsymbol{n}$ denotes the outward unit normal on the boundary $\partial\Omega$. For the first variational formulation of the MHD equations (Sections 3.3.1 and 3.4.1), it is assumed in addition that Ω is either convex polyhedral or with a $\mathcal{C}^{1,1}$ boundary. When Ω is a non-convex polyhedron (namely has re-entrant corners), specific difficulties appear for the approximation of the magnetic field. This point motivates the second variational formulation which, contrarily to the first one, is valid on non-convex polyhedra (Sections 3.3.2 and 3.4.2).

3.1 A general framework for problems with constraints

In this section, we introduce the mathematical tools that will be needed to treat the divergence-free constraints in the MHD equations.

3.1.1 A model problem: the Stokes equations

The Stokes problem consists in looking for two functions $\boldsymbol{u} \in \mathbb{H}^1(\Omega)$ and $p \in L^2_0(\Omega)$ such that

$$\begin{cases} -\Delta \boldsymbol{u} + \boldsymbol{\nabla} p = \boldsymbol{f} & \text{in } \Omega, \\ \operatorname{div} \boldsymbol{u} = 0 & \text{in } \Omega, \\ \boldsymbol{u} = 0 & \text{on } \partial\Omega, \end{cases} \tag{3.1}$$

where $\boldsymbol{f}$ is given in $\mathbb{H}^{-1}(\Omega)$.

First variational formulation Following the same line as in Chapter 2 for the theoretical analysis of the hydrodynamics equations, we multiply formally these equations by $\boldsymbol{v} \in V = \{\boldsymbol{w} \in \mathbb{H}^1_0(\Omega), \operatorname{div} \boldsymbol{w} = 0\}$ to obtain a first variational formulation of (3.1), identical to (2.33):

Find $\boldsymbol{u} \in V$ such that, for all $\boldsymbol{v} \in V$,

$$\int_\Omega \boldsymbol{\nabla u} : \boldsymbol{\nabla v} = \langle \boldsymbol{f}, \boldsymbol{v} \rangle. \tag{3.2}$$

The pressure term is absent, eliminated in this formulation because the divergence of $\boldsymbol{v}$ vanishes. It is very convenient from the theoretical viewpoint to look for the solution in the space V. The resulting variational problem is then coercive and can be straightforwardly studied with the Lax–Milgram theorem (see Chapter 2). Nevertheless this formulation has two major drawbacks from a practical viewpoint: it does not provide the pressure and it is based on a space which is not convenient in a classical finite element framework (the usual finite element spaces typically provide approximations of $\mathbb{H}^1(\Omega)$ not of V).

Second variational formulation Practical implementations use a second variational formulation of the Stokes problem obtained by multiplying the first equation of (3.1) by $\boldsymbol{v} \in \mathbb{H}^1_0(\Omega)$ and the second one by $q \in L^2_0(\Omega)$:

Find $(\boldsymbol{u}, p) \in \mathbb{H}^1_0(\Omega) \times L^2_0(\Omega)$ such that for all $(\boldsymbol{v}, q) \in \mathbb{H}^1_0(\Omega) \times L^2_0(\Omega)$,

$$\begin{aligned} \int_\Omega \boldsymbol{\nabla u} : \boldsymbol{\nabla v} - \int_\Omega p \, \mathit{div} \boldsymbol{v} &= \langle \boldsymbol{f}, \boldsymbol{v} \rangle, \\ \int_\Omega q \, \mathit{div} \boldsymbol{u} &= 0. \end{aligned} \tag{3.3}$$

In contrast to (3.2), this formulation does provide the pressure and is posed in spaces which are *a priori* easier to approximate with finite elements. Nevertheless, if we try to apply the Lax–Milgram theorem, we discover a difficulty in (3.3). Indeed, denoting by $\mathcal{X}$ the space $\mathbb{H}^1_0(\Omega) \times L^2_0(\Omega)$ endowed with the norm $\|(\boldsymbol{v}, q)\|^2_{\mathcal{X}} = \|\boldsymbol{v}\|^2_1 + \|q\|^2_0$, introducing on $\mathcal{X} \times \mathcal{X}$ the bilinear form

$$\Psi((\boldsymbol{u}, p), (\boldsymbol{v}, q)) = \int_\Omega \boldsymbol{\nabla u} : \boldsymbol{\nabla v} - \int_\Omega p \operatorname{div} \boldsymbol{v} + \int_\Omega q \operatorname{div} \boldsymbol{u},$$

and defining $F \in \mathcal{X}'$ by

$$\langle F, (\boldsymbol{v}, q)\rangle = \langle \boldsymbol{f}, \boldsymbol{v}\rangle,$$

we may rephrase formulation (3.3) as:

Find $(\boldsymbol{u}, p) \in \mathcal{X}$ such that for all $(\boldsymbol{v}, q) \in \mathcal{X}$

$$\Psi((\boldsymbol{u}, p), (\boldsymbol{v}, q)) = \langle F, (\boldsymbol{v}, q)\rangle.$$

Investigating the coercivity of $\Psi(\cdot,\cdot)$, we compute:

$$\Psi((\boldsymbol{u}, p), (\boldsymbol{u}, p)) = \int_\Omega |\boldsymbol{\nabla} \boldsymbol{u}|^2,$$

which can clearly not be bounded from below by $\|(\boldsymbol{u}, p)\|^2_{\mathcal{X}}$, because $\|p\|_0$ is absent. In conclusion, formulation (3.3) cannot be studied in the usual framework of the Lax–Milgram theorem. The theoretical framework for such a formulation is presented in the next section.

3.1.2 Abstract framework for a linear problem

The main results we now present date back to the seminal works by I. Babuška [11] and F. Brezzi [31]. Many existing textbooks offer a comprehensive presentation of this theory. We refer in particular to [106] and [75] which have inspired our presentation. Our purpose is to provide a brief and self-contained exposition to supply the reader with the basic tools necessary to understand the discretization of the MHD equations.

Let $(X, \|\cdot\|_X)$ and $(M, \|\cdot\|_M)$ be two real Hilbert spaces endowed with the scalar products $(\cdot,\cdot)_X$ and $(\cdot,\cdot)_M$. Their dual spaces are denoted as usual by X' and M'. We introduce two continuous bilinear forms $a : X \times X \to \mathbb{R}$ and $b : X \times M \to \mathbb{R}$ associated to the operators $A : X \to X'$ and $B : X \to M'$ defined by

$$\langle Au, v\rangle = a(u, v), \quad \forall\, (u, v) \in X \times X, \tag{3.4}$$

$$\langle Bv, q\rangle = b(v, q), \quad \forall\, (v, q) \in X \times M. \tag{3.5}$$

The dual operator of B is denoted by B^T and is defined by $B^T : M \to X'$, $\langle B^T q, v\rangle = b(v, q) = \langle Bv, q\rangle$, for all $(v, q) \in X \times M$.

For $f \in X'$ and $g \in M'$, we consider the following problem.

Find $(u, p) \in X \times M$ such that, for all $(v, q) \in X \times M$

$$\begin{cases} a(u, v) + b(v, p) = \langle f, v\rangle, \\ \qquad\qquad b(u, q) = \langle g, q\rangle, \end{cases} \tag{3.6}$$

or, in an equivalent form:

Find $(u, p) \in X \times M$ *such that*

$$\begin{cases} Au + B^T p = f, \\ \qquad Bu = g. \end{cases} \tag{3.7}$$

We now search for necessary and sufficient conditions ensuring that this problem has a unique solution. For this purpose, we introduce the kernel of B,

$$V = \operatorname{Ker} B = \{u \in X, Bu = 0\} = \{u \in X, b(u, q) = 0, \forall\, q \in M\},$$

along with its polar set,

$$V^o = (\operatorname{Ker} B)^o = \{h \in X', \langle h, w\rangle = 0, \forall\, v \in V\},$$

and we denote by Π the canonical injection of X' into V': for $h \in X'$, we define Πh in V' by

$$\langle \Pi h, v\rangle = \langle h, v\rangle, \text{ for all } v \in V.$$

In other words, if h is a continuous linear form defined on X then Πh is its restriction to V. Note that $\|\Pi h\|_{V'} \leq \|h\|_{X'}$ and that $V^o = \operatorname{Ker} \Pi$. We then have:

Theorem 3.1 *Problem (3.7) has a unique solution if and only if*

(i) $\Pi \circ A$ *is an isomorphism from* $V = \mathit{Ker} B$ *onto* $V' = (\mathit{Ker} B)'$;
(ii) $B : X \to M'$ *is surjective.*

Proof We first suppose that (3.7) has a unique solution.

Let us prove that B is surjective. Let h in M'. Then problem (3.7) has a unique solution corresponding to $f = 0$ and $g = h$, thus there exists $u \in X$ such that $Bu = h$.

We now prove that $\Pi \circ A$ is surjective from V onto V'. Let $f \in V'$. As a linear continuous form on $V \subset X$, f can be extended on X (Hahn–Banach theorem). Let us denote this extension by $\tilde{f}$ (by construction $\Pi\tilde{f} = f$). There exists a unique $(u, p) \in X \times M$ such that

$$\begin{cases} Au + B^T p = \tilde{f}, \\ \qquad Bu = 0. \end{cases}$$

Thus, in particular for $v \in V$,

$$\langle Au, v\rangle + \langle B^T p, v\rangle = \langle \tilde{f}, v\rangle = \langle f, v\rangle,$$

but $\langle B^T p, v\rangle = \langle p, Bv\rangle = 0$, thus $\Pi A u = f$.

It remains to show that $\Pi \circ A$ is injective. Let $u \in V$ be such that $\Pi A u = 0$. For all $v \in V$,

$$\langle Au, v\rangle = 0, \tag{3.8}$$

thus $Au \in V^o = (\operatorname{Ker} B)^o$.

We have just proved that B is surjective, thus $\operatorname{Im} B = M'$ which is obviously closed in M'. Thus (3.8) and the closed range theorem (see Remark 3.1.1) yields $Au \in \operatorname{Im} B^T$, which means there exists $p \in M$ such that $B^T p = -Au$. The pair (u, p) therefore satisfies

$$\begin{cases} Au + B^T p = 0, \\ \qquad\quad Bu = 0. \end{cases}$$

By assumption, this problem has a unique solution which is clearly $(0, 0)$. Thus $u = 0$, which proves that $\Pi \circ A$ is injective.

We now assume that assumptions (i) and (ii) of Theorem 3.1 hold true. To show the existence of u, we first use assumption (ii) to define $u_g \in X$ such that $Bu_g = g$. Then, $\Pi f - \Pi A u_g$ being an element of V', assumption (i) gives the existence of $u_0 \in V$ such that

$$\Pi A u_0 = \Pi f - \Pi A u_g.$$

Defining $u = u_0 + u_g$, we have $\Pi A u = \Pi f$, which means

$$\langle f - Au, v \rangle = 0, \quad \forall v \in V.$$

Thus $f - Au \in (\operatorname{Ker} B)^o$. The operator B being surjective, the closed range theorem ensures that $(\operatorname{Ker} B)^o = \operatorname{Im} B^T$. Thus there exists $p \in M$ such that $f - Au = B^T p$. In conclusion, we have

$$\begin{aligned} Au + B^T p &= f, \\ Bu &= g, \end{aligned}$$

which shows the existence of a solution. To prove uniqueness, we consider (u, p) such that

$$\begin{aligned} Au + B^T p &= 0, \\ Bu &= 0. \end{aligned}$$

This implies that $\Pi A u + \Pi B^T p = 0$. But, as seen above, the restriction of $B^T p$ to V vanishes, thus $\Pi A u = 0$. With assumption (i) this yields $u = 0$. We finally notice that B being surjective, B^T is injective, thus $B^T p = 0$ implies $p = 0$. Problem (3.7) therefore has a unique solution. □

Remark 3.1.1 [Closed range theorem] In the above proof, we have used the closed range theorem (see, *e.g.* H. Brezis [30]) which states that the following four proposition are equivalent:

(i) $\operatorname{Im} B$ is closed;
(ii) $\operatorname{Im} B^T$ is closed;
(iii) $(\operatorname{Ker} B)^o = \operatorname{Im} B^T$;
(iv) $(\operatorname{Ker} B^T)^o = \operatorname{Im} B$.

Note that in infinite-dimensional spaces, $h \in (\operatorname{Ker} B)^o$ does not yield in general $h \in \operatorname{Im} B^T$. ◇

3.1.3 Application to the Stokes problem

We may apply the above result to the Stokes problem on Ω with $X = \mathbb{H}_0^1(\Omega)$, $M = L_0^2(\Omega)$, $A = -\Delta$, $B = -\mathrm{div}$, $B^T = \boldsymbol{\nabla}$, $V = \{\boldsymbol{v} \in X, \mathrm{div}\, \boldsymbol{v} = 0\}$, $a(\boldsymbol{u}, \boldsymbol{v}) = \int_\Omega \boldsymbol{\nabla u} : \boldsymbol{\nabla v}$, $b(\boldsymbol{v}, q) = -\int_\Omega q \mathrm{div}\, \boldsymbol{v}$. Assumption (i) of Theorem 3.1 is a direct consequence of the coercivity of $(\boldsymbol{\nabla}\cdot, \boldsymbol{\nabla}\cdot)$ on $X \times X$ (where $(\cdot,\cdot)$ denotes the scalar product on $L^2(\Omega)$ or $(L^2(\Omega))^d$, $d \in \mathbb{N}$, according to the context). The proof of assumption (ii) is much more involved. The key result is

Theorem 3.2 *Let Ω be an open bounded set with a Lipschitz boundary. Then $\boldsymbol{\nabla} : L^2(\Omega) \to \mathbb{H}^{-1}(\Omega)$ has a closed range.*

The proof of this theorem is very technical. We refer the reader to [106, p. 20].

Corollary 3.3 *Let Ω be an open bounded connected set with a Lipschitz boundary. Then the operator div is surjective from $\mathbb{H}_0^1(\Omega)$ onto $L_0^2(\Omega)$.*

Proof The operator $B^T = \boldsymbol{\nabla}$ defined on M has a closed range and is injective (if $\boldsymbol{\nabla} p = 0$ on a connected domain then p is constant, and this constant is zero since it has a zero mean value). Thus the closed range theorem implies that $B = -\mathrm{div}$ is surjective. □

As a consequence, we can state the following existence and uniqueness result for the Stokes problem.

Corollary 3.4 *Let Ω be an open bounded connected set with a Lipschitz boundary. If $\boldsymbol{f} \in \mathbb{H}^{-1}(\Omega)$, there exists a unique $(\boldsymbol{u}, p) \in \mathbb{H}_0^1(\Omega) \times L_0^2(\Omega)$ solution of problem (3.1).*

Remark 3.1.2 Let us recall the De Rham theorem which is sometimes used in the mathematical analysis of the Stokes equations (see, *e.g.* R. Temam [227] and Chapter 2).

> If $\boldsymbol{f} \in (\mathcal{D}'(\Omega))^d$ and $\langle \boldsymbol{f}, \phi \rangle = 0$ for all $\phi \in (\mathcal{D}(\Omega))^d$ such that $\mathrm{div}\, \phi = 0$ then there exists $p \in \mathcal{D}'(\Omega)$ such $\boldsymbol{f} = \boldsymbol{\nabla} p$.

Note that Theorem 3.2 gives directly a simplified version of the De Rahm theorem sufficient for our purpose. Indeed, since $\mathrm{Im}\,(\boldsymbol{\nabla})$ is closed, the closed range theorem yields $\mathrm{Im}\,(\boldsymbol{\nabla}) = (\mathrm{Ker}\,(\mathrm{div}))^o$, which means

> If $\boldsymbol{f} \in \mathbb{H}^{-1}(\Omega)$, and $\langle \boldsymbol{f}, \boldsymbol{v} \rangle = 0$ for all $\boldsymbol{v} \in \mathbb{H}_0^1(\Omega)$ such that $\mathrm{div}\, \boldsymbol{v} = 0$ then there exists $p \in L_0^2(\Omega)$ such that $\boldsymbol{f} = \boldsymbol{\nabla} p$. ◊

3.1.4 The inf–sup condition

In this section, we return to the abstract framework of Section 3.1. We give a standard equivalent formulation of assumption (ii) in Theorem 3.1, which is important for applications.

Definition 3.5 *The property:*

$$\exists\, \beta > 0,\ \inf_{q \in M} \sup_{v \in X} \frac{b(v, q)}{\|v\|_X \|q\|_M} \geq \beta, \tag{3.9}$$

is known as the inf–sup condition or the Babuška–Brezzi condition or the LBB condition (for Ladyzhenskaya–Babuška–Brezzi)[22].

The inf–sup condition (3.9) can be conveniently rewritten as

$$\exists\,\beta > 0, \forall\, q \in M, \sup_{v \in X} \frac{b(v,q)}{\|v\|_X} \geq \beta \|q\|_M, \tag{3.10}$$

or as

$$\exists\,\beta > 0, \forall\, q \in M, \|B^T q\|_{X'} \geq \beta \|q\|_M. \tag{3.11}$$

The following theorem shows that the inf–sup condition (3.9) is a necessary and sufficient condition to ensure property (ii) of Theorem 3.1. In particular, we will show below that this is the practical criterion to check if the problem is well-posed after discretization.

Definition 3.6 *We define the orthogonal space of V by*

$$V^\perp = \{v \in X, (v,u)_X = 0, \forall\, u \in V\}.$$

Note that the spaces V^o and $V^\perp$ can be identified through a norm-preserving isomorphism (the Riesz representation operator).

Theorem 3.7 *For $\beta > 0$, the following three statements are equivalent:*

(i) The inf–sup condition (3.9) is satisfied with the constant β.
(ii) B^T is an isomorphism from M onto V^o and

$$\forall\, q \in M, \|B^T q\|_{X'} \geq \beta \|q\|_M.$$

(iii) B is an isomorphism from $V^\perp$ onto M' and

$$\forall\, u \in V^\perp, \|Bu\|_{M'} \geq \beta \|u\|_X.$$

Proof The assertion (ii) $\Longrightarrow$ (i) is trivial (see (3.11)).

We suppose that (i) holds true and we prove (ii). The form (3.11) of the inf–sup condition gives, for all $q \in M$,

$$\|B^T q\|_{X'} \geq \beta \|q\|_M,$$

which shows that B^T is injective and therefore bijective from M on $\operatorname{Im} B^T$. Moreover, it also yields that $(B^T)^{-1}$ is continuous. Indeed, let $f \in \operatorname{Im} B^T$. Then there exists $q \in M$ such that $f = B^T q$, and

$$\|(B^T)^{-1} f\|_M \leq \frac{1}{\beta} \|f\|_{X'}.$$

Thus $\operatorname{Im} B^T$ is closed (as the inverse range of M by the continuous mapping $(B^T)^{-1}$). Thus, with the closed range theorem, $\operatorname{Im} B^T = (\operatorname{Ker} B)^o = V^o$, which proves that B^T is bijective from M onto V^o.

[22]Notice that we should rigorously write $\inf_{q \in M \setminus \{0\}}$ and $\sup_{v \in X \setminus \{0\}}$. Here and below, to simplify these expressions, we implicitly exclude the value 0 each time it appears in a denominator.

The equivalence between (ii) and (iii) is straightforward by duality. For example, assertion (ii) implies that B, identified to $(B^T)^T$, is an isomorphism from $(V^o)'$ onto $M'' = M$. Using the isometric identification of $(V^o)'$ and $V^\perp$, we then deduce assertion (iii). The other implication can be proved with analogous arguments. □

The following result can be easily deduced from Theorem 3.7.

Corollary 3.8 *The following assertions are equivalent;*

(i) the inf–sup condition (3.9) is satisfied;
(ii) $B^T : M \to X'$ is injective and B^T has a closed range;
(iii) $B : X \to M'$ is surjective.

Corollaries 3.3 and 3.8 yield the following inf–sup condition for the operator $B = -\mathrm{div}$, which is useful for the analysis of the Stokes problem and the MHD equations.

Corollary 3.9 *There exists $\beta > 0$ such that*

$$\inf_{q \in L_0^2(\Omega)} \sup_{\boldsymbol{v} \in \mathbb{H}_0^1(\Omega)} \frac{\displaystyle\int_\Omega q\, div\, \boldsymbol{v}}{\|\boldsymbol{v}\|_1 \|q\|_0} \geq \beta. \tag{3.12}$$

We can now state the following result which gives necessary and sufficient conditions for problem (3.6) to be well-posed.

Theorem 3.10 *Problem (3.6) admits a unique solution if and only if*

(i) $\Pi \circ A$ is an isomorphism from $V = Ker\, B$ on V';
(ii) the inf–sup condition (3.9) is satisfied.

When these conditions are fulfilled, we denote by β the best constant in the inf–sup condition and we define the positive constant[23]

$$\alpha = \inf_{v \in V} \frac{\|\Pi A v\|}{\|v\|}.$$

Then, the unique solution (u, p) of Problem (3.6) satisfies:

$$\|u\|_X \leq \frac{1}{\beta}\left(1 + \frac{\|a\|}{\alpha}\right)\|g\|_{M'} + \frac{1}{\alpha}\|f\|_{X'}, \tag{3.13}$$

$$\|p\|_X \leq \frac{\|a\|}{\beta^2}\left(1 + \frac{\|a\|}{\alpha}\right)\|g\|_{M'} + \frac{1}{\beta}\left(1 + \frac{\|a\|}{\alpha}\right)\|f\|_{X'}. \tag{3.14}$$

Proof (sketch) The first part of the theorem is a straightforward consequence of Theorem 3.1 and Corollary 3.8. For the second part, estimates (3.13) and (3.14) can be established with standard tools using some arguments of the proof of Theorem 3.1. □

[23]Recall that the existence of a constant $\alpha > 0$ such that $\alpha\|u\|_V \leq \|\Pi A u\|_{V'}$ results from the open mapping theorem.

3.1.5 The mixed Galerkin method

We are interested in this section in the approximation of problem (3.6) with a Galerkin method. To simplify the presentation, we assume that $a(\cdot,\cdot)$ is coercive on $V \times V$,

$$\exists\,\alpha > 0, \quad a(v,v) \geq \alpha\|v\|^2, \quad \forall\, v \in V, \tag{3.15}$$

which implies assumption (i) of Theorem 3.10 (by the Lax–Milgram theorem). In addition, we assume that the inf–sup condition (3.9) holds true on $X \times M$. Thus, assumptions (i) and (ii) of Theorem 3.10 are fulfilled. This ensures that problem (3.6) is well-posed.

Now, as usual in Galerkin methods, we consider *finite-dimensional subspaces* X_h of X and M_h of M, and we introduce the discrete problem, which is the *mixed Galerkin approximation* of problem (3.6):

Find $(u_h, p_h) \in X_h \times M_h$*, such that for all* $(v_h, q_h) \in X_h \times M_h$*,*

$$\begin{cases} a(u_h,v_h) + b(v_h,p_h) = \langle f, v_h\rangle, \\ \qquad\qquad\quad b(u_h,q_h) = \langle g, q_h\rangle. \end{cases} \tag{3.16}$$

Note that if a problem can be studied with the Lax–Milgram theorem on a space X, then any of its finite-dimensional internal approximations on $X_h \subset X$ can also be treated by the Lax–Milgram theorem. The well-posedness of the discrete problem is therefore straightforward. In contrast, the fact that problem (3.6) is well-posed does not imply in general that its discrete counterpart (3.16) is also well-posed. The reason for this is twofold:

- First, if we define

$$V_h = \{u_h \in X_h, \forall\, q_h \in M_h, b(u_h,q_h) = 0\}. \tag{3.17}$$

  Then V_h is not necessarily included in V (for example, for the Stokes problem, $u_h \in V_h$ does not imply that u_h is divergence free). Thus, the fact that the continuous problem satisfies property (i) of Theorem 3.1 on V does not imply that the discrete problem satisfies an analogous property on V_h.
- Second, the inf–sup condition on $X \times M$,

$$\exists\,\beta > 0, \inf_{q\in M} \sup_{v\in X} \frac{b(v,q)}{\|v\|_X\|q\|_M} \geq \beta,$$

  only implies

$$\exists\,\beta > 0, \quad \inf_{q_h\in M_h} \sup_{v\in X} \frac{b(v,q_h)}{\|v\|_X\|q_h\|_M} \geq \beta,$$

  which is an inf–sup condition on $X \times M_h$. The latter does not imply in general an inf–sup condition on $X_h \times M_h$, since $X \supset X_h$.

It is therefore necessary to assume that assumptions (i) and (ii) of Theorem 3.10 are satisfied by the discrete problem on $X_h \times M_h$ itself. Then, the following theorem gives an estimate of the error of the Galerkin method by the interpolation error. In other words, it is the analog of the well-known Céa Lemma for elliptic operators (*e.g.* A. Quarteroni and A. Valli [192]).

Theorem 3.11 *Assume that the coercivity hypothesis on V (3.15) and the inf–sup condition (3.9) hold and let (u,p) be the solution of (3.6). Assume in addition that*[24]

(i) $\exists\, \alpha_h > 0$ *such that* $\forall\, v_h \in V_h,\ a(v_h, v_h) \geq \alpha_h \|v_h\|_X^2$;

(ii) $\exists\, \beta_h > 0$, *such that* $\displaystyle\inf_{q_h \in M_h} \sup_{v_h \in X_h} \frac{b(v_h, q_h)}{\|v_h\|_X \|q_h\|_M} \geq \beta_h$.

Then problem (3.16) admits a unique solution which satisfies

$$\|u - u_h\|_X \leq \left(1 + \frac{\|a\|}{\alpha_h}\right)\left(1 + \frac{\|b\|}{\beta_h}\right) \inf_{v_h \in X_h} \|u - u_h\|_X + \frac{\|b\|}{\alpha_h} \inf_{q_h \in M_h} \|p - q_h\|_M, \tag{3.18}$$

and

$$\|p - p_h\|_X \leq \frac{\|a\|}{\beta_h}\left(1 + \frac{\|a\|}{\alpha_h}\right)\left(1 + \frac{\|b\|}{\beta_h}\right) \inf_{v_h \in X_h} \|u - u_h\|_X \left(1 + \frac{\|b\|}{\beta_h} + \frac{\|a\|\|b\|}{\alpha_h \beta_h}\right) \inf_{q_h \in M_h} \|p - q_h\|_M. \tag{3.19}$$

The proof of this theorem being somehow technical, we skip it and refer the reader to [75] or [106]. Considering the classical estimations of the interpolation errors, this result can readily give the convergence of the mixed Galerkin method. For optimal convergence rates, α_h and β_h are typically required to be independent of h.

3.1.6 Algebraic aspects

Let us consider a basis $(\varphi_i)_{i=1,\dots,N_u}$ (resp. $(\psi_i)_{i=1,\dots,N_p}$) of X_h (resp. of M_h). Any element $u_h \in X_h$ and $p_h \in M_h$ can be decomposed on these bases:

$$u_h = \sum_{i=1}^{N_u} U_i \varphi_i, \qquad \text{and} \qquad p_h = \sum_{i=1}^{N_p} P_i \psi_i.$$

We denote by U the vector $(U_1, \dots, U_{N_u})^T \in \mathbb{R}^{N_u}$, by P the vector $(P_1, \dots, P_{N_p})^T \in \mathbb{R}^{N_p}$, by F the vector $(\langle f, \varphi_1\rangle, \dots, \langle f, \varphi_{N_u}\rangle)^T \in \mathbb{R}^{N_u}$ and we define the following matrices:

$$\mathrm{A} = [a(\varphi_j, \varphi_i)]_{i,j=1,\dots,N_u}, \qquad \mathrm{B} = [b(\varphi_j, \psi_i)]_{i=1,\dots,N_p, j=1,\dots,N_u}, \tag{3.20}$$

where the index i indicates the rows and j the columns.

[24] Recall that V_h is defined in (3.17).

Assuming for simplicity that $g = 0$, problem (3.16) readily takes the following algebraic form:

Find $(\mathrm{U}, \mathrm{P}) \in \mathbb{R}^{\mathrm{N_u}} \times \mathbb{R}^{\mathrm{N_p}}$ *such that*

$$\begin{bmatrix} \mathrm{A} & \mathrm{B}^T \\ \mathrm{B} & 0 \end{bmatrix} \begin{bmatrix} \mathrm{U} \\ \mathrm{P} \end{bmatrix} = \begin{bmatrix} \mathrm{F} \\ 0 \end{bmatrix}. \tag{3.21}$$

If the unknown U is eliminated from system (3.21), we obtain the so-called *dual problem*[25]:

$$\mathrm{BA}^{-1}\mathrm{B}^T\mathrm{P} = \mathrm{BA}^{-1}\mathrm{F}. \tag{3.22}$$

Once the dual problem is solved, U may be recovered by solving

$$\mathrm{AU} = \mathrm{F} - \mathrm{B}^T\mathrm{P}.$$

For conciseness, let us denote

$$\mathrm{S} = \begin{bmatrix} A & \mathrm{B}^T \\ \mathrm{B} & 0 \end{bmatrix}.$$

We collect in the following proposition a few simple algebraic properties.

Proposition 3.12 *Assume that properties (i) and (ii) of Theorem 3.11 hold and assume in addition that $a(\cdot,\cdot)$ is symmetric. Then*

(i) The matrix S *is invertible and has* N_u *positive and* N_p *negative eigenvalues.*
(ii) The matrix $\mathrm{BA}^{-1}\mathrm{B}^T$ *is symmetric positive definite.*

When the dual problem is solved with a gradient method ($\mathrm{BA}^{-1}\mathrm{B}^T$ being symmetric positive definite), one obtains the well-known Uzawa algorithm.

At the algebraic level, the inf–sup condition on $X_h \times M_h$ is simply equivalent to

$$\mathrm{Ker}\,\mathrm{B}^T = \{0\}.$$

If the inf–sup condition does not hold, the approximation by X_h and M_h is said to be *unstable*: there exists $\mathrm{P}^* \in \mathbb{R}^{\mathrm{N_p}} - \{0\}$ such that $\mathrm{B}^T\mathrm{P}^* = 0$, and the linear systems (3.21) and (3.22) are ill-posed. Indeed, if (U, P) is a solution of (3.21), one can build an infinity of solutions $(\mathrm{U}, \mathrm{P} + k\mathrm{P}^*)$, $k \in \mathbb{R}$. Such a vector P^* is often called a *spurious mode* (or a *spurious pressure* in the context of hydrodynamics equations).

3.1.7 Mixed finite element for the Stokes problem

We assume that the reader is familiar with the basic results on the finite element method, typically for discretization of Lax–Milgram type problems. In particular, the classical theory of Lagrange interpolation is assumed to be known. Elementary as well as advanced results can be found for example in [192], [39] or [75].

[25]The matrix $\mathrm{BA}^{-1}\mathrm{B}^T$ is the *Schur complement* with respect to P.

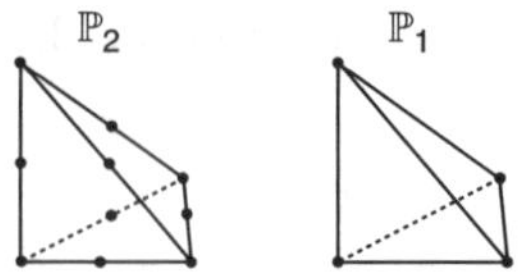

Fig. 3.1 The Taylor–Hood ($\mathbb{P}_2/\mathbb{P}_1$) finite element.

Section 3.1.5 has provided the theoretical results to analyze the convergence of a Galerkin discretization of problem (3.6). In this section, we briefly present a few finite element spaces adapted to the discretization of the Stokes equations (3.3). The simplicity of problem (3.3) is an advantage for the clarity of the exposition. In addition to this, we shall see in Section 3.4 that the finite element spaces for the Stokes problem are also well-suited for the MHD equations.

The domain $\Omega \subset \mathbb{R}^3$ is supposed to be polyhedral and partitioned in a mesh $\mathcal{T}_h$, which consists of tetrahedral elements K. Each tetrahedron is assumed to be the image by an affine map T_K of a reference tetrahedron $\hat{K}$. We define $h = \max_{K\in\mathcal{T}_h}\{h_K\}$ where h_K denotes the diameter of an element K. The family of meshes $(\mathcal{T}_h)_{h>0}$ is assumed to be regular. We denote by $\mathbb{P}_k(\hat{K})$ the space of polynomials of degree k defined on $\hat{K}$. We introduce the classical Lagrange finite element space

$$Y_h^r = \{v_h \in \mathcal{C}^0(\overline{\Omega}), v_h \circ T_K \in \mathbb{P}_r(\hat{K}), \forall\, K \in \mathcal{T}_h\}. \tag{3.23}$$

When $X_h = (Y_h^r)^3 \cap \mathbb{H}_0^1(\Omega)$ and $M_h = Y_h^s \cap L_0^2(\Omega)$, one says that the Stokes problem is solved with the $\mathbb{P}_r/\mathbb{P}_s$ finite element[26]: this means that the velocity (resp. the pressure) is approximated in a space of continuous functions, piecewise polynomials of degree r (resp. s). As mentioned above, the main difficulty is to find spaces X_h and M_h that satisfy the inf–sup condition:

$$\exists\,\beta_h > 0,\ \inf_{q_h\in M_h}\ \sup_{\boldsymbol{v}_h\in X_h} \frac{\displaystyle\int_\Omega q_h \operatorname{div} \boldsymbol{v}_h}{\|\boldsymbol{v}_h\|_1 \|q_h\|_0} \geq \beta_h, \tag{3.24}$$

For example, the pairs $\mathbb{P}_r/\mathbb{P}_r$, $r \geq 1$, or $\mathbb{P}_1/\mathbb{P}_0$ are known to be *unstable*. This means that the spaces X_h and M_h built on these finite elements do not satisfy the inf–sup condition (3.24). The resulting linear system is therefore singular.

A simple example of a stable pair using the standard spaces Y_h^r is $\mathbb{P}_2/\mathbb{P}_1$ which is known as the Taylor–Hood finite element (Figure 3.1). The inf–sup constant β_h of this element is in addition independent of h which ensures an optimal convergence rate. The following result can be proved (see for example [75, Section 4.2.5]).

[26]Note that in practical implementations, it is not necessary to enforce a zero mean value for the pressure. It is possible to get rid of the undetermined constant by fixing the value of the pressure at one point or simply by using an iterative method to solve the linear systems.

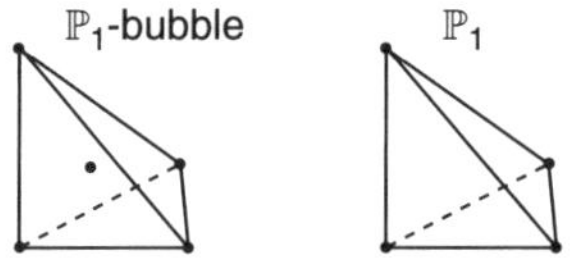

Fig. 3.2 The mini ($\mathbb{P}_1$-bubble/$\mathbb{P}_1$) finite element.

Proposition 3.13 *Assume that the solution of the Stokes problem satisfies $\boldsymbol{u} \in \mathbb{H}^3(\Omega) \cap \mathbb{H}_0^1(\Omega)$ and $p \in H^2(\Omega) \cap L_0^2(\Omega)$. Assume moreover that each tetrahedron of the mesh has at least three edges within Ω. Then there exists $c > 0$ such that the solution $(\boldsymbol{u}_h, p_h)$ computed with the $\mathbb{P}_2/\mathbb{P}_1$ finite element satisfies for all $h > 0$:*

$$\|\boldsymbol{u} - \boldsymbol{u}_h\|_{\mathbb{H}^1(\Omega)} + \|p - p_h\|_{L^2(\Omega)} \leq ch^2(\|\boldsymbol{u}\|_{\mathbb{H}^3(\Omega)} + \|p\|_{H^2(\Omega)}).$$

Let us present another popular element, the so-called $\mathbb{P}_1$-bubble/$\mathbb{P}_1$ pair, also known as the *mini-element.* It consists in adding to the $\mathbb{P}_1/\mathbb{P}_1$ element one degree of freedom for each component of the velocity on the barycenters of the tetrahedra (Figure 3.2). Let $\hat{b} \in H_0^1(\hat{K})$ taking the value 1 at the barycenter of $\hat{K}$ and such that $0 \leq \hat{b} \leq 1$. The function $\hat{b}$ is called a "bubble function." We then define the space

$$\mathcal{P}_{1,h}^b = \left\{ v_h \in \mathcal{C}^0(\bar{\Omega}), v_h \circ T_K \in \mathbb{P}_1(\hat{K}) \oplus \operatorname{span}\{\hat{b}\}, \forall\, K \in \mathcal{T}_h \right\}.$$

The Stokes problem is said to be solved with the mini-element, or the $\mathbb{P}^1$-bubble/$\mathbb{P}^1$ finite element, when one uses the spaces $X_h = (\mathcal{P}_{1,h}^b)^3 \cap \mathbb{H}_0^1(\Omega)$ to approximate the velocity and $M_h = Y_h^1 \cap L_0^2(\Omega)$ to approximate the pressure. As for the Taylor–Hood element, the inf–sup constant β_h of the mini-element is independent of h which ensures an optimal convergence rate. We then have the following result (see for example [75, Section 4.2.4]).

Proposition 3.14 *Assume that the solution of the Stokes problem satisfies $\boldsymbol{u} \in \mathbb{H}^2(\Omega) \cap \mathbb{H}_0^1(\Omega)$ and $p \in H^1(\Omega) \cap L_0^2(\Omega)$. Then there exists $c > 0$ such that the solution $(\boldsymbol{u}_h, p_h)$ computed with the $\mathbb{P}_1$-bubble/$\mathbb{P}_1$ finite element satisfies for all $h > 0$:*

$$\|\boldsymbol{u} - \boldsymbol{u}_h\|_{\mathbb{H}^1(\Omega)} + \|p - p_h\|_{L^2(\Omega)} \leq ch(\|\boldsymbol{u}\|_{\mathbb{H}^2(\Omega)} + \|p\|_{\mathbb{H}^1(\Omega)}).$$

Many other examples of stable pairs can be found in the literature. We refer the interested reader to V. Girault and P.A. Raviart [106, Chapter 2], F. Brezzi and M. Fortin [32, Chapter 4] or A. Ern and J.-L. Guermond [75, Chapter 4].

3.1.8 Extension to nonlinear problems

We now need to extend the previous results, specific to *linear* problems, to the class of nonlinear problems we will meet when considering the MHD equations. For this purpose, we define a nonlinear form

$$a(\cdot;\cdot,\cdot) : X \times X \times X \to \mathbb{R},$$

where for any w in X, $a(w;\cdot,\cdot)$ is a bilinear continuous form on $X \times X$. Note that the continuity on X with respect to w is not assumed. Let $f \in X'$. We consider the following nonlinear problem:

Find $(u,p) \in X \times M$ *such that for all* $(v,q) \in X \times M$,

$$\begin{cases} a(u;u,v) + b(v,p) = \langle f, v\rangle, \\ \qquad\qquad\quad b(u,q) = 0. \end{cases} \tag{3.25}$$

It is built from (3.6) replacing $a(u,v)$ by $a(u;u,v)$. We define the operator B as in (3.5) and we again denote by V the kernel of B. The linear theory presented above is of course instrumental in addressing this nonlinear problem: loosely speaking, the linear theory provides the existence of the Lagrange multiplier p as soon as the nonlinear problem "restricted to the kernel of B" has a solution u (see below, problem (3.26)). The existence of the solution u in V will be obtained in the sequel by compactness arguments as in Chapter 2. But for the moment, let us just consider the question of the existence of p. We introduce the problem:

Find $u \in V$ *such that for all* $v \in V$,

$$a(u;u,v) = \langle f, v\rangle. \tag{3.26}$$

If u is a solution of (3.26) and if the inf–sup condition (3.9) holds, then it can be proved exactly as in the linear case that there exists $p \in M$ such that

$$b(v,p) = \langle f, v\rangle - a(u;u,v),$$

for all $v \in V$. Indeed, the inf–sup condition yields that B is surjective, thus $\operatorname{Im} B$ is closed, and the closed range theorem implies $\operatorname{Im} B^T = (\operatorname{Ker} B)^o = V^o$. As a consequence, if $u \in V$ is a solution of (3.26), then $\langle f,\cdot\rangle - a(u;u,\cdot)$ is a linear form which belongs to V^o and thus to $\operatorname{Im} B^T$. This implies the existence and the uniqueness of $p \in M$ such that (u,p) is a solution of (3.25). We see in this argument that $a(u;u,\cdot)$ plays the same role as $a(u,\cdot)$ in the linear theory. Thus, the only remaining task to study (3.25) is to analyze problem (3.26). This is the purpose of the following theorem.

Theorem 3.15 *We assume that*

(i) there exists $\alpha > 0$, *such that for all* $v \in V$,

$$a(v;v,v) \geq \alpha \|v\|_X^2;$$

(ii) the space V *is separable and that for any sequence* v_n *that weakly converges to* v *in* V, $a(v_n;v_n,w)$ *converges to* $a(v;v,w)$ *for all* $w \in V$.

Then there exists at least one $u \in V$ *solution of problem (3.26). If, in addition we assume that:*

(iii) the ellipticity property (i) holds uniformly with respect to the first variable, i.e. there exists $\alpha > 0$ such that for all $v, w \in V$,

$$a(w; v, v) \geq \alpha \|v\|_X^2,$$

(iv) there exists a constant $\gamma > 0$ such that, for all u_1, u_2, v, w in V,

$$|a(u_2; v, w) - a(u_1; v, w)| \leq \gamma \|u_2 - u_1\|_X \|v\|_X \|w\|_X,$$

then problem (3.26) has a unique solution $u \in V$ provided that

$$\frac{\gamma \|f\|_{V'}}{\alpha^2} < 1. \tag{3.27}$$

For the proof, we refer to [106]. Therein, [106, Theorem IV.1.2] gives the existence of u and [106, Theorem IV.1.3] establishes its uniqueness, the latter under more general assumptions.

3.2 A glance at stabilized finite elements

In this section, we briefly present stabilized finite element methods. These techniques are in particular a way to discretize problems with constraints, such as (3.1). In the case of Stokes equations, they allow us to circumvent the compatibility condition for the velocity and pressure finite element spaces (3.24). They, however, have a much wider range of application since they can also improve the efficiency of finite element methods in other type of problems, such as, *e.g.* advection–diffusion equations. Both aspects of the stabilization techniques will be exploited below (see Section 3.5). Our presentation is very brief, our aim being just to introduce the basic ideas that will be used in Section 3.5 for the MHD equations.

We first consider the discretization of the Stokes problem (3.1) with a right-hand side $\boldsymbol{f} \in \mathbb{L}^2(\Omega)$. Let $\mathcal{T}_h$ be a finite element mesh over Ω, defined as in Section 3.1.7. Starting from the weak form of the Stokes equations (3.3) and discretizing them as the abstract problem (3.16), we obtain:

Find $(\boldsymbol{u}_h, p_h) \in X_h \times M_h$ such that for all $(\boldsymbol{v}_h, q_h) \in X_h \times M_h$,

$$\Psi((\boldsymbol{u}_h, p_h), (\boldsymbol{v}_h, q_h)) = \langle F, (\boldsymbol{v}_h, q_h) \rangle,$$

with

$$\Psi((\boldsymbol{u}_h, p_h), (\boldsymbol{v}_h, q_h)) = \int_\Omega \boldsymbol{\nabla} \boldsymbol{u}_h : \boldsymbol{\nabla} \boldsymbol{v}_h - \int_\Omega p_h \operatorname{div} \boldsymbol{v}_h + \int_\Omega q_h \operatorname{div} \boldsymbol{u}_h,$$

and

$$\langle F, (\boldsymbol{v}_h, q_h) \rangle = \int_\Omega \boldsymbol{f} \cdot \boldsymbol{v}_h.$$

It has been shown in Section 3.1.5 that this problem is well-posed if and only if the spaces X_h and M_h satisfy the inf–sup condition (3.9) on $X_h \times M_h$. This is a consequence of the non-coercivity of this problem on $X_h \times M_h$ (see Section 3.1). Basically, *stabilization* techniques consist in adding mesh-dependent terms to the classical Galerkin formulation in order to recover coercivity. Such terms are chosen *strongly consistent*, which means that they vanish for the exact solution.

For example, we can replace $a(\cdot,\cdot)$ by $a_h(\cdot,\cdot)$ defined by

$$\begin{aligned} a_h((\boldsymbol{u}_h, p_h), (\boldsymbol{v}_h, q_h)) &= a((\boldsymbol{u}_h, p_h), (\boldsymbol{v}_h, q_h)) \\ &\quad + \sum_{K \in \mathcal{T}_h} \int_K \tau(-\Delta\, \boldsymbol{u}_h + \boldsymbol{\nabla} p_h) \cdot (-\xi \Delta\, \boldsymbol{v}_h + \boldsymbol{\nabla} q_h), \end{aligned}$$

and F by F_h:

$$\langle F_h, (\boldsymbol{v}_h, q_h) \rangle = \langle F, (\boldsymbol{v}_h, q_h) \rangle + \sum_{K \in \mathcal{T}_h} \int_K \tau \boldsymbol{f}.(-\xi \Delta\, \boldsymbol{v}_h + \boldsymbol{\nabla} q_h),$$

where $\tau|_K = \lambda h_K^2$ is the stabilization coefficient. The constant λ has to be selected in order to ensure coercivity. The value of ξ depends on the variant of the method: -1 (Douglas–Wang method, DWG), 0 (Streamline Upwind Petrov Galerkin, SUPG) or 1 (Galerkin Least Square, GLS).

The convergence of the stabilized method may then be proved irrespective of any inf–sup condition. In particular, equal order elements $\mathbb{P}_r/\mathbb{P}_r$ for the velocity and the pressure can be used. We refer to J. Douglas and J. Wang [70], T.J.R. Hughes, L.P. Franca and M. Balestra [129] for the original papers and to A. Quarteroni and A. Valli [192] for a pedagogical presentation.

As mentioned above, the idea of adding "strongly consistent" terms to the Galerkin formulation can also improve the classical finite element methods to solve the advection–diffusion equation:

$$-\eta \Delta\, u + \boldsymbol{a} \cdot \boldsymbol{\nabla} u = f \text{ in } \Omega, \tag{3.28}$$

where η is a positive given constant. It is indeed well-known that the finite element method is not well-suited when the diffusion is overtaken by the convection at the element scale, in other words when $\|\boldsymbol{a}\|_{\infty,K} h_K/\eta$ is large. The stabilization methods can improve the convergence in that case. For simplicity, we suppose that $f \in L^2(\Omega)$, $u = 0$ on $\partial\Omega$ and $\operatorname{div} \boldsymbol{a} = 0$. For a more complete study we refer to [192] for example.

We choose for X_h a standard finite element space, for example consisting of continuous piecewise polynomials of degree k. We define the bilinear form

$$\Phi(w, v) = \eta \int_\Omega \boldsymbol{\nabla} w \cdot \boldsymbol{\nabla} v + \int_\Omega \boldsymbol{a} \cdot \boldsymbol{\nabla} w\, v,$$

and

$$\langle F, v \rangle = \int_\Omega f v.$$

In the sequel, the solution of the continuous problem (3.28) is denoted by u, the interpolate of u in X_h is denoted by $\tilde{u}_h$ and the solution of the discrete Galerkin problem is denoted by u_h. Thus we have:

$$\Phi(u_h, v_h) = \langle F, v_h \rangle, \forall\, v_h \in X_h. \tag{3.29}$$

The interpolation error is denoted by $\pi_h = u - \tilde{u}_h$, the approximation error $e_h = \tilde{u}_h - u_h$ and the global error $\epsilon_h = u - u_h$. We have $\epsilon_h = \pi_h + e_h$.

By classical estimates, the interpolation error is shown to be bounded as follows

$$\|\boldsymbol{\nabla} e_h\|_{\mathbb{L}^2} \leq C \sqrt{1 + \frac{\|\boldsymbol{a}\|_\infty^2 h^2}{\eta^2}}\ |u|_{k+1}\ h^k, \tag{3.30}$$

where C is a constant that does not depend on h, $\boldsymbol{a}$, u and η. For advection-dominated flows, the right-hand side of this estimate is large so that the method may be quite inaccurate. Numerical observations show that results exhibit possibly large oscillations.

The way to estimate the error of the Galerkin method is based on four properties: linearity, strong consistency, coercivity and continuity. The stabilized methods, in the vein of those introduced above for the Stokes problem, are generalized Galerkin methods where the bilinear form Φ of the continuous formulation is replaced by a bilinear form Φ_h depending on the mesh and still satisfying these four properties.

For example, we may define Φ_h on $X_h \times X_h$ by

$$\Phi_h(w, v) = \Phi(w, v) + \sum_{K \in \mathcal{T}_h} \int_K \tau(-\eta \Delta\, w + \boldsymbol{a} \cdot \boldsymbol{\nabla} w)(\xi \eta \Delta\, v + \boldsymbol{a} \cdot \boldsymbol{\nabla} v), \tag{3.31}$$

and

$$\langle F_h, v \rangle = \int_\Omega f v + \sum_{K \in \mathcal{T}_h} \int_K \tau f(\xi \eta \Delta\, v + \boldsymbol{a} \cdot \boldsymbol{\nabla} v). \tag{3.32}$$

The stabilization coefficient τ can be defined on Ω by:

$$\tau|_K = \tau_K \text{ with } \tau_K = \frac{\lambda h_K}{\|\boldsymbol{a}\|_{\infty,K}}, \forall K \in \mathcal{T}_h$$

where λ is a non-negative constant that has to be fixed to ensure the coercivity of the approximate problem, and $\xi = -1, 0$ or 1 depending on the chosen variant. For example, for $\xi = 1$, it can be proved that the error estimate for the stabilization methods is

$$\|\boldsymbol{\nabla} e_h\|_{\mathbb{L}^2} \leq C \sqrt{1 + \frac{\|\boldsymbol{a}\|_\infty h}{\eta}}\ |u|_{k+1}\ h^k, \tag{3.33}$$

where C is a constant independent of h, $\boldsymbol{a}$, u and η. Comparing (3.30) and (3.33), this shows that the stabilized methods may improve the results when $\|\boldsymbol{a}\|_\infty h/\eta$

is large. The improvement of the constant in the estimations is not the only advantage: it can be shown that the stabilization terms also allows a control of the convection derivative

$$\int_\Omega |\boldsymbol{a} \cdot \boldsymbol{\nabla} e_h|^2,$$

independent of the Peclet number $\|\mathbf{a}\|_\infty h/\eta$. In the case when $\|\boldsymbol{a}\|_\infty h/\eta < 1$ (diffusion-dominated flow), the estimate (3.33) is worse than that obtained with the classical Galerkin approximation (3.30). To avoid this we can choose $\tau_K = \lambda h_K^2/\eta$ in the cells K where diffusion dominates. We then recover an estimation like (3.30) and the stabilization coefficient τ remains continuous when $\|\boldsymbol{a}\|_\infty h/\eta$ takes the value 1.

In Section 3.5, we will present a stabilized finite method for the linearized MHD equations which circumvents the inf–sup condition and stabilizes the advection terms.

3.3 Mixed formulations of the stationary MHD equations

We present in this section two variational formulations of the stationary MHD equations which are both adapted to the discretization by mixed finite elements that will be presented in Section 3.4.

The first formulation, extensively studied by M.D. Gunzburger *et al.* in [118], makes use of standard Lagrange finite element spaces and is therefore the easiest to implement, but it is restricted to domains which are either convex polyhedral or with a $\mathcal{C}^{1,1}$ boundary. Note that, after discretization, the boundary of the computational domain usually does not have the $\mathcal{C}^{1,1}$ regularity. In practice, this formulation is thus limited to convex domains and may therefore not be convenient for some applications.

The second formulation, proposed by D. Schötzau in [205], is also valid on non-convex polyhedra. In such domains, re-entrant corners may yield singularities in the magnetic field that are missed by classical $\mathbb{H}^1$-conforming finite element. This formulation is based on the $H(\mathrm{div})$ space for the magnetic field and leads to a discretization by the Nédélec finite element.

As announced before, the abstract theory presented in Section 3.1 is the key ingredient for the study of these formulations at the continuous and the discrete levels.

Let $\boldsymbol{f} \in \mathbb{H}^{-1}(\Omega)$, $\boldsymbol{h} \in \mathbb{L}^2(\Omega)$. We denote by Re, Rm and S three constant positive parameters. We consider the stationary MHD problem, as formulated in Chapter 1, in its non-dimensionalized form:

Find $(\boldsymbol{u}, \boldsymbol{B}, p) \in \mathbb{H}^1(\Omega) \times \mathbb{H}^1(\Omega) \times L_0^2(\Omega)$ *such that*

$$\begin{cases} -\dfrac{1}{Re}\Delta \boldsymbol{u} + \boldsymbol{\nabla} p + \boldsymbol{u} \cdot \boldsymbol{\nabla} \boldsymbol{u} + S\, \boldsymbol{B} \times \mathbf{curl}\, \boldsymbol{B} = \boldsymbol{f}, \\ \mathit{div}\, \boldsymbol{u} = 0, \\ \dfrac{1}{Rm}\mathbf{curl}\,(\mathbf{curl}\, \boldsymbol{B}) - \mathbf{curl}\,(\boldsymbol{u} \times \boldsymbol{B}) = \boldsymbol{h}, \\ \mathit{div}\, \boldsymbol{B} = 0, \end{cases} \tag{3.34}$$

in $\mathcal{D}'(\Omega)$, and satisfying the following homogeneous boundary conditions:

$$\begin{cases} \dfrac{1}{Rm}\mathbf{curl}\,\boldsymbol{B} \times \boldsymbol{n} = 0, \\ \boldsymbol{B} \cdot \boldsymbol{n} = 0, \\ \boldsymbol{u} = 0. \end{cases} \tag{3.35}$$

The right-hand side $\boldsymbol{h}$ in (3.34) is not present in the original magnetic equation. Such a term appears (among others) when a lifting of non-homogeneous magnetic conditions is introduced (see Sections 3.8 and 2.2.5). We shall suppose that

$$\int_\Omega \boldsymbol{h} \cdot \boldsymbol{\nabla v} = 0 \tag{3.36}$$

for all $\boldsymbol{v} \in \mathbb{H}^1_n(\Omega)$. This assumption is compatible with the physical equations (see Section 3.8).

We recall that the Poincaré inequality reads

$$\exists\, c_{\Omega,1} > 0,\ \ \forall \boldsymbol{v} \in \mathbb{H}^1_0(\Omega),\ \ c_{\Omega,1}\|\boldsymbol{v}\|_1^2 \le |\boldsymbol{v}|_1^2. \tag{3.37}$$

We recall the definition of the spaces

$$\mathbb{H}^1_n(\Omega) = \{\boldsymbol{C} \in \mathbb{H}^1(\Omega), \boldsymbol{C} \cdot \boldsymbol{n}|_{\partial\Omega} = 0\},$$

and we introduce

$$\begin{aligned} \mathcal{H}(\Omega) &= H(\mathbf{curl}\,;\Omega) \cap H_0(\mathrm{div}\,;\Omega) \\ &= \{\boldsymbol{C} \in \mathbb{L}^2(\Omega), \mathbf{curl}\,\boldsymbol{C} \in \mathbb{L}^2(\Omega), \mathrm{div}\,\boldsymbol{C} \in L^2(\Omega), \boldsymbol{C} \cdot \boldsymbol{n}|_{\partial\Omega} = 0\}. \end{aligned} \tag{3.38}$$

3.3.1 A formulation for convex polyhedra and regular domains

To derive a first variational formulation of problem (3.34)–(3.35) we make, throughout this section, the following additional assumption on the domain:

$$\Omega \text{ either is a convex polyhedron or has a } \mathcal{C}^{1,1} \text{ boundary.} \tag{3.39}$$

Note that this hypothesis is very strong and may be not satisfied in practical applications. Before stating a property of $\mathcal{H}(\Omega)$ when (3.39) holds, we recall that for a general domain Ω, bounded, simply connected and only Lipschitz, we have

$$\exists\, c_{\Omega,2} > 0,\ \ \forall\, \boldsymbol{B} \in \mathcal{H}(\Omega),\ \ c_{\Omega,2}\|\boldsymbol{B}\|_0^2 \le \|\mathbf{curl}\,\boldsymbol{B}\|_0^2 + \|\mathrm{div}\,\boldsymbol{B}\|_0^2. \tag{3.40}$$

Proposition 3.16 *[106, Section I.3.5] If assumption (3.39) holds then $\mathcal{H}(\Omega)$ is continuously embedded in $\mathbb{H}^1(\Omega)$. Thus,*

$$\exists\, c_{\Omega,3} > 0,\ \ \forall\, \boldsymbol{B} \in \mathcal{H}(\Omega),\ \ c_{\Omega,3}\|\boldsymbol{B}\|_1^2 \le \|\mathbf{curl}\,\boldsymbol{B}\|_0^2 + \|\,div\,\boldsymbol{B}\|_0^2. \tag{3.41}$$

Assumption (3.39) on Ω also allows for the following lemma. The result will be useful to recover the divergence-free constraint on $\boldsymbol{B}$ in the variational formulation that is proposed below.

Lemma 3.17 *If $\boldsymbol{B} \in \mathbb{H}^1_n(\Omega)$ then there exists $\psi \in H^2(\Omega)$ such that*

$$\begin{cases} -\Delta \psi = div \boldsymbol{B} & in\ \Omega, \\ \dfrac{\partial \psi}{\partial n} = 0 & on\ \partial\Omega, \end{cases} \tag{3.42}$$

In particular $\boldsymbol{\nabla}\psi$ is in $\mathbb{H}^1_n(\Omega)$.

Proof First note that this Neumann problem is well-posed since

$$\int_\Omega \operatorname{div} \boldsymbol{B} = \int_{\partial\Omega} \boldsymbol{B} \cdot \boldsymbol{n} = 0.$$

The right-hand side $\operatorname{div} \boldsymbol{B}$ being in $L^2(\Omega)$, and the domain satisfying (3.39), it is well-known that the solution to the Neumann problem exists and is in $H^2(\Omega)$. Moreover $\boldsymbol{\nabla}\psi \cdot \boldsymbol{n} = 0$ on $\partial\Omega$, thus $\boldsymbol{\nabla}\psi \in \mathbb{H}^1_n(\Omega)$. □

To be consistent with the notation used in Section 3.1, we denote by V the kernel of the divergence operator in $\mathbb{H}^1_0$:

$$V = \{\boldsymbol{v} \in \mathbb{H}^1_0(\Omega), \operatorname{div} \boldsymbol{v} = 0\}. \tag{3.43}$$

We introduce the spaces

$$\mathcal{X} = \mathbb{H}^1_0(\Omega) \times \mathbb{H}^1_n(\Omega) \qquad \text{and} \qquad \mathcal{V} = V \times \mathbb{H}^1_n(\Omega), \tag{3.44}$$

endowed with the norm $\|(\boldsymbol{v}, \boldsymbol{C})\|_{\mathcal{X}} = (\|\boldsymbol{v}\|_1^2 + \|\boldsymbol{C}\|_1^2)^{1/2}$. We also introduce

$$\mathcal{M} = L^2_0(\Omega), \tag{3.45}$$

endowed with the norm $\|q\|_{\mathcal{M}} = \|q\|_0$.

We then consider the variational formulation

Find $((\boldsymbol{u}, \boldsymbol{B}), p) \in \mathcal{X} \times \mathcal{M}$ such that for all $((\boldsymbol{v}, \boldsymbol{C}), q) \in \mathcal{X} \times \mathcal{M}$

$$\begin{aligned} \mathcal{A}((\boldsymbol{u}, \boldsymbol{B}), (\boldsymbol{v}, \boldsymbol{C})) + \mathcal{C}((\boldsymbol{u}, \boldsymbol{B}), (\boldsymbol{u}, \boldsymbol{B}), (\boldsymbol{v}, \boldsymbol{C})) + \mathcal{B}((\boldsymbol{v}, \boldsymbol{C}), p) &= \mathcal{L}((\boldsymbol{v}, \boldsymbol{C})), \\ \mathcal{B}((\boldsymbol{u}, \boldsymbol{B}), q) &= 0, \end{aligned} \tag{3.46}$$

where the following operators, $\mathcal{A}, \mathcal{B}, \mathcal{C}$ and $\mathcal{L}$, have been introduced:

$$\mathcal{A}((\boldsymbol{u}, \boldsymbol{B}), (\boldsymbol{v}, \boldsymbol{C})) = \frac{1}{Re}\int_\Omega \boldsymbol{\nabla u} : \boldsymbol{\nabla v} + \frac{S}{Rm}\int_\Omega (\operatorname{div} \boldsymbol{B} \operatorname{div} \boldsymbol{C} + \mathbf{curl}\, \boldsymbol{B} \cdot \mathbf{curl}\, \boldsymbol{C}), \tag{3.47}$$

$$\mathcal{B}((\boldsymbol{v}, \boldsymbol{C}), q) = \int_\Omega q \operatorname{div} \boldsymbol{v}, \tag{3.48}$$

$$\begin{aligned} &\mathcal{C}((\boldsymbol{v}_1, \boldsymbol{C}_1), (\boldsymbol{v}_2, \boldsymbol{C}_2), (\boldsymbol{v}_3, \boldsymbol{C}_3)) = \\ &\quad \int_\Omega (\boldsymbol{v}_1 \cdot \boldsymbol{\nabla v}_2 \cdot \boldsymbol{v}_3 - S\, \mathbf{curl}\, \boldsymbol{C}_2 \times \boldsymbol{C}_1 \cdot \boldsymbol{v}_3 - S\, \boldsymbol{v}_2 \times \boldsymbol{C}_1 \cdot \mathbf{curl}\, \boldsymbol{C}_3), \end{aligned} \tag{3.49}$$

$$\mathcal{L}((\boldsymbol{v}, \boldsymbol{C})) = \langle \boldsymbol{f}, \boldsymbol{v} \rangle + \langle S\, \boldsymbol{h}, \boldsymbol{C} \rangle. \tag{3.50}$$

Proposition 3.18 *If (3.36) holds and if* $((\boldsymbol{u}, \boldsymbol{B}), p) \in \mathcal{X} \times \mathcal{M}$ *is a solution of problem (3.46) then* $((\boldsymbol{u}, \boldsymbol{B}), p)$ *is a solution to (3.34)–(3.35).*

Proof Let $((\boldsymbol{u}, \boldsymbol{B}), p)$ be a solution of (3.46), and let us prove that $((\boldsymbol{u}, \boldsymbol{B}), p)$ is a solution to (3.34)–(3.35). The first equation of (3.34) can be classically recovered by taking $\boldsymbol{C} = 0$ in (3.46) and the second equation of (3.34) comes straightforwardly from the second equation of (3.46). Next, $\boldsymbol{v} = 0$ in (3.46) gives

$$\frac{S}{Rm}\int_\Omega (\mathbf{curl}\,\boldsymbol{B} \cdot \mathbf{curl}\,\boldsymbol{C} + \operatorname{div}\boldsymbol{B}\operatorname{div}\boldsymbol{C}) - S\int_\Omega \boldsymbol{u} \times \boldsymbol{B} \cdot \mathbf{curl}\,\boldsymbol{C} = S\int_\Omega \boldsymbol{h} \cdot \boldsymbol{C}.$$

Let ψ be given by Lemma 3.17. Taking $\boldsymbol{C} = \boldsymbol{\nabla}\psi$ in this equation, we obtain

$$\int_\Omega (\operatorname{div}\boldsymbol{B})^2 = 0,$$

which yields $\operatorname{div}\boldsymbol{B} = 0$ a.e., and thus, for all $\boldsymbol{C} \in \mathbb{H}^1_n(\Omega)$,

$$\frac{S}{Rm}\int_\Omega \mathbf{curl}\,\boldsymbol{B} \cdot \mathbf{curl}\,\boldsymbol{C} - S\int_\Omega \boldsymbol{u} \times \boldsymbol{B} \cdot \mathbf{curl}\,\boldsymbol{C} = S\int_\Omega \boldsymbol{h} \cdot \boldsymbol{C}.$$

Taking $\boldsymbol{C} \in \mathbb{H}^1_0(\Omega)$, we readily obtain

$$\frac{1}{Rm}\mathbf{curl}\,(\mathbf{curl}\,\boldsymbol{B}) - \mathbf{curl}\,(\boldsymbol{u} \times \boldsymbol{B}) = \boldsymbol{h}.$$

Then, since $\boldsymbol{h}$ is in $\mathbb{L}^2(\Omega)$, this equation provides additional regularity on $\boldsymbol{B}$. Next, when inserted in the variational formulation, it yields the first boundary condition in (3.35). □

Lemma 3.19 *The following properties hold:*

(i) The bilinear form $\mathcal{A}(\cdot,\cdot)$ *is continuous and coercive on* $\mathcal{X} \times \mathcal{X}$ *with a coercivity constant given by*

$$\alpha = \min\left(\frac{c_{\Omega,1}}{Re}, \frac{S c_{\Omega,3}}{Rm}\right). \tag{3.51}$$

(ii) The trilinear form $\mathcal{C}(\cdot,\cdot,\cdot)$ *is continuous on* $\mathcal{X} \times \mathcal{X} \times \mathcal{X}$.

(iii) If $((\boldsymbol{u}, \boldsymbol{B}), (\boldsymbol{v}, \boldsymbol{C}), (\boldsymbol{w}, \boldsymbol{D})) \in \mathcal{X} \times \mathcal{X} \times \mathcal{X}$ *with* $\operatorname{div}\boldsymbol{u} = 0$, *then*

$$\mathcal{C}((\boldsymbol{u}, \boldsymbol{B}), (\boldsymbol{v}, \boldsymbol{C}), (\boldsymbol{w}, \boldsymbol{D})) = -\mathcal{C}((\boldsymbol{u}, \boldsymbol{B}), (\boldsymbol{w}, \boldsymbol{D}), (\boldsymbol{v}, \boldsymbol{C})).$$

Proof (i) Continuity of $\mathcal{A}(\cdot,\cdot)$: let $(\boldsymbol{u}, \boldsymbol{B})$ and $(\boldsymbol{v}, \boldsymbol{C})$ be in $\mathcal{X}$. Then

$$\begin{aligned}\mathcal{A}((\boldsymbol{u}, \boldsymbol{B}), (\boldsymbol{v}, \boldsymbol{C})) &= \frac{1}{Re}\int_\Omega \boldsymbol{\nabla}\boldsymbol{u} : \boldsymbol{\nabla}\boldsymbol{v} + \frac{S}{Rm}\int_\Omega (\mathbf{curl}\,\boldsymbol{B} \cdot \mathbf{curl}\,\boldsymbol{C} + \operatorname{div}\boldsymbol{B}\operatorname{div}\boldsymbol{C}) \\ &\le \frac{1}{Re}\|\boldsymbol{u}\|_1\|\boldsymbol{v}\|_1 + \frac{2S}{Rm}\|\boldsymbol{B}\|_1\|\boldsymbol{C}\|_1 + \frac{3S}{Rm}\|\boldsymbol{B}\|_1\|\boldsymbol{C}\|_1 \\ &\le \max\left(\frac{1}{Re}, \frac{3S}{Rm}\right)\|(\boldsymbol{u}, \boldsymbol{B})\|_{\mathcal{X}}\|(\boldsymbol{v}, \boldsymbol{C})\|_{\mathcal{X}}.\end{aligned}$$

Coercivity of $\mathcal{A}(\cdot,\cdot)$: let $(\boldsymbol{v},\boldsymbol{C})$ be in $\mathcal{X}$. Then

$$\mathcal{A}((\boldsymbol{v},\boldsymbol{C}),(\boldsymbol{v},\boldsymbol{C})) = \frac{1}{Re}\int_\Omega |\boldsymbol{\nabla v}|^2 + \frac{S}{Rm}\int_\Omega |\mathbf{curl}\,\boldsymbol{C}|^2 + \frac{S}{Rm}\int_\Omega |\mathrm{div}\,\boldsymbol{C}|^2$$
$$\geq \min\left(\frac{c_{\Omega,1}}{Re}, \frac{Sc_{\Omega,3}}{Rm}\right) \|(\boldsymbol{v},\boldsymbol{C})\|^2_{\mathcal{X}},$$

where $c_{\Omega,1}$ and $c_{\Omega,3}$ are respectively defined in (3.37) and (3.41).

(ii) Continuity of $\mathcal{C}(\cdot,\cdot,\cdot)$: let $(\boldsymbol{u},\boldsymbol{B})$, $(\boldsymbol{v},\boldsymbol{C})$ and $(\boldsymbol{w},\boldsymbol{D})$ be in $\mathcal{X}$. Then

$$\begin{aligned}
&\mathcal{C}((\boldsymbol{u},\boldsymbol{B}),(\boldsymbol{v},\boldsymbol{C}),(\boldsymbol{w},\boldsymbol{D}))\\
&\quad\leq \int_\Omega (|\boldsymbol{u}|\,|\boldsymbol{\nabla v}|\,|\boldsymbol{w}| + S|\boldsymbol{B}|\,|\mathbf{curl}\,\boldsymbol{C}|\,|\boldsymbol{w}| + S|\boldsymbol{v}|\,|\boldsymbol{B}|\,|\mathbf{curl}\,\boldsymbol{D}|)\\
&\quad\leq c\max(1,S)(\|\boldsymbol{u}\|_1\|\boldsymbol{v}\|_1\|\boldsymbol{w}\|_1 + \|\boldsymbol{C}\|_1\|\boldsymbol{B}\|_1\|\boldsymbol{w}\|_1 + \|\boldsymbol{v}\|_1\|\boldsymbol{B}\|_1\|\boldsymbol{D}\|_1)\\
&\quad\leq c\max(1,S)[\|\boldsymbol{u}\|_1\|\boldsymbol{v}\|_1\|\boldsymbol{w}\|_1 + \|\boldsymbol{B}\|_1(\|(\boldsymbol{v},\boldsymbol{C})\|_{\mathcal{X}}\|(\boldsymbol{w},\boldsymbol{D})\|_{\mathcal{X}})]\\
&\quad\leq c\max(1,S)\|(\boldsymbol{u},\boldsymbol{B})\|_{\mathcal{X}}\|(\boldsymbol{v},\boldsymbol{C})\|_{\mathcal{X}}\|(\boldsymbol{w},\boldsymbol{D})\|_{\mathcal{X}}
\end{aligned}$$

where c denotes various different positive constants only depending on Ω. The Cauchy–Schwarz inequality in $\mathbb{R}^2$ (namely, for all (v,w) and (e,f) in $\mathbb{R}^2$, $ve + wf \leq \sqrt{v^2+w^2}\sqrt{e^2+f^2}$) has been used several times.

(iii) Let $((\boldsymbol{u},\boldsymbol{B}),(\boldsymbol{v},\boldsymbol{C})) \in \mathcal{X}\times\mathcal{X}$. Then

$$\mathcal{C}((\boldsymbol{u},\boldsymbol{B}),(\boldsymbol{v},\boldsymbol{C}),(\boldsymbol{v},\boldsymbol{C})) = \int_\Omega (\boldsymbol{u}\cdot\boldsymbol{\nabla}\frac{\boldsymbol{v}^2}{2} + S(\boldsymbol{B}\times\mathbf{curl}\,\boldsymbol{C}\cdot\boldsymbol{v} - \boldsymbol{v}\times\boldsymbol{B}\cdot\mathbf{curl}\,\boldsymbol{C})).$$

We first note that the last two terms vanish since $\boldsymbol{B}\times\mathbf{curl}\,\boldsymbol{C}\cdot\boldsymbol{v} = \boldsymbol{v}\times\boldsymbol{B}\cdot\mathbf{curl}\,\boldsymbol{C}$. In addition,

$$\int_\Omega \boldsymbol{u}\cdot\boldsymbol{\nabla v}\cdot\boldsymbol{v} = -\frac{1}{2}\int_\Omega |\boldsymbol{v}|^2\mathrm{div}\,\boldsymbol{u} + \frac{1}{2}\int_{\partial\Omega} |\boldsymbol{v}|^2\boldsymbol{u}\cdot\boldsymbol{n},$$

thus $\mathcal{C}((\boldsymbol{u},\boldsymbol{B}),(\boldsymbol{v},\boldsymbol{C}),(\boldsymbol{v},\boldsymbol{C})) = 0$ as long as $\mathrm{div}\,\boldsymbol{u} = 0$ and $\boldsymbol{u}\cdot\boldsymbol{n} = 0$ on $\partial\Omega$. The skew-symmetry property follows by linearity. □

Remark 3.3.1 Let us make precise here what remains of the above results if the domain Ω does not satisfy (3.39). More specifically, we suppose here that Ω is a polyhedron, not necessarily convex. In this case, we also have the coercivity of $\mathcal{A}$ on $\mathcal{X}\times\mathcal{X}$. In particular, the following problem: find $\boldsymbol{B}\in\mathbb{H}^1_n(\Omega)$ such that, $\forall\,\boldsymbol{C}\in\mathbb{H}^1_n(\Omega)$,

$$\int_\Omega \mathbf{curl}\,\boldsymbol{B}\cdot\mathbf{curl}\,\boldsymbol{C} + \int_\Omega \mathrm{div}\,\boldsymbol{B}\,\mathrm{div}\,\boldsymbol{C} = \int_\Omega \boldsymbol{h}\cdot\boldsymbol{C},$$

still admits a solution.

The importance of the convexity (or the regularity of the boundary) comes from the following fact: for a general domain Ω, $\mathcal{H}(\Omega)$ is embedded in a space of functions which are *locally* in $\mathbb{H}^1$, but $\mathcal{H}(\Omega)$ is not embedded in $\mathbb{H}^1(\Omega)$. Actually, one can prove that for a polyhedral domain Ω, $\mathcal{H}(\Omega)$ is embedded in $\mathbb{H}^1(\Omega)$ if and only if Ω is convex. Therefore, if Ω is not convex, the solution of the above problem may not be the physical solution of the real problem, namely: find $\boldsymbol{B} \in \mathcal{H}(\Omega)$ such that, $\forall\, \boldsymbol{C} \in \mathcal{H}(\Omega)$,

$$\int_\Omega \mathbf{curl}\,\boldsymbol{B}\cdot\mathbf{curl}\,\boldsymbol{C} + \int_\Omega \operatorname{div}\boldsymbol{B}\cdot\operatorname{div}\boldsymbol{C} = \int_\Omega \boldsymbol{h}\cdot\boldsymbol{C}.$$

Notice that this problem is well-defined since the bilinear form is coercive on $\mathcal{H}(\Omega)$. As mentioned in Proposition 3.16, another possible assumption to ensure that $\mathcal{H}(\Omega)$ is embedded in $\mathbb{H}^1(\Omega)$ is that Ω has a $\mathcal{C}^{1,1}$ boundary. Note that all these results also hold for other types of boundary conditions (namely those corresponding to the space $H_0(\mathbf{curl}\,;\Omega)\cap H(\operatorname{div}\,;\Omega)$; see Section 3.8). We refer to M. Costabel [41], M. Costabel and M. Dauge [42] and C. Amrouche et al. [4] and to Section 3.4.2. ♢

To apply the abstract theory of Section 3.1, we need an inf–sup condition on the operator $\mathcal{B}$. We show in the next lemma that it is a straightforward consequence of the classical inf–sup condition for the Stokes equations.

Lemma 3.20 *The bilinear form $\mathcal{B}(\cdot,\cdot)$ is continuous on $\mathcal{X}\times\mathcal{M}$ and satisfies the inf–sup condition:*

$$\exists\,\beta>0,\ \ \forall\, q\in\mathcal{M},\ \sup_{(\boldsymbol{v},\boldsymbol{C})\in\mathcal{X}} \frac{\mathcal{B}((\boldsymbol{v},\boldsymbol{C}),q)}{\|(\boldsymbol{v},\boldsymbol{C})\|_{\mathcal{X}}} \geq \beta\|q\|_{\mathcal{M}}.$$

Proof First, let us check the continuity: let $(\boldsymbol{v},\boldsymbol{C})\in\mathcal{X}$ and $q\in\mathcal{M}$. Then

$$|\mathcal{B}((\boldsymbol{v},\boldsymbol{C}),q)| = \left|\int_\Omega q\operatorname{div}\boldsymbol{v}\right| \leq \|q\|_0\|\operatorname{div}\boldsymbol{v}\|_0 \leq c\|q\|_{\mathcal{M}}\|(\boldsymbol{v},\boldsymbol{C})\|_{\mathcal{X}}.$$

Next, we establish the inf–sup condition: let $q\in\mathcal{M}$. Then

$$\begin{aligned}\sup_{(\boldsymbol{v},\boldsymbol{C})\in\mathcal{X}} \frac{\mathcal{B}((\boldsymbol{v},\boldsymbol{C}),q)}{\|(\boldsymbol{v},\boldsymbol{C})\|} &\geq \sup_{(\boldsymbol{v},\boldsymbol{C})\in\mathbb{H}^1_0(\Omega)\times\{0\}} \frac{\mathcal{B}((\boldsymbol{v},\boldsymbol{C}),q)}{\|(\boldsymbol{v},\boldsymbol{C})\|} = \sup_{\boldsymbol{v}\in\mathbb{H}^1_0(\Omega)} \frac{\int_\Omega q\operatorname{div}\boldsymbol{v}}{\|\boldsymbol{v}\|_1}\\ &\geq \beta\|q\|_{\mathcal{M}},\end{aligned}$$

the last inequality coming from the standard inf–sup condition (3.12). □

The last step of the analysis is to establish the coercivity and compactness properties required by Theorem 3.15.

Lemma 3.21 *We introduce the notation, consistent with Section 3.1.8,*

$$a((\boldsymbol{u},\boldsymbol{B});(\boldsymbol{v},\boldsymbol{C}),(\boldsymbol{w},\boldsymbol{D})) = \mathcal{A}((\boldsymbol{v},\boldsymbol{C}),(\boldsymbol{w},\boldsymbol{D})) + \mathcal{C}((\boldsymbol{u},\boldsymbol{B}),(\boldsymbol{v},\boldsymbol{C}),(\boldsymbol{w},\boldsymbol{D})).$$

Then, the following properties hold:

(i) $\forall (\boldsymbol{v},\boldsymbol{C}) \in \mathcal{V}, a((\boldsymbol{v},\boldsymbol{C});(\boldsymbol{v},\boldsymbol{C}),(\boldsymbol{v},\boldsymbol{C})) \geq \alpha\|(\boldsymbol{v},\boldsymbol{C})\|^2_{\mathcal{X}}$, *with* α *defined in (3.51).*

(ii) If $(\boldsymbol{u}_n,\boldsymbol{B}_n)$ *weakly converges to* $(\boldsymbol{u},\boldsymbol{B})$ *in* $\mathcal{V}$ *as* $n\to\infty$ *then for all* $(\boldsymbol{v},\boldsymbol{C}) \in \mathcal{X}\times\mathcal{X}$,

$$a((\boldsymbol{u}_n,\boldsymbol{B}_n);(\boldsymbol{u}_n,\boldsymbol{B}_n),(\boldsymbol{v},\boldsymbol{C})) \longrightarrow a((\boldsymbol{u},\boldsymbol{B});(\boldsymbol{u},\boldsymbol{B}),(\boldsymbol{v},\boldsymbol{C})) \text{ as } n\to\infty.$$

(iii) There exists a constant $\gamma > 0$, *which only depends on the domain* Ω, *such that, for all* $(\boldsymbol{u}_1,\boldsymbol{B}_1),(\boldsymbol{u}_2,\boldsymbol{B}_2),(\boldsymbol{v},\boldsymbol{C})$ *and* $(\boldsymbol{w},\boldsymbol{D})$ *in* $\mathcal{X}$,

$$\begin{aligned}&|a((\boldsymbol{u}_2,\boldsymbol{B}_2);(\boldsymbol{v},\boldsymbol{C}),(\boldsymbol{w},\boldsymbol{D})) - a((\boldsymbol{u}_1,\boldsymbol{B}_1);(\boldsymbol{v},\boldsymbol{C}),(\boldsymbol{w},\boldsymbol{D}))|\\ &\leq \gamma\max(1,S)\|(\boldsymbol{u}_2,\boldsymbol{B}_2)-(\boldsymbol{u}_1,\boldsymbol{B}_1)\|_{\mathcal{X}}\|(\boldsymbol{v},\boldsymbol{C})\|_{\mathcal{X}}\|(\boldsymbol{w},\boldsymbol{D})\|_{\mathcal{X}}.\end{aligned}$$

Proof (i) With the properties proved in Lemma 3.19, there exists $\alpha > 0$ such that for all $(\boldsymbol{v},\boldsymbol{C}) \in \mathcal{V}$, $a((\boldsymbol{v},\boldsymbol{C});(\boldsymbol{v},\boldsymbol{C}),(\boldsymbol{v},\boldsymbol{C})) = \mathcal{A}((\boldsymbol{v},\boldsymbol{C}),(\boldsymbol{v},\boldsymbol{C})) \geq \alpha\|(\boldsymbol{v},\boldsymbol{C})\|^2_{\mathcal{X}}$.

(ii) Let $(\boldsymbol{u}_n,\boldsymbol{B}_n)$ be a sequence which weakly converges to $(\boldsymbol{u},\boldsymbol{B})$ in $\mathcal{V}$. Then

$$\begin{aligned}&|a((\boldsymbol{u}_n,\boldsymbol{B}_n);(\boldsymbol{u}_n,\boldsymbol{B}_n),(\boldsymbol{v},\boldsymbol{C})) - a((\boldsymbol{u},\boldsymbol{B});(\boldsymbol{u},\boldsymbol{B}),(\boldsymbol{v},\boldsymbol{C}))|\\ &\leq |\mathcal{A}((\boldsymbol{u}_n-\boldsymbol{u},\boldsymbol{B}_n-\boldsymbol{B}),(\boldsymbol{v},\boldsymbol{C}))| + |\mathcal{C}((\boldsymbol{u}_n-\boldsymbol{u},\boldsymbol{B}_n-\boldsymbol{B}),(\boldsymbol{u}_n,\boldsymbol{B}_n),(\boldsymbol{v},\boldsymbol{C}))|\\ &\quad + |\mathcal{C}((\boldsymbol{u},\boldsymbol{B}),(\boldsymbol{u}_n-\boldsymbol{u},\boldsymbol{B}_n-\boldsymbol{B}),(\boldsymbol{v},\boldsymbol{C}))|.\end{aligned} \tag{3.52}$$

The first term in the right-hand side of this inequality reads

$$\begin{aligned}&\mathcal{A}((\boldsymbol{u}_n-\boldsymbol{u},\boldsymbol{B}_n-\boldsymbol{B}),(\boldsymbol{v},\boldsymbol{C}))\\ &= \frac{1}{Re}\int_\Omega \boldsymbol{\nabla}(\boldsymbol{u}_n-\boldsymbol{u}) : \boldsymbol{\nabla}\boldsymbol{v}\\ &\quad + \frac{S}{Rm}\int_\Omega \mathbf{curl}\,(\boldsymbol{B}_n-\boldsymbol{B})\cdot\mathbf{curl}\,\boldsymbol{C} + \frac{S}{Re}\int_\Omega \mathrm{div}\,(\boldsymbol{B}_n-\boldsymbol{B})\mathrm{div}\,\boldsymbol{C}.\end{aligned}$$

The weak convergence of $(\boldsymbol{u}_n,\boldsymbol{B}_n)$ is enough to pass to the limit in these linear terms. The second term of the right-hand side of the inequality is slightly more involved due to nonlinearities.

$$\begin{aligned}&\mathcal{C}\,((\boldsymbol{u}_n-\boldsymbol{u},\boldsymbol{B}_n-\boldsymbol{B}),(\boldsymbol{u}_n,\boldsymbol{B}_n),(\boldsymbol{v},\boldsymbol{C}))\\ &= +\int_\Omega (\boldsymbol{u}_n-\boldsymbol{u})\cdot\boldsymbol{\nabla}\boldsymbol{u}_n\cdot\boldsymbol{v}\\ &\quad + S\int_\Omega (\boldsymbol{B}_n-\boldsymbol{B})\times\mathbf{curl}\,\boldsymbol{B}_n\cdot\boldsymbol{v} + S\int_\Omega (\boldsymbol{B}_n-\boldsymbol{B})\times\boldsymbol{u}_n\cdot\mathbf{curl}\,\boldsymbol{C}.\end{aligned}$$

Using the compact embedding of $\mathbb{H}^1(\Omega)$ in $\mathbb{L}^4(\Omega)$, we infer from the weak convergence of $(\boldsymbol{u}_n,\boldsymbol{B}_n)$ to $(\boldsymbol{u},\boldsymbol{B})$, the strong convergence (up to an extraction)

$$(\boldsymbol{u}_n,\boldsymbol{B}_n) \to (\boldsymbol{u},\boldsymbol{B}) \text{ in } \mathbb{L}^4(\Omega)\times\mathbb{L}^4(\Omega).$$

Thus, using the following inequalities,

$$\left|\int_\Omega (\boldsymbol{u}-\boldsymbol{u}_n)\cdot\boldsymbol{\nabla}\boldsymbol{u}_n\cdot\boldsymbol{v}\right| \le \|\boldsymbol{u}-\boldsymbol{u}_n\|_{\mathbb{L}^4}\|\boldsymbol{\nabla}\boldsymbol{u}_n\|_0\|\boldsymbol{v}\|_{\mathbb{L}^4},$$

$$\left|\int_\Omega (\boldsymbol{B}_n-\boldsymbol{B})\times\mathbf{curl}\,\boldsymbol{B}_n\cdot\boldsymbol{v}\right| \le C\|\boldsymbol{B}_n-\boldsymbol{B}\|_{\mathbb{L}^4}\|\boldsymbol{\nabla}\boldsymbol{B}_n\|_0\|\boldsymbol{v}\|_{\mathbb{L}^4},$$

$$\left|\int_\Omega (\boldsymbol{B}_n-\boldsymbol{B})\times\boldsymbol{u}_n\cdot\mathbf{curl}\,\boldsymbol{C}\right| \le \|\boldsymbol{B}_n-\boldsymbol{B}\|_{\mathbb{L}^4}\|\boldsymbol{u}_n\|_{\mathbb{L}^4}\|\mathbf{curl}\,\boldsymbol{C}\|_0,$$

we can conclude that the second term of the right-hand side of (3.52) goes to zero. The third term,

$$\begin{aligned}&\mathcal{C}\left((\boldsymbol{u},\boldsymbol{B}),(\boldsymbol{u}_n-\boldsymbol{u},\boldsymbol{B}_n-\boldsymbol{B}),(\boldsymbol{v},\boldsymbol{C})\right)\\ &=\int_\Omega \boldsymbol{u}\cdot\boldsymbol{\nabla}(\boldsymbol{u}-\boldsymbol{u}_n)\cdot\boldsymbol{v}\\ &\quad+S\int_\Omega \boldsymbol{B}\times\mathbf{curl}\,(\boldsymbol{B}_n-\boldsymbol{B})\cdot\boldsymbol{v}+S\int_\Omega \boldsymbol{B}\times(\boldsymbol{u}_n-\boldsymbol{u})\cdot\mathbf{curl}\,\boldsymbol{C},\end{aligned}$$

can be treated with similar arguments:

$$\left|\int_\Omega \boldsymbol{u}\cdot\boldsymbol{\nabla}(\boldsymbol{u}-\boldsymbol{u}_n)\cdot\boldsymbol{v}\right| = \left|\int_\Omega \boldsymbol{u}\cdot\boldsymbol{\nabla}\boldsymbol{v}\cdot(\boldsymbol{u}-\boldsymbol{u}_n)\right| \le \|\boldsymbol{u}\|_{\mathbb{L}^4}\|\boldsymbol{u}-\boldsymbol{u}_n\|_{\mathbb{L}^4}\|\boldsymbol{\nabla}\boldsymbol{v}\|_0,$$

$$\left|\int_\Omega \boldsymbol{B}\times(\boldsymbol{u}_n-\boldsymbol{u})\cdot\mathbf{curl}\,\boldsymbol{C}\right| \le \|\boldsymbol{B}\|_{\mathbb{L}^4}\|\boldsymbol{u}_n-\boldsymbol{u}\|_{\mathbb{L}^4}\|\mathbf{curl}\,\boldsymbol{C}\|_0.$$

The remaining term $S\int_\Omega \boldsymbol{B}\times\mathbf{curl}\,(\boldsymbol{B}_n-\boldsymbol{B})\cdot\boldsymbol{v}$ goes to zero because of the weak convergence of $\boldsymbol{B}_n$ to $\boldsymbol{B}$ in $\mathbb{H}^1(\Omega)$.

(iii) Let $(\boldsymbol{u}_1,\boldsymbol{B}_1),(\boldsymbol{u}_2,\boldsymbol{B}_2),(\boldsymbol{v},\boldsymbol{C})$ and $(\boldsymbol{w},\boldsymbol{D})$ be in $\mathcal{X}$. Using again the continuous embedding of $\mathbb{H}^1(\Omega)$ in $\mathbb{L}^4(\Omega)$,

$$\begin{aligned}&|a((\boldsymbol{u}_2,\boldsymbol{B}_2);(\boldsymbol{v},\boldsymbol{C}),(\boldsymbol{w},\boldsymbol{D}))-a((\boldsymbol{u}_1,\boldsymbol{B}_1);(\boldsymbol{v},\boldsymbol{C}),(\boldsymbol{w},\boldsymbol{D}))|\\ &=\left|\int_\Omega [(\boldsymbol{u}_2-\boldsymbol{u}_1)\cdot\boldsymbol{\nabla}\boldsymbol{v}\cdot\boldsymbol{w}-S\mathbf{curl}\,\boldsymbol{C}\times(\boldsymbol{B}_2-\boldsymbol{B}_1)\cdot\boldsymbol{w}-S\boldsymbol{v}\times(\boldsymbol{B}_2-\boldsymbol{B}_1)\cdot\mathbf{curl}\,\boldsymbol{D}]\right|\\ &\le c(\|\boldsymbol{u}_2-\boldsymbol{u}_1\|_1\|\boldsymbol{v}\|_1\|\boldsymbol{w}\|_1+S\|\boldsymbol{C}\|_1\|\boldsymbol{B}_2-\boldsymbol{B}_1\|_1\|\boldsymbol{w}\|_1+S\|\boldsymbol{v}\|_1\|\boldsymbol{B}_2-\boldsymbol{B}_1\|_1\|\boldsymbol{D}\|_1)\\ &\le\gamma\max(1,S)\|(\boldsymbol{u}_2,\boldsymbol{B}_2)-(\boldsymbol{u}_1,\boldsymbol{B}_1)\|_{\mathcal{X}}\|(\boldsymbol{v},\boldsymbol{C})\|_{\mathcal{X}}\|(\boldsymbol{w},\boldsymbol{D})\|_{\mathcal{X}},\end{aligned}$$

where c and γ denote constants depending only on Ω. □

We now state the main result of this section (this result has already been briefly sketched in Section 2.2.4.3).

Theorem 3.22 *The variational formulation (3.46) of the stationary MHD system admits a solution* $(\boldsymbol{u}, \boldsymbol{B}, p) \in \mathbb{H}_0^1(\Omega) \times \mathbb{H}_n^1(\Omega) \times L_0^2(\Omega)$ *that satisfies in particular*

$$\|(\boldsymbol{u}, \boldsymbol{B})\|_{\mathcal{X}} \leq \frac{\|\mathcal{L}\|_{\mathcal{V}'}}{\alpha},$$

where $\mathcal{L}$ *is defined in (3.50) and* α *in (3.51). If, in addition,*

$$\frac{\gamma \max(1, S)\|\mathcal{L}\|_{\mathcal{V}'}}{\alpha^2} < 1,$$

where γ *is a constant depending only on* Ω *defined in Lemma 3.21, then the solution is unique.*

Proof This theorem straightforwardly results from Lemmata 3.19, 3.20, 3.21, and Theorem 3.15. □

3.3.2 A formulation for non-convex polyhedra

In the previous section we have presented a formulation for the MHD equations in domains which are either convex or with a boundary of class $\mathcal{C}^{1,1}$. The purpose of this section is to present a formulation which is valid under the following assumption (irrespective of the convexity of the domain):

$$\Omega \text{ is a Lipschitz polyhedron.} \tag{3.53}$$

In the cases, relevant for applications, when Ω is a non-convex polyhedron, the difficulty comes from the fact that the magnetic field is in general not in $\mathbb{H}^1(\Omega)$ (see Remark 3.3.1 above). The approximation of the magnetic field with classical $\mathbb{H}^1$-conforming Lagrange finite elements may miss singularities, present in the physical solution, due to re-entrant corners. It is worth emphasizing that, in such domains, the numerical approximation of the magnetic field with standard Lagrange finite element may *look stable* (since the $\int \mathbf{curl} \cdot \mathbf{curl} \cdot + \int \operatorname{div} \cdot \operatorname{div} \cdot$ operator is coercive, see Remark 3.3.1) even if it is *not*! This problem is the main motivation of the formulation (due to D. Schötzau [205]) that we now present and that will be discretized in Section 3.4.2 with the Nédélec finite elements.

Let us just mention that a formulation, based on a technique called *weight regularization*, has recently been proposed by M. Costabel and M. Dauge [43] to overcome this difficulty while keeping Lagrange finite elements. We also refer to U. Hasler *et al.* [120] for an application of these ideas in the context of MHD.

We begin with the following technical result that will be useful in the sequel.

Proposition 3.23 *There exists a parameter* $\varepsilon_1 > 0$*, which only depends on* Ω*, such that* $\mathcal{H}(\Omega)$ *is compactly embedded into* $\mathbb{L}^{3+\varepsilon_1}(\Omega)$.

Proof The result, which can be found in [205], is a direct consequence of the Sobolev embedding theorem and of the following property proved in [4]: there exists $s > 1/2$ such that $\mathcal{H}(\Omega)$ is continuously embedded into $\mathbb{H}^s(\Omega)$.

Recall that the spaces $\mathcal{H}(\Omega)$ and V are defined in (3.38) and (3.43), respectively. We introduce the following spaces:

$$\tilde{\mathcal{X}} = \mathbb{H}_0^1(\Omega) \times H(\mathbf{curl}\,,\Omega), \quad \text{and} \quad \tilde{\mathcal{V}} = V \times (\mathcal{H}(\Omega) \cap H(\mathrm{div}^{\,0},\Omega)), \tag{3.54}$$

endowed with the norm $\|(\boldsymbol{v},\boldsymbol{C})\|_{\tilde{\mathcal{X}}} = (\|\boldsymbol{v}\|_1^2 + \|\boldsymbol{C}\|_{\mathrm{curl}}^2)^{1/2}$, where $\|\boldsymbol{C}\|_{\mathrm{curl}}^2 = \|\boldsymbol{C}\|_0^2 + \|\mathbf{curl}\,\boldsymbol{C}\|_0^2$, and

$$\tilde{\mathcal{M}} = L_0^2(\Omega) \times H^1(\Omega)/\mathbb{R}, \tag{3.55}$$

endowed with the norm $\|(q,s)\|_{\tilde{\mathcal{M}}} = (\|q\|_0^2 + \|s\|_1^2)^{1/2}$. Comparing those definitions with (3.44) and (3.45), we notice in particular that the magnetic field how belongs to $H(\mathbf{curl}\,,\Omega)$ instead of $\mathbb{H}_n^1(\Omega)$.

We then consider the following problem (analogous to (3.46)):

Find $((\boldsymbol{u},\boldsymbol{B}),(p,r)) \in \tilde{\mathcal{X}} \times \tilde{\mathcal{M}}$ *such that for all* $((\boldsymbol{v},\boldsymbol{C}),(q,s)) \in \tilde{\mathcal{X}} \times \tilde{\mathcal{M}}$,

$$\begin{aligned} \tilde{\mathcal{A}}((\boldsymbol{u},\boldsymbol{B}),(\boldsymbol{v},\boldsymbol{C})) + \mathcal{C}((\boldsymbol{u},\boldsymbol{B}),(\boldsymbol{u},\boldsymbol{B}),(\boldsymbol{v},\boldsymbol{C})) + \tilde{\mathcal{B}}((\boldsymbol{v},\boldsymbol{C}),(p,r)) &= \mathcal{L}(\boldsymbol{v},\boldsymbol{C}),\\ \tilde{\mathcal{B}}((\boldsymbol{u},\boldsymbol{B}),(q,s)) &= 0, \end{aligned} \tag{3.56}$$

where the following operators have been introduced:

$$\tilde{\mathcal{A}}((\boldsymbol{u},\boldsymbol{B});(\boldsymbol{v},\boldsymbol{C})) = \frac{1}{Re}\int_\Omega \boldsymbol{\nabla}\boldsymbol{u} : \boldsymbol{\nabla}\boldsymbol{v} + \frac{S}{Rm}\int_\Omega \mathbf{curl}\,\boldsymbol{B}\cdot\mathbf{curl}\,\boldsymbol{C}, \tag{3.57}$$

$$\tilde{\mathcal{B}}((\boldsymbol{v},\boldsymbol{C}),(q,s)) = -\int_\Omega q\,\mathrm{div}\,\boldsymbol{v} - \int_\Omega \boldsymbol{\nabla} s\cdot\boldsymbol{C}. \tag{3.58}$$

The symbol ˜ aims to distinguish these operators from their counterparts defined in the previous section for convex domains (see (3.47) and (3.48)). The operators $\mathcal{C}$ and $\mathcal{L}$ are defined as in the previous case by (3.49) and (3.50). Note that, contrary to the formulation of Section 3.3.1, a Lagrange multiplier r corresponding to the divergence-free constraint on $\boldsymbol{B}$ has been introduced. In other words, r plays for $\boldsymbol{B}$ the role of p for $\boldsymbol{u}$.

Before analyzing this formulation, we first check that it indeed defines a solution of the stationary MHD equations. □

Proposition 3.24 *If (3.36) holds and if* $((\boldsymbol{u},\boldsymbol{B}),(p,r)) \in \tilde{\mathcal{X}} \times \tilde{\mathcal{M}}$ *is a solution of problem (3.56) then* $((\boldsymbol{u},\boldsymbol{B}),p)$ *is a solution of (3.34)–(3.35).*

Proof It is standard to recover the first and the second equations of (3.34) from (3.56) by taking $\boldsymbol{C} = 0$ and $s = 0$. Next, taking $\boldsymbol{v} = 0$, we obtain:

$$\frac{S}{Rm}\int_\Omega \mathbf{curl}\,\boldsymbol{B}\cdot\mathbf{curl}\,\boldsymbol{C} - S\int_\Omega \boldsymbol{u}\times\boldsymbol{B}\cdot\mathbf{curl}\,\boldsymbol{C} - \int_\Omega \boldsymbol{\nabla} r\cdot\boldsymbol{C} = S\int_\Omega \boldsymbol{h}\cdot\boldsymbol{C},$$

for all $\boldsymbol{C} \in H(\mathbf{curl}\,,\Omega)$. In particular, since $r \in H^1(\Omega)/\mathbb{R}$, we can take $\boldsymbol{C} = \boldsymbol{\nabla} r$ as a test function. This gives:

$$-\int_\Omega |\boldsymbol{\nabla} r|^2 = \int_\Omega \boldsymbol{h}\cdot\boldsymbol{\nabla} r = 0$$

which yields $r = 0$ (in $H^1(\Omega)/\mathbb{R}$). In other words, the "pseudo-pressure" which appears in the magnetic equation as the Lagrange multiplier corresponding the constraint on $\operatorname{div} \boldsymbol{B}$, is in fact zero. We then easily obtain the third equation of (3.34). Note that without the assumption (3.36) on $\boldsymbol{h}$, this non-physical "pressure" would remain in the magnetic equation.

It remains to prove that $\boldsymbol{B}$ is divergence-free. For this purpose, we recall the Helmholtz decomposition (see, *e.g.* [106]):

$$\mathbb{L}^2(\Omega) = H_0(div^0, \Omega) \oplus \boldsymbol{\nabla} H^1(\Omega).$$

This allows us to write

$$\boldsymbol{B} = \boldsymbol{B}_0 + \boldsymbol{\nabla} s_0,$$

with $\operatorname{div} \boldsymbol{B}_0 = 0$ and $s_0 \in H^1(\Omega)$. Taking $q = 0$, the second equation of (3.56) gives

$$\int_\Omega \boldsymbol{B} \cdot \boldsymbol{\nabla} s = 0,$$

for all $s \in H^1(\Omega)/\mathbb{R}$. In particular,

$$\int_\Omega \boldsymbol{B}_0 \cdot \boldsymbol{\nabla} s_0 + \int_\Omega |\boldsymbol{\nabla} s_0|^2 = 0.$$

The first term of the left-hand side vanishes by integration by parts. Thus $\boldsymbol{\nabla} s_0 = 0$ a.e. and $\boldsymbol{B} = \boldsymbol{B}_0$, which proves that $\operatorname{div} \boldsymbol{B} = 0$ and $\boldsymbol{B} \cdot \boldsymbol{n} = 0$.

We now proceed to the proof of the existence of a solution for the formulation (3.56). For this purpose, following the same lines as in the previous section, we collect in three lemmata the properties of the forms $\tilde{\mathcal{A}}, \tilde{\mathcal{B}}$ and $\mathcal{C}$ that will allow us to apply Theorem 3.15. □

Lemma 3.25 *The following properties hold:*

(i) The bilinear form $\tilde{\mathcal{A}}(\cdot,\cdot)$ is continuous on $\tilde{\mathcal{X}} \times \tilde{\mathcal{X}}$ and coercive on $\tilde{\mathcal{V}} \times \tilde{\mathcal{V}}$ with a coercivity constant given by

$$\alpha = \min\left(\frac{c_{\Omega,1}}{Re}, \frac{(1+c_{\Omega,2})S}{2Rm}\right). \tag{3.59}$$

(ii) Let $(\boldsymbol{u}, \boldsymbol{B}) \in \tilde{\mathcal{X}}$. The bilinear form $\mathcal{C}((\boldsymbol{u}, \boldsymbol{B}), \cdot, \cdot)$ is continuous on $\tilde{\mathcal{X}} \times \tilde{\mathcal{X}}$. If moreover $div \boldsymbol{B} = 0$, there exists a constant $c > 0$ such that

$$|\mathcal{C}((\boldsymbol{u}, \boldsymbol{B}), (\boldsymbol{v}, \boldsymbol{C}), (\boldsymbol{w}, \boldsymbol{D}))| \leq c \|(\boldsymbol{u}, \boldsymbol{B})\|_{\tilde{\mathcal{X}}} \|(\boldsymbol{v}, \boldsymbol{C})\|_{\tilde{\mathcal{X}}} \|(\boldsymbol{w}, \boldsymbol{D})\|_{\tilde{\mathcal{X}}}.$$

(iii) If $((\boldsymbol{u}, \boldsymbol{B}), (\boldsymbol{v}, \boldsymbol{C}), (\boldsymbol{w}, \boldsymbol{D})) \in \tilde{\mathcal{X}} \times \tilde{\mathcal{X}} \times \tilde{\mathcal{X}}$ with $div \boldsymbol{u} = 0$, then

$$\mathcal{C}((\boldsymbol{u}, \boldsymbol{B}), (\boldsymbol{v}, \boldsymbol{C}), (\boldsymbol{w}, \boldsymbol{D})) = -\mathcal{C}((\boldsymbol{u}, \boldsymbol{B}), (\boldsymbol{w}, \boldsymbol{D}), (\boldsymbol{v}, \boldsymbol{C})).$$

Remark 3.3.2 Notice that $\tilde{\mathcal{A}}(\cdot,\cdot)$ is coercive on $\tilde{\mathcal{V}} \times \tilde{\mathcal{V}}$ but not on $\tilde{\mathcal{X}} \times \tilde{\mathcal{X}}$ in general. This property is sufficient to apply the general theory presented in Section 3.1. In contrast, the bilinear form $\mathcal{A}(\cdot,\cdot)$ chosen for in convex domains is coercive on the whole space $\mathcal{X} \times \mathcal{X}$ (Lemma 3.19). ◇

Proof (i) We first prove the continuity of $\tilde{\mathcal{A}}(\cdot,\cdot)$: let $(\boldsymbol{u},\boldsymbol{B})$ and $(\boldsymbol{v},\boldsymbol{C})$ be in $\tilde{\mathcal{X}}$. Then

$$\begin{aligned}\tilde{\mathcal{A}}((\boldsymbol{u},\boldsymbol{B}),(\boldsymbol{v},\boldsymbol{C})) &= \frac{1}{Re}\int_\Omega \boldsymbol{\nabla u} : \boldsymbol{\nabla v} + \frac{S}{Rm}\int_\Omega \mathbf{curl}\,\boldsymbol{B}\cdot\mathbf{curl}\,\boldsymbol{C}\\ &\le \frac{1}{Re}\|\boldsymbol{u}\|_1\|\boldsymbol{v}\|_1 + \frac{S}{Rm}\|\boldsymbol{B}\|_{\mathrm{curl}}\|\boldsymbol{C}\|_{\mathrm{curl}}\\ &\le \max\left(\frac{1}{Re},\frac{S}{Rm}\right)\|(\boldsymbol{u},\boldsymbol{B})\|_{\tilde{\mathcal{X}}}\|(\boldsymbol{v},\boldsymbol{C})\|_{\tilde{\mathcal{X}}}.\end{aligned}$$

Let us now check the coercivity of $\tilde{\mathcal{A}}(\cdot,\cdot)$: let $(\boldsymbol{v},\boldsymbol{C})\in\tilde{\mathcal{V}}$. Then

$$\begin{aligned}\tilde{\mathcal{A}}((\boldsymbol{v},\boldsymbol{C}),(\boldsymbol{v},\boldsymbol{C})) &= \frac{1}{Re}\int_\Omega |\boldsymbol{\nabla v}|^2 + \frac{S}{Rm}\int_\Omega |\mathbf{curl}\,\boldsymbol{C}|^2\\ &\ge \min\left(\frac{c_{\Omega,1}}{Re},\frac{(1+c_{\Omega,2})S}{2Rm}\right)\|(\boldsymbol{v},\boldsymbol{C})\|^2_{\tilde{\mathcal{X}}},\end{aligned}$$

where $c_{\Omega,1}$ and $c_{\Omega,2}$ are, respectively, defined in (3.37) and (3.40). Note that we have used the fact that $\boldsymbol{C}\in\tilde{\mathcal{V}}$ is divergence-free.

(ii) Let $(\boldsymbol{u},\boldsymbol{B}),(\boldsymbol{v},\boldsymbol{C})$ and $(\boldsymbol{w},\boldsymbol{D})$ be in $\tilde{\mathcal{X}}$. Using the Hölder inequality and the embedding $\mathbb{H}^1(\Omega)\subset\mathbb{L}^6(\Omega)$,

$$\begin{aligned}&\mathcal{C}((\boldsymbol{u},\boldsymbol{B}),(\boldsymbol{v},\boldsymbol{C}),(\boldsymbol{w},\boldsymbol{D}))\\ &\le \int_\Omega (|\boldsymbol{u}|\,|\boldsymbol{\nabla v}|\,|\boldsymbol{w}| + S|\boldsymbol{B}|\,|\mathbf{curl}\,\boldsymbol{C}|\,|\boldsymbol{w}| + S|\boldsymbol{v}|\,|\boldsymbol{B}|\,|\mathbf{curl}\,\boldsymbol{D}|)\\ &\le c(\|\boldsymbol{u}\|_1\|\boldsymbol{v}\|_1\|\boldsymbol{w}\|_1 + S\|\boldsymbol{C}\|_{\mathrm{curl}}\|\boldsymbol{B}\|_{L^{3+\varepsilon_1}}\|\boldsymbol{w}\|_{L^{6-\varepsilon_2}}\\ &\quad + S\|\boldsymbol{v}\|_{L^{6-\varepsilon_2}}\|\boldsymbol{B}\|_{L^{3+\varepsilon_1}}\|\boldsymbol{D}\|_{\mathrm{curl}})\\ &\le c\max(1,S)(\|\boldsymbol{u}\|_1\|\boldsymbol{v}\|_1\|\boldsymbol{w}\|_1 + \|\boldsymbol{C}\|_{\mathrm{curl}}\|\boldsymbol{B}\|_{L^{3+\varepsilon_1}}\|\boldsymbol{w}\|_1\\ &\quad + \|\boldsymbol{v}\|_1\|\boldsymbol{B}\|_{L^{3+\varepsilon_1}}\|\boldsymbol{D}\|_{\mathrm{curl}}),\end{aligned}$$

where c denotes various different positive constants only depending on Ω. The small parameter $\varepsilon_1>0$ is given by Proposition 3.23, and $\varepsilon_2>0$ is such that

$$\frac{1}{3+\varepsilon_1}+\frac{1}{6-\varepsilon_2}=\frac{1}{2}.$$

Thus, there exists $c_{\boldsymbol{u},\boldsymbol{B}}>0$ such that

$$\mathcal{C}((\boldsymbol{u},\boldsymbol{B}),(\boldsymbol{v},\boldsymbol{C}),(\boldsymbol{w},\boldsymbol{D}))\le c_{\boldsymbol{u},\boldsymbol{B}}\|(\boldsymbol{v},\boldsymbol{C})\|_{\tilde{\mathcal{X}}}\|(\boldsymbol{w},\boldsymbol{D})\|_{\tilde{\mathcal{X}}},$$

which proves the continuity of the bilinear form $\mathcal{C}((\boldsymbol{u},\boldsymbol{B}),\cdot,\cdot)$ on $\tilde{\mathcal{X}}\times\tilde{\mathcal{X}}$. If we moreover assume that $\operatorname{div}\boldsymbol{B}=0$ (as it is the case in particular when $(\boldsymbol{u},\boldsymbol{B})\in\tilde{\mathcal{V}}$), then, using again the embedding $\mathcal{H}(\Omega)\subset\mathbb{L}^{3+\varepsilon_1}(\Omega)$ (see Proposition 3.23),

$$\begin{aligned}
&\mathcal{C}((\boldsymbol{u},\boldsymbol{B}),(\boldsymbol{v},\boldsymbol{C}),(\boldsymbol{w},\boldsymbol{D}))\\
&\le c\max(1,S)(\|\boldsymbol{u}\|_1\|\boldsymbol{v}\|_1\|\boldsymbol{w}\|_1+\|\boldsymbol{C}\|_{\text{curl}}\|\boldsymbol{B}\|_{\text{curl}}\|\boldsymbol{w}\|_1+\|\boldsymbol{v}\|_1\|\boldsymbol{B}\|_{\text{curl}}\|\boldsymbol{D}\|_{\text{curl}})\\
&\le c\max(1,S)\|(\boldsymbol{u},\boldsymbol{B})\|_{\tilde{\mathcal{X}}}\|(\boldsymbol{v},\boldsymbol{C})\|_{\tilde{\mathcal{X}}}\|(\boldsymbol{w},\boldsymbol{D})\|_{\tilde{\mathcal{X}}}.
\end{aligned}$$

Note that the assumption $\operatorname{div}\boldsymbol{B}=0$ is essential for this proof.

(iii) The proof of this property follows the lines of that in Lemma 3.19. □

We now establish an inf–sup condition for the bilinear form $\tilde{\mathcal{B}}$. Note that this property is more involved than for the form $\mathcal{B}$ of the previous section. Indeed, due to the introduction of a magnetic Lagrange multiplier, it does not only rely upon the sole inf–sup condition of the Stokes problem.

Lemma 3.26 *The bilinear form $\tilde{\mathcal{B}}(\cdot,\cdot)$ is continuous on $\tilde{\mathcal{X}}\times\tilde{\mathcal{M}}$ and satisfies the inf–sup condition:*

$$\exists\beta>0,\ \forall(q,s)\in\tilde{\mathcal{M}},\ \sup_{(\boldsymbol{v},\boldsymbol{C})\in\tilde{\mathcal{X}}}\frac{\tilde{\mathcal{B}}((\boldsymbol{v},\boldsymbol{C}),(q,s))}{\|(\boldsymbol{v},\boldsymbol{C})\|_{\tilde{\mathcal{X}}}}\ge\beta\|(q,s)\|_{\tilde{\mathcal{M}}}. \tag{3.60}$$

Proof We first prove the continuity of $\tilde{\mathcal{B}}$: let $(\boldsymbol{v},\boldsymbol{C})\in\tilde{\mathcal{X}}$ and $(q,s)\in\tilde{\mathcal{M}}$. Then

$$\begin{aligned}
\left|\tilde{\mathcal{B}}((\boldsymbol{v},\boldsymbol{C}),(q,s))\right|&\le\left|\int_\Omega q\operatorname{div}\boldsymbol{v}\right|+\left|\int_\Omega\boldsymbol{C}\cdot\boldsymbol{\nabla}s\right|\\
&\le\|q\|_0\|\boldsymbol{v}\|_1+\|\boldsymbol{C}\|_0\|s\|_1\le c\|(q,s)\|_{\tilde{\mathcal{M}}}\|(\boldsymbol{v},\boldsymbol{C})\|_{\tilde{\mathcal{X}}}.
\end{aligned}$$

We next establish the inf–sup condition. Let $(q,s)\in\tilde{\mathcal{M}}$. Using assertion (iii) of Theorem 3.7, it is easy to check that the standard inf–sup condition (see (3.12)),

$$\inf_{q\in L^2_0(\Omega)}\sup_{\boldsymbol{v}\in\mathbb{H}^1_0(\Omega)}\frac{\displaystyle\int_\Omega q\operatorname{div}\boldsymbol{v}}{\|\boldsymbol{v}\|_1\|q\|_0}\ge\beta_1>0, \tag{3.61}$$

implies the existence of $\boldsymbol{v}_0\in\mathbb{H}^1_0(\Omega)$ such that $\int_\Omega q\operatorname{div}\boldsymbol{v}_0=\|q\|_0^2$, and $\beta_1\|\boldsymbol{v}_0\|_1\le\|q\|_0$. We admit for the moment that we also have:

$$\inf_{s\in H^1(\Omega)/\mathbb{R}}\sup_{\boldsymbol{C}\in H(\text{curl},\Omega)}\frac{\displaystyle\int_\Omega\boldsymbol{C}\cdot\boldsymbol{\nabla}s}{\|\boldsymbol{C}\|_{\text{curl}}\|s\|_1}\ge\beta_2>0, \tag{3.62}$$

thus, by the same argument, there exists $\boldsymbol{C}_0\in H(\mathbf{curl},\Omega)$ such that $\int_\Omega\boldsymbol{\nabla}s\cdot\boldsymbol{C}_0=\|s\|_1^2$ and $\beta_2\|\boldsymbol{C}_0\|_{\text{curl}}\le\|s\|_1$. Thus, for all $(q,s)\in\tilde{\mathcal{M}}$, there exists $(\boldsymbol{v}_0,\boldsymbol{C}_0)\in\tilde{\mathcal{X}}$ such that

$$\mathcal{B}((\boldsymbol{v}_0,\boldsymbol{C}_0),(q,s))\ge\|(q,s)\|^2_{\tilde{\mathcal{M}}},$$

with $\|(\boldsymbol{v}_0, \boldsymbol{C}_0)\|^2_{\tilde{X}} \leq \max(\frac{1}{\beta_1^2}, \frac{1}{\beta_2^2})\|(q,s)\|^2_{\tilde{\mathcal{M}}}$. This implies the inf–sup condition (3.60) with

$$\beta = \frac{1}{\sqrt{\max(\beta_1^{-2}, \beta_2^{-2})}}.$$

It remains to prove (3.62). This can be readily done by noticing that if $s \in H^1(\Omega)/\mathbb{R}$ then $\boldsymbol{\nabla} s \in H(\mathbf{curl}, \Omega)$. Thus, using the Poincaré inequality in $H^1(\Omega)/\mathbb{R}$, there exists $\beta_2 > 0$, depending only on Ω, such that

$$\sup_{\boldsymbol{C} \in H(\mathrm{curl}, \Omega)} \frac{\displaystyle\int_\Omega \boldsymbol{C} \cdot \boldsymbol{\nabla} s}{\|s\|_1 \|\boldsymbol{C}\|_{\mathrm{curl}}} \geq \frac{\displaystyle\int_\Omega |\boldsymbol{\nabla} s|^2}{\|s\|_1 \|\boldsymbol{\nabla} s\|_{\mathrm{curl}}} \geq \beta_2,$$

which proves (3.62). □

The coercivity and compactness properties needed to apply Theorem 3.15 are provided by the following lemma.

Lemma 3.27 *We introduce the notation, consistent with Section 3.1.8,*

$$\tilde{a}((\boldsymbol{u}, \boldsymbol{B}); (\boldsymbol{v}, \boldsymbol{C}), (\boldsymbol{w}, \boldsymbol{D})) = \tilde{\mathcal{A}}((\boldsymbol{v}, \boldsymbol{C}), (\boldsymbol{w}, \boldsymbol{D})) + \mathcal{C}((\boldsymbol{u}, \boldsymbol{B}), (\boldsymbol{v}, \boldsymbol{C}), (\boldsymbol{w}, \boldsymbol{D})).$$

Then the following properties hold:

(i) $\forall (\boldsymbol{v}, \boldsymbol{C}) \in \tilde{\mathcal{V}}, \tilde{a}((\boldsymbol{v}, \boldsymbol{C}); (\boldsymbol{v}, \boldsymbol{C}), (\boldsymbol{v}, \boldsymbol{C})) \geq \alpha\|(\boldsymbol{v}, \boldsymbol{C})\|^2_{\tilde{\mathcal{X}}}$, *with* α *defined in (3.59).*

(ii) If $(\boldsymbol{u}_n, \boldsymbol{B}_n)$ *weakly converges to* $(\boldsymbol{u}, \boldsymbol{B})$ *in* $\tilde{\mathcal{V}}$ *then*

$$\tilde{a}((\boldsymbol{u}_n, \boldsymbol{B}_n); (\boldsymbol{u}_n, \boldsymbol{B}_n), (\boldsymbol{v}, \boldsymbol{C})) \longrightarrow \tilde{a}((\boldsymbol{u}, \boldsymbol{B}); (\boldsymbol{u}, \boldsymbol{B}), (\boldsymbol{v}, \boldsymbol{C})), \text{ as } n \to +\infty.$$

(iii) There exists a constant $\gamma > 0$, *which only depends on the domain* Ω, *such that, for all* $(\boldsymbol{u}_1, \boldsymbol{B}_1), (\boldsymbol{u}_2, \boldsymbol{B}_2), (\boldsymbol{v}, \boldsymbol{C})$ *and* $(\boldsymbol{w}, \boldsymbol{D})$ *in* $\tilde{\mathcal{V}}$,

$$|\tilde{a}((\boldsymbol{u}_2, \boldsymbol{B}_2); (\boldsymbol{v}, \boldsymbol{C}), (\boldsymbol{w}, \boldsymbol{D})) - \tilde{a}((\boldsymbol{u}_1, \boldsymbol{B}_1); (\boldsymbol{v}, \boldsymbol{C}), (\boldsymbol{w}, \boldsymbol{D}))|$$
$$\leq \gamma \max(1, S)\|(\boldsymbol{u}_2, \boldsymbol{B}_2) - (\boldsymbol{u}_1, \boldsymbol{B}_1)\|_{\tilde{\mathcal{X}}}\|(\boldsymbol{v}, \boldsymbol{C})\|_{\tilde{\mathcal{X}}}\|(\boldsymbol{w}, \boldsymbol{D})\|_{\tilde{\mathcal{X}}}.$$

Proof (i) With the properties (i) and (iii) of Lemma 3.25, there exists $\alpha > 0$ such that, for all $(\boldsymbol{v}, \boldsymbol{C}) \in \tilde{\mathcal{V}}$,

$$\tilde{a}((\boldsymbol{v}, \boldsymbol{C}); (\boldsymbol{v}, \boldsymbol{C}), (\boldsymbol{v}, \boldsymbol{C})) = \tilde{\mathcal{A}}((\boldsymbol{v}, \boldsymbol{C}), (\boldsymbol{v}, \boldsymbol{C})) \geq \alpha\|(\boldsymbol{v}, \boldsymbol{C})\|^2_{\tilde{\mathcal{X}}}.$$

(ii) The proof follows the same lines as for Lemma 3.19. Let $(\boldsymbol{u}_n, \boldsymbol{B}_n)$ be a sequence which weakly converges to $(\boldsymbol{u}, \boldsymbol{B})$ in $\tilde{\mathcal{V}}$.

$$|\tilde{a}((\boldsymbol{u}_n, \boldsymbol{B}_n); (\boldsymbol{u}_n, \boldsymbol{B}_n), (\boldsymbol{v}, \boldsymbol{C})) - \tilde{a}((\boldsymbol{u}, \boldsymbol{B}); (\boldsymbol{u}, \boldsymbol{B}), (\boldsymbol{v}, \boldsymbol{C}))|$$
$$\leq |\tilde{\mathcal{A}}((\boldsymbol{u}_n - \boldsymbol{u}, \boldsymbol{B}_n - \boldsymbol{B}), (\boldsymbol{v}, \boldsymbol{C}))||\mathcal{C}((\boldsymbol{u}_n - \boldsymbol{u}, \boldsymbol{B}_n - \boldsymbol{B}), (\boldsymbol{u}_n, \boldsymbol{B}_n), (\boldsymbol{v}, \boldsymbol{C}))|$$
$$+ |\mathcal{C}((\boldsymbol{u}, \boldsymbol{B}), (\boldsymbol{u}_n - \boldsymbol{u}, \boldsymbol{B}_n - \boldsymbol{B}), (\boldsymbol{v}, \boldsymbol{C}))|. \tag{3.63}$$

The first term in the right-hand side of this inequality reads

$$\mathcal{A}((\boldsymbol{u}_n - \boldsymbol{u}, \boldsymbol{B}_n - \boldsymbol{B}), (\boldsymbol{v}, \boldsymbol{C})) = \frac{1}{Re}\int_\Omega \nabla(\boldsymbol{u}_n - \boldsymbol{u}) : \nabla \boldsymbol{v} + \frac{S}{Rm}\int_\Omega \mathbf{curl}\,(\boldsymbol{B}_n - \boldsymbol{B}) \cdot \mathbf{curl}\,\boldsymbol{C}.$$

The weak convergence of $(\boldsymbol{u}_n, \boldsymbol{B}_n)$ is sufficient to pass to the limit in these linear terms. The passage to the limit in the second term of the right-hand side of inequality (3.63)

$$\begin{aligned}&\mathcal{C}\,((\boldsymbol{u}_n - \boldsymbol{u}, \boldsymbol{B}_n - \boldsymbol{B}), (\boldsymbol{u}_n, \boldsymbol{B}_n), (\boldsymbol{v}, \boldsymbol{C}))\\ &= \int_\Omega (\boldsymbol{u}_n - \boldsymbol{u}) \cdot \nabla \boldsymbol{u}_n \cdot \boldsymbol{v}\\ &\quad + S\int_\Omega (\boldsymbol{B}_n - \boldsymbol{B}) \times \mathbf{curl}\,\boldsymbol{B}_n \cdot \boldsymbol{v} + S\int_\Omega (\boldsymbol{B}_n - \boldsymbol{B}) \times \boldsymbol{u}_n \cdot \mathbf{curl}\,\boldsymbol{C}\end{aligned}$$

relies on the compact embedding of $\mathcal{H}(\Omega)$ in $\mathbb{L}^{3+\varepsilon_1}(\Omega)$, where $\varepsilon_1 > 0$ is given by Proposition 3.23, along with the compact embedding $\mathbb{H}^1(\Omega)$ in $\mathbb{L}^{6-\varepsilon_3}(\Omega)$, where ε_3 is such that

$$\frac{1}{3+\varepsilon_1} + \frac{1}{6-\varepsilon_3} = \frac{1}{2}.$$

These properties yield (up to an extraction):

$$(\boldsymbol{u}_n, \boldsymbol{B}_n) \to (\boldsymbol{u}, \boldsymbol{B}) \text{ in } \mathbb{L}^{6-\varepsilon_3}(\Omega) \times \mathbb{L}^{3+\varepsilon_1}(\Omega).$$

Thus, using the inequalities,

$$\left|\int_\Omega (\boldsymbol{u} - \boldsymbol{u}_n) \cdot \nabla \boldsymbol{u}_n \cdot \boldsymbol{v}\right| \le \|\boldsymbol{u} - \boldsymbol{u}_n\|_{\mathbb{L}^4} \|\nabla \boldsymbol{u}_n\|_0 \|\boldsymbol{v}\|_{\mathbb{L}^4},$$

$$\left|\int_\Omega (\boldsymbol{B}_n - \boldsymbol{B}) \times \mathbf{curl}\,\boldsymbol{B}_n \cdot \boldsymbol{v}\right| \le c\|\boldsymbol{B}_n - \boldsymbol{B}\|_{\mathbb{L}^{3+\varepsilon_1}} \|\mathbf{curl}\,\boldsymbol{B}_n\|_0 \|\boldsymbol{v}\|_{\mathbb{L}^{6-\varepsilon_3}},$$

$$\left|\int_\Omega (\boldsymbol{B}_n - \boldsymbol{B}) \times \boldsymbol{u}_n \cdot \mathbf{curl}\,\boldsymbol{C}\right| \le c\|\boldsymbol{B}_n - \boldsymbol{B}\|_{\mathbb{L}^{3+\varepsilon_1}} \|\boldsymbol{u}_n\|_{\mathbb{L}^{6-\varepsilon_3}} \|\mathbf{curl}\,\boldsymbol{C}\|_{\mathbb{L}^2},$$

we may conclude that the second term of the right-hand side of (3.63) goes to zero. The other terms can be treated with similar arguments as those of Lemma 3.21.

(iii) Let $(\boldsymbol{u}_1, \boldsymbol{B}_1), (\boldsymbol{u}_2, \boldsymbol{B}_2), (\boldsymbol{v}, \boldsymbol{C})$ and $(\boldsymbol{w}, \boldsymbol{D})$ be in $\tilde{\mathcal{V}}$. Using the same arguments as in (ii) of Lemma 3.25,

$$\begin{aligned}&|a((\boldsymbol{u}_2, \boldsymbol{B}_2); (\boldsymbol{v}, \boldsymbol{C}), (\boldsymbol{w}, \boldsymbol{D})) - a((\boldsymbol{u}_1, \boldsymbol{B}_1); (\boldsymbol{v}, \boldsymbol{C}), (\boldsymbol{w}, \boldsymbol{D}))|\\ &= \left|\int_\Omega [(\boldsymbol{u}_2 - \boldsymbol{u}_1) \cdot \nabla \boldsymbol{v} \cdot \boldsymbol{w} - S\mathbf{curl}\,\boldsymbol{C} \times (\boldsymbol{B}_2 - \boldsymbol{B}_1) \cdot \boldsymbol{w} - S\boldsymbol{v} \times (\boldsymbol{B}_2 - \boldsymbol{B}_1) \cdot \mathbf{curl}\,\boldsymbol{D}]\right|\\ &\le c(\|\boldsymbol{u}_2 - \boldsymbol{u}_1\|_1 \|\boldsymbol{v}\|_1 \|\boldsymbol{w}\|_1 + S\|\boldsymbol{C}\|_{\text{curl}} \|\boldsymbol{B}_2 - \boldsymbol{B}_1\|_{\text{curl}} \|\boldsymbol{w}\|_1\\ &\quad + S\|\boldsymbol{v}\|_1 \|\boldsymbol{B}_2 - \boldsymbol{B}_1\|_{\text{curl}} \|\boldsymbol{D}\|_{\text{curl}})\\ &\le \gamma \max(1, S) \|(\boldsymbol{u}_2, \boldsymbol{B}_2) - (\boldsymbol{u}_1, \boldsymbol{B}_1)\|_{\tilde{\mathcal{X}}} \|(\boldsymbol{v}, \boldsymbol{C})\|_{\tilde{\mathcal{X}}} \|(\boldsymbol{w}, \boldsymbol{D})\|_{\tilde{\mathcal{X}}}.\end{aligned}$$

where c and γ denote constants depending only on Ω. We have used in particular the fact that $\operatorname{div} \boldsymbol{B} = 0$ when $(\boldsymbol{u}, \boldsymbol{B}) \in \tilde{\mathcal{V}}$, and the embedding given in Proposition 3.23. Note that this property holds true in $\tilde{\mathcal{V}}$, not in $\tilde{\mathcal{X}}$, contrary to the analogous property in the previous section (see point (iii) of Lemma 3.21). □

We now state the main result of this section.

Theorem 3.28 *The variational formulation (3.56) of the stationary MHD system admits a solution* $(\boldsymbol{u}, \boldsymbol{B}, p, r) \in \mathbb{H}_0^1(\Omega) \times H(\mathbf{curl}, \Omega) \times L_0^2(\Omega) \times H^1(\Omega)/\mathbb{R}$ *which satisfies in particular*

$$\|(\boldsymbol{u}, \boldsymbol{B})\|_{\tilde{\mathcal{X}}} \leq \frac{\|\mathcal{L}\|_{\tilde{\mathcal{V}}'}}{\alpha},$$

where $\mathcal{L}$ *is defined in (3.50) and* α *in (3.59). If, in addition,*

$$\frac{\gamma \max(1, S) \|\mathcal{L}\|_{\tilde{\mathcal{V}}'}}{\alpha^2} < 1,$$

where γ *is a constant depending only on* Ω *defined in Lemma 3.27, then the solution is unique.*

Proof This theorem is a straightforward corollary of Lemmata 3.25, 3.26, 3.27, and Theorem 3.15. □

Remark 3.3.3 In the above formulation, valid for non-convex polyhedra, the relation $\operatorname{div} \boldsymbol{B} = 0$ is enforced with a Lagrange multiplier r. A very similar approach, using an analogous functional setting, has been proposed by F. Kikuchi [133, 134] in the context of electrostatic and magnetostatic problems and by F. Assous et al. [9] for solving the time-dependent Maxwell equations. Using different function spaces (typically involving classical Lagrange finite elements), N. Ben Salah et al. [19, 20] also used a Lagrange multiplier to enforce the divergence free constraint on $\boldsymbol{B}$. On the contrary, in the formulation (3.46) presented in Section 3.3.1 for convex polyhedron or regular domains, $\operatorname{div} \boldsymbol{B} = 0$ was recovered as a by-product of the variational formulation, without any Lagrange multiplier. Note that formulation (3.46) should not be confused with a penalization: at the continuous level, the divergence of $\boldsymbol{B}$ is *exactly* zero, and after discretization, it goes to zero with the *discretization step*. ◊

3.4 Mixed finite elements for MHD

We present in this section two finite element discretizations of the stationary MHD equation. The first one uses classical Lagrange finite elements for the magnetic field whereas the second one is based on the Nédélec (edge) finite element. The reader interested in a comparison between Lagrange finite elements and edge elements is referred to J.-L. Guermond and P.D. Minev [114].

3.4.1 Mixed finite elements on convex polyhedra and regular domains

We consider the formulation (3.46) defined in Section 3.3.1 for convex polyhedra or domains with a $\mathcal{C}^{1,1}$ boundary. We introduce three finite element spaces

$$X_h \subset \mathbb{H}_0^1(\Omega), \quad W_h \subset \mathbb{H}_n^1(\Omega), \quad M_h \subset L_0^2(\Omega), \tag{3.64}$$

to approximate $\boldsymbol{u}, \boldsymbol{B}$ and p, respectively. We assume that the velocity space X_h and the pressure space M_h satisfy a classical inf–sup condition (3.24) for the Stokes problem with a constant β_h *independent* of h. We also assume that these spaces satisfy the standard approximation property (see, *e.g.* P.G. Ciarlet [39]):

$$\inf_{\boldsymbol{v}_h \in X_h} \|\boldsymbol{v} - \boldsymbol{v}_h\|_1 + \inf_{q_h \in M_h} \|q - q_h\|_0 \leq c\, h^{\min(s,k)} \left[\|\boldsymbol{v}\|_{s+1} + \|q\|_s\right], \tag{3.65}$$

for $\boldsymbol{v} \in \mathbb{H}^{s+1}(\Omega)$ and $p \in H^s(\Omega)$ with a regularity exponent $s > 1/2$, and an approximation order $k \geq 1$. There are many spaces satisfying these assumptions. Among the examples mentioned in Section 3.1.7, we may quote the Taylor–Hood element (for which $k = 2$) and the mini-element (for which $k = 1$). Other examples may be found, *e.g.* in [106] and [32].

We assume in addition that the space W_h for the magnetic field satisfies an analogous approximation property:

$$\inf_{\boldsymbol{C}_h \in W_h} \|\boldsymbol{C} - \boldsymbol{C}_h\|_1 \leq c\, h^{\min(s,l)} \|\boldsymbol{C}\|_{s+1}, \tag{3.66}$$

for $\boldsymbol{C} \in \mathbb{H}^{s+1}(\Omega)$ with a regularity exponent $s > 1/2$, and an approximation order $l \geq 1$. Then the following theorem can be proved (see [118] for a close result).

Theorem 3.29 *We assume that*

$$\frac{\gamma \max(1, S)(\|\boldsymbol{f}\|_{-1}^2 + \|\boldsymbol{h}\|_0^2)^{1/2}}{\alpha^2} < \frac{1}{2},$$

where γ is a constant depending only on Ω defined in Lemma 3.21. We assume that the solution $(\boldsymbol{u}, \boldsymbol{B}, p)$ of the MHD system (3.46) satisfies

$$\boldsymbol{u} \in \mathbb{H}^{s+1}(\Omega), p \in H^s(\Omega), \boldsymbol{B} \in \mathbb{H}^{s+1}(\Omega),$$

for $s > 1/2$. Let $(\boldsymbol{u}_h, \boldsymbol{B}_h, p_h)$ denote the approximate solution in the space $X_h \times W_h \times M_h$ defined above. Then we have

$$\begin{aligned} &\|(\boldsymbol{u} - \boldsymbol{u}_h, \boldsymbol{B} - \boldsymbol{B}_h)\|_{\mathcal{X}} + \|p - p_h\|_{\mathcal{M}} \\ &\leq c\left[h^{\min(s,k)}\left(\|\boldsymbol{u}\|_{s+1} + \|p\|_s\right) + h^{\min(s,l)}\|\boldsymbol{B}\|_{s+1}\right]. \end{aligned}$$

Two conclusions may be drawn from this result. First, in this formulation, there is much freedom in the choice of the approximation space for $\boldsymbol{B}$. For example standard Lagrange finite element spaces are convenient. The only difficulty comes from the spaces for $\boldsymbol{u}$ and p which are constrained by the same inf–sup condition as in the Stokes problem. Second, once X_h and M_h are chosen, the convergence estimate indicates that the optimal choice is to take a space W_h such that $l = k$. In other words, the best choice is also the simplest one: the optimal convergence rate is achieved when $\boldsymbol{u}$ and $\boldsymbol{B}$ are approximated with the same finite element. It is therefore quite easy to modify a standard incompressible Navier–Stokes solver to implement this formulation.

3.4.2 Mixed finite elements on non-convex polyhedra

We now present a stable discretization, proposed by D. Schötzau [205], of the mixed formulation (3.56) valid for general polyhedra, in particular non-convex. We suppose that Ω satisfies (3.53). We recall that the mesh $\mathcal{T}_h$ is assumed to be made of tetrahedra. For the velocity and the pressure, we choose as above spaces satisfying the standard approximation property (3.65) and the Stokes inf–sup condition (3.24) (for example the Taylor–Hood or the mini finite element, see Section 3.1.7). For the magnetic field, the discretization is based on Nédélec (or edge) finite elements. Let $\tilde{\mathbb{P}}_k(K)$ be the space of homogeneous polynomials of degree k defined on K. We define, for $k \geq 1$,

$$\mathcal{D}_k(K) = \{\boldsymbol{p} \in [\tilde{\mathbb{P}}_k(K)]^3, \text{ such that } \boldsymbol{p}(\boldsymbol{x}) \cdot \boldsymbol{x} = 0 \text{ on } K\},$$

$$\mathcal{N}_k(K) = [\mathbb{P}_{k-1}(K)]^3 \oplus \mathcal{D}_k(K).$$

In the case $k = 1$, which is the most used in practice, we have the following equivalent definition:

$$\mathcal{N}_1(K) = [\mathbb{P}_0(K)]^3 \oplus (\boldsymbol{x} \times [\mathbb{P}_0(K)]^3).$$

For the associated finite element, a degree of freedom of a vector field consists of the integral along an edge of the component of the vector parallel to the edge. This is the conventional meaning of the arrows on the representation of this element in Figure 3.3.

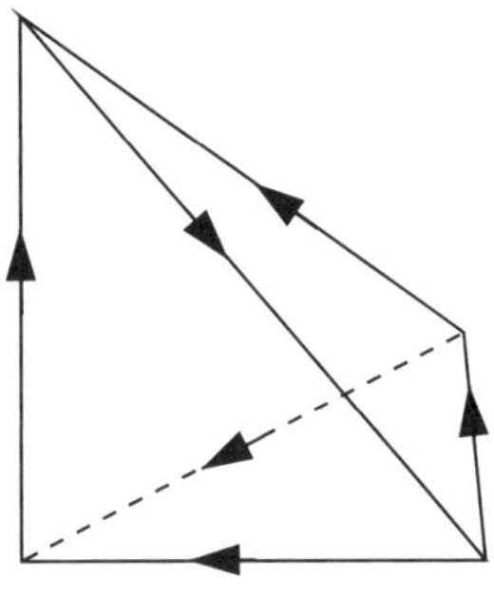

Fig. 3.3 The Nédélec finite element of lowest order (for the magnetic field).

The magnetic field is then approximated in the space

$$W_h = \{\boldsymbol{C}_h \in H(\mathbf{curl}\,,\Omega) \text{ such that } \boldsymbol{C}_h|_K \in \mathcal{N}_k(K), K \in \mathcal{T}_h\}, \tag{3.67}$$

and the Lagrange multiplier r associated to the divergence constraint on the magnetic field is approximated in

$$S_h = \{s_h \in H^1(\Omega)/\mathbb{R} \text{ such that } s_h|_K \in \mathbb{P}_k(K), K \in \mathcal{T}_h\}. \tag{3.68}$$

We have $\boldsymbol{\nabla} S_h \subset W_h$. Therefore, using the same argument as in the continuous case, it can be checked that the inf–sup condition (3.62) holds true on $W_h \times S_h$ with a constant β independent of h. Moreover, the following approximation result holds:

$$\inf_{\boldsymbol{C}_h \in W_h} \|\boldsymbol{C} - \boldsymbol{C}_h\|_{\text{curl}} + \inf_{s_h \in S_h} \|s - s_h\|_1 \leq c\, h^{\min(s,k)} \left[\|\boldsymbol{C}\|_s + \|\mathbf{curl}\,\boldsymbol{C}\|_s + \|r\|_{s+1}\right], \tag{3.69}$$

for $\boldsymbol{C} \in \mathbb{H}^s(\Omega)$, $\mathbf{curl}\,\boldsymbol{C} \in \mathbb{H}^s(\Omega)$ and $r \in H^{s+1}(\Omega)$ for $s > 1/2$. We then have the following results whose proof can be found in [205].

Theorem 3.30 *[205] We assume that*

$$\frac{\gamma \max(1, S)(\|\boldsymbol{f}\|_{-1}^2 + \|\boldsymbol{h}\|_0^2)^{1/2}}{\alpha^2} < \frac{1}{2},$$

where γ is a constant, depending only on Ω, defined in Lemma 3.27. The spaces X_h and M_h are assumed to satisfy the approximation property (3.65) and the inf–sup condition (3.24) with a constant β_h independent of h. The spaces W_h and S_h are defined by (3.67) and (3.68). We assume in addition that the solution $(\boldsymbol{u}, \boldsymbol{B}, p, r)$ of the MHD system (3.56) satisfies

$$\boldsymbol{u} \in \mathbb{H}^{s+1}(\Omega),\ p \in H^s(\Omega),\ \boldsymbol{B} \in \mathbb{H}^s(\Omega),\ \mathbf{curl}\,\boldsymbol{B} \in \mathbb{H}^s(\Omega),\ r \in H^{s+1}(\Omega)/\mathbb{R},$$

for $s > 1/2$. Let $(\boldsymbol{u}_h, \boldsymbol{B}_h, p_h, r_h)$ be the solution to the problem (3.56) approximated in the space $X_h \times W_h \times M_h \times S_h$. Then we have

$$\begin{aligned} &\|(\boldsymbol{u} - \boldsymbol{u}_h, \boldsymbol{B} - \boldsymbol{B}_h)\|_{\tilde{\mathcal{X}}} + \|(p - p_h, r - r_h)\|_{\tilde{\mathcal{M}}} \\ &\leq c\, h^{\min(s,k)} \left[\|\boldsymbol{u}\|_{s+1} + \|p\|_s + \|\boldsymbol{B}\|_s + \|\mathbf{curl}\,\boldsymbol{B}\|_s + \|r\|_{s+1}\right]. \end{aligned}$$

In conclusion, a possible choice to discretize formulation (3.56) consists in approximating the velocity and the pressure with the $\mathbb{P}^1$-bubble/$\mathbb{P}^1$ finite element, the magnetic field with the Nédélec finite element of lowest order, and the Lagrange multiplier associated with the divergence-free constraint on $\boldsymbol{B}$ with the $\mathbb{P}_1$ finite element. With these spaces, Theorem 3.30 ensures an optimal convergence rate as long as the solution is regular enough.

3.5 Stabilized finite elements for MHD

We consider in this section the linearized stationary MHD equations. We present the analysis of a stabilized finite element formulation proposed in [89]. As explained in Section 3.2, stabilized finite elements may be useful in advection-dominated problems. In many industrial situations (such as aluminum electrolysis, electromagnetic pumping, MHD generator, see, *e.g.* R. Berton [21]), the hydrodynamic Reynolds number Re being high (typically 10^4), the hydrodynamic equation is typically advection-dominated. It is therefore natural to use some stabilizing techniques for these equations. In contrast, the magnetic Reynolds number Rm being moderate or low (from 10^{-2} to 1), the magnetic equation is *a priori* dominated by the diffusion. Nevertheless, we will show in this section that it may be useful to *also* stabilize the magnetic equation, due to the coupling with hydrodynamics.

In the formulation proposed in this section, the velocity, the magnetic field and the pressure may be approximated with the same finite element. The implementation of this formulation is therefore simpler than those presented in Sections 3.4.1 and 3.4.2.

For simplicity, we assume that

$$\Omega \text{ is a bounded regular convex open subset of } \mathbb{R}^3, \tag{3.70}$$

and we consider a linearized version of the MHD system:

$$\boldsymbol{a}\cdot\boldsymbol{\nabla}\boldsymbol{u} - \frac{1}{Re}\Delta\,\boldsymbol{u} + \boldsymbol{\nabla}p + S\,\boldsymbol{b}\times\mathbf{curl}\,\boldsymbol{B} = \boldsymbol{f}, \tag{3.71}$$

$$\operatorname{div}\boldsymbol{u} = 0, \tag{3.72}$$

$$\frac{1}{Rm}\,\mathbf{curl}\,(\mathbf{curl}\,\boldsymbol{B}) - \mathbf{curl}\,(\boldsymbol{u}\times\boldsymbol{b}) = \boldsymbol{h}, \tag{3.73}$$

$$\operatorname{div}\boldsymbol{B} = 0. \tag{3.74}$$

This simplified system is of practical interest since it appears in particular when the original nonlinear system is solved with the Picard algorithm (see Section 3.6.1). In such a case $\boldsymbol{a}$ and $\boldsymbol{b}$ stand for $\boldsymbol{u}^n$ and $\boldsymbol{B}^n$ in the Picard iterations and are supposed to be known and smooth. Again for simplicity, we suppose that $\operatorname{div}\boldsymbol{a} = 0$ (this is not necessary, see Remark 3.5.3), and we deal with homogeneous boundary conditions on $\partial\Omega$:

$$\boldsymbol{u} = 0, \tag{3.75}$$

$$\boldsymbol{B}\cdot\boldsymbol{n} = 0, \tag{3.76}$$

$$\mathbf{curl}\,\boldsymbol{B}\times\boldsymbol{n} = 0. \tag{3.77}$$

The velocity $\boldsymbol{u}$ is approximated in

$$X_h = (Y_h^k)^3 \cap \mathbb{H}_0^1(\Omega),$$

the pressure p in

$$M_h = Y_h^m \cap L_0^2(\Omega),$$

and the magnetic field $\boldsymbol{B}$ in

$$W_h = (Y_h^l)^3 \cap \mathbb{H}_n^1(\Omega),$$

where the space Y_h^k is a classical Lagrange finite element space defined in (3.23). We have the following classical property of approximation:

$$\begin{aligned} \|u - \Pi_h u\|_{L^2(K)} + h_K \|\nabla(u - \Pi_h u)\|_{L^2(K)} + h_K^2 \|\Delta\,(u - \Pi_h u)\|_{L^2(K)} \\ \leq C h_K^{k+1} |u|_{k+1,K}, \end{aligned} \tag{3.78}$$

where $\Pi_h u$ denotes the interpolate in Y_h^k of $u \in H^{k+1}(K)$. We will use the inverse inequality (see, *e.g.* P.G. Ciarlet [39])

$$\sum_{K \in \mathcal{T}_h} h_K^2 \int_K |\Delta\, v_h|^2 \leq d_0 \int_\Omega |\nabla v_h|^2, \qquad \forall\, v_h \in X_h, \tag{3.79}$$

where d_0 is a non-negative constant independent of h.

With the assumption (3.70) made on Ω, we have the following inequality for $\boldsymbol{B} \in \mathbb{H}_n^1(\Omega)$ (V. Girault and P.A. Raviart [106], Theorem 3.9):

$$\int_\Omega |\nabla \boldsymbol{B}|^2 \leq \int_\Omega |\mathbf{curl}\, \boldsymbol{B}|^2 + |\mathrm{div}\, \boldsymbol{B}|^2. \tag{3.80}$$

We introduce the notation

$$\eta = 1/Re \qquad \text{and} \qquad \alpha = 1/Rm.$$

To simplify the presentation, we assume that the hydrodynamics is globally advection-dominated:

$$\frac{|\boldsymbol{a}(x)| h_K}{\eta} > 1, \qquad \forall\, x \in K, \tag{3.81}$$

for all elements K of the mesh. Let $\lambda_u > 0$ and $\lambda_B > 0$ be two constant parameters to be fixed. We define the stabilization coefficients by:

$$\tau_u|_K = \frac{\lambda_u h_K}{|\boldsymbol{a}(x)|},$$

and

$$\tau_B|_K = \frac{\lambda_B h_K^2}{\alpha}.$$

Note that (3.81) implies

$$\eta \tau_u|_K \leq \lambda_u h_K^2. \tag{3.82}$$

The resolution of the linearized MHD equations (3.71)–(3.77) can be done by a mixed Galerkin method (as in Sections 3.4.1 for formulation (3.46)): find $\boldsymbol{u}_h \in X_h$, $\boldsymbol{B}_h \in W_h$ and $p_h \in M_h$ such that for all $(\boldsymbol{v}_h, \boldsymbol{C}_h, q_h) \in X_h \times W_h \times M_h$

$$\Phi_G(\boldsymbol{u}_h, \boldsymbol{B}_h, p_h; \boldsymbol{v}_h, \boldsymbol{C}_h, q_h) = \langle F_G; \boldsymbol{v}_h, \boldsymbol{C}_h, q_h \rangle,$$

with

$$\Phi_G(\boldsymbol{u},\boldsymbol{B},p;\boldsymbol{v},\boldsymbol{C},q) = \int_\Omega (\eta\boldsymbol{\nabla}\boldsymbol{u}\cdot\boldsymbol{\nabla}\boldsymbol{v} + \boldsymbol{a}\cdot\boldsymbol{\nabla}\boldsymbol{u}\cdot\boldsymbol{v} - p\,\mathrm{div}\,\boldsymbol{v} + S\,\boldsymbol{b}\times\mathbf{curl}\,\boldsymbol{B}\cdot\boldsymbol{v})$$
$$+\int_\Omega q\,\mathrm{div}\,\boldsymbol{u}$$
$$+\int_\Omega (\alpha S\,\mathbf{curl}\,\boldsymbol{B}\cdot\mathbf{curl}\,C + \alpha S\,\mathrm{div}\,\boldsymbol{B}\mathrm{div}\,C$$
$$-\ S\,\boldsymbol{u}\times\boldsymbol{b}\cdot\mathbf{curl}\,\boldsymbol{C}),$$

and

$$\langle F_G; v, C, q\rangle = \int_\Omega \boldsymbol{f}\cdot\boldsymbol{v} + \int_\Omega S\,\boldsymbol{h}\cdot\boldsymbol{C}.$$

Let us now define the stabilization terms:

$$\Phi_S(\boldsymbol{u},\boldsymbol{B},p;\boldsymbol{v},\boldsymbol{C},q) = \sum_{K\in\mathcal{T}_h}\int_K \tau_u\,(\boldsymbol{a}\cdot\boldsymbol{\nabla}\boldsymbol{u} - \eta\Delta\,\boldsymbol{u} + \boldsymbol{\nabla}p + S\,\boldsymbol{b}\times\mathbf{curl}\,\boldsymbol{B})\cdot$$
$$(\boldsymbol{a}\cdot\boldsymbol{\nabla}\boldsymbol{v} + \xi\eta\Delta\,\boldsymbol{v} + \boldsymbol{\nabla}q + S\,\boldsymbol{b}\times\mathbf{curl}\,C)$$
$$+\sum_{K\in\mathcal{T}_h}\int_K \tau_B\,(-\alpha S\Delta\,\boldsymbol{B} - S\mathbf{curl}\,(\boldsymbol{u}\times\boldsymbol{b}))\cdot$$
$$(\alpha\xi\Delta\,C - \mathbf{curl}\,(\boldsymbol{v}\times\boldsymbol{b})),$$

and

$$\langle F_S;\boldsymbol{v},\boldsymbol{C},q\rangle = \sum_{K\in\mathcal{T}_h}\int_K \tau_u\,\boldsymbol{f}\cdot(\boldsymbol{a}\cdot\boldsymbol{\nabla}\boldsymbol{v} + \xi\eta\Delta\,\boldsymbol{v} + \boldsymbol{\nabla}q + S\,\boldsymbol{b}\times\mathbf{curl}\,C)$$
$$+\sum_{K\in\mathcal{T}_h}\int_K \tau_B\,S\,\boldsymbol{h}\cdot(\alpha\xi\Delta\,C - \mathbf{curl}\,(\boldsymbol{v}\times\boldsymbol{b})),$$

where ξ is a constant equal to -1, 0 or 1 depending on the variant of the method (see Section 3.2). Notice that the three variants coincide when $\mathbb{P}_1$ elements are used. To simplify the exposition, we will assume in the sequel that $\xi = 1$.

The stabilized problem that we now propose to analyze reads: find $\boldsymbol{u}_h \in X_h$, $\boldsymbol{B}_h \in W_h$ and $p_h \in M_h$ such that for all $(\boldsymbol{v}_h,\boldsymbol{C}_h,q_h) \in X_h\times W_h\times M_h$,

$$\Phi(\boldsymbol{u}_h,\boldsymbol{B}_h,p_h;\boldsymbol{v}_h,\boldsymbol{C}_h,q_h) = \langle F;\boldsymbol{v}_h,\boldsymbol{C}_h,q_h\rangle, \tag{3.83}$$

with $\Phi = \Phi_G + \Phi_S$ and $F = F_G + F_S$.

The first step in the analysis of this method consists in proving its stability, *i.e.* its coercivity in a mesh-dependent norm defined by

$$|||(\boldsymbol{v}_h,\boldsymbol{C}_h,q_h)|||_h = \Big(\frac{1}{2}\int_\Omega \eta|\boldsymbol{\nabla}\boldsymbol{v}_h|^2 + \frac{1}{2}\int_\Omega S\alpha|\mathbf{curl}\,\boldsymbol{C}_h|^2$$
$$+\frac{1}{2}\int_\Omega S\alpha|\mathrm{div}\,\boldsymbol{C}_h|^2 + \int_\Omega \tau_B S|\mathbf{curl}\,(\boldsymbol{v}_h\times\boldsymbol{b})|^2 \tag{3.84}$$
$$+\int_\Omega \tau_u|\boldsymbol{a}\cdot\boldsymbol{\nabla}\boldsymbol{v}_h + \boldsymbol{\nabla}q_h + S\,\boldsymbol{b}\times\mathbf{curl}\,\boldsymbol{C}_h|^2\Big)^{1/2},$$

for $(\boldsymbol{v}_h, \boldsymbol{C}_h, q_h) \in X_h \times W_h \times M_h$. The following lemma gives the conditions on λ_u and λ_B that ensure the stability when $\xi = 1$. The cases $\xi = 0$ and $\xi = -1$ are treated in [89].

Lemma 3.31 *If $\xi = 1$ and if the positive parameters λ_u and λ_B are such that*

$$1 - d_0\lambda_B \geq \frac{1}{2} \qquad \text{and} \qquad 1 - d_0\lambda_u \geq \frac{1}{2}, \tag{3.85}$$

where d_0 is the constant defined in the inverse inequality (3.79), then

$$\Phi(\boldsymbol{v}_h, \boldsymbol{C}_h, q_h; \boldsymbol{v}_h, \boldsymbol{C}_h, q_h) \geq |||(\boldsymbol{v}_h, \boldsymbol{C}_h, q_h)|||_h^2, \tag{3.86}$$

for all $(\boldsymbol{v}_h, \boldsymbol{C}_h, q_h) \in X_h \times W_h \times M_h$.

Proof To start with, we compute $\Phi_G(\boldsymbol{v}_h, \boldsymbol{C}_h, q_h; \boldsymbol{v}_h, \boldsymbol{C}_h, q_h)$ which corresponds to the (non-stabilized) Galerkin method. The advection term $\int \boldsymbol{a} \cdot \boldsymbol{\nabla} \boldsymbol{v}_h \cdot \boldsymbol{v}_h$ cancels out owing to the fact that $\boldsymbol{a}$ is divergence-free. Simple manipulations, in particular based on the relation

$$\boldsymbol{b} \times \mathbf{curl}\, \boldsymbol{C}_h \cdot \boldsymbol{v}_h = \boldsymbol{v}_h \times \boldsymbol{b} \cdot \mathbf{curl}\, \boldsymbol{C}_h,$$

lead to the expression

$$\Phi_G(\boldsymbol{v}_h, \boldsymbol{C}_h, q_h; \boldsymbol{v}_h, \boldsymbol{C}_h, q_h) = \int_\Omega \left(\eta|\boldsymbol{\nabla}\boldsymbol{v}_h|^2 + \alpha S|\mathbf{curl}\, \boldsymbol{C}_h|^2 + \alpha S|\mathrm{div}\, \boldsymbol{C}_h|^2\right). \tag{3.87}$$

Let us next consider the stabilization terms $\Phi(\boldsymbol{v}_h, \boldsymbol{C}_h, q_h; \boldsymbol{v}_h, \boldsymbol{C}_h, q_h)$. Using (3.82) and the inverse inequality (3.79), we have:

$$\begin{aligned}
&\Phi_S(\boldsymbol{v}_h, \boldsymbol{C}_h, q_h; \boldsymbol{v}_h, \boldsymbol{C}_h, q_h) \\
&= \sum_{K \in \mathcal{T}_h} \int_K \left(\tau_u|\boldsymbol{a} \cdot \boldsymbol{\nabla}\boldsymbol{v}_h + \boldsymbol{\nabla} q_h + S\,\boldsymbol{b} \times \mathbf{curl}\, \boldsymbol{C}_h|^2 - \tau_u \eta^2 |\Delta\, \boldsymbol{v}_h|^2\right) \\
&\quad + \sum_{K \in \mathcal{T}_h} \int_K \left(\tau_B S|\mathbf{curl}\,(\boldsymbol{v}_h \times \boldsymbol{b})|^2 - \tau_B S\alpha^2 |\Delta\, \boldsymbol{C}_h|^2\right) \\
&\geq \int_\Omega \tau_u |\boldsymbol{a} \cdot \boldsymbol{\nabla}\boldsymbol{v}_h + \boldsymbol{\nabla} q_h + S\,\boldsymbol{b} \times \mathbf{curl}\, \boldsymbol{C}_h|^2 - d_0\lambda_u \int_\Omega \eta|\boldsymbol{\nabla}\boldsymbol{v}_h|^2 \\
&\quad + \int_\Omega \tau_B S|\mathbf{curl}\,(\boldsymbol{v}_h \times \boldsymbol{b})|^2 - d_0\lambda_B \int_\Omega S\alpha|\boldsymbol{\nabla}\boldsymbol{C}_h|^2.
\end{aligned}$$

Collecting this last inequality, (3.80) and (3.87), we deduce

$$\begin{aligned}
\Phi(\boldsymbol{v}_h, \boldsymbol{C}_h, q_h; \boldsymbol{v}_h, \boldsymbol{C}_h, q_h) \geq{}& (1 - d_0\lambda_u) \int_\Omega \eta|\boldsymbol{\nabla}\boldsymbol{v}_h|^2 + (1 - d_0\lambda_B) \int_\Omega S\alpha|\mathbf{curl}\, \boldsymbol{C}_h|^2 \\
&+ (1 - d_0\lambda_B) \int_\Omega S\alpha|\mathrm{div}\, \boldsymbol{C}_h|^2 + \int_\Omega \tau_B S|\mathbf{curl}\,(\boldsymbol{v}_h \times \boldsymbol{b})|^2 \\
&+ \int_\Omega \tau_u |\boldsymbol{a} \cdot \boldsymbol{\nabla}\boldsymbol{v}_h + \boldsymbol{\nabla} q_h + S\,\boldsymbol{b} \times \mathbf{curl}\, \boldsymbol{C}_h|^2.
\end{aligned}$$

With definition (3.84) and assumptions (3.85), this concludes the proof of the lemma. □

Let us denote by $(\boldsymbol{u}, \boldsymbol{B}, p)$ the exact solution of the MHD equations (3.71)–(3.77), by $(\tilde{\boldsymbol{u}}_h, \tilde{\boldsymbol{B}}_h, \tilde{p}_h)$ the interpolate of $(\boldsymbol{u}, \boldsymbol{B}, p)$ in $X_h \times W_h \times M_h$ and by $(\boldsymbol{u}_h, \boldsymbol{B}_h, p_h)$ the solution obtained by the stabilized finite element method (3.83) with $\xi = 1$. Note that the existence and uniqueness of this approximate solution is ensured by the Lax–Milgram theorem as long as assumptions of Lemma 3.31 are fulfilled. The convergence of the stabilized formulation (3.83) is established in the following theorem.

Theorem 3.32 *Under hypotheses (3.81) and (3.85), the approximation error* $(\boldsymbol{e}_u, \boldsymbol{e}_b, e_p) = (\tilde{\boldsymbol{u}}_h - \boldsymbol{u}_h, \tilde{\boldsymbol{B}}_h - \boldsymbol{B}_h, \tilde{p}_h - p_h)$ *can be estimated as follows:*

$$
\begin{aligned}
|||(\boldsymbol{e}_u, \boldsymbol{e}_b, e_p)|||_h^2 \leq C_{\lambda_u,\lambda_B} \Bigg[& \left(\frac{\lambda_u h}{\|\boldsymbol{a}\|_\infty} + \frac{h^2}{\eta} \right) h^{2m} |p|_{m+1}^2 \\
& + \left(\lambda_u (\|\boldsymbol{a}\|_\infty h + \eta) + \frac{S^2 h^2}{\alpha} (\|\boldsymbol{b}\|_\infty^2 + h^2 \|\boldsymbol{\nabla} \boldsymbol{b}\|_\infty^2) \right) h^{2k} |\boldsymbol{u}|_{k+1}^2 \\
& + \left(\lambda_u \frac{S^2 \|\boldsymbol{b}\|_\infty^2}{\|\boldsymbol{a}\|_\infty} h + \lambda_B \alpha \right) h^{2l} |\boldsymbol{B}|_{l+1}^2 \Bigg].
\end{aligned} \tag{3.88}
$$

Proof We denote the interpolation error for the velocity

$$\boldsymbol{\pi}_u = \boldsymbol{u} - \tilde{\boldsymbol{u}}_h,$$

and the global error by

$$\boldsymbol{\epsilon}_u = \boldsymbol{u} - \boldsymbol{u}_h.$$

We define in the same way $\boldsymbol{\pi}_b$, $\boldsymbol{\epsilon}_b$ for the magnetic field and π_p, ϵ_p for the pressure. Note that we have the relation $\boldsymbol{\epsilon}_u = \boldsymbol{\pi}_u + \boldsymbol{e}_u$, $\boldsymbol{\epsilon}_b = \boldsymbol{\pi}_b + \boldsymbol{e}_b$, $\epsilon_p = \pi_p + e_p$. □

The linearity and the strong consistency of the stabilized formulation implies:

$$
\begin{aligned}
\Phi(\boldsymbol{e}_u, \boldsymbol{e}_b, e_p; \boldsymbol{e}_u, \boldsymbol{e}_b, e_p) &= \Phi(\boldsymbol{\epsilon}_u - \boldsymbol{\pi}_u, \boldsymbol{\epsilon}_b - \boldsymbol{\pi}_b, \epsilon_p - \pi_p; \boldsymbol{e}_u, \boldsymbol{e}_b, e_p) \\
&= -\Phi(\boldsymbol{\pi}_u, \boldsymbol{\pi}_b, \pi_p; \boldsymbol{e}_u, \boldsymbol{e}_b, e_p).
\end{aligned}
$$

The stability property proved in Lemma 3.31 yields

$$\Phi(\boldsymbol{e}_u, \boldsymbol{e}_b, e_p; \boldsymbol{e}_u, \boldsymbol{e}_b, e_p) \geq |||(\boldsymbol{e}_u, \boldsymbol{e}_b, e_p)|||_h^2,$$

the norm $|||\cdot|||_h$ being defined in (3.84). Thus, we have

$$
\begin{aligned}
& \int_\Omega (\eta |\boldsymbol{\nabla} \boldsymbol{e}_u|^2 + S\alpha |\mathbf{curl}\, \boldsymbol{e}_b|^2 + S\alpha |\mathrm{div}\, \boldsymbol{e}_b|^2) + 2 \int_\Omega \tau_B S |\mathbf{curl}\, (\boldsymbol{e}_u \times \boldsymbol{b})|^2 \\
& + 2 \int_\Omega \tau_u |\boldsymbol{a} \cdot \boldsymbol{\nabla} \boldsymbol{e}_u + \boldsymbol{\nabla} e_p + S\, \boldsymbol{b} \times \mathbf{curl}\, \boldsymbol{e}_b|^2 \\
& \leq 2 |\Phi(\boldsymbol{\pi}_u, \boldsymbol{\pi}_b, \pi_p; \boldsymbol{e}_u, \boldsymbol{e}_b, e_p)|.
\end{aligned} \tag{3.89}
$$

We now estimate the right-hand side of this inequality. The hope is that the quantities involving $\boldsymbol{e}_u, \boldsymbol{e}_b$ and e_p will appear with an order smaller than those of the left-hand side. As far as Φ_G is concerned, we have

$$
\begin{aligned}
&2\,\Phi_G(\boldsymbol{\pi}_u, \boldsymbol{\pi}_b, \pi_p; \boldsymbol{e}_u, \boldsymbol{e}_b, e_p) \\
&= 2\int_\Omega e_p \operatorname{div} \boldsymbol{\pi}_u \\
&\quad + 2\int_\Omega (\eta \boldsymbol{\nabla}\boldsymbol{\pi}_u \cdot \boldsymbol{\nabla}\boldsymbol{e}_u + \boldsymbol{a}\cdot\boldsymbol{\nabla}\boldsymbol{\pi}_u \cdot \boldsymbol{e}_u - \pi_p \operatorname{div} \boldsymbol{e}_u + S\,\boldsymbol{b}\times \mathbf{curl}\,\boldsymbol{\pi}_b \cdot \boldsymbol{e}_u) \\
&\quad + 2\int_\Omega (\alpha S\,\mathbf{curl}\,\boldsymbol{\pi}_b \cdot \mathbf{curl}\,\boldsymbol{e}_b + \alpha S \operatorname{div} \boldsymbol{\pi}_b \operatorname{div} \boldsymbol{e}_b - S\,\boldsymbol{\pi}_u \times \boldsymbol{b}\cdot \mathbf{curl}\,\boldsymbol{e}_b).
\end{aligned}
$$

The diffusion terms are treated as follows:

$$
2\eta \int_\Omega |\boldsymbol{\nabla}\boldsymbol{\pi}_u||\boldsymbol{\nabla}\boldsymbol{e}_u| \le \gamma_1 \eta \int_\Omega |\boldsymbol{\nabla}\boldsymbol{\pi}_u|^2 + \frac{\eta}{\gamma_1}\int_\Omega |\boldsymbol{\nabla}\boldsymbol{e}_u|^2,
$$

$$
\begin{aligned}
&2\alpha S \int_\Omega (|\mathbf{curl}\,\boldsymbol{\pi}_b||\mathbf{curl}\,\boldsymbol{e}_b| + |\operatorname{div} \boldsymbol{\pi}_b||\operatorname{div} \boldsymbol{e}_b|) \\
&\qquad \le \gamma_2 \alpha S \int_\Omega (|\mathbf{curl}\,\boldsymbol{\pi}_b|^2 + |\operatorname{div} \boldsymbol{\pi}_b|^2) + \frac{\alpha S}{\gamma_2}\int_\Omega (|\mathbf{curl}\,\boldsymbol{e}_b|^2 + |\operatorname{div} \boldsymbol{e}_b|^2).
\end{aligned}
$$

We use the stabilization in order to control the convection and the pressure terms:

$$
\begin{aligned}
&2\left|\int_\Omega \boldsymbol{a}\cdot\boldsymbol{\nabla}\boldsymbol{\pi}_u \cdot \boldsymbol{e}_u + e_p \operatorname{div} \boldsymbol{\pi}_u - S\boldsymbol{\pi}_u \times \boldsymbol{b}\cdot \mathbf{curl}\,\boldsymbol{e}_b\right| \\
&\qquad \le \frac{1}{\gamma_3}\int_\Omega \tau_u |\boldsymbol{a}\cdot\boldsymbol{\nabla}\boldsymbol{e}_u + \boldsymbol{\nabla} e_p + S\,\boldsymbol{b}\times \mathbf{curl}\,\boldsymbol{e}_b|^2 + \gamma_3 \int_K \frac{1}{\tau_u}|\boldsymbol{\pi}_u|^2.
\end{aligned}
$$

We see here the interest of stabilizing both the magnetic and hydrodynamics parts: on the one hand, the term $\boldsymbol{\pi}_u \times \boldsymbol{b}\cdot \mathbf{curl}\,\boldsymbol{e}_b$ which comes from the magnetic equations is controlled by the stabilization term of the Navier–Stokes equations; on the other hand, $\boldsymbol{b}\times \mathbf{curl}\,\boldsymbol{\pi}_b \cdot \boldsymbol{e}_u$ which comes from the Lorentz force of the Navier–Stokes equations is controlled by the stabilization term of the magnetic equations (see Remark 3.5.2). Integrating by parts, we actually have:

$$
2\left|\int_\Omega S\,\boldsymbol{b}\times \mathbf{curl}\,\boldsymbol{\pi}_b \cdot \boldsymbol{e}_u\right| \le \frac{S}{\gamma_4}\int_\Omega \tau_B |\mathbf{curl}\,(\boldsymbol{e}_u \times \boldsymbol{b})|^2 + S\gamma_4 \int_\Omega \frac{1}{\tau_B}|\boldsymbol{\pi}_b|^2.
$$

The pressure term is bounded with:

$$
2\int_\Omega |\pi_p||\operatorname{div} \boldsymbol{e}_u| \le \frac{3\eta}{\gamma_5}\int_\Omega |\boldsymbol{\nabla}\boldsymbol{e}_u|^2 + \frac{\gamma_5}{\eta}\int_\Omega |\pi_p|^2.
$$

The estimation of the stabilization term $\Phi_S(\boldsymbol{\pi}_u, \boldsymbol{\pi}_b, \pi_p; \boldsymbol{e}_u, \boldsymbol{e}_b, e_p)$ is rather technical. We refer to [89] for the details. After computations, and inserting all the estimations in (3.89), we eventually obtain:

$$\begin{aligned}|||(\boldsymbol{e}_u, \boldsymbol{e}_b, e_p)|||_h^2 \leq\ & c_1 \int_\Omega \tau_u |\boldsymbol{a}\cdot \boldsymbol{\nabla}\boldsymbol{\pi}_u|^2 + c_2 \sum_{K\in\mathcal{T}_h} \int_K \tau_u \eta^2 |\Delta\, \boldsymbol{\pi}_u|^2 \\ & + c_3 \int_\Omega \tau_u |\boldsymbol{\nabla}\pi_p|^2 + \frac{c_4}{\eta}\int_\Omega |\pi_p|^2 + c_5 \int_\Omega \tau_u S^2 |\boldsymbol{b}\times \mathbf{curl}\, \boldsymbol{\pi}_b|^2 \\ & + c_6 \sum_{K\in\mathcal{T}_h} \int_K \tau_B \alpha^2 |\Delta\, \boldsymbol{\pi}_b|^2 + c_7 \int_K S^2 \tau_B |\mathbf{curl}\,(\boldsymbol{\pi}_u \times \boldsymbol{b})|^2,\end{aligned} \tag{3.90}$$

where the constants c_i do not depend on h_K, η, α, $\boldsymbol{a}$ and $\boldsymbol{b}$. In order to complete the proof, we establish the following estimates. Using (3.82), we have:

$$c_1 \int_K \tau_u |\boldsymbol{a}\cdot \boldsymbol{\nabla}\boldsymbol{\pi}_u|^2 + c_2 \int_K \tau_u \eta^2 |\Delta\, \boldsymbol{\pi}_u|^2 \leq C(\|a\|_\infty h_K + \eta)\lambda_u h_K^{2k} |u|_{k+1,K}^2,$$

$$c_3 \int_K \tau_u |\boldsymbol{\nabla}\pi_p|^2 + \frac{c_4}{\eta}\int_K |\pi_p|^2 \leq C\left(\frac{\lambda_u}{\|\boldsymbol{a}\|_\infty} + \frac{h_K}{\eta}\right) h_K^{2m+1} |p|_{m+1,K}^2,$$

$$\begin{aligned} c_5 \int_K \tau_u S^2 |\boldsymbol{b}\times \mathbf{curl}\, \boldsymbol{\pi}_b|^2 + c_6 \int_K \tau_B \alpha^2 |\Delta\, \boldsymbol{\pi}_b|^2 & \\ \leq C\left(\lambda_u \frac{S^2 \|\boldsymbol{b}\|_\infty^2}{\|\boldsymbol{a}\|_\infty} h_K + \lambda_B \alpha\right) h_K^{2l} |B|_{l+1,K}^2, & \end{aligned}$$

$$\begin{aligned} c_7 \int_K S^2 \tau_B |\mathbf{curl}\,(\boldsymbol{\pi}_u \times \boldsymbol{b})|^2 &= c_7 \int_K S^2 \tau_B |\boldsymbol{\pi}_u|^2 (|\mathrm{div}\, \boldsymbol{b}|^2 + |\boldsymbol{\nabla}\boldsymbol{b}|^2) \\ &\quad + c_7 \int_K S^2 \tau_B |b|^2 (|\mathrm{div}\, \boldsymbol{\pi}_u|^2 + |\boldsymbol{\nabla}\boldsymbol{\pi}_u|^2) \\ &\leq C(\|\boldsymbol{b}\|_\infty^2 + h_K^2 \|\boldsymbol{\nabla b}\|_\infty^2) \frac{S^2 h_K^2}{\alpha} h_K^{2k} |u|_{k+1,K}^2. \end{aligned}$$

Inserting these inequalities in (3.90), we obtain (3.88). This concludes the proof of Theorem 3.32.

Remark 3.5.1 Notice that the stabilization terms in the Navier–Stokes equations are necessary to prove the convergence of the pressure but the stabilization terms in the magnetic equation are *not* needed to prove the convergence of $\boldsymbol{B}_h$. In other words, formulation (3.83) is still valid with $\lambda_B = 0$, but it is necessary to have $\lambda_u > 0$. However, it may be useful to choose $\lambda_B > 0$ to improve the quality of the results, as explained in the next remark. ◇

Remark 3.5.2 It has been noticed in the proof of Theorem 3.32 that the quantity $\boldsymbol{\pi}_u \times \boldsymbol{b} \cdot \mathbf{curl}\, \boldsymbol{e}_b$ coming from the magnetic equations was controlled using the stabilization terms of the Navier–Stokes equations, and conversely for

$\boldsymbol{b} \times \mathbf{curl}\, \boldsymbol{\pi}_b \cdot \boldsymbol{e}_u$. This is a consequence of the stabilization of the magnetic equation (*i.e.* $\lambda_B > 0$). It is worth mentioning how one would control $\boldsymbol{b} \times \mathbf{curl}\, \boldsymbol{\pi}_b \cdot \boldsymbol{e}_u$ if the magnetic equations were not stabilized (*i.e.* if $\lambda_B = 0$):

$$\begin{aligned}\left|\int_\Omega \boldsymbol{b} \times \mathbf{curl}\, \boldsymbol{\pi}_b \cdot \boldsymbol{e}_u\right| &= \left|\int_\Omega \mathbf{curl}\,(\boldsymbol{e}_u \times \boldsymbol{b}) \cdot \boldsymbol{\pi}_b\right| \\ &\leq 2C(\|\boldsymbol{b}\|_\infty + \|\nabla \boldsymbol{b}\|_\infty)\|\nabla \boldsymbol{e}_u\|_{L^2}\|\boldsymbol{\pi}_b\|_{L^2} \\ &\leq \frac{\eta}{\gamma_1}\int_\Omega |\nabla \boldsymbol{e}_u|^2 + C\frac{\gamma_1\|\boldsymbol{b}\|^2_{W^{1,\infty}}}{\eta}\int_\Omega |\boldsymbol{\pi}_b|^2.\end{aligned}$$

Thus, the following term would occur in the error estimate on $\boldsymbol{B}$:

$$\frac{h^2\|b\|^2_{W^{1,\infty}}}{\eta}h^{2l}|\boldsymbol{B}|^2_{l+1}.$$

This yields the convergence of the approximation, but, with the typical values of h, η and S encountered in practical 3D computations, this estimate is less accurate than that of Theorem 3.32. Numerical simulations confirm that it is indeed useful to stabilize the magnetic equations. ♢

Remark 3.5.3 As mentioned above, the fields $\boldsymbol{a}$ and $\boldsymbol{b}$ may be seen as the velocity $\boldsymbol{u}_h^n$ and magnetic field $\boldsymbol{B}_h^n$ occurring in Picard iterations (see (3.91) below). In this case $\boldsymbol{a} \in X_h$ and $\boldsymbol{b} \in W_h$. The above proof has thus to be adapted. First, the field $\boldsymbol{a} = \boldsymbol{u}_h^n$ is not divergence-free. Nevertheless, in the proof of Theorem 3.32, this assumption is only needed to ensure the skew-symmetry of the trilinear advection term. Thus, it suffices to replace in the definition of Φ

$$\int_\Omega \boldsymbol{a} \cdot \nabla \boldsymbol{v} \cdot \boldsymbol{w}$$

by

$$\int_\Omega \boldsymbol{a} \cdot \nabla \boldsymbol{v} \cdot \boldsymbol{w} + \frac{1}{2}\int_\Omega \boldsymbol{w} \cdot \boldsymbol{v}\, \mathrm{div}\, \boldsymbol{a},$$

(see R. Temam [227] for this classical trick). Let us now see what implies the assumption $\boldsymbol{b} = \boldsymbol{B}_h^n \in W_h$. We have the following inverse inequality (see P.G. Ciarlet [39, Theorem 17.2] for example):

$$\|\nabla \boldsymbol{b}\|_\infty \leq \frac{c}{h}\|\boldsymbol{b}\|_\infty,$$

where c is a constant independent of h. Thus, $(\|\boldsymbol{b}\|^2_\infty + h^2\|\nabla \boldsymbol{b}\|^2_\infty)$ could be replaced by $C\|\boldsymbol{b}\|^2_\infty$ in estimate (3.88). ♢

Remark 3.5.4 For the sake of simplicity, it has been assumed in this proof that the regime was advection-dominated on the whole domain. Nevertheless, for the numerical simulations, one should adapt the stabilization coefficient to the regime of the flow, as explained in Section 3.2 for the advection–diffusion equation. We refer, for example, to L.P. Franca and S.L. Frey in [82] and to L. Tobiska and R. Verfürth [233]. ♢

Remark 3.5.5 Theorem 3.32 provides an error estimate of the velocity in the $H^1(\Omega)$ seminorm. One could also establish an error estimate in the $L^2(\Omega)$ norm using standard arguments (see for example [75]). ◊

3.6 Solution strategy and algebraic aspects

We present in this section some algorithms, commonly used in practice, to solve the stationary and transient MHD equations. To avoid an excess of formalism, we choose to work on the strong formulation (3.34) and we omit the boundary conditions. In practice, these algorithms are of course implemented on the discretized versions of the variational formulations (3.46) or (3.56).

3.6.1 Fully coupled iterations for stationary problems

Suppose that we have an initial guess $(\boldsymbol{u}^0, \boldsymbol{B}^0)$, and that $(\boldsymbol{u}^n, \boldsymbol{B}^n)$ is known at iteration $n \geq 0$. Then the Picard algorithm consists in searching for $(\boldsymbol{u}^{n+1}, \boldsymbol{B}^{n+1}, p^{n+1})$, a solution to:

$$\left\{\begin{aligned} \boldsymbol{u}^n.\boldsymbol{\nabla}\boldsymbol{u}^{n+1} - \frac{1}{Re}\Delta\,\boldsymbol{u}^{n+1} + \boldsymbol{\nabla}p^{n+1} + S\,\boldsymbol{B}^n \times \mathbf{curl}\,\boldsymbol{B}^{n+1} &= \boldsymbol{f}, \\ \operatorname{div}\boldsymbol{u}^{n+1} &= 0, \\ \frac{1}{Rm}\mathbf{curl}\,(\mathbf{curl}\,\boldsymbol{B}^{n+1}) - \mathbf{curl}\,(\boldsymbol{u}^{n+1} \times \boldsymbol{B}^n) &= \boldsymbol{h}, \\ \operatorname{div}\boldsymbol{B}^{n+1} &= 0. \end{aligned}\right. \tag{3.91}$$

On the other hand, the Newton–Raphson algorithm consists in first solving for the increment $(\delta\boldsymbol{u}, \delta\boldsymbol{B}, \delta p)$ such that

$$\left\{\begin{aligned} &\boldsymbol{u}^n.\boldsymbol{\nabla}\delta\boldsymbol{u} + \delta\boldsymbol{u}.\boldsymbol{\nabla}\boldsymbol{u}^n - \frac{1}{Re}\Delta\,\delta\boldsymbol{u} + \boldsymbol{\nabla}\delta p + S\,\boldsymbol{B}^n \times \mathbf{curl}\,\delta\boldsymbol{B} \\ &\quad + S\,\delta\boldsymbol{B} \times \mathbf{curl}\,\boldsymbol{B}^n = \boldsymbol{f} - \boldsymbol{u}^n.\boldsymbol{\nabla}\boldsymbol{u}^n + \frac{1}{Re}\Delta\,\boldsymbol{u}^n - \boldsymbol{\nabla}p^n - S\,\boldsymbol{B}^n \times \mathbf{curl}\,\boldsymbol{B}^n, \\ &\qquad \operatorname{div}\delta\boldsymbol{u} = 0, \\ &\frac{1}{Rm}\mathbf{curl}\,(\mathbf{curl}\,\delta\boldsymbol{B}) - \mathbf{curl}\,(\delta\boldsymbol{u} \times \boldsymbol{B}^n) \\ &\quad - \mathbf{curl}\,(\boldsymbol{u}^n \times \delta\boldsymbol{B}) = \boldsymbol{h} - \frac{1}{Rm}\mathbf{curl}\,(\mathbf{curl}\,\boldsymbol{B}^n) + \mathbf{curl}\,(\boldsymbol{u}^n \times \boldsymbol{B}^n), \\ &\qquad \operatorname{div}\delta\boldsymbol{B} = 0. \end{aligned}\right. \tag{3.92}$$

The new iterate is then defined by

$$(\boldsymbol{u}^{n+1}, \boldsymbol{B}^{n+1}, p^{n+1}) = (\boldsymbol{u}^n, \boldsymbol{B}^n, p^n) + \lambda(\delta\boldsymbol{u}, \delta\boldsymbol{B}, \delta p).$$

The convergence of these methods is studied by M.D. Gunzburger et al. in [118]. In the original Newton–Raphson algorithm, the value of λ is 1. But it is usually preferred to determine λ with an adaptive linesearch procedure (see for example C.T. Kelley [131]).

As is often observed for many other contexts, an efficient strategy for the MHD equations is to first perform a few iterations of fixed point for robustness and then switch to the Newton algorithm for efficiency.

In both Picard iterations and Newton iterations, the hydrodynamic and the magnetic equations are coupled: in three dimensions, seven scalar unknowns have to be solved simultaneously (actually eight of them if a magnetic Lagrange multiplier is introduced as in formulation (3.56)). This results in very large sparse linear systems. Efficient solvers for sparse non-definite linear system have therefore to be used (typically GMRES, see Y. Saad and M.H. Schultz [200]), in combination with appropriate preconditioners (for example incomplete LU factorization).

It can be easily checked, proceeding as in Section 3.1.6, that the mixed finite element method applied to formulation (3.46) has the following matrix form:

$$\begin{bmatrix} A_f^n & -(L^n)^T & D_f^T \\ L^n & A_m & 0 \\ D_f & 0 & 0 \end{bmatrix} \begin{bmatrix} U^{n+1} \\ B^{n+1} \\ P^{n+1} \end{bmatrix} = \begin{bmatrix} F_f \\ F_m \\ 0 \end{bmatrix}.$$

Here A_f^n comes from the advection and viscous terms in the Navier–Stokes equation (it depends on $\boldsymbol{u}^n$), D_f^T from the gradient of the pressure, D_f from the incompressibility constraint, A_m from the magnetic diffusive terms and L^n from the Lorentz force (it depends on $\boldsymbol{B}^n$).

If we alternatively consider formulation (3.56), also valid on non-convex polyhedra, the algebraic form reads:

$$\begin{bmatrix} A_f^n & -(L^n)^T & D_f^T & 0 \\ L^n & \tilde{A}_m & 0 & D_m^T \\ D_f & 0 & 0 & 0 \\ 0 & D_m & 0 & 0 \end{bmatrix} \begin{bmatrix} U^{n+1} \\ B^{n+1} \\ P^{n+1} \\ R^{n+1} \end{bmatrix} = \begin{bmatrix} F_f \\ F_m \\ 0 \\ 0 \end{bmatrix},$$

where $\tilde{A}_m$ indicates the new form of the magnetic diffusive terms and D_m is the block related to the divergence of $\boldsymbol{B}$.

The stabilized finite element formulation (3.83) yields the following matrix form:

$$\begin{bmatrix} A_{f,s}^n & -(L_s^n)^T & B_s^T \\ L_s^n & A_{m,s} & C_{2,s}^n \\ B_s & C_{3,s}^n & -C_{1,s}^n \end{bmatrix} \begin{bmatrix} U^{n+1} \\ B^{n+1} \\ P^{n+1} \end{bmatrix} = \begin{bmatrix} F_f \\ F_m \\ 0 \end{bmatrix}.$$

We notice in particular that the entry located on the last row and last column is now $-C_{1,s}$ (instead of 0 in the previous formulations). The presence of this block allows the matrix to be invertible even when $\operatorname{Ker} B_s^T \neq \{0\}$, *i.e.* when the inf–sup condition for the velocity and the pressure spaces is not satisfied.

3.6.2 Decoupled iterations for stationary problems

The above-mentioned fully coupled algorithms are stable for a wide range of parameters but are computationally expensive since all the unknowns are solved for simultaneously. It is therefore tempting to consider segregated schemes. An instance is:

$$\begin{cases} -\dfrac{1}{Re}\Delta\,\boldsymbol{u}^{n+1} + \boldsymbol{\nabla} p^{n+1} = \boldsymbol{f} - \boldsymbol{u}^n.\boldsymbol{\nabla}\boldsymbol{u}^n + S\,\mathbf{curl}\,\boldsymbol{B}^n \times \boldsymbol{B}^n, \\ \operatorname{div}\boldsymbol{u}^{n+1} = 0, \end{cases} \tag{3.93}$$

and

$$\begin{cases} \dfrac{1}{Rm}\mathbf{curl}\,(\mathbf{curl}\,\boldsymbol{B}^{n+1}) = \mathbf{curl}\,(\boldsymbol{u}^n \times \boldsymbol{B}^n) + \boldsymbol{h}, \\ \operatorname{div}\boldsymbol{B}^{n+1} = 0. \end{cases} \tag{3.94}$$

The advantage of such a scheme is obvious: the two subproblems to be solved at each iteration are uncoupled and an efficient dedicated solver can be used for each of them. There are many instances of such solvers for the hydrodynamics equations (see, *e.g.* O. Pironneau [186] or R. Glowinski [109]). Regarding the magnetic equation, we refer for example to [94]. Nevertheless, as it is often the case for segregated schemes, the decoupling comes at a price: stability properties deteriorate. Although some theoretical convergence results exist in academic settings (M.D. Gunzburger et al. [118]), the algorithm is usually unstable when the physical parameters have realistic values.

A slightly more implicit, decoupled scheme may be considered:

$$\begin{cases} -\dfrac{1}{Re}\Delta\,\boldsymbol{u}^{n+1} + \boldsymbol{u}^n.\boldsymbol{\nabla}\boldsymbol{u}^{n+1} + \boldsymbol{\nabla} p^{n+1} = \boldsymbol{f} + S\,\mathbf{curl}\,\boldsymbol{B}^n \times \boldsymbol{B}^n, \\ \operatorname{div}\boldsymbol{u}^{n+1} = 0, \end{cases} \tag{3.95}$$

and $\boldsymbol{B}^{n+1}$ is found by solving

$$\begin{cases} \dfrac{1}{Rm}\mathbf{curl}\,\mathbf{curl}\,\boldsymbol{B}^{n+1} - \mathbf{curl}\,(\boldsymbol{u}^{n+1} \times \boldsymbol{B}^n) = \boldsymbol{h}, \\ \operatorname{div}\boldsymbol{B}^{n+1} = 0. \end{cases} \tag{3.96}$$

The semi-implicit treatment of the advection in (3.95) *a priori* improves the stability of the hydrodynamics equation compared to (3.93). Nevertheless, this algorithm is still unsatisfactory in practical MHD problems. One may improve its stability by introducing relaxation: Let $0 \leq \theta < 1$. Then the velocity $\boldsymbol{u}^{n+1}$ is computed as follows:

$$\begin{cases} -\dfrac{1}{Re}\Delta\,\boldsymbol{u}^{n+1/2} + \boldsymbol{u}^n.\boldsymbol{\nabla}\boldsymbol{u}^{n+1/2} + \boldsymbol{\nabla} p^{n+1/2} = \boldsymbol{f} + S\,\mathbf{curl}\,\boldsymbol{B}^n \times \boldsymbol{B}^n, \\ \operatorname{div}\boldsymbol{u}^{n+1/2} = 0, \end{cases} \tag{3.97}$$

$$\boldsymbol{u}^{n+1} = \theta\boldsymbol{u}^n + (1-\theta)\boldsymbol{u}^{n+1/2}, \tag{3.98}$$

and $\boldsymbol{B}^{n+1}$ is found by solving

$$\begin{cases} \dfrac{1}{Rm}\mathbf{curl}\,\mathbf{curl}\,\boldsymbol{B}^{n+1/2} - \mathbf{curl}\,(\boldsymbol{u}^{n+1} \times \boldsymbol{B}^{n+1/2}) = 0, \\ \mathrm{div}\,\boldsymbol{B}^{n+1/2} = 0. \end{cases} \tag{3.99}$$

$$\boldsymbol{B}^{n+1} = \theta \boldsymbol{B}^n + (1-\theta)\boldsymbol{B}^{n+1/2}. \tag{3.100}$$

For Hartmann flows (see Section 3.7), it can be shown (see [88, Chapter 7]), that with a Hartmann number (defined in Section 1.6) typically greater than 10, the convergence is only achieved with a very strong relaxation ($\theta \approx 1$) and thus is very slow.

One may conjecture that the poor behavior observed in the simulation of the stationary MHD equations using decoupled schemes is mainly due to the explicit treatment of the Lorentz force in the right-hand side of (3.95). An alternative formulation, proposed in [88], consists in treating a part of this force implicitly. For this purpose, we use Ohm's law (see (1.36) in Chapter 1),

$$\boldsymbol{j} = Rm(\boldsymbol{E} + \boldsymbol{u} \times \boldsymbol{B}),$$

to write the Lorentz force:

$$S\,\boldsymbol{j} \times \boldsymbol{B} = S\,Rm\,(\boldsymbol{E} \times \boldsymbol{B} + (\boldsymbol{u} \times \boldsymbol{B}) \times \boldsymbol{B}).$$

In this form, the part of the Lorentz force involving $\boldsymbol{u}$ can be treated implicitly while preserving the equations uncoupled. More precisely, the approach consists of the following iterations.

First, solve

$$\begin{cases} \dfrac{1}{Rm}\mathbf{curl}\,(\mathbf{curl}\,\boldsymbol{B}^{n+1}) = \mathbf{curl}\,(\boldsymbol{u}^n \times \boldsymbol{B}^n), \\ \mathrm{div}\,\boldsymbol{B}^{n+1} = 0. \end{cases} \tag{3.101}$$

Second, using the fact that for stationary problems the electric field $\boldsymbol{E}$ is the gradient of a potential Φ, solve the scalar Poisson equation with a Neumann boundary condition:

$$\begin{cases} \mathrm{div}\,(Rm\boldsymbol{\nabla}\Phi^{n+1}) = \mathrm{div}\,(Rm\,\boldsymbol{u}^n \times \boldsymbol{B}^{n+1}), \\ Rm\,\dfrac{\partial \Phi^{n+1}}{\partial \boldsymbol{n}} = Rm\,\boldsymbol{u}^n \times \boldsymbol{B}^{n+1} \cdot \boldsymbol{n} - \mathbf{curl}\,\boldsymbol{B}^{n+1} \cdot \boldsymbol{n}. \end{cases} \tag{3.102}$$

Third, solve the Navier–Stokes equations

$$\begin{cases} \boldsymbol{u}^n \cdot \boldsymbol{\nabla}\boldsymbol{u}^{n+1} - \dfrac{1}{Re}\Delta\,\boldsymbol{u}^{n+1} + \boldsymbol{\nabla}p^{n+1} \\ \quad - S\,Rm(\boldsymbol{u}^{n+1} \times \boldsymbol{B}^{n+1}) \times \boldsymbol{B}^{n+1} = \boldsymbol{f} - S\,Rm\boldsymbol{\nabla}\Phi^{n+1} \times \boldsymbol{B}^{n+1}, \\ \mathrm{div}\,\boldsymbol{u}^{n+1} = 0. \end{cases} \tag{3.103}$$

We are not aware of any theoretical analysis of this scheme. Nevertheless, numerical observations show that it is efficient and stable in various configurations. Comparisons with fully coupled methods are presented in Section 3.7.

3.6.3 Fully coupled iterations for transient problems

Among the numerous possible time discretization algorithms, we mention the following semi-implicit Euler scheme that will be used in Chapter 5. Let us denote by δt the timestep and $(\boldsymbol{u}^n, p^n, \boldsymbol{B}^n)$ the approximate value at time t^n. The unknowns fields at time t^{n+1} are obtained by solving:

$$\left\{\begin{aligned} \frac{\boldsymbol{u}^{n+1}-\boldsymbol{u}^n}{\delta t} + \boldsymbol{u}^n.\boldsymbol{\nabla}\boldsymbol{u}^{n+1} - \frac{1}{Re}\Delta\,\boldsymbol{u}^{n+1} + \boldsymbol{\nabla}p^{n+1} + S\,\boldsymbol{B}^n \times \mathbf{curl}\,\boldsymbol{B}^{n+1} &= \boldsymbol{f},\\ \operatorname{div}\boldsymbol{u}^{n+1} &= 0,\\ \frac{\boldsymbol{B}^{n+1}-\boldsymbol{B}^n}{\delta t} + \frac{1}{Rm}\mathbf{curl}\,(\mathbf{curl}\,\boldsymbol{B}^{n+1}) - \mathbf{curl}\,(\boldsymbol{u}^{n+1}\times\boldsymbol{B}^n) &= \boldsymbol{h},\\ \operatorname{div}\boldsymbol{B}^{n+1} &= 0. \end{aligned}\right.$$

This scheme has been studied by F. Armero and J.C. Simo [8], with an emphasis on dissipative properties. The resulting linear system is very similar to the Picard iterations, in particular it couples the seven scalar unknowns. It is only first-order accurate in time but has a nice stability property in the energy norm: multiplying the first equation by $\boldsymbol{u}^{n+1}$, and the third one by $\boldsymbol{B}^{n+1}$, and integrating (assuming that the boundary conditions and the right-hand side are zero to simplify), it is a matter of routine to check that:

$$\begin{aligned} \frac{1}{2\delta t}\int_\Omega \left(|\boldsymbol{u}^{n+1}|^2 + S\,|\boldsymbol{B}^{n+1}|^2\right) &+ \frac{1}{Re}\int_\Omega |\boldsymbol{\nabla}\boldsymbol{u}^{n+1}|^2\\ &+ \frac{S}{Rm}\int_\Omega |\mathbf{curl}\,\boldsymbol{B}^{n+1}|^2 \le \frac{1}{2\delta t}\int_\Omega \left(|\boldsymbol{u}^n|^2 + S\,|\boldsymbol{B}^n|^2\right). \end{aligned}$$

Note that this property mainly relies on the fact that owing to the specific linearization performed on the coupling terms, the integral $\displaystyle\int_\Omega \boldsymbol{B}^n \times \mathbf{curl}\,\boldsymbol{B}^{n+1} \cdot \boldsymbol{u}^{n+1}$ exactly compensates for the integral $\displaystyle\int_\Omega \mathbf{curl}\,(\boldsymbol{u}^{n+1}\times\boldsymbol{B}^n)\cdot\boldsymbol{B}^{n+1}$. Thus, the discretization does not induce any spurious energy transfer between hydrodynamics and electromagnetism. This cancellation effect is the main advantage of solving the system in a fully coupled way.

3.6.4 MHD versus Navier–Stokes solvers

To conclude this section, we address the following practical question: how to develop a MHD solver from an existing finite element Navier–Stokes solver. Of course this question is very "code-dependent" and cannot be addressed in details in a textbook. We therefore restrict ourselves to two general comments, which we however believe to be useful practically.

First, the simplest strategy typically consists in choosing a Lagrange finite element for the magnetic field since such elements are usually present in a Navier–Stokes solver. The formulation presented in Section 3.4.1 is therefore the most convenient. Nevertheless, we recall that it is limited to convex polyhedra. The formulation of Section 3.4.2 is valid on "arbitrary" domains but is also more demanding from the implementation viewpoint since the Nédélec finite elements are generally not present in a Navier–Stokes code. An interesting alternative may be to consider the regularization techniques recently proposed by M. Costabel and M. Dauge [43] and used in MHD by U. Hasler et al. [120].

Second, it is well-known (see, for example, R. Moreau [172]) that the MHD equations can be rewritten in the following way (the so-called Helmholtz formulation):

$$\begin{cases} \rho\,\dfrac{\partial \boldsymbol{u}}{\partial t} + \rho\,\boldsymbol{u}\cdot\nabla\,\boldsymbol{u} - \eta\,\Delta\boldsymbol{u} + \nabla\left(p + \dfrac{|\boldsymbol{B}|^2}{2\mu}\right) = \dfrac{1}{\mu}\,\boldsymbol{B}\cdot\nabla\boldsymbol{B} + \rho\mathbf{g}, \\ \operatorname{div}\boldsymbol{u} = 0, \end{cases} \tag{3.104}$$

$$\begin{cases} \dfrac{\partial \boldsymbol{B}}{\partial t} - \dfrac{1}{\mu\sigma}\Delta\boldsymbol{B} + \boldsymbol{u}\cdot\nabla\boldsymbol{B} - \boldsymbol{B}\cdot\nabla\boldsymbol{u} + \nabla q = 0, \\ \operatorname{div}\boldsymbol{B} = 0, \end{cases} \tag{3.105}$$

where the following identities have been used:

$$(\mathbf{curl}\,\boldsymbol{B})\times\boldsymbol{B} = \boldsymbol{B}\cdot\nabla\boldsymbol{B} - \frac{1}{2}\nabla(|\boldsymbol{B}|^2),$$

$$\mathbf{curl}\,(\boldsymbol{u}\times\boldsymbol{B}) = \boldsymbol{B}\cdot\nabla\boldsymbol{u} - \boldsymbol{B}\operatorname{div}\boldsymbol{u} - \boldsymbol{u}\cdot\nabla\boldsymbol{B} + \boldsymbol{u}\operatorname{div}\boldsymbol{B}.$$

The q variable is in fact a dummy Lagrange multiplier: by taking the divergence of the equation on $\boldsymbol{B}$, one obtains that $\Delta q = 0$, and therefore $q = 0$, at least if q is taken in a suitable functional space. Notice that if $(\boldsymbol{u}, p, \boldsymbol{B})$ is a solution to (1.40), then $(\boldsymbol{u}, p, \boldsymbol{B}, q = 0)$ is a solution to (3.104)–(3.105). The Helmholtz formulation is very similar to the Navier–Stokes equations (in fact it is exactly the vorticity equation) which makes it attractive for an implementation in a Navier–Stokes solver. Nevertheless, if we use typical Stokes finite elements (in primitive variables $(\boldsymbol{u}, p)$) to discretize the variational formulation which results from (3.104)–(3.105), the functional spaces are not well-suited to implement the natural magnetic boundary conditions (see Section 3.8). In conclusion, even if more demanding, the variational formulations based on the **curl curl** formulation of the MHD equations seems more convenient than those resulting from (3.105).

3.7 Examples of test cases and simulations

We present in this section a few numerical experiments. Some of them directly come from the classical MHD literature (see, for example, R. Moreau [172] or W.F. Hughes and F.J. Young [130]) and can be used as benchmarks to validate an MHD solver. In the whole section, we consider the stationary MHD equations

$$\begin{cases} -\dfrac{1}{Re}\Delta\,\boldsymbol{u} + \boldsymbol{\nabla} p + \boldsymbol{u}\cdot\boldsymbol{\nabla}\boldsymbol{u} + S\,\boldsymbol{B}\times\mathbf{curl}\,\boldsymbol{B} = 0, \\ \operatorname{div}\boldsymbol{u} = 0, \\ \dfrac{1}{Rm}\mathbf{curl}\,(\mathbf{curl}\,\boldsymbol{B}) - \mathbf{curl}\,(\boldsymbol{u}\times\boldsymbol{B}) = 0, \\ \operatorname{div}\boldsymbol{B} = 0. \end{cases}$$

3.7.1 Hartmann flows

Hartmann flows are the MHD version of the classical Poiseuille flows. They consist of the internal flow of an incompressible fluid in the presence of an external transverse magnetic field $\boldsymbol{B}^d$. The geometry is either 2D or 3D, the walls of the duct are either conducting or insulating.

We impose no-slip boundary condition on the wall

$$\boldsymbol{u} = 0, \text{ on the wall } \Gamma_{\text{wall}},$$

and Neumann boundary conditions on the inlet and the outlet

$$-p\boldsymbol{n} + \frac{1}{Re}\boldsymbol{\nabla u} \cdot \boldsymbol{n} = -p_d\boldsymbol{n} \text{ on the inlet } \Gamma_{\text{in}} \text{ and the outlet } \Gamma_{\text{out}}. \tag{3.106}$$

The pressure p_d in (3.106) is given. The magnetic boundary conditions are discussed below.

3.7.1.1 Hartmann flows in 2D We consider the 2D domain $\Omega = [0, L]\times[-1, 1]$ and the external magnetic field $\boldsymbol{B}^d = (0, 1)$. It can be checked that the velocity and the magnetic fields have the following form: $\boldsymbol{u} = (u(y), 0)$ and $\boldsymbol{B} = (b(y), 1)$. The pressure is given by:

$$p_d(x, y) = -Gx - S\,b^2(y)/2 + p_0.$$

The unknown functions $u(y)$ and $b(y)$ are solutions to the following system:

$$\begin{aligned} u''(y) + Re\,S\,b'(y) &= -G\,Re, \\ b''(y) + Rm\,u'(y) &= 0. \end{aligned} \tag{3.107}$$

The wall is assumed to be perfectly insulating. On the boundary, we impose the tangential part of the magnetic field:

$$\boldsymbol{B} \times \boldsymbol{n} = \boldsymbol{B}^d \times \boldsymbol{n}. \tag{3.108}$$

The explicit solution of (3.107) is

$$u(y) = \frac{G\,Re}{Ha\tanh(Ha)}\left(1 - \frac{\cosh(y\,Ha)}{\cosh(Ha)}\right),$$

and

$$b(y) = \frac{G}{S}\left(\frac{\sinh(y\,Ha)}{\sinh(Ha)} - y\right),$$

where the Hartmann number Ha is defined by $BL\sqrt{\sigma/\eta}$ (B and L being the reference magnetic field and length, respectively).

The numerical solution is obtained by solving the formulation (3.46) presented in Section 3.3.1 with the Taylor–Hood finite element (see Section 3.1.7). The resolution of the nonlinear system is performed with a combination of Picard iterations and Newton algorithm, as explained in Section 3.6.1. Figure 3.4 show the velocity and the magnetic field. The analytical solutions along with the numerical ones are represented on Figures 3.5 and 3.6.

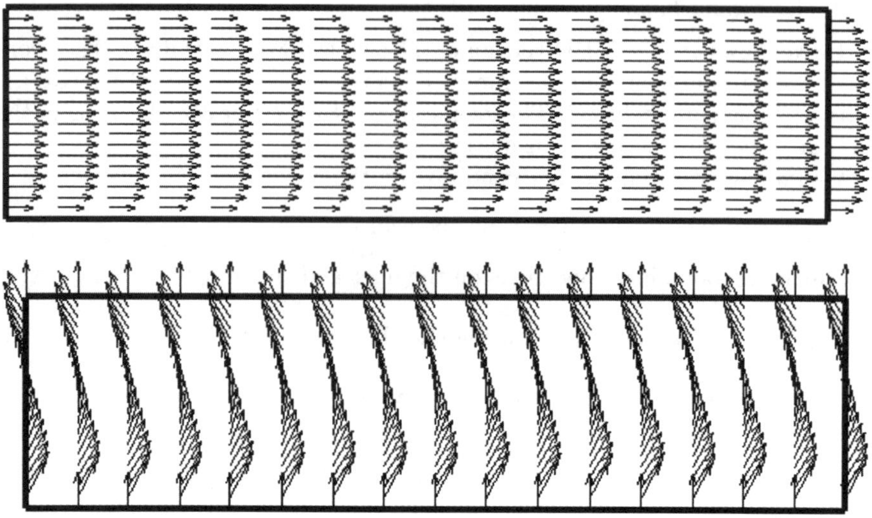

Fig. 3.4 Velocity and magnetic field in a 2D Hartmann flow with insulating walls ($Re = 10$, $Rm = 10$, $S = 1$, $G = 1$).

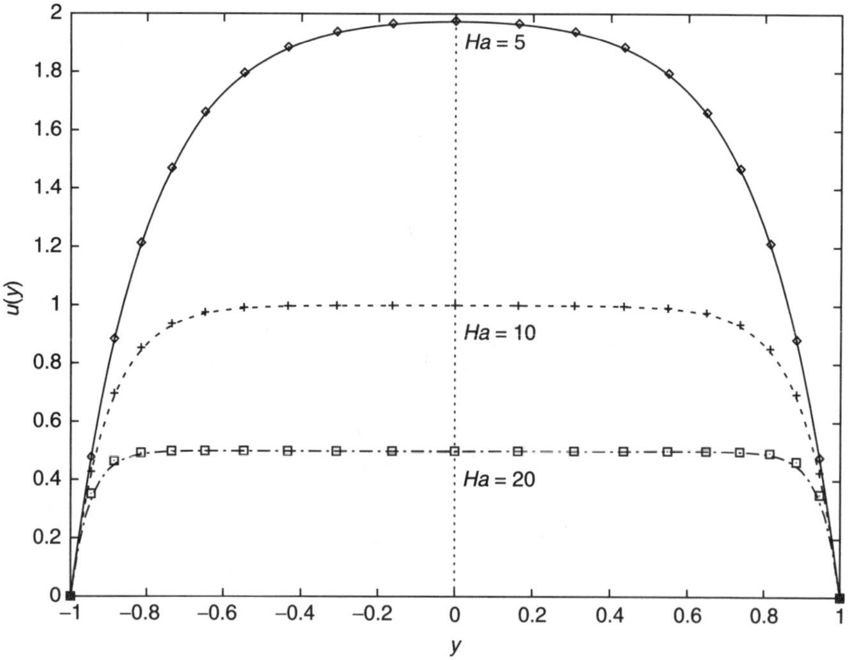

Fig. 3.5 Hartmann flow in 2D with insulating walls: computed (points) and theoretical (line) $\boldsymbol{u}$.

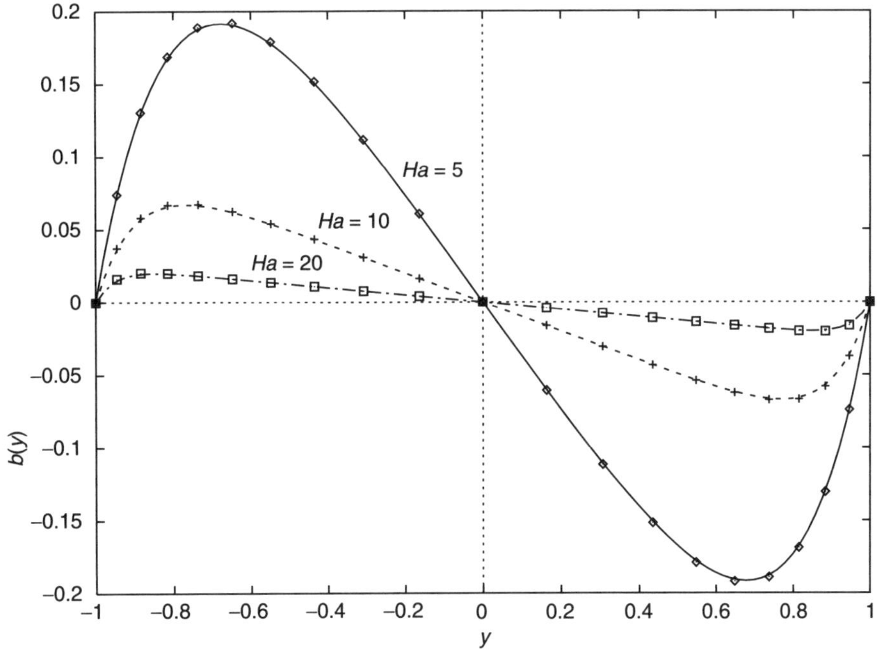

Fig. 3.6 Hartmann flow in 2D with insulating walls: computed (points) and theoretical (line) $\boldsymbol{B}$.

3.7.1.2 Hartmann flow in 3D We now consider a Hartmann flow in a 3D rectangular duct. The axis is Ox, the section in the plane (Oy, Oz) is a rectangle of length $2y_0$ and width $2z_0$. The value of the transverse magnetic field is given by $\boldsymbol{B}^d = (0, 1, 0)$. The velocity and the magnetic field have the following form: $\boldsymbol{u} = (u(y, z), 0, 0)$ and $\boldsymbol{B} = (b(y, z), 1, 0)$. The functions u and b are solutions to:

$$\begin{cases} \dfrac{\partial^2 u}{\partial y^2} + \dfrac{\partial^2 u}{\partial z^2} + Re\, S \dfrac{\partial b}{\partial y} = -GRe, \\ \dfrac{\partial^2 b}{\partial y^2} + \dfrac{\partial^2 b}{\partial z^2} + Rm \dfrac{\partial u}{\partial y} = 0. \end{cases} \tag{3.109}$$

Various cases are considered in the literature depending on whether each of the walls is considered perfectly conducting or perfectly insulating. We refer to W.F. Hughes and F.J. Young [130] for an exhaustive presentation. For conciseness, we restrict ourselves to the case of insulating walls. We therefore have $u(\pm y_0, z) = u(y, \pm z_0) = 0$ and $b(\pm y_0, z) = b(y, \pm z_0) = 0$. The solution of (3.107) can be obtained with elementary arguments but tedious computations. We refer for example to [88] or [130].

For the numerical simulations, we choose $y_0 = 2$, $z_0 = 1$ and we consider the same boundary conditions as in 2D: a pressure drop is imposed with (3.106), no-slip is assumed on the wall, and the magnetic boundary conditions are given

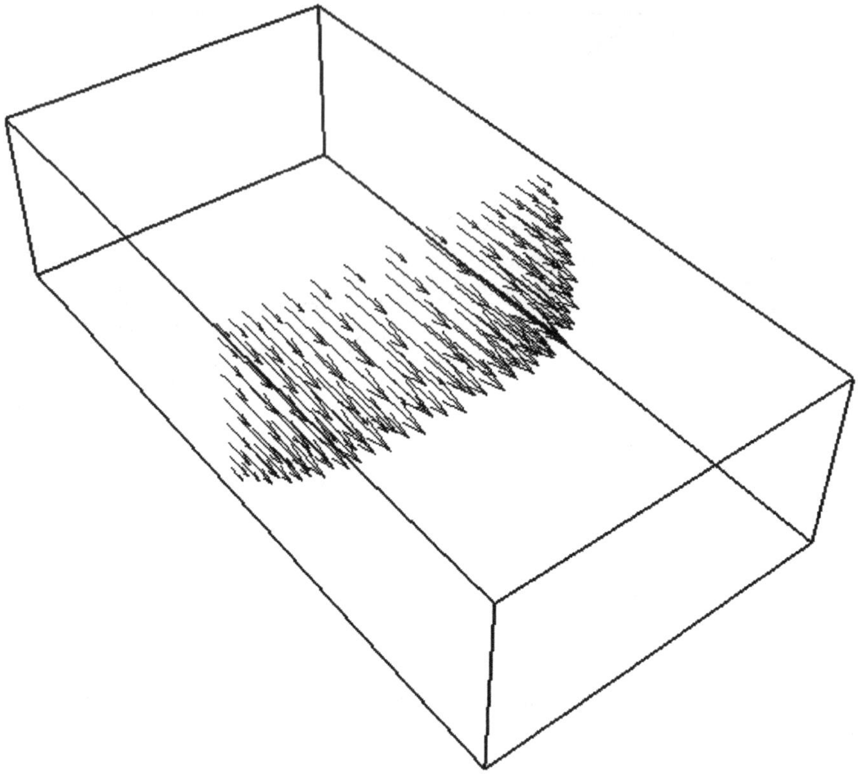

Fig. 3.7 Velocity field in a 3D Hartmann flow (insulating walls).

by (3.108). We use the stabilized formulation (3.83) with $\mathbb{P}_1/\mathbb{P}_1$ finite elements and a combination of Picard and Newton iterations. Figures 3.7 and 3.8 show the velocity and the magnetic field, respectively. The solution being invariant along the axis of the duct, we have represented the vector on a plane orthogonal to the flow. As in 2D, it can be checked that analytical and numerical solutions are in very good agreement [88].

3.7.2 A fluid carrying current in the presence of a magnetic field

In this simulation, an electric current flows through a fluid. The fluid is enclosed within a parallelepiped whose top and bottom are assumed to be perfectly conducting and whose sides are perfectly insulating. External electric conductors surround the parallelepiped and create a magnetic field (see Figure 3.9).

The magnetic field created by the external conductors and the internal homogeneous current (the fluid being at rest) are computed by the Biot–Savart formula. This field is used to enforce the boundary conditions on $\boldsymbol{B}$ in order to perform the MHD computations.

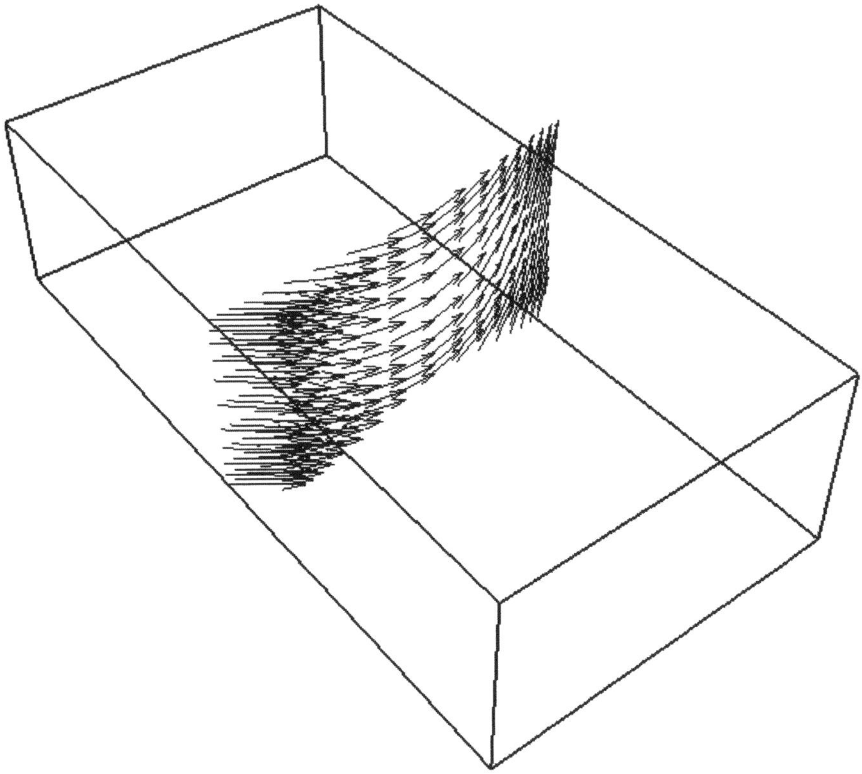

Fig. 3.8 Magnetic field in a 3D Hartmann flow (insulating walls).

Figure 3.10 displays the results obtained with four or five external conductors. Other configurations can be considered (see [89]). These numerical results are in good qualitative agreement with the predictions by J.M. Blanc and P. Entner [22].

3.7.3 Convergence of nonlinear algorithms

We now compare, for two test cases, the convergence rate of some nonlinear algorithms presented in Section 3.6 for the solution of the stationary MHD equations. In Figure 3.11, P denotes the Picard algorithm (3.91), P/NR denotes the Newton–Raphson algorithm (3.92) initialized by three Picard iterations, "relax" denotes the uncoupled iterations with relaxation (3.97)–(3.100), and "Ohm" denotes the uncoupled formulation (3.101)–(3.103) based on the splitting of Ohm's law.

We first consider a 2D Hartmann flow with insulating walls and $Ha = Re = Rm = 10$, $S = 1$. As announced in Section 3.6, the convergence rate of Newton–Raphson and Picard algorithms is higher than the uncoupled iterations with relaxation. Of course one fully coupled iteration is more expensive than a

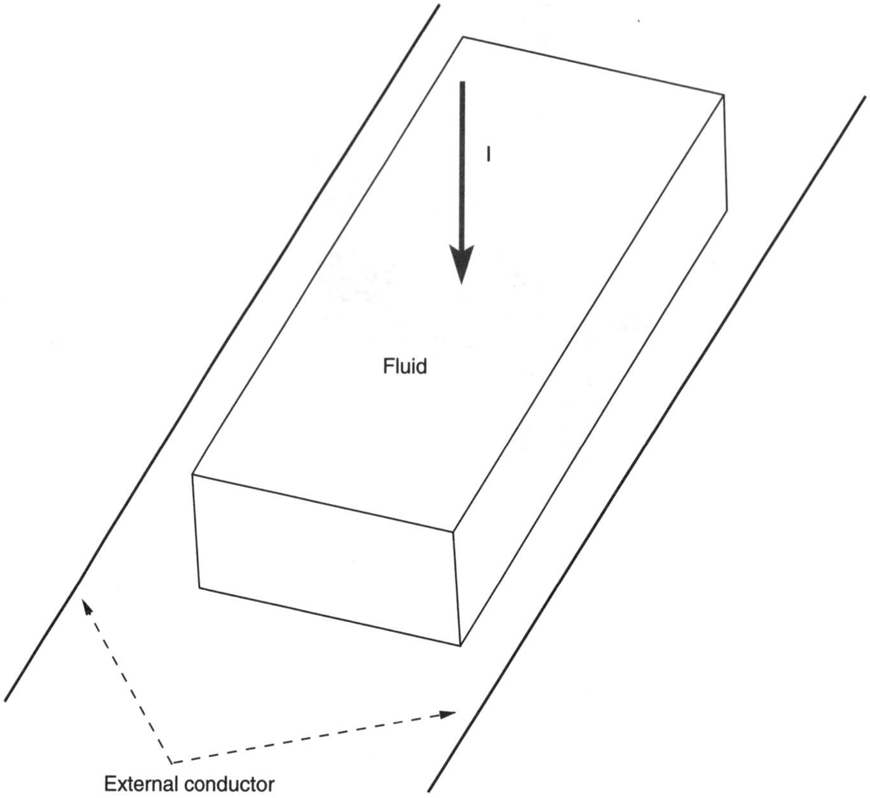

Fig. 3.9 A vertical electric current flows through a fluid enclosed in a box surrounded by external conductors.

decoupled one. But *in fine*, the high convergence rate of fully coupled methods make them competitive in term of CPU time. Moreover, with Picard iterations or Newton–Raphson algorithm, the convergence is achieved in a robust way, without tuning any relaxation parameter. The convergence rate obtained with the uncoupled "Ohm" algorithm (3.101)–(3.103) is very satisfactory. Nevertheless, in those special cases of Hartmann flows, some coupling terms vanish (the right-hand side of $(3.102)_1$), which artificially favors this scheme.

We compare the same schemes in Figure 3.12 for the 3D experiment described in Section 3.7.2 (two external conductors, $Re = 100$, $Rm = S = 1$). The physical configuration is now complex enough to ensure that none of the coupling or nonlinear terms degenerate. The above conclusions are confirmed. In particular, it seems that the uncoupled "Ohm" scheme (3.101)–(3.103) is an interesting alternative to the fully coupled scheme. We nevertheless recall that it is only defined for stationary (magnetic) problems since it essentially relies upon the fact that the electric field $\boldsymbol{E}$ is derived from a potential Φ.

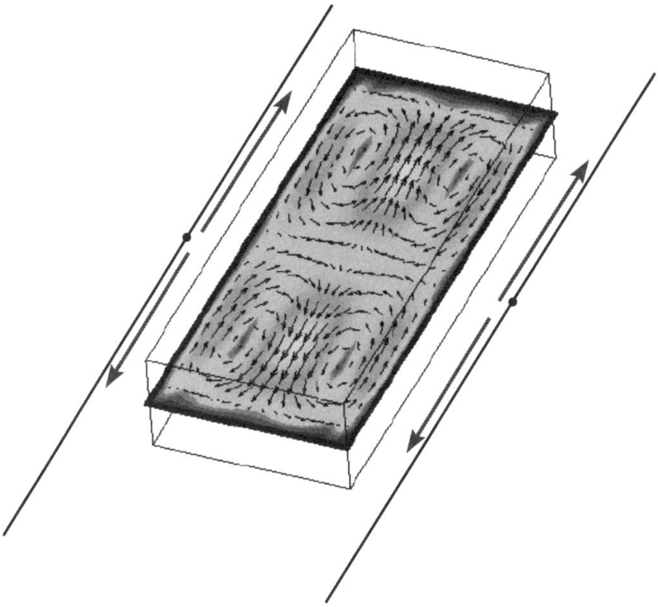

(i) Configuration with four external conductors (four symmetric vortices).

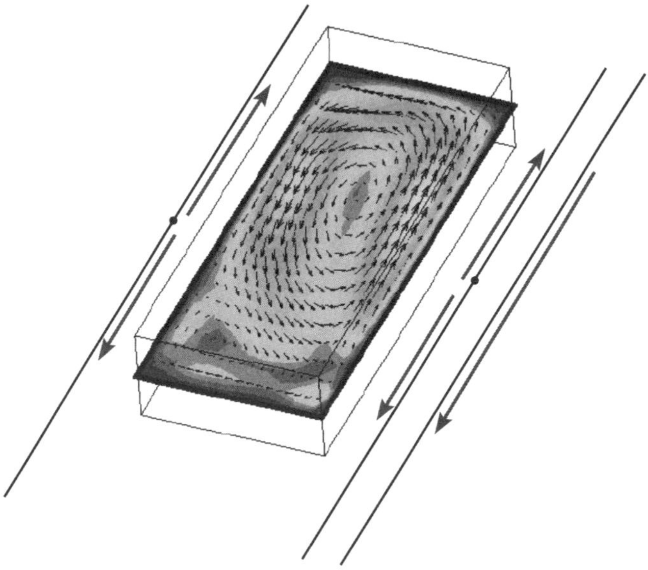

(ii) Configuration with five external conductors (asymmetric vortices).

Fig. 3.10 Simulations performed in the setting of Figure 3.9. **See plate 1.**

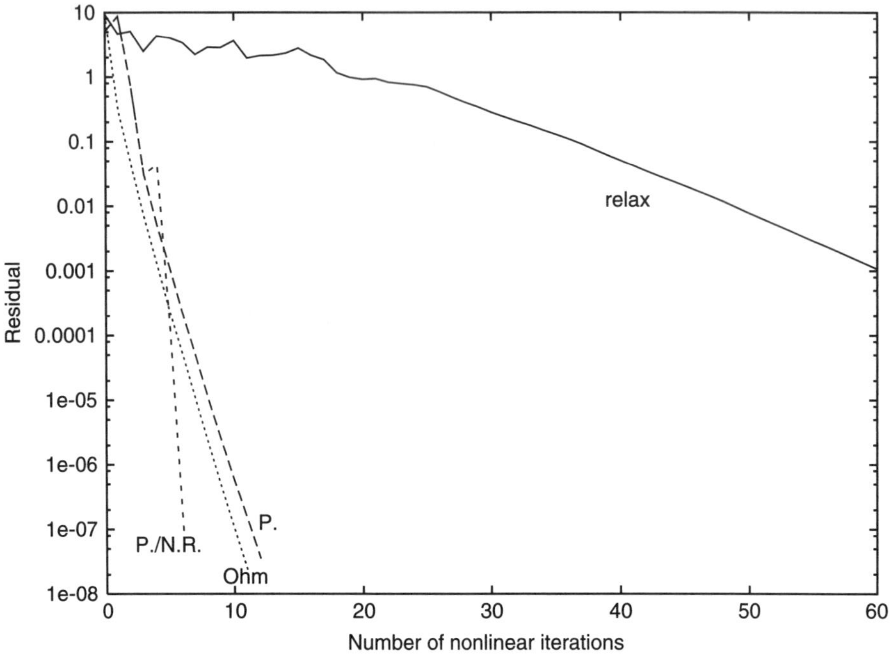

Fig. 3.11 Evolution of the nonlinear residual with various algorithms in the case of a 2D Hartman flow ($Ha = 10$).

3.8 About the boundary conditions

Let Ω be a bounded regular domain of $\mathbb{R}^3$ whose boundary is denoted by Γ, and let Ω' be the exterior of Ω, *i.e.* $\mathbb{R}^3 \backslash \bar{\Omega}$. We denote by $\boldsymbol{n}$ the outward-pointing normal to Ω.

The solution of the Maxwell equations over the whole space $\mathbb{R}^3$ satisfies the following transmission relations across Γ

$$\begin{cases} (i) & (\boldsymbol{D}_0 - \boldsymbol{D}) \cdot \boldsymbol{n} = \rho_\Gamma, \\ (ii) & (\boldsymbol{H}_0 - \boldsymbol{H}) \times \boldsymbol{n} = -\boldsymbol{j}_\Gamma, \\ (iii) & (\boldsymbol{B}_0 - \boldsymbol{B}) \cdot \boldsymbol{n} = 0, \\ (iv) & (\boldsymbol{E}_0 - \boldsymbol{E}) \times \boldsymbol{n} = 0, \end{cases} \tag{3.110}$$

where ρ_Γ and $\boldsymbol{j}_\Gamma$ are the surface density of charge and current, respectively (see R. Dautray and J.-L. Lions [45] for example), $(\boldsymbol{B}, \boldsymbol{D}, \boldsymbol{E}, \boldsymbol{H})$ and $(\boldsymbol{B}_0, \boldsymbol{D}_0, \boldsymbol{E}_0, \boldsymbol{H}_0)$ denote the solution in Ω and Ω', respectively.

We now assume that Ω and Ω' are both perfect media, in the sense that the following relations hold:

$$\boldsymbol{D} = \epsilon \boldsymbol{E} \text{ and } \boldsymbol{H} = \frac{\boldsymbol{B}}{\mu},$$

and we assume that μ and ϵ are constant over $\mathbb{R}^3$.

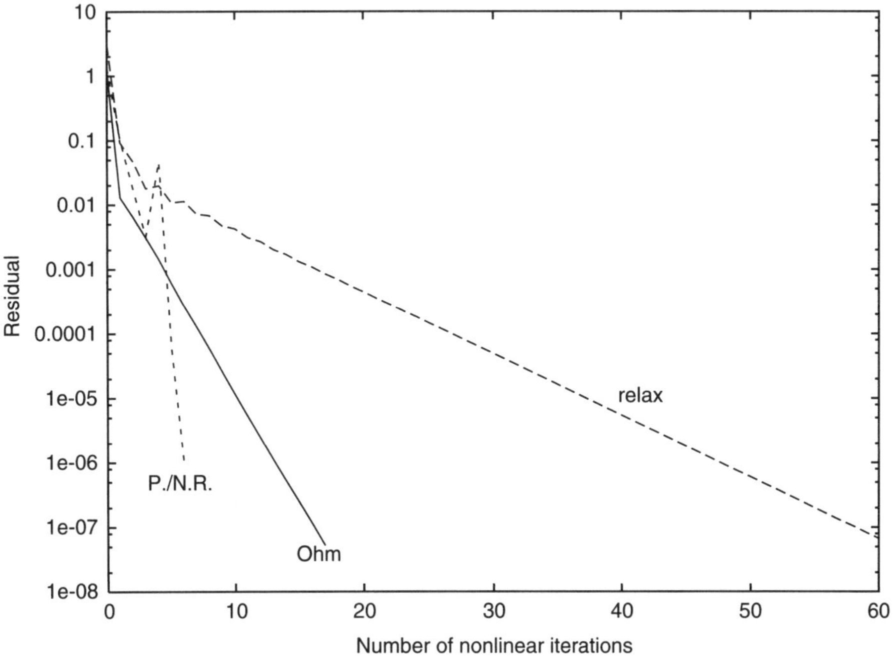

Fig. 3.12 Nonlinear residual with various algorithms in the 3D experiment of Section 3.7.2 (with two external conductors, $Re = 100$, $Rm = S = 1$).

When solving electromagnetism problems on bounded domains, two sets of boundary conditions are typically used: either keeping (*iii*) and (*iv*) in (3.110) as, for example, in F. Kikuchi [134],

$$\begin{cases} \boldsymbol{B} \cdot \boldsymbol{n} = \boldsymbol{B}_0 \cdot \boldsymbol{n}, \\ \boldsymbol{E} \times \boldsymbol{n} = \boldsymbol{E}_0 \times \boldsymbol{n}, \end{cases} \tag{3.111}$$

or keeping (*i*) and (*ii*) in (3.110) as, for example, in J.-L. Guermond and P.D. Minev [114],

$$\begin{cases} \boldsymbol{E} \cdot \boldsymbol{n} = \boldsymbol{E}_0 \cdot \boldsymbol{n} - \dfrac{1}{\epsilon}\, \rho_\Gamma, \\ \boldsymbol{B} \times \boldsymbol{n} = \boldsymbol{B}_0 \times \boldsymbol{n} + \mu \boldsymbol{j}_\Gamma. \end{cases} \tag{3.112}$$

In the MHD problems we are interested in, Ohm's law (1.36) allows us to eliminate the electric field $\boldsymbol{E}$. The above two sets of boundary conditions therefore become: either

$$\begin{cases} \boldsymbol{B} \cdot \boldsymbol{n} = \boldsymbol{B}_0 \cdot \boldsymbol{n}, \\ \left(\dfrac{1}{\sigma\mu}\mathbf{curl}\, \boldsymbol{B} - \boldsymbol{u} \times \boldsymbol{B}\right) \times \boldsymbol{n} = \boldsymbol{E}_0 \times \boldsymbol{n}. \end{cases} \tag{3.113}$$

or

$$\boldsymbol{B} \times \boldsymbol{n} = \boldsymbol{B}_0 \times \boldsymbol{n} + \mu \boldsymbol{j}_\Gamma. \tag{3.114}$$

Note that in the second case, the boundary conditions on $\boldsymbol{E}$ are not necessary to compute $\boldsymbol{B}$. We now make precise the functional setting for these boundary conditions.

3.8.1 First set of boundary conditions

Starting from (3.113), a first possible set of boundary conditions for the MHD equations on Ω is

$$\begin{cases} \boldsymbol{B}\cdot\boldsymbol{n} = q, \\ \left(\dfrac{1}{\sigma\mu}\,\mathbf{curl}\,\boldsymbol{B} - \boldsymbol{u}\times\boldsymbol{B}\right)\times\boldsymbol{n} = \boldsymbol{k}, \end{cases} \tag{3.115}$$

where $q \in H^{1/2}(\partial\Omega)$ and $\boldsymbol{k} \in \mathbb{H}^{-1/2}(\partial\Omega)$ are assumed to satisfy the following compatibility conditions:

$$\begin{cases} \displaystyle\int_{\partial\Omega} q = 0 \\ \boldsymbol{k}\cdot\boldsymbol{n} = 0. \end{cases} \tag{3.116}$$

The first relation comes from $\operatorname{div}\boldsymbol{B} = 0$, the second one from $(\boldsymbol{E}\times\boldsymbol{n})\cdot\boldsymbol{n} = 0$. To recover $\operatorname{div}\boldsymbol{B} = 0$ from the variational formulation presented in Section 3.3.1, we would also make the following technical assumption (see [118]):

$$\langle \boldsymbol{k}, \boldsymbol{\nabla}\phi\rangle_{\partial\Omega} = 0, \quad \forall\,\phi \in H^2(\Omega). \tag{3.117}$$

From the physical viewpoint it is in particular satisfied as long as $\mathbf{curl}\,\boldsymbol{E} = 0$.

The second relation of (3.115) will appear as a *natural* (or Neumann) boundary condition in the variational formulation of the equations. It will therefore modify the right-hand side of the variational equations. The first one will appear as an *essential* (or Dirichlet) boundary condition. With a "lifting" procedure, detailed below, q can be assumed to be 0. The appropriate functional space to search for the solution $\boldsymbol{B}$ is therefore

$$H(\mathbf{curl}\,;\Omega) \cap H_0(\operatorname{div},\Omega).$$

When Ω is either convex or has a $\mathcal{C}^{1,1}$ boundary, this space is a subspace of $\mathbb{H}^1$ and the appropriate space is simply $\mathbb{H}^1_n(\Omega)$.

It remains to describe the lifting of q. For the sake of simplicity, we assume that Ω has a $\mathcal{C}^{1,1}$ boundary, and we refer to Section 4.1 of [118] for more general assumptions. Suppose that we search for $\boldsymbol{B}$ such that

$$\begin{cases} \mathbf{curl}\,(\mathbf{curl}\,\boldsymbol{B}) - \mathbf{curl}\,(\boldsymbol{u}\times\boldsymbol{B}) = 0, & \text{in } \Omega, \\ \operatorname{div}\boldsymbol{B} = 0, & \text{in } \Omega, \\ \boldsymbol{B}\cdot\boldsymbol{n} = q, & \text{on } \partial\Omega, \\ \mathbf{curl}\,\boldsymbol{B}\times\boldsymbol{n} = \boldsymbol{k}, & \text{on } \partial\Omega. \end{cases}$$

We solve the preliminary problem

$$\begin{cases} -\Delta\, b = 0, & \text{in } \Omega, \\ \dfrac{\partial b}{\partial \boldsymbol{n}} = q, & \text{on } \partial\Omega, \end{cases}$$

whose solution is in $H^2(\Omega)$ and satisfies $\|b\|_{H^2(\Omega)} \leq c\|q\|_{H^{1/2}(\partial\Omega)}$. Let $\overline{\boldsymbol{B}} = \boldsymbol{\nabla} b$. Then we have:

$$\overline{\boldsymbol{B}} \in \mathbb{H}^1(\Omega),\ \operatorname{div}\overline{\boldsymbol{B}} = 0,\ \mathbf{curl}\,\overline{\boldsymbol{B}} = 0,\ \overline{\boldsymbol{B}}\cdot\boldsymbol{n} = q, \tag{3.118}$$

$$\|\overline{\boldsymbol{B}}\|_{\mathbb{H}^1} \leq c\|q\|_{H^{1/2}(\partial\Omega)}. \tag{3.119}$$

Then, $\tilde{\boldsymbol{B}} = \boldsymbol{B} - \overline{\boldsymbol{B}}$ is in $\mathbb{H}^1_n(\Omega)$ and is a solution of

$$\begin{cases} \mathbf{curl}\,(\mathbf{curl}\,\tilde{\boldsymbol{B}}) - \mathbf{curl}\,(\boldsymbol{u}\times\tilde{\boldsymbol{B}}) = \mathbf{curl}\,(\boldsymbol{u}\times\overline{\boldsymbol{B}}), & \text{in } \Omega,\\ \operatorname{div}\tilde{\boldsymbol{B}} = 0, & \text{in } \Omega,\\ \tilde{\boldsymbol{B}}\cdot\boldsymbol{n} = 0, & \text{on } \partial\Omega,\\ \mathbf{curl}\,\tilde{\boldsymbol{B}}\times\boldsymbol{n} = \boldsymbol{k}, & \text{on } \partial\Omega. \end{cases}$$

In conclusion, when we reduce the problem with $\boldsymbol{B}\cdot\boldsymbol{n} = q$ to a problem with homogeneous boundary conditions $\boldsymbol{B}\cdot\boldsymbol{n} = 0$, we introduce a term $\mathbf{curl}\,(\boldsymbol{u}\times\overline{\boldsymbol{B}})$ on the right-hand side of the magnetic equations (as well as other terms in the Navier–Stokes equations). These modifications cannot be considered as trivial. Nevertheless, with estimate (3.119), the analysis done with $\overline{\boldsymbol{B}} = 0$ can be extended to treat the case $\overline{\boldsymbol{B}} \neq 0$ by slightly modifying some arguments (see Section 2.2.5 and [118]).

3.8.2 Second set of boundary conditions

Starting from (3.114), a second possible set of boundary conditions for the MHD equations on Ω is

$$\boldsymbol{B}\times\boldsymbol{n} = \boldsymbol{k}, \tag{3.120}$$

where $\boldsymbol{k} \in \mathbb{H}^{1/2}(\partial\Omega)$ is assumed to satisfy the compatibility condition

$$\boldsymbol{k}\cdot\boldsymbol{n} = 0, \tag{3.121}$$

since of course $(\boldsymbol{B}\times\boldsymbol{n})\cdot\boldsymbol{n} = 0$.

Reduction to homogeneous boundary conditions can be done in a similar way as above. We can look for a $\overline{\boldsymbol{B}}$ satisfying

$$\overline{\boldsymbol{B}} \in \mathbb{H}^1(\Omega),\ \operatorname{div}\overline{\boldsymbol{B}} = 0,\ \mathbf{curl}\,\overline{\boldsymbol{B}} = 0,\ \overline{\boldsymbol{B}}\times\boldsymbol{n} = \boldsymbol{k}, \tag{3.122}$$

$$\|\overline{\boldsymbol{B}}\|_{\mathbb{H}^1} \leq c\|\boldsymbol{k}\|_{\mathbb{H}^{1/2}(\partial\Omega)}. \tag{3.123}$$

This can be done if we assume that $\boldsymbol{n}\times\boldsymbol{k} \in \mathbb{H}^{1/2}(\partial\Omega)$ and that

$$\int_{\partial\Omega} \boldsymbol{k}\cdot\boldsymbol{\nabla}\phi = 0,\ \ \forall\,\phi \in H^2(\Omega).$$

When this last compatibility condition is not fulfilled, a lifting $\overline{\boldsymbol{B}}$ which does not satisfy $\mathbf{curl}\,\overline{\boldsymbol{B}} = 0$ can nevertheless be built and, assuming $\boldsymbol{k}$ sufficiently small, the analysis proceeds as before. We refer to [118] (Section 5) for the details.

When considering homogeneous boundary conditions of type (3.120), the natural space to look for the magnetic field is

$$H_0(\mathbf{curl}\,;\Omega) \cap H(\mathrm{div}\,,\Omega),$$

and we recall that when Ω is either convex or has a $\mathcal{C}^{1,1}$ boundary, this space is a subspace of $\mathbb{H}^1$ and the appropriate space is simply $\mathbb{H}^1_\tau(\Omega)$.

3.8.3 Practical implementation of the boundary conditions

In practice, the essential boundary conditions on the magnetic field ($\boldsymbol{B}\cdot\boldsymbol{n} = q$ or $\boldsymbol{B}\times\boldsymbol{n} = \boldsymbol{k}$) can be incorporated directly in the functional space at the discrete level. But the simplest way is to use a penalization procedure, by adding to the variational formulation (for example (3.46)) the term $\frac{1}{\epsilon}\int_{\partial\Omega}(\boldsymbol{B}\cdot\boldsymbol{n} - q)\,\boldsymbol{C}\cdot\boldsymbol{n}$ or the term $\frac{1}{\epsilon}\int_{\partial\Omega}(\boldsymbol{B}\times\boldsymbol{n} - \boldsymbol{k})\cdot\boldsymbol{C}\times\boldsymbol{n}$, where $\epsilon > 0$. Other procedures can be found in J. Zhu et al. [94] and [239].

In view of (3.113) or (3.114), we see that q and $\boldsymbol{k}$ in (3.115) or (3.120) can be determined assuming a value for the external electromagnetic field $(\boldsymbol{E}_0, \boldsymbol{B}_0)$.

In the case of a perfectly conducting wall, since $\boldsymbol{E}$ and $\boldsymbol{B}$ are zero in the wall, we have $\boldsymbol{B}_0\cdot\boldsymbol{n} = 0$ and $\boldsymbol{E}_0\times\boldsymbol{n} = 0$, so that the normal boundary conditions (3.113) are particularly well adapted.

In the more general cases, the tangential components on $\partial\Omega$ of the internal field $\boldsymbol{B}$ have to match those of the exterior field $\boldsymbol{B}_0$ (assuming for simplicity that there is no surface current). A possible strategy is the following: first, compute an "initial guess" $\boldsymbol{B}_0^{\mathrm{guess}}$ in the whole space $\mathbb{R}^3$ by solving a pure electromagnetism problem (for example solving an electric problem inside Ω and applying the Biot–Savart law); next, use the tangential boundary conditions (3.114) to match the field $\boldsymbol{B}$, a solution of the MHD problem with Ω, and the field $\boldsymbol{B}_0^{\mathrm{guess}}$ on the boundary. This strategy is illustrated in the example of Section 3.7.2. Of course, this provides an approximate solution of the problem, but fixed point iterations can be performed in order to converge to a solution of the problem on the whole space.

4

MATHEMATICAL ANALYSIS OF TWO-FLUID PROBLEMS

We are now ready to begin the specific study of the multifluid MHD system (1.49). The additional difficulty with respect to the three former chapters lies in the *multifluid* nature, or in more general terms, the fact that the density is *not* homogeneous.

Throughout this chapter, the equation that will play a crucial role is the equation of the conservation of mass (1.1), included in system (1.49), and that we rewrite here for convenience:

$$\frac{\partial \rho}{\partial t} + \operatorname{div}(\rho \boldsymbol{u}) = 0 \quad \text{on } \Omega. \tag{4.1}$$

In the homogeneous case, this equation coincides with the incompressibility condition (1.10), but this is no longer the case here.

The heterogeneity of ρ causes several difficulties. Even if the solution $\rho(t, \mathbf{x})$ to (4.1) were known in advance, which is of course not the case, the fact that $\rho(t, \mathbf{x})$ may take different values at different points of the domain Ω implies that two other important parameters of (1.49), namely the density-dependent viscosity η and electric conductivity σ, may also take different values. If these values $\eta(\rho)$ and $\sigma(\rho)$ vary continuously with respect to the space variable, then the situation is relatively simple, and like that of the constant case. But for the relevant situation here, η and σ have *discontinuities*, or *jumps*, because they non-trivially depend on ρ and the latter has jumps (think of our prototypical case of two different homogeneous fluids).

Why is it that discontinuities of η and σ make the mathematical analysis of (1.49) more difficult? We shall see this formally in Section 4.1.1, and in more details in the subsequent sections. Let us only mention now that when both η and σ are constant, we have $-\operatorname{div}(\eta(\rho)(\boldsymbol{\nabla u} + \boldsymbol{\nabla u}^T)) = -\overline{\eta}\Delta \boldsymbol{u}$ and $\mathbf{curl}(\frac{1}{\sigma(\rho)}\,\mathbf{curl}\,\boldsymbol{B}) = -\frac{1}{\overline{\sigma}}\Delta \boldsymbol{B}$. The presence of the Laplacian operator in the equations has a regularizing effect. A better regularity of the fields $\boldsymbol{u}$ and $\boldsymbol{B}$ follows. The question of existence of a solution becomes easier.

In addition to this, ρ is of course not known in advance. Therefore we have to understand how to *construct* ρ. At this stage of our exposition, the reader is now familiar with the usual issues of *extracting compactness* from *a priori* estimates. This is useful to pass to the limit in the nonlinear terms for a sequence of approximate solutions. Here, the difficulty is that (4.1) does not *a priori* provide any compactness on ρ as a function of the space variable (or, in other words, any bounds on the derivatives of ρ with respect to the space variable $\mathbf{x}$). For illustration, the reader may think of the parabolic equation

$$\frac{\partial \rho}{\partial t} + \operatorname{div}(\rho \boldsymbol{u}) - \Delta \rho = 0$$

which, in contrast, has a regularizing effect. Starting from an initial $\rho_0 \in L^2$, and assuming, say, that $\boldsymbol{u}$ is divergence-free, we readily obtain that $\rho(t, \cdot) \in H^1$ for almost all times. To some extent, the situation is natural, since the case under study embodies the case of discontinuous densities, for which derivatives in space are only measures. Extracting some compactness from equation (4.1) is much more intricate. In particular, we shall see that, owing to an argument based on the theory of renormalized solutions and due to R. DiPerna and P.-L. Lions [68], the transport equation (4.1) *propagates* in time the compactness of ρ (see below, Theorem 4.1). A converging sequence of initial conditions ρ_0 and a weakly convergent sequence of velocity fields $\boldsymbol{u}$ give rise to a converging sequence of solutions ρ.

Regarding the mathematical analysis we conduct in the present chapter, there are a number of differences compared to Chapter 2. First we will not be so elementary in our exposition: taking advantage of our introductory survey of Chapter 2, we will concentrate on the additional difficulties, and essentially outline the techniques, rather than presenting all the details.

Second, again relying on our exposition of Chapter 2, we shall directly proceed to the *three-dimensional* density-dependent MHD system, skipping any two-dimensional setting and any purely hydrodynamic case. Our proofs and results apply to the latter, often with significant simplifications.

Third, we address more advanced issues. We present in Section 4.3 the study of the long-time dynamics of our model system. This study has not only a mathematical purpose; it is very much related to the practical questions we shall examine in Chapter 6.

4.1 The difficulties of the non-homogeneous case

For convenience, let us again recall the two-fluid MHD system (1.49):

$$\begin{cases} \begin{cases} \dfrac{\partial}{\partial t}(\rho \boldsymbol{u}) + \operatorname{div}(\rho \boldsymbol{u} \otimes \boldsymbol{u}) - \operatorname{div}(\eta(\rho)(\boldsymbol{\nabla} \boldsymbol{u} + \boldsymbol{\nabla} \boldsymbol{u}^T)) = -\boldsymbol{\nabla} p + \rho \boldsymbol{g} \\ \qquad\qquad\qquad\qquad\qquad\qquad\qquad\qquad + \dfrac{1}{\bar{\mu}} \mathbf{curl}\, \boldsymbol{B} \times \boldsymbol{B}, \\ \qquad\qquad\qquad\qquad\qquad\qquad \operatorname{div} \boldsymbol{u} = 0, \end{cases} \\ \dfrac{\partial \rho}{\partial t} + \operatorname{div}(\rho \boldsymbol{u}) = 0, \\ \begin{cases} \dfrac{\partial \boldsymbol{B}}{\partial t} + \dfrac{1}{\bar{\mu}} \mathbf{curl} \left(\dfrac{1}{\sigma(\rho)} \mathbf{curl}\, \boldsymbol{B} \right) = \mathbf{curl}(\boldsymbol{u} \times \boldsymbol{B}), \\ \qquad\qquad \operatorname{div} \boldsymbol{B} = 0, \end{cases} \end{cases}$$

$$\begin{cases} \boldsymbol{u}|_{\partial\Omega} = 0, \\ \mathbf{curl}\,\boldsymbol{B} \times \boldsymbol{n}|_{\partial\Omega} = 0, \\ \boldsymbol{B} \cdot \boldsymbol{n}|_{\partial\Omega} = 0, \end{cases}$$

$$\begin{cases} \boldsymbol{u}(t=0,\mathbf{x}) = \boldsymbol{u}_0, \\ \boldsymbol{B}(t=0,\mathbf{x}) = \boldsymbol{B}_0, \\ \rho(t=0,\mathbf{x}) = \begin{cases} \overline{\rho}_1 > 0, & \text{constant on}\,\Omega_1, \\ \overline{\rho}_2 > 0, & \text{constant on}\,\Omega_2, \\ \text{with} \quad \overline{\Omega}_1 \cup \overline{\Omega}_2 = \overline{\Omega}, & \text{meas}(\Omega_i) > 0, \quad i = 1, 2. \end{cases} \end{cases}$$

It is a special case of the analogous *multifluid* system, and, beyond, a special case of the *density-dependent* MHD equations (1.42).

Henceforth, we fix the constant value $\overline{\mu}$ of the magnetic permeability to the value 1, in order to lighten the notation.

4.1.1 A formal mathematical argument

As mentioned above, the discontinuities of η and σ make the mathematical analysis of (1.49) more difficult. Let us formally argue on the parabolic magnetic equation, for clarity of exposition. A straightforward formal argument shows that when $\operatorname{div} \boldsymbol{B} = 0$, we have $\mathbf{curl}\,\mathbf{curl}\,\boldsymbol{B} = -\Delta\,\boldsymbol{B}$, thus the case of a constant conductivity σ simplifies into

$$\frac{\partial \boldsymbol{B}}{\partial t} - \frac{1}{\overline{\sigma}} \Delta\,\boldsymbol{B} = \boldsymbol{f} := \mathbf{curl}\,(\boldsymbol{u} \times \boldsymbol{B})$$

which coincides with the three-dimensional heat equation on $\boldsymbol{B}$. Consequently, the field $\boldsymbol{B}$ is very regular. Of course, this depends on the regularity of the right-hand side, thus on that of $\boldsymbol{u}$. We omit this in this very vague argument. The regularity of $\boldsymbol{B}$ improves the regularity in the hydrodynamics equation, and thus simplifies the analysis. It is all the more true as the same formal argument applies to the hydrodynamics equation because η is also constant, thus (using $\operatorname{div} \boldsymbol{u} = 0$) we have $\operatorname{div}(\eta(\rho)\,(\boldsymbol{\nabla u} + \boldsymbol{\nabla u}^T)) = \overline{\eta}\Delta\,\boldsymbol{u}$. Again we argue formally, and skip some more involved arguments.

It remains now to understand why such additional regularity on $\boldsymbol{u}$ indeed simplifies the specific difficulty related to the heterogeneity of ρ. For this purpose, we simply recall that, for a *regular* velocity field $\boldsymbol{u}$, we may indeed solve the ordinary differential equation

$$\begin{cases} \dfrac{dX}{ds} = \boldsymbol{u}(X(s;x,t),s) \\ X(t;x,t) = x. \end{cases} \tag{4.2}$$

This is the famous *characteristics method.* In other words, we replace the Eulerian viewpoint by the Lagrangian viewpoint. The solution of (4.1) is

$$\rho(x,t) = \rho_0(X(0;x,t)),$$

and we may consider the system coupling (1.49) and (4.2) where the only unknown fields are $\boldsymbol{u}$ and $\boldsymbol{B}$. The reader can thus understand that the case when η

and σ are both constant is easier to analyze. Our observation that the regularity of the velocity field $\boldsymbol{u}$ readily simplifies the problem of the knowledge of the density field ρ, is the bottom line for the proof of our main result in Section 4.2.2. We shall proceed by *regularization* of the velocity (see in particular the setting considered in Section 4.2.2.2).

In addition to the above questions related to the heterogeneity of ρ, there is the possible vanishing of ρ in some parts of the domain. It can be formally understood on the first line of (1.49). There is no more information on $\boldsymbol{u}$ in such zones, since the derivative in time concerns $\rho\boldsymbol{u}$. Difficulties are thus expected, and the analysis is indeed much more difficult when ρ is allowed to vanish at initial time (a fact that propagates forward in time, owing to (4.1)). In Section 4.2.1, we shall explain why we do not allow ρ to vanish in our study.

Actually, the proof outlined above is the essence of the work by A.V. Kazhikov, S.N. Antontsev and A.V. Monakhov in [7]. This work indeed shows the well-posedness of the density-dependent Navier–Stokes equations

$$\begin{cases} \dfrac{\partial}{\partial t}(\rho\boldsymbol{u}) + \operatorname{div}(\rho\boldsymbol{u}\otimes\boldsymbol{u}) - \overline{\eta}\,\Delta\,\boldsymbol{u} = -\boldsymbol{\nabla}p + \boldsymbol{f}, \\ \operatorname{div}\boldsymbol{u} = 0, \\ \dfrac{\partial\rho}{\partial t} + \operatorname{div}(\rho\boldsymbol{u}) = 0, \end{cases}$$

when

- the viscosity is constant;
- the initial condition ρ_0 is bounded away from zero;
- the problem is set in two dimensions.

We refer to the original reference for more details.

In contrast to the situation studied in [7], the landscape completely changes when η and σ are not constant, which is *generically* the case if ρ is not homogeneous. The fields $\boldsymbol{u}$ and $\boldsymbol{B}$ solving (1.49) are not *a priori* as regular as above. More appropriately stated, they are *necessarily* not so regular, and huge difficulties arise, mainly related to the treatment of ρ.

4.1.2 The major ingredient

The fundamental ingredient for our analysis is a compactness result based upon the theory of renormalized solutions by R. DiPerna and P.-L. Lions [68]. For consistency, we now cite it under the form proved by P.-L. Lions in [155]:

Theorem 4.1 *[155, Theorem 2.4, p. 41 and Remark 2.4, 3, p. 42]* **[P.-L. Lions]** *Let ρ_n and $\boldsymbol{u}_n$ satisfy $\rho_n \in \mathcal{C}([0,T], L^1(\Omega))$, $0 \le \rho_n \le C$ a.e on $\Omega\times(0,T)$, $\boldsymbol{u}_n \in L^2(0,T;\mathbb{H}^1_0(\Omega))$, $\|\boldsymbol{u}_n\|_{L^2(0,T;\mathbb{H}^1(\Omega))} \le C$ and $div\,\boldsymbol{u}_n = 0$ (C denotes various constants independent of n). We note $\rho_{0n} = \rho_n(0)$ and we assume*

$$\frac{\partial\rho_n}{\partial t} + div(\rho_n\boldsymbol{u}_n) = 0 \text{ in } \mathcal{D}'(\Omega\times(0,T)), \tag{4.3}$$

$$\rho_{0n} \to \rho_0 \text{ in } L^1(\Omega) \text{ and } \boldsymbol{u}_n \rightharpoonup \boldsymbol{u} \text{ weakly in } L^2(0,T;\mathbb{H}^1(\Omega)). \tag{4.4}$$

Then:

(1) ρ_n *converges in* $\mathcal{C}([0,T], L^p(\Omega))$ *for all* $1 \leq p < \infty$ *to the unique function* ρ *bounded on* $\Omega \times (0,T)$*, a solution to*

$$\begin{cases} \dfrac{\partial \rho}{\partial t} + div(\rho \boldsymbol{u}) = 0 \text{ in } \mathcal{D}'(\Omega \times (0,T)), \\ \\ \rho(0) = \rho_0 \text{ in } \Omega. \end{cases} \tag{4.5}$$

(2) We assume in addition that $\rho_n |\boldsymbol{u}_n|^2$ *is bounded in* $L^\infty(0,T; L^1(\Omega))$ *and that we have for some* $m \geq 1$*, and for all* n*,*

$$\left| \left\langle \frac{\partial(\rho_n \boldsymbol{u}_n)}{\partial t}, \phi \right\rangle \right| \leq C \|\phi\|_{L^2(0,T;\mathbb{H}^m(\Omega))}$$

for all $\phi \in \mathcal{C}_0^\infty(\Omega \times (0,T))^3$ *such that* $div \phi = 0$ *on* $\Omega \times (0,T)$*. Then* $\sqrt{\rho_n} \boldsymbol{u}_n$ *converges to* $\sqrt{\rho} \boldsymbol{u}$ *in* $L^p(0,T; \mathbb{L}^r(\Omega))$ *for* $2 < p < \infty$, $1 \leq r < 6p/(3p-4)$ *and* $\boldsymbol{u}_n$ *converges to* $\boldsymbol{u}$ *in* $L^\theta(0,T;\mathbb{L}^{3\theta}(\Omega))$ *for* $1 \leq \theta < 2$ *on the set* $\{\rho > 0\}$*.*

The above theorem is instrumental in the sequel. Loosely speaking, it settles the question of the time-dependence of the density, by replacing the density-dependent problem by a problem where the density is "known." By this, we mean that in fact the density is known to converge strongly for the sequence of approximate solutions, which, from the mathematical standpoint, is almost the same as being known! This vague statement is clarified in Section 4.2.2. The approximate densities (strongly) converge at initial time (the approximation is constructed for this purpose), and the equation *propagates* this compactness *in time*. This is the power of the above theorem.

For consistency, we find it useful to now outline the proof given in [155] of this theorem. Actually, we shall concentrate on the proof of Assertion 1.

Proof of Theorem 4.1, Assertion 1 The proof of Assertion 1 falls in five steps, steps 2 and 3 being the major ones.

Step 1: Some preliminary convergences In view of the L^∞ bounds on ρ_n, we know that an extraction of ρ_n weakly converges in $L^p(0,T;L^p(\Omega))$, to some ρ, for all $1 < p < +\infty$. In the sequel, we do not make the extraction explicit, since we will eventually show in Step 4 that our result concerns not only one extraction but the whole sequence itself. In addition, again because ρ_n is bounded in L^∞, and $\boldsymbol{u}_n$ is bounded in $L^2(0,T;\mathbb{H}^1(\Omega))$ (thus in $L^2(0,T;\mathbb{L}^2(\Omega))$), we deduce from (4.3) that $\partial \rho_n/\partial t$ is bounded in, say, $L^2(0,T,\mathbb{H}^{-1}(\Omega))$. From the latter, along with the $L^2(0,T;\mathbb{L}^2(\Omega))$ bound on ρ_n, we obtain, by application of Lemma 2.8 in Chapter 2, that ρ_n converges to ρ strongly in $L^2(0,T;\mathbb{H}^{-1}(\Omega))$. Actually, we also have convergence in $\mathcal{C}(0,T;L^2_w(\Omega))$ (see the definition of this space in Section 4.2.1 below). In turn, this shows the convergence of $\rho_n \boldsymbol{u}_n$ to $\rho \boldsymbol{u}$ in the sense of distributions at least, by taking the product of the strongly convergent sequence ρ_n in $L^2(0,T;\mathbb{H}^{-1}(\Omega))$, and the weakly convergent sequence $\boldsymbol{u}_n$ in $L^2(0,T;\mathbb{H}^1(\Omega))$ (in the vein of Proposition 2.3 of Chapter 2). We may therefore pass to the limit in (4.3) and obtain (4.5), at least in the sense of distributions. At this stage, we do not know yet that ρ lies in $\mathcal{C}([0,T],L^1(\Omega))$, nor that ρ_n strongly converges to ρ in that space. Likewise, we do not know the uniqueness of ρ, nor the convergence of the whole sequence ρ_n. Only the convergence of an extraction holds, and the limit ρ *a priori* depends on the extraction.

Step 2: A formal argument Assume momentarily that we are allowed to multiply (4.3) by ρ_n and integrate over Ω. Then we obtain

$$\frac{1}{2}\frac{d}{dt}\int_\Omega \rho_n^2 = 0, \tag{4.6}$$

since

$$\int_\Omega \operatorname{div}(\rho_n \boldsymbol{u}_n)\,\rho_n = \int_\Omega \frac{\rho_n^2}{2}\operatorname{div}\boldsymbol{u}_n + \int_{\partial\Omega}\frac{\rho_n^2}{2}\boldsymbol{u}_n\cdot\boldsymbol{n} = 0,$$

using that $\boldsymbol{u}_n$ is both divergence-free and vanishes on the boundary. Likewise, multiplying (4.5) by ρ,

$$\frac{1}{2}\frac{d}{dt}\int_\Omega \rho^2 = 0. \tag{4.7}$$

It follows that

$$\int_\Omega \rho_n^2 = \int_\Omega \rho_{0n}^2 \overset{n\to+\infty}{\longrightarrow} \int_\Omega \rho_0^2 = \int_\Omega \rho^2. \tag{4.8}$$

This shows that for all time, $\rho_n(t,\cdot)$ not only converges weakly in L^2 to $\rho(t,\cdot)$, but also strongly (recall that in a reflexive Banach space such as L^2, weak convergence and convergence of the norm imply strong convergence). Likewise, writing (4.5) for two solutions ρ_1 and ρ_2, multiplying both equations by $\rho_1-\rho_2$ and substracting the two, we obtain

$$\frac{1}{2}\frac{d}{dt}\int_\Omega (\rho_1-\rho_2)^2 = 0. \tag{4.9}$$

This proves uniqueness since ρ_1 and ρ_2 agree at initial time.

Unfortunately, such calculations are only formal. We need to use a regularization argument: regularize the solution, perform some analogous computations on the regularized function, and pass to the limit. This is the purpose of the subsequent steps. □

Step 3: A regularization result We introduce the following lemma (often called the *commutation lemma* for reasons that will soon be clear):

Lemma 4.2 *[155,* Lemma 2.3, p. 43*] Let $\boldsymbol{v}\in \mathbb{W}^{1,\alpha}(\mathbb{R}^d)$, $g\in L^\beta(\mathbb{R}^d)$, with $1\le\alpha,\beta\le\infty$, $\frac{1}{\alpha}+\frac{1}{\beta}\le 1$. Let $\omega\in\mathcal{C}_0^\infty(\mathbb{R}^d)$, $\int\omega=1$, $\omega\ge 0$, and set $\omega_\varepsilon=\frac{1}{\varepsilon^d}\omega(\frac{\cdot}{\varepsilon})$. Then we have, for $\frac{1}{\gamma}=\frac{1}{\alpha}+\frac{1}{\beta}$, and for some C independent of $\varepsilon\in(0,1)$,*

$$\left\|div(g\boldsymbol{v})\star\omega_\varepsilon - div((g\star\omega_\varepsilon)\boldsymbol{v})\right\|_{L^\gamma(\mathbb{R}^d)} \le C\left\|\boldsymbol{v}\right\|_{\mathbb{W}^{1,\alpha}(\mathbb{R}^d)}\left\|g\right\|_{L^\beta(\mathbb{R}^d)} \tag{4.10}$$

where $\star$ denotes the convolution product. In addition, if $\gamma < +\infty$,

$$div(g\boldsymbol{v}) \star \omega_\varepsilon - div((g \star \omega_\varepsilon)\boldsymbol{v}) \quad \stackrel{\varepsilon \to 0}{\longrightarrow} 0, \qquad \text{in} \quad L^\gamma(\mathbb{R}^d). \tag{4.11}$$

Again, owing to the practical importance of this lemma, we wish to briefly mention now why such a result is likely to hold. Note that proving (4.10) is sufficient. Then (4.11) follows by a classical argument based on linearity and density. For (4.10), we remark that

$$s_\varepsilon(\mathbf{x}) = \operatorname{div}(g\boldsymbol{v}) \star \omega_\varepsilon - \operatorname{div}((g \star \omega_\varepsilon)\boldsymbol{v}) + (g \star \omega_\varepsilon)\operatorname{div}\boldsymbol{v}$$

can be written

$$s_\varepsilon(\mathbf{x}) = \int \frac{1}{\varepsilon}(\boldsymbol{v}(\mathbf{y}) - \boldsymbol{v}(\mathbf{x})) \,.\, (\boldsymbol{\nabla}\omega)\left(\frac{\mathbf{x}-\mathbf{y}}{\varepsilon}\right) \frac{1}{\varepsilon^d}\, g(\mathbf{y})d\mathbf{y}.$$

On this formula, we remark that, because of the support of the function $(\boldsymbol{\nabla}\omega)\left(\frac{\mathbf{x}-\mathbf{y}}{\varepsilon}\right)$, the integral can be roughly restricted to a ball of radius ε around $\mathbf{x}$. Then the integral is estimated using the Hölder inequality and classical arguments.

We now use Lemma 4.2 as follows. For $\varepsilon > 0$ fixed, and g a solution in $L^\infty(0, T; L^\infty(\Omega))$ to

$$\frac{\partial g}{\partial t} + \operatorname{div}(g\,\boldsymbol{u}) = 0 \tag{4.12}$$

we convolute this equation with ω_ε, and obtain

$$\frac{\partial g_\varepsilon}{\partial t} + \operatorname{div}(g_\varepsilon\,\boldsymbol{u}) = -\operatorname{div}(g\boldsymbol{u}) \star \omega_\varepsilon + \operatorname{div}(g_\varepsilon \boldsymbol{u}), \tag{4.13}$$

denoting $g_\varepsilon = g \star \omega_\varepsilon$. Next, for ε fixed, we perform on g_ε the formal calculations of Step 2. For $\beta \in C^1(\mathbb{R}, \mathbb{R})$, we multiply the equation by $\beta'(g_\varepsilon)$ and obtain (using $\operatorname{div}\boldsymbol{u} = 0$)

$$\frac{\partial \beta(g_\varepsilon)}{\partial t} + \operatorname{div}(\beta(g_\varepsilon)\,\boldsymbol{u}) = r_\varepsilon := (-\operatorname{div}(g\boldsymbol{u}) \star \omega_\varepsilon + \operatorname{div}(g_\varepsilon \boldsymbol{u}))\,\beta'(g_\varepsilon). \tag{4.14}$$

Letting ε now go to zero, and using Lemma 4.2, this shows that $\beta(g)$ satisfies the equation

$$\frac{\partial \beta(g)}{\partial t} + \operatorname{div}(\beta(g)\,\boldsymbol{u}) = 0. \tag{4.15}$$

We will repeatedly use this fact in the next two steps.

Step 4: Uniqueness Uniqueness of ρ is a simple consequence of the above regularization argument. Considering ρ_1 and ρ_2 two solutions to (4.5), both with initial condition ρ_0, we know by linearity that $g = \rho_1 - \rho_2$ also satisfies equation (4.12) with zero initial value function. Using a sequence of functions $\beta \in C^1(\mathbb{R}, \mathbb{R})$ that converges in C^0 to the absolute value, we obtain from (4.15):

$$\frac{\partial |g|}{\partial t} + \operatorname{div}(|g|\, \boldsymbol{u}) = 0,$$

which we integrate over Ω to obtain $\frac{d}{dt} \int |g| = 0$, and thus $g \equiv 0$, since it is zero at initial time. The uniqueness of ρ follows, and, from this, the convergence of the whole sequence ρ_n. This convergence is only weak for the moment.

Step 5: Strong convergence A second application of the regularization argument allows us to prove that $\rho \in \mathcal{C}(0, T; L^p(\Omega))$. Indeed, choosing $g = \rho$ in (4.12), we make the regularization for two values ε and ε'. Subtracting the two equations (4.13) obtained in this manner, and multiplying next by $\beta'(\rho_\varepsilon - \rho_{\varepsilon'})$, for some functions β that will eventually approximate $|\cdot|^p$, it is easy to show on (4.14) that

$$\frac{d}{dt} \int_\Omega |\rho_\varepsilon - \rho_{\varepsilon'}|^p \leq \left(\int_\Omega |r_\varepsilon - r_{\varepsilon'}|^{p'} \right)^{\frac{1}{p'}} \left(\int_\Omega |\rho_\varepsilon - \rho_{\varepsilon'}|^p \right)^{\frac{1}{p}}.$$

An application of the Gronwall lemma then shows that the sequence ρ_ε is a Cauchy sequence in $\mathcal{C}(0, T; L^p(\Omega))$. By construction it converges to ρ, thus $\rho \in \mathcal{C}(0, T; L^p(\Omega))$.

The third and final application of the regularization argument is the following. From the formal calculations performed in Step 2, we know it suffices to justify (4.6)–(4.7). This in turn yields the two conservations $\int_\Omega \rho^2 = \int_\Omega \rho_0^2$ and $\int_\Omega \rho_n^2 = \int_\Omega \rho_{0n}^2$ (for all n) used in (4.8). To this end, we use the regularization step with $g = \rho$ and $\beta(t) = t^2$, obtain from (4.15) that ρ^2 is the solution with the initial condition ρ_0^2, and deduce the above conservation $\int_\Omega \rho^2 = \int_\Omega \rho_0^2$ using equation (4.15) integrated over Ω. We argue similarly with ρ_n.

This concludes the outline of the proof of Assertion 1 of Theorem 4.1. For more details, along with the proof of Assertion 2, we refer to the original reference.

Some comments upon the previous proof are now in order.

The proof we have just outlined is prototypical of proofs of well-posedness for equations using the notion of renormalized solutions, *à la* Di Perna-Lions (see R.J. DiPerna and P.-L. Lions [68] for an original reference on transport equations). The idea is to obtain an appropriate notion of solution. The notion should be:

- sufficiently weak so that equations with irregular data still have solutions (typically, equations of the type (4.5) for irregular vector fields $\boldsymbol{u}$);
- sufficiently strong so that uniqueness can still be proved (notice that if the notion of solutions is too much weakened "almost" every function is a solution, and uniqueness fails).

Often, the point is thus to obtain uniqueness, while existence is "easy".

Now, as seen in the formal calculations of Step 2, uniqueness is ensured as long as appropriate "conservation laws" (or in other words *a priori* estimates) can be established. The latter basically rely on being allowed to *multiply* the equation by sufficiently many test functions, and typically by the solution itself (like in Step 2), or functions of the solution (like the $\beta'(g)$ of Step 3). The notion of renormalized solutions precisely aims to treat general nonlinear functions $\beta(\rho)$ instead of ρ itself: ρ is said to be a renormalized solution if $\beta(\rho)$ for a large class of (nonlinear) functions β is a "classical" solution (compare (4.15) to (4.12)). To some extent, the fact that the *a priori* estimates hold true is encoded in the actual notion of solution. More rigorously, the latter are shown to be valid, by the same regularizing tool as above, and uniqueness follows. The proof of well-posedness therefore amounts to a two-fold strategy: *a priori* estimates and regularization. For the *a priori* estimates, global estimates on the data are necessary: here, it is the divergence-free constraint $\operatorname{div} \boldsymbol{u} = 0$ that allows us to establish (4.6), (4.7), (4.9). For the regularization, the local regularity of the data is needed: here we use that $\boldsymbol{u}(t,\cdot) \in \mathbb{H}^1$ for Lemma 4.2 to apply in Step 3.

In the particular case addressed here, the proof follows the same lines, but the specific renormalization procedure (replacing ρ by $\beta(\rho)$) is not needed for the notion of solution itself because ρ is assumed to be bounded. It is only useful for the uniqueness, and, further, the regularity of the solution. □

4.1.3 Short overview of the state of the art for the hydrodynamic case

We have already mentioned that some global existence and regularity results have been established by A.V. Kazhikov, S.N. Antontsev and A.V. Monakhov in [7] in the two-dimensional case. They suppose that the viscosity is constant in the whole domain and that the initial density is bounded from below by a positive constant.

A. Nouri and F. Poupaud consider in [182] the transport equation for both the density and the viscosity and they use the concept of renormalized solutions of R.J. DiPerna and P.-L. Lions. This allows them to prove the existence of a global weak solution for several fluids with various viscosities and various densities bounded from below by a positive constant. On the other hand, when the viscosity is constant but the density is no longer bounded from below, the existence of global-in-time weak solutions is due to J. Simon [210, 211].

But to date, the most complete study of the density-dependent Navier–Stokes equations is due to P.-L. Lions in [155]. In this approach, the viscosity is a function of the density. The initial density is assumed to be non-negative, but not necessarily bounded from below by a positive constant, which also allows one to consider free-surface problems. The main result proved in [155] in this setting is the global existence of a weak solution. Moreover, as long as a strong solution exists, then any weak solution is equal to it (see [155] and also B. Desjardins [65] for a proof of existence of a strong solution under particular assumptions).

4.2 Weak solutions of the multifluid MHD system

We are now in a position to establish the existence of weak solutions to system (1.49). We begin with the mathematical setting (functional spaces and definition of the notion of solutions in Definition 4.3). We next enter the three steps of the proof of our main result (Theorem 4.4) based on an appropriate regularization and a linearization.

4.2.1 Mathematical setting of the equations

We recall that Ω is a simply connected, fixed bounded domain in $\mathbb{R}^3$ enclosed in a $\mathcal{C}^\infty$ boundary $\partial\Omega$. We again denote by $\boldsymbol{n}$ the outward-pointing normal to Ω.

The density-dependent MHD problem we consider is to find two vector-valued functions, the velocity $\boldsymbol{u}$ and the magnetic field $\boldsymbol{B}$, and two scalar functions, the density ρ and the pressure p, defined on $\Omega \times [0, T]$, solving (1.49).

Initially, ρ is piecewise constant as prescribed in (1.49). Our proof would indeed carry through for an initial condition of the type

$$\rho|_{t=0} = \rho_0 \text{ on } \Omega, \tag{4.16}$$

with

$$\text{ess inf}_\Omega \rho_0 > 0. \tag{4.17}$$

In fact, our argument would also apply to an initial condition ρ_0 satisfying only

$$\rho_0 \geq 0, \tag{4.18}$$

but in that latter case we encounter a specific difficulty related to the MHD modeling. Indeed, a zone where the density vanishes is a zone of a vacuum. We should extend our magnetic equations to the vacuum. The difficulty is that the parabolic form of the magnetic equation used in (1.49) is no longer valid, because we have derived it from the general equations (1.25)–(1.28) using Ohm's law. Therefore, for consistency, we should replace in such a zone of the vacuum the magnetic equation in (1.49) by the original Maxwell equations (1.25)–(1.28). Then the mathematical analysis is out of reach. An alternative would be to keep system (1.49) even when ρ_0 only satisfies (4.18), while giving to the vacuum ($\rho = 0$) an artifical electric conductivity $\sigma(\rho = 0) > 0$: the analysis below then goes through, but the physical relevance of the system is questionable.

We therefore prefer to concentrate on the case (4.17). A slight improvement of this condition is

$$\rho_0 > 0, \tag{4.19}$$

(and indeed this is the condition used in J.-F. Gerbeau and C. Le Bris [90]), but the improvement from (4.17) to (4.19) is not significant from the application viewpoint. Therefore we stay with (4.17).

For brevity, we will make use of the notation from Chapter 2. We will use the functional spaces (2.18)–(2.21), and (2.79)–(2.81), including $\mathcal{V}$, H, V, $\|\boldsymbol{v}\|_V$, $\mathcal{W}$, W, $\|\boldsymbol{C}\|_W$, and also the related notation $\mathbb{L}^2(\Omega)$, $\mathbb{H}^1(\Omega)$.

Regarding the regularity of the data (initial and boundary conditions), we shall impose:

$$\rho_0 \in L^\infty(\Omega), \tag{4.20}$$

$$\boldsymbol{u}_0 \in H, \tag{4.21}$$

$$\boldsymbol{B}_0 \in H. \tag{4.22}$$

In addition, we shall suppose in the sequel, unless otherwise mentioned, that

$$\boldsymbol{f} \in L^2(0,T;\mathbb{L}^2(\Omega)) \tag{4.23}$$

and that η and σ are continuous functions on $[0,+\infty)$ such that

$$0 < \overline{\eta}_- \leq \eta(\xi) \leq \overline{\eta}_+ \quad \text{for } \xi \in (0,\infty), \tag{4.24}$$

$$0 < \overline{\sigma}_- \leq \sigma(\xi) \leq \overline{\sigma}_+ \quad \text{for } \xi \in (0,\infty). \tag{4.25}$$

Considering the above setting, we now introduce the notion of solutions we shall manipulate. A straightforward formal calculation on (1.49) shows that we may expect the following *a priori* estimate to hold:

$$\frac{d}{dt}\left(\int_\Omega \rho\frac{\boldsymbol{u}^2}{2} + \int_\Omega \rho\frac{\boldsymbol{B}^2}{2\overline{\mu}}\right) + 2\int_\Omega \eta(\rho)\,|\boldsymbol{D}(\boldsymbol{u})|^2 + \int_\Omega \frac{1}{\overline{\mu}\,\sigma(\rho)}\,|\mathbf{curl}\,\boldsymbol{B}|^2 = \int_\Omega \rho\boldsymbol{g}\cdot\boldsymbol{u} \tag{4.26}$$

together with the fact that, owing to the maximum principle, ρ remains valued in $\{\overline{\rho}_1,\overline{\rho}_2\}$, which are the two values taken by the piecewise-constant initial density. Therefore, it is natural to expect a solution in the following functional spaces:

$$\begin{cases} \rho \in L^\infty(0,T;L^\infty(\Omega)), \\ \boldsymbol{u} \in L^\infty(0,T;\mathbb{L}^2(\Omega)) \cap L^2(0,T;\mathbb{H}^1(\Omega)), \\ \boldsymbol{B} \in L^\infty(0,T;\mathbb{L}^2(\Omega)) \cap L^2(0,T;\mathbb{H}^1(\Omega)), \end{cases} \tag{4.27}$$

along with the divergence-free constraints on $\boldsymbol{u}$ and $\boldsymbol{B}$. Note that to establish the formal bound on $\boldsymbol{u}$, we have used the fact that

$$2\int_\Omega \eta|\boldsymbol{D}(\boldsymbol{u})|^2\,dx \geq \frac{\overline{\eta}_-}{2}\int_\Omega |\boldsymbol{\nabla u} + \boldsymbol{\nabla u}^T|^2\,dx = \overline{\eta}_-\int_\Omega |\boldsymbol{\nabla u}|^2\,dx \tag{4.28}$$

since $\operatorname{div}\boldsymbol{u} = 0$.

We thus introduce the following.

Definition 4.3 *For $T > 0$, we shall say that $(\rho,\boldsymbol{u},\boldsymbol{B})$ is a weak solution on $\Omega \times [0,T]$ to the problem (1.49)–(1.48) with the assumptions (4.20), (4.21), (4.22), (4.23), (4.24), (4.25), if*

$$\rho \in L^\infty(\Omega \times (0,T)) \cap \mathcal{C}(0,T;L^p(\Omega)), \quad \forall p \geq 1, \tag{4.29}$$

$$\boldsymbol{u} \in L^2(0,T;V) \cap L^\infty(0,T;H) \cap \mathcal{C}(0,T;H_w), \tag{4.30}$$

$$\boldsymbol{B} \in L^2(0,T;W) \cap L^\infty(0,T;H) \cap \mathcal{C}([0,T],H_w) \tag{4.31}$$

and $(\rho, \boldsymbol{u}, \boldsymbol{B})$ *are such that (4.1) holds in the sense of distributions in* $\Omega \times (0, T)$ *and*

$$\iint_{\Omega\times(0,\infty)} -\rho\boldsymbol{u}\cdot\frac{\partial\boldsymbol{\phi}}{\partial t} - \rho\boldsymbol{u}\otimes\boldsymbol{u} : \boldsymbol{\nabla}\boldsymbol{\phi} + 2\eta\boldsymbol{D}(\boldsymbol{u}) : \boldsymbol{D}(\boldsymbol{\phi})\,dxdt$$
$$= \iint_{\Omega\times(0,\infty)} \left(\rho\boldsymbol{g} + \frac{1}{\bar{\mu}}(\mathbf{curl}\,\boldsymbol{B})\times\boldsymbol{B}\right)\cdot\boldsymbol{\phi}\,dxdt + \int_\Omega \rho_0\boldsymbol{u}_0\cdot\boldsymbol{\phi}(x,0)\,dx, \tag{4.32}$$

$$\iint_{\Omega\times(0,\infty)} -\boldsymbol{B}\frac{\partial\boldsymbol{\phi}}{\partial t} + \frac{1}{\sigma\bar{\mu}}\mathbf{curl}\,\boldsymbol{B}\cdot\mathbf{curl}\,\boldsymbol{\phi}\,dxdt = \iint_{\Omega\times(0,\infty)} \mathbf{curl}\,(\boldsymbol{u}\times\boldsymbol{B})\cdot\boldsymbol{\phi}\,dxdt + \int_\Omega \boldsymbol{B}_0\cdot\boldsymbol{\phi}(x,0)\,dx, \tag{4.33}$$

for all $\boldsymbol{\phi} \in \mathcal{D}(\Omega \times [0, T))^3$.

In the above statement, we need to clarify the meaning of the functional space $\mathcal{C}([0,T], H_w)$, called the space of *weakly continuous functions* valued in H. The fact that $\boldsymbol{B}$ belongs to this space means $\forall \boldsymbol{C} \in H$, $t \to \int_\Omega \boldsymbol{B}(t)\cdot\boldsymbol{C}\,dx$ is a continuous scalar function. Another equivalent way of stating this is to say that $\boldsymbol{B} \in \mathcal{C}([0,T], H_w)$ when first $\boldsymbol{B} \in L^\infty(0,T;H)$, and second $\boldsymbol{B} \in \mathcal{C}([0,T],\mathcal{B})$ where $\mathcal{B}$ is the bounded domain in H, where $\boldsymbol{B}(t,\cdot)$ varies, equipped with a norm defining the same topology as the topology endowed by the weak L^2 topology.

4.2.2 Existence of a weak solution

This section is devoted to the statement and proof of our main result.

Theorem 4.4 *Under the regularity assumptions on the data (4.20), (4.21), (4.22), (4.23), (4.24), (4.25), there exists a weak solution (in the sense of Definition 4.3)* $(\rho, \boldsymbol{u}, \boldsymbol{B})$ *to the multifluid MHD equations (1.49).*

In addition,

$$\left.\begin{array}{l}\text{meas}\{x \in \Omega/\alpha \le \rho(x,t) \le \beta\}\\ \textit{is independent of } t \ge 0 \textit{ for all } 0 \le \alpha \le \beta < \infty.\end{array}\right\} \tag{4.34}$$

Remark 4.2.1 Of course, (4.34) implies that, when the density is initially piecewise constant, taking only two values $\rho_0 \in \{\overline{\rho_1}, \overline{\rho_2}\}$, then, for any later time $t > 0$, the density $\rho(t,\cdot)$ still takes value $\overline{\rho_1}$ or $\overline{\rho_2}$. In addition, the volume of the two phases is preserved forward in time. ◇

Remark 4.2.2 Notice that, like for the standard Navier–Stokes equation and *a fortiori* for the density-dependent equation with given forces treated in [155], we do not know if a weak solution is unique. We do not know either if a strong solution exists for all times. ◇

For the proof of Theorem 4.4, the idea is to introduce a regularized problem (namely (4.63) in Section 4.2.2.2 below) for which the velocity field, denoted by $r_\varepsilon(\boldsymbol{u})$, is sufficiently regular. Then

$$\frac{\partial \rho}{\partial t} + \operatorname{div}(\rho r_\varepsilon(\boldsymbol{u})) = 0$$

makes sense as a classical transport equation.

At the same time, we regularize the conductivity σ, the viscosity η, and the nonlinear terms in the hydrodynamics and the magnetic equations:

$$\begin{cases} \dfrac{\partial(\rho\boldsymbol{u})}{\partial t} + \operatorname{div}(\rho r_\varepsilon(\boldsymbol{u}) \otimes \boldsymbol{u}) - \operatorname{div}(2\eta_\varepsilon \boldsymbol{D}(\boldsymbol{u})) + \boldsymbol{\nabla} p = \rho \boldsymbol{f}_\varepsilon + \mathbf{curl}\, \boldsymbol{B} \times s_\varepsilon(\boldsymbol{B}), \\ \operatorname{div} \boldsymbol{u} = 0, \\ \dfrac{\partial \rho}{\partial t} + \operatorname{div}(r_\varepsilon(\boldsymbol{u})\rho) = 0, \\ \dfrac{\partial \boldsymbol{B}}{\partial t} + \mathbf{curl}\left(\dfrac{1}{\sigma_\varepsilon}\mathbf{curl}\, \boldsymbol{B}\right) = \mathbf{curl}\,(\boldsymbol{u} \times s_\varepsilon(\boldsymbol{B})), \\ \operatorname{div} \boldsymbol{B} = 0. \end{cases}$$

Showing the existence of a solution to this regularized problem is the purpose of our first two steps. We linearize the problem in Subsection 4.2.2.1 and then use a fixed point argument in Subsection 4.2.2.2. Proving the theorem then amounts to passing to the limit $\varepsilon \to 0$ in the regularized problem. In the third step, we make use of the compactness Theorem 4.1 above.

In comparison with the case studied in [155], the new difficulty is that we have to check that the force term $\mathbf{curl}\, \boldsymbol{B} \times \boldsymbol{B}$ does not introduce any perturbation on the estimates on the velocity $\boldsymbol{u}$ and the density ρ. In addition, we have to recover some compactness on $\boldsymbol{B}$ through the parabolic equation in order to pass to the limit in the nonlinear terms $\mathbf{curl}\, \boldsymbol{B} \times \boldsymbol{B}$ and $\mathbf{curl}\,(\boldsymbol{u} \times \boldsymbol{B})$. This program has been successfully performed in Chapter 2 in the one-fluid case (see Section 2.2) and we now need to extend it to cover the multifluid case.

4.2.2.1 First step: a linear coupled problem Consider $\rho, \boldsymbol{w}$ and $\boldsymbol{h}$ arbitrarily fixed such that

$$\rho \in \mathcal{C}([0,T], \mathcal{C}^k(\overline{\Omega})),\ \forall k \geq 0, \text{such that } 0 < \rho_1 \leq \rho(x,t) \leq \rho_2, \tag{4.35}$$

$$\frac{\partial \rho}{\partial t} \in L^2(0,T; \mathcal{C}^k(\overline{\Omega})),\ \forall k \geq 0, \tag{4.36}$$

$$\boldsymbol{w} \in L^2(0,T; \mathbb{L}^\infty(\Omega)), \text{with } \operatorname{div} \boldsymbol{w} = 0 \text{ and } \frac{\partial \rho}{\partial t} + \operatorname{div}(\rho \boldsymbol{w}) = 0, \tag{4.37}$$

$$\boldsymbol{h} \in L^2(0,T; \mathbb{L}^\infty(\Omega) \cap \mathbb{W}^{1,3}(\Omega)) \text{ with } \operatorname{div} \boldsymbol{h} = 0. \tag{4.38}$$

We introduce the following linearized MHD system with prescribed density: find $(\boldsymbol{u}, \boldsymbol{B}, p)$ on $\Omega \times [0,T]$, solving

$$\begin{cases} \rho \dfrac{\partial \boldsymbol{u}}{\partial t} + \rho(\boldsymbol{w} \cdot \boldsymbol{\nabla})\boldsymbol{u} - \operatorname{div}(2\eta \boldsymbol{D}(\boldsymbol{u})) + \boldsymbol{\nabla} p = \rho \boldsymbol{f} + \mathbf{curl}\, \boldsymbol{B} \times \boldsymbol{h}, \\ \operatorname{div} \boldsymbol{u} = 0, \\ \dfrac{\partial \boldsymbol{B}}{\partial t} + \mathbf{curl} \left(\dfrac{1}{\sigma} \mathbf{curl}\, \boldsymbol{B} \right) = \mathbf{curl}\,(\boldsymbol{u} \times \boldsymbol{h}), \\ \operatorname{div} \boldsymbol{B} = 0. \end{cases} \tag{4.39}$$

The boundary and initial conditions on $\boldsymbol{u}$ and $\boldsymbol{B}$ are those of (1.49).

We assume here that the viscosity and the conductivity satisfy, in addition to (4.24) and (4.25), the following regularity properties:

$$\eta \in \mathcal{C}^\infty([0,\infty)), \tag{4.40}$$

$$\sigma \in \mathcal{C}^\infty([0,\infty)). \tag{4.41}$$

On the other hand, we momentarily only assume that

$$\boldsymbol{f} \in L^2(0,T;\mathbb{H}^{-1}(\Omega)), \tag{4.42}$$

$$\boldsymbol{u}_0, \boldsymbol{B}_0 \in H. \tag{4.43}$$

Although we shall use a strong solution of this problem in the sequel, it is useful for the proof of the following proposition to define a notion of weak solution for (4.39): we shall say that $(\boldsymbol{u}, \boldsymbol{B})$ is a weak solution of (4.39) if this pair is a solution to

Find $\boldsymbol{u} \in L^2(0,T;V)$ and $\boldsymbol{B} \in L^2(0,T;W)$, satisfying the initial conditions, such that

$$\int_\Omega \rho \left(\frac{\partial \boldsymbol{u}}{\partial t} + (\boldsymbol{w} \cdot \boldsymbol{\nabla})\boldsymbol{u} \right) \cdot \boldsymbol{v}\, dx + \int_\Omega 2\eta \boldsymbol{D}(\boldsymbol{u}) : \boldsymbol{D}(\boldsymbol{v})\, dx = \langle \rho \boldsymbol{f}, \boldsymbol{v} \rangle + \int_\Omega \mathbf{curl}\, \boldsymbol{B} \times \boldsymbol{h} \cdot \boldsymbol{v}\, dx \tag{4.44}$$

$$\int_\Omega \frac{\partial \boldsymbol{B}}{\partial t} \cdot \boldsymbol{C}\, dx + \int_\Omega \frac{1}{\sigma} \mathbf{curl}\, \boldsymbol{B} \cdot \mathbf{curl}\, \boldsymbol{C}\, dx = \int_\Omega \mathbf{curl}\,(\boldsymbol{u} \times \boldsymbol{h}) \cdot \boldsymbol{C}\, dx \tag{4.45}$$

for all $\boldsymbol{v} \in V$ and for all $\boldsymbol{C} \in W$.

Notice that we have made use of the regularity (4.37) of $\boldsymbol{w}$ to define this problem. Then we have:

Proposition 4.5 *Assume (4.35)–(4.38) and (4.40)–(4.43). Then, there exists a unique pair $(\boldsymbol{u}, \boldsymbol{B}) \in L^2(0,T;V) \times L^2(0,T;W)$ weak solution to the problem (4.39) and a distribution $p \in \mathcal{D}'(\Omega \times (0,T))$, unique up to an additive constant, satisfying (4.39). In addition, $\boldsymbol{u}$ and $\boldsymbol{B}$ belong to $\mathcal{C}(0,T;H)$.*

If in addition we assume $\boldsymbol{f} \in L^2(0,T;\mathbb{L}^2(\Omega))$, $\boldsymbol{u}_0 \in V$ and $\boldsymbol{B}_0 \in W$, then we have the following additional regularities:

$$\boldsymbol{u} \in L^2(0,T;\mathbb{H}^2(\Omega)) \cap \mathcal{C}(0,T;V), \tag{4.46}$$

$$\boldsymbol{B} \in L^2(0,T;\mathbb{H}^2(\Omega)) \cap \mathcal{C}(0,T;W), \tag{4.47}$$

$$\frac{\partial \boldsymbol{u}}{\partial t} \in L^2(0,T;H), \tag{4.48}$$

$$\frac{\partial \boldsymbol{B}}{\partial t} \in L^2(0,T;H), \tag{4.49}$$

$$p \in L^2(0,T;H^1(\Omega)). \tag{4.50}$$

It is a simple matter to prove Proposition 4.5, adapting the proof of Proposition 2.19 in Chapter 2. No new argument is needed. We therefore only sketch the proof.

Sketch of proof for Proposition 4.5 We first approximate (4.39) by a Galerkin method. The finite-dimensional fields $\boldsymbol{u}_n, \boldsymbol{B}_n$ solve the associated finite dimensional variational formulation readily deduced from (4.44)–(4.45). It is easily checked (in particular using (4.28)) that $\boldsymbol{u}_n$ and $\boldsymbol{B}_n$ satisfy the first energy estimate[27]:

$$\begin{aligned}\frac{d}{dt}\int_\Omega \rho|\boldsymbol{u}_n|^2 + |\boldsymbol{B}_n|^2\,dx + \overline{\eta}_-\int_\Omega |\boldsymbol{\nabla} \boldsymbol{u}_n|^2\,dx + \frac{2}{\overline{\sigma}_+}\int_\Omega |\mathbf{curl}\,\boldsymbol{B}_n|^2\,dx \\ \leq C\|\rho\|^2_{\mathcal{C}^1(\overline{\Omega})}\|\boldsymbol{f}\|^2_{H^{-1}(\Omega)}.\end{aligned}$$

Using (4.35), we deduce by the Gronwall lemma that

$$\boldsymbol{u}_n \text{ is bounded in } L^2(0,T;V) \cap L^\infty(0,T;H),$$

$$\boldsymbol{B}_n \text{ is bounded in } L^2(0,T;W) \cap L^\infty(0,T;H).$$

So, there exists $\boldsymbol{u} \in L^2(0,T;V) \cap L^\infty(0,T;H)$ such that $\boldsymbol{u}_n$ converges to $\boldsymbol{u}$ (up to the extraction of subsequences) for the weak-$\star$ topology of $L^\infty(0,T;H)$ and for the weak topology of $L^2(0,T;V)$. In the same way, there exists $\boldsymbol{B} \in L^2(0,T;W) \cap L^\infty(0,T;H)$ such that $\boldsymbol{B}_n$ converges to $\boldsymbol{B}$ for the weak-$\star$ topology of $L^\infty(0,T;H)$ and for the weak topology of $L^2(0,T;W)$. Clearly, the pair $(\boldsymbol{u}, \boldsymbol{B})$ is a solution of (4.39). By linearity, the uniqueness of the solution to (4.39) is clear. The existence of a convenient pressure field p is deduced by standard arguments, already seen in Chapter 2.

In addition, both $\partial\boldsymbol{u}/\partial t$ and $\partial\boldsymbol{B}/\partial t$ belong to $L^2(0,T;\mathbb{H}^{-1})$ (at least). Therefore, since $\boldsymbol{u}$ and $\boldsymbol{B}$ belong to $L^2(0,T;\mathbb{H}^1)$, we deduce, by an argument already seen in Chapter 2, that $\boldsymbol{u}$ and $\boldsymbol{B}$ belong to $\mathcal{C}(0,T;H)$.

[27] Note that the linearization of the terms $\mathbf{curl}\,\boldsymbol{B} \times \boldsymbol{B}$ and $\mathbf{curl}\,(\boldsymbol{u} \times \boldsymbol{B})$ is chosen on purpose, for this *a priori* estimate to hold.

The additional assumptions of regularity for $\boldsymbol{f}$, $\boldsymbol{u}_0$ and $\boldsymbol{B}_0$ enable us to obtain the following two estimates for the approximate solution $(\boldsymbol{u}_n, \boldsymbol{B}_n)$

$$\frac{1}{2}\int_\Omega \rho \left|\frac{\partial \boldsymbol{u}_n}{\partial t}\right|^2 dx + \int_\Omega \rho \boldsymbol{w}\cdot\boldsymbol{\nabla}\boldsymbol{u}_n \cdot \frac{\partial \boldsymbol{u}_n}{\partial t}\,dx + \int_\Omega 2\eta \boldsymbol{D}(\boldsymbol{u}_n) : \frac{\partial \boldsymbol{D}}{\partial t}(\boldsymbol{u}_n)\,dx = \int_\Omega (\rho \boldsymbol{f} + \mathbf{curl}\,\boldsymbol{B}_n \times \boldsymbol{h})\cdot \frac{\partial \boldsymbol{u}_n}{\partial t}\,dx, \tag{4.51}$$

$$\frac{1}{2}\int_\Omega \left|\frac{\partial \boldsymbol{B}_n}{\partial t}\right|^2 dx + \int_\Omega \frac{1}{\sigma}\mathbf{curl}\,\boldsymbol{B}_n \cdot \frac{\partial \mathbf{curl}\,\boldsymbol{B}_n}{\partial t}\,dx = \int_\Omega \mathbf{curl}\,(\boldsymbol{u}_n \times \boldsymbol{h})\cdot \frac{\partial \boldsymbol{B}_n}{\partial t}\,dx, \tag{4.52}$$

and therefore also

$$\begin{aligned} &\frac{1}{2}\frac{d}{dt}\left(\int_\Omega \eta\,|\boldsymbol{D}(\boldsymbol{u}_n)|^2\,dx + \int_\Omega \frac{1}{\sigma}\,|\mathbf{curl}\,\boldsymbol{B}_n|^2\,dx\right) \\ &\le \gamma_0(t)\left(\int_\Omega \eta\,|\boldsymbol{D}(\boldsymbol{u}_n)|^2\,dx + \int_\Omega \frac{1}{\sigma}\,|\mathbf{curl}\,\boldsymbol{B}_n|^2\,dx\right) + \gamma_1(t), \end{aligned} \tag{4.53}$$

where

$$\begin{aligned} \gamma_0(t) = &\frac{3\rho_2^2}{\rho_1\overline{\eta}_-}\|\boldsymbol{w}\|^2_{\mathbb{L}^\infty(\Omega)} + \left(\frac{4}{\overline{\eta}_-} + \frac{3\overline{\sigma}_+}{2\rho_1}\right)\|\boldsymbol{h}\|^2_{\mathbb{L}^\infty(\Omega)} \\ &+ \frac{4c_0^2}{\overline{\eta}_-}\|\boldsymbol{\nabla h}\|^2_{L^3(\Omega)} + \frac{1}{\overline{\eta}_-}\left\|\frac{\partial \eta}{\partial t}\right\|_{L^\infty(\Omega)} + 2\overline{\sigma}_+\left\|\frac{\partial\left(\frac{1}{\sigma}\right)}{\partial t}\right\|_{L^\infty(\Omega)}, \end{aligned}$$

and

$$\gamma_1(t) = \frac{3\rho_2^2}{2\rho_1}\|\boldsymbol{f}\|^2_{\mathbb{L}^2(\Omega)}.$$

This is obtained using standard manipulations on the equations. The assumptions (4.35)–(4.38) and (4.40)–(4.43) imply that $\gamma_0 \in L^1(0,T)$. Moreover $\gamma_1 \in L^1(0,T)$ since $\boldsymbol{f} \in L^2(0,T;\mathbb{L}^2(\Omega))$. Using that $\boldsymbol{u}_0 \in V$, $\boldsymbol{B}_0 \in W$, and the Gronwall lemma, we obtain:

$$\boldsymbol{u}_n \text{ is bounded in } L^\infty(0,T;V),$$

$$\boldsymbol{B}_n \text{ is bounded in } L^\infty(0,T;W).$$

We then deduce from (4.51)–(4.52) that

$$\frac{\partial \boldsymbol{u}_n}{\partial t} \text{ and } \frac{\partial \boldsymbol{B}_n}{\partial t} \text{ are bounded in } L^2(0,T;H).$$

Passing to the limit, we obtain

$$\frac{\partial \boldsymbol{u}}{\partial t} \in L^2(0,T;H), \tag{4.54}$$

$$\frac{\partial \boldsymbol{B}}{\partial t} \in L^2(0,T;H). \tag{4.55}$$

The regularities (4.46) and (4.47) are consequences of (4.54), (4.55) and classical regularity argument on the Stokes problem and the magnetostatic problem already seen in the preceding chapters. Here, we omit those for brevity. This concludes our sketch of the proof of Proposition 4.5. □

4.2.2.2 Second step: an approximate nonlinear problem In this section, we solve a regularized MHD problem by using the Schauder fixed point theorem and the results of step 1.

Let $\boldsymbol{u} \in L^2(0,T;V)$. We define a regularized velocity $r_\varepsilon(\boldsymbol{u})$ as in [155]: $r_\varepsilon(\boldsymbol{u}) \in L^2(0,T;\mathcal{C}^\infty(\Omega)^3)$, $\operatorname{div} r_\varepsilon(\boldsymbol{u}) = 0$ and $r_\varepsilon(\boldsymbol{u}) \in \mathbb{L}^\infty(\Omega)$ for all time, $r_\varepsilon(\boldsymbol{u})$ vanishes near $\partial\Omega$, and

$$\lim_{\varepsilon \to 0} r_\varepsilon(\boldsymbol{u}) = \boldsymbol{u} \quad \text{in } \mathbb{L}^p(\Omega), \tag{4.56}$$

for all p such that $\boldsymbol{u} \in \mathbb{L}^p(\Omega)$. For $\boldsymbol{B} \in L^2(0,T;W)$, we also build a regularization $s_\varepsilon(\boldsymbol{B})$ as follows. We first extend $\boldsymbol{B}$ to $\mathbb{R}^3$ by 0. We next define $s_\varepsilon(\boldsymbol{B}) = \boldsymbol{B} \star \omega_\varepsilon$ (ω_ε is a regularizing kernel). Let us notice that $s_\varepsilon(\boldsymbol{B}) \in L^2(0,T;\mathcal{C}^\infty(\Omega))$ and $\operatorname{div} s_\varepsilon(\boldsymbol{B}) = 0$ (since $\boldsymbol{B} \cdot \boldsymbol{n} = 0$) but $s_\varepsilon(\boldsymbol{B}) \cdot \boldsymbol{n} \neq 0$. We have in particular

$$\lim_{\varepsilon \to 0} s_\varepsilon(\boldsymbol{B}) = \boldsymbol{B} \quad \text{in } \mathbb{L}^p(\Omega), \tag{4.57}$$

for all p convenient. We set $\boldsymbol{f}_\varepsilon = (\boldsymbol{f}\, 1_{(d>2\varepsilon)}) \star \omega_\varepsilon$ where $d = \operatorname{dist}(x, \partial\Omega)$.

Without loss of generality, we may assume that $\eta(\xi)$ is constant for ξ large enough (since ρ remains in $[0, \|\rho_0\|_{L^\infty(\Omega)}]$). We denote by $\eta^\varepsilon \in \mathcal{C}^\infty([0,\infty))$ a function bounded away from 0, and such that $\sup_{[0,\infty)} |\eta^\varepsilon - \eta| \leq \varepsilon$. Moreover, $\eta^\varepsilon(\xi)$ is supposed to be constant for ξ large enough. Then, we define $\eta_\varepsilon = \overline{\eta}(\rho) \star \omega_\varepsilon|_\Omega$ with $\overline{\eta}(\rho) = \eta^\varepsilon(\rho)$ in Ω and $= 1$ in Ω^c. We define σ_ε from σ like η_ε from η. The initial data are also regularized like in [155]. Let us just recall that

$$\varepsilon \leq \rho_0^\varepsilon \leq \overline{\rho}_2, \tag{4.58}$$

$$\lim_{\varepsilon \to 0} \rho_0^\varepsilon = \rho_0 \quad \text{in } L^p(\Omega) \quad (1 \leq p < \infty), \tag{4.59}$$

$$\lim_{\varepsilon \to 0} \boldsymbol{u}_0^\varepsilon = \boldsymbol{u}_0 \quad \text{in } \mathbb{L}^2(\Omega). \tag{4.60}$$

Finally, $\boldsymbol{B}_0 \in H$ is regularized as follows. We extend $\boldsymbol{B}_0$ on $\mathbb{R}^3$ by 0 and we define $\boldsymbol{B}_0^\varepsilon = (\boldsymbol{B}_0\, 1_{(d>2\varepsilon)}) \star \omega_\varepsilon$. Note that $\boldsymbol{B}_0$ vanishes near $\partial\Omega$ and that we have

$$\lim_{\varepsilon \to 0} \boldsymbol{B}_0^\varepsilon = \boldsymbol{B}_0 \quad \text{in } \mathbb{L}^p(\Omega) \quad (1 \leq p < \infty). \tag{4.61}$$

We now consider the following transport equation with the regularized velocity field $r_\varepsilon(\boldsymbol{u})$:

$$\frac{\partial \rho}{\partial t} + \operatorname{div}\left(r_\varepsilon(\boldsymbol{u})\rho\right) = 0, \tag{4.62}$$

and couple the latter with the other equations (also appropriately regularized). We obtain the following problem:

$$\left\{\begin{aligned} &\frac{\partial(\rho\boldsymbol{u})}{\partial t} + \operatorname{div}\left(\rho r_\varepsilon(\boldsymbol{u}) \otimes \boldsymbol{u}\right) - \operatorname{div}\left(2\eta_\varepsilon \boldsymbol{D}(\boldsymbol{u})\right) + \boldsymbol{\nabla} p = \rho \boldsymbol{f}_\varepsilon + \mathbf{curl}\, \boldsymbol{B} \times s_\varepsilon(\boldsymbol{B}),\\ &\operatorname{div} \boldsymbol{u} = 0,\\ &\frac{\partial \rho}{\partial t} + \operatorname{div}\left(r_\varepsilon(\boldsymbol{u})\rho\right) = 0,\\ &\frac{\partial \boldsymbol{B}}{\partial t} + \mathbf{curl}\left(\frac{1}{\sigma_\varepsilon}\mathbf{curl}\, \boldsymbol{B}\right) = \mathbf{curl}\left(\boldsymbol{u} \times s_\varepsilon(\boldsymbol{B})\right),\\ &\operatorname{div} \boldsymbol{B} = 0, \end{aligned}\right. \tag{4.63}$$

supplied with the initial conditions

$$\rho|_{t=0} = \rho_0^\varepsilon, \tag{4.64}$$

$$\boldsymbol{u}|_{t=0} = \boldsymbol{u}_0^\varepsilon, \tag{4.65}$$

$$\boldsymbol{B}|_{t=0} = \boldsymbol{B}_0^\varepsilon, \tag{4.66}$$

and the usual boundary conditions. We have:

Proposition 4.6 *The above regularized problem (4.63) has a solution* $(\rho, \boldsymbol{u}, \boldsymbol{B}) \in \mathcal{C}^\infty(\overline{\Omega} \times [0, +\infty))^7$.

Proof (1) First, we prove by a fixed point argument that the regularized problem has a solution in $\mathcal{C}(\overline{\Omega} \times [0, T]) \times L^2(0, T; V) \times L^2(0, T; W)$.

Let us consider the convex set C_ε in $\mathcal{C}(\overline{\Omega} \times [0, T]) \times L^2(0, T; V) \times L^2(0, T; W)$ defined by

$$\begin{aligned} C_\varepsilon = \{(\overline{\rho}, \overline{\boldsymbol{u}}, \overline{\boldsymbol{B}}) \in \mathcal{C}(\overline{\Omega} \times [0, T]) \times L^2(0, T; V) \times L^2(0, T; W), \text{such that}\\ \varepsilon \le \overline{\rho} \le \rho_2 \text{ in } \overline{\Omega} \times [0, T], \|\overline{\boldsymbol{u}}\|_{L^2(0,T;V)} \le R_0, \|\overline{\boldsymbol{B}}\|_{L^2(0,T;W)} \le R_0\} \end{aligned}$$

where R_0 is a constant to be determined.

For $(\overline{\rho}, \overline{\boldsymbol{u}}, \overline{\boldsymbol{B}}) \in C_\varepsilon$ we define $F(\overline{\rho}, \overline{\boldsymbol{u}}, \overline{\boldsymbol{B}}) = (\rho, \boldsymbol{u}, \boldsymbol{B})$ as follows. First of all, we solve

$$\left\{\begin{aligned} &\frac{\partial \rho}{\partial t} + \operatorname{div}\left(\rho r_\varepsilon(\overline{\boldsymbol{u}})\right) = 0 \quad \text{in } \Omega \times (0, T),\\ &\rho|_{t=0} = \rho_0^\varepsilon \quad \text{in } \Omega. \end{aligned}\right. \tag{4.67}$$

This is a classical transport equation since, by construction, $r_\varepsilon(\overline{\boldsymbol{u}})$ is regular, divergence-free and vanishes near $\partial\Omega$. Thus ρ is given by

$$\rho(x, t) = \rho_0^\varepsilon(X(0; x, t)), \quad \forall (x, t) \in \overline{\Omega} \times [0, T],$$

where X is the solution of the ordinary differential equation

$$\begin{cases} \dfrac{dX}{ds} = r_\varepsilon(\overline{\boldsymbol{u}})(X(s;x,t),s), \\ X(t;x,t) = x. \end{cases}$$

□

We deduce from (4.58) that $\varepsilon \le \rho \le \rho_2$ in $\overline{\Omega} \times [0,T]$. Thus $\rho \in \mathcal{C}([0,T];\mathcal{C}^k(\overline{\Omega}))$ for all $k \ge 0$, and is bounded in this space uniformly in $(\overline{\rho}, \overline{\boldsymbol{u}})$. Furthermore, we deduce from (4.67) that $\partial\rho/\partial t$ is bounded in $L^2(0,T;\mathcal{C}^k(\overline{\Omega}))$ for all $k \ge 0$ uniformly in $(\overline{\rho}, \overline{\boldsymbol{u}})$. Therefore the set of ρ (such that $(\rho, \boldsymbol{u}, \boldsymbol{B}) = F(\overline{\rho}, \overline{\boldsymbol{u}}, \overline{\boldsymbol{B}})$ for $(\overline{\rho}, \overline{\boldsymbol{u}}, \overline{\boldsymbol{B}}) \in C_\varepsilon$) is compact in $\mathcal{C}(\overline{\Omega} \times [0,T])$.

Next, we set $\boldsymbol{w} = r_\varepsilon(\overline{\boldsymbol{u}})$ and $\boldsymbol{h} = s_\varepsilon(\overline{\boldsymbol{B}})$ and we use Proposition 4.5 to define $(\boldsymbol{u}, \boldsymbol{B})$ as the unique solution to:

$$\begin{cases} \dfrac{\partial(\rho\boldsymbol{u})}{\partial t} + \operatorname{div}(\rho r_\varepsilon(\overline{\boldsymbol{u}}) \otimes \boldsymbol{u}) - \operatorname{div}(2\eta_\varepsilon \boldsymbol{D}(\boldsymbol{u})) + \boldsymbol{\nabla} p = \rho \boldsymbol{f}_\varepsilon + \mathbf{curl}\,\boldsymbol{B} \times s_\varepsilon(\overline{\boldsymbol{B}}), \\ \dfrac{\partial \boldsymbol{B}}{\partial t} + \mathbf{curl}\left(\dfrac{1}{\sigma_\varepsilon}\mathbf{curl}\,\boldsymbol{B}\right) = \mathbf{curl}\,(\boldsymbol{u} \times s_\varepsilon(\overline{\boldsymbol{B}})), \\ \operatorname{div} \boldsymbol{u} = 0, \\ \operatorname{div} \boldsymbol{B} = 0, \end{cases} \tag{4.68}$$

with the usual boundary conditions on $\boldsymbol{B}$ and the initial conditions (4.65)–(4.66). We recall that $\boldsymbol{u} \in L^2(0,T;\mathbb{H}^2(\Omega)) \cap \mathcal{C}(0,T;V)$ and $\boldsymbol{B} \in L^2(0,T;\mathbb{H}^2(\Omega)) \cap \mathcal{C}(0,T;W)$ which justifies the manipulations hereafter.

Now, let us choose R_0 in such a way that $(\rho, \boldsymbol{u}, \boldsymbol{B})$ is in C_ε. By a now standard argument, we have the energy identity:

$$\frac{1}{2}\frac{d}{dt}\int_\Omega \rho|\boldsymbol{u}|^2 + |\boldsymbol{B}|^2\,dx + \int_\Omega 2\eta_\varepsilon|\boldsymbol{D}(\boldsymbol{u})|^2 + \frac{1}{\sigma_\varepsilon}|\mathbf{curl}\,\boldsymbol{B}|^2\,dx = \int_\Omega \rho \boldsymbol{f}_\varepsilon \cdot \boldsymbol{u}\,dx. \tag{4.69}$$

Then, the Cauchy–Schwarz inequality and $\|\boldsymbol{u}\|_{L^2(\Omega)} \le c(\Omega)\|\boldsymbol{\nabla}\boldsymbol{u}\|_{L^2(\Omega)}$ lead to

$$\frac{d}{dt}\int_\Omega \rho|\boldsymbol{u}|^2 + |\boldsymbol{B}|^2\,dx + \int_\Omega \frac{\overline{\eta}_-}{2}|\boldsymbol{\nabla}\boldsymbol{u}|^2 + \frac{2}{\overline{\sigma}_+}|\mathbf{curl}\,\boldsymbol{B}|^2\,dx \le \frac{2\rho_2 c(\Omega)^2}{\overline{\eta}_-}\|\boldsymbol{f}_\varepsilon\|^2_{L^2(\Omega)}.$$

Finally, using $0 < \varepsilon \le \rho$ we obtain by the Gronwall lemma:

$$\sup_{t\in[0,T]} \|\boldsymbol{u}(t)\|_{L^2(\Omega)} + \sup_{t\in[0,T]} \|\boldsymbol{B}(t)\|_{L^2(\Omega)} + \|\boldsymbol{u}\|_{L^2(0,T;V)} + \|\boldsymbol{B}\|_{L^2(0,T;W)} \le c_0$$

where c_0 is a constant which is independent of R_0, $\overline{\boldsymbol{u}}$, $\overline{\boldsymbol{B}}$. Hence, with $R_0 = c_0$, we have $F(\overline{\rho}, \overline{\boldsymbol{u}}, \overline{\boldsymbol{B}}) \in C_\varepsilon$.

In order to apply the Schauder theorem, we still have to prove that the mapping F is compact on C. Replacing $\boldsymbol{w}$ by $r_\varepsilon(\overline{\boldsymbol{u}})$ and $\boldsymbol{h}$ by $s_\varepsilon(\overline{\boldsymbol{B}})$ in the proof of Proposition 4.5, part 2, we see that:

$$\frac{\partial \boldsymbol{B}}{\partial t} \text{ and } \frac{\partial \boldsymbol{u}}{\partial t} \text{ are bounded in } L^2(0,T;\mathbb{L}^2(\Omega)), \text{ and}$$

$$\boldsymbol{B} \text{ and } \boldsymbol{u} \text{ are bounded in } L^2(0,T;\mathbb{H}^2(\Omega)).$$

We deduce that the set of $\boldsymbol{u}$ (resp. $\boldsymbol{B}$) built above is relatively compact in $L^2(0,T;\mathbb{H}^1(\Omega))$. Since V and W are closed subsets of $\mathbb{H}^1(\Omega)$, the set of $\boldsymbol{u}$ (resp. $\boldsymbol{B}$) is relatively compact in $L^2(0,T;V)$ (resp. in $L^2(0,T;W)$). Let us recall that the set of ρ is compact in $\mathcal{C}(\overline{\Omega}\times[0,T])$. Hence the mapping F is compact on C and has a fixed point $(\rho,\boldsymbol{u},\boldsymbol{B})$ which is a solution of (4.63).

(2) The solution $(\rho,\boldsymbol{u},\boldsymbol{B})$ built above satisfies $\rho\in\mathcal{C}([0,T];\mathcal{C}^k(\overline{\Omega}))$, $\boldsymbol{u}\in L^2(0,T;\mathbb{H}^2(\Omega))\cap\mathcal{C}(0,T;V)$, $\boldsymbol{B}\in L^2(0,T;\mathbb{H}^2(\Omega))\cap\mathcal{C}(0,T;W)$, $\partial\boldsymbol{u}/\partial t$ and $\partial\boldsymbol{B}/\partial t\in L^2(0,T;H)$.

The smoothness of $r_\varepsilon(\boldsymbol{u})$, $s_\varepsilon(\boldsymbol{B})$, $\boldsymbol{u}_0^\varepsilon$ and $\boldsymbol{B}_0^\varepsilon$ allows us to apply the same regularity arguments as in part 2 of Proposition 4.5 which provides more regularity on $(\boldsymbol{u},\boldsymbol{B})$ and therefore on ρ. By bootstrapping we conclude that ρ, $\boldsymbol{u}$ and $\boldsymbol{B}$ are in $\mathcal{C}^\infty(\overline{\Omega}\times[0,+\infty))$. □

4.2.2.3 Third step: passage to the limit We denote by $(\rho^\varepsilon,\boldsymbol{u}^\varepsilon,\boldsymbol{B}^\varepsilon)$ the smooth approximate solution provided by Proposition 4.6. We have:

$$\frac{\partial\rho^\varepsilon}{\partial t}+\operatorname{div}(r_\varepsilon(\boldsymbol{u}^\varepsilon)\rho^\varepsilon)=0. \tag{4.70}$$

Let β_n be a function of class $\mathcal{C}^1(\mathbb{R},\mathbb{R})$. Multiplying (4.70) by $\beta_n'(\rho)$ and using $\operatorname{div} r_\varepsilon(\boldsymbol{u}^\varepsilon)=0$ we have

$$\frac{\partial\beta_n(\rho^\varepsilon)}{\partial t}+r_\varepsilon(\boldsymbol{u}^\varepsilon)\cdot\boldsymbol{\nabla}\beta_n(\rho^\varepsilon)=0.$$

We integrate this equation on $\Omega\times[0,T]$ and we use again that $r_\varepsilon(\boldsymbol{u}^\varepsilon)$ is divergence-free and vanishes on the boundary to obtain

$$\int_\Omega \beta_n(\rho^\varepsilon(x,t))\,dx=\int_\Omega\beta_n(\rho_0^\varepsilon(x))\,dx. \tag{4.71}$$

For $0\le\alpha\le\beta<\infty$ we choose (for n large enough) $0\le\beta_n\le 1$ such that $\beta_n(\xi)=0$ if $\xi\notin[\alpha,\beta]$, $\beta_n(\xi)=1$ if $\xi\in[\alpha+1/n,\beta-1/n]$. Letting n go to $+\infty$ in (4.71) we deduce that (4.34) holds with ρ^ε, *i.e.*

$$\int_\Omega 1_{[\alpha,\beta]}(\rho^\varepsilon(x,t))\,dx=\int_\Omega 1_{[\alpha,\beta]}(\rho_0^\varepsilon(x))\,dx \tag{4.72}$$

where $\chi_{[\alpha,\beta]}(\xi)=1$ on $[\alpha,\beta]$ and 0 elsewhere. In particular, with $\alpha=0$ and $\beta=\|\rho_0\|_{L^\infty(\Omega)}$ this yields the following L^∞-estimate on ρ:

$$0\le\rho^\varepsilon\le\|\rho_0\|_{L^\infty(\Omega)}.$$

In addition, we have the energy identity (4.69):

$$\frac{1}{2}\frac{d}{dt}\int_\Omega \rho^\varepsilon|\boldsymbol{u}^\varepsilon|^2+|\boldsymbol{B}^\varepsilon|^2\,dx+\int_\Omega 2\eta_\varepsilon|\boldsymbol{D}(\boldsymbol{u}^\varepsilon)|^2+\frac{1}{\sigma_\varepsilon}|\mathbf{curl}\,\boldsymbol{B}^\varepsilon|^2\,dx=\int_\Omega \rho^\varepsilon \boldsymbol{f}_\varepsilon\cdot\boldsymbol{u}^\varepsilon\,dx$$

which implies, using the Gronwall lemma:

$$\|\boldsymbol{u}^\varepsilon\|_{L^2(0,T;V)}\le c, \tag{4.73}$$

$$\sup_{t\in[0,T]}\|\rho^\varepsilon|\boldsymbol{u}^\varepsilon|^2\|_{L^1(\Omega)}\le c, \tag{4.74}$$

$$\left\|\frac{1}{\sqrt{\sigma_\varepsilon}}\mathbf{curl}\,\boldsymbol{B}^\varepsilon\right\|_{L^2(0,T;\mathbb{L}^2(\Omega))}\le c, \tag{4.75}$$

$$\sup_{t\in[0,T]}\|\boldsymbol{B}^\varepsilon\|_{\mathbb{L}^2(\Omega)}\le c, \tag{4.76}$$

where c denotes various constants independent of ε.

In view of these estimates, and using Theorem 4.1, our goal is now to pass to the limit in the weak formulation. Extracting subsequences if necessary and using (4.73) and (4.76), we may define $\boldsymbol{u}$ as the weak limit of $\boldsymbol{u}^\varepsilon$ in $L^2(0,T;V)$ and $\boldsymbol{B}$ as the limit of $\boldsymbol{B}^\varepsilon$ for the weak-$\star$ topology of $L^\infty(0,T;\mathbb{L}^2(\Omega))$.

Let us remark that $0\le\sigma\le\overline{\sigma}_+$ and (4.75) imply that $\boldsymbol{B}\in\mathbb{H}^1(\Omega)$ and $\mathbf{curl}\,\boldsymbol{B}^\varepsilon$ converges to $\mathbf{curl}\,\boldsymbol{B}$ weakly in $L^2(0,T;\mathbb{L}^2(\Omega))$.

In view of (4.59) and (4.73), the first assertion of Theorem 4.1 implies that ρ^ε converges to some $\rho\in\mathcal{C}([0,T];L^p(\Omega))$ with $1\le p<\infty$ and

$$\frac{\partial\rho}{\partial t}+\mathrm{div}\,(\rho\boldsymbol{u})=0.$$

Passing to the limit in (4.72), we deduce that for $0\le\alpha\le\beta<\infty$

$$\int_\Omega \chi_{[\alpha,\beta]}(\rho(x,t))\,dx=\int_\Omega \chi_{[\alpha,\beta]}(\rho_0(x))\,dx,$$

which proves (4.34).

The convergence of ρ^ε as $\varepsilon\to0$ implies that

$$\lim_{\varepsilon\to0}\eta_\varepsilon=\eta(\rho)\quad\text{in }\mathcal{C}([0,T];L^p(\Omega))\text{ for }1\le p<\infty \tag{4.77}$$

$$\lim_{\varepsilon\to0}\sigma_\varepsilon=\sigma(\rho)\quad\text{in }\mathcal{C}([0,T];L^p(\Omega))\text{ for }1\le p<\infty \tag{4.78}$$

$$\lim_{\varepsilon\to0}\rho^\varepsilon\boldsymbol{f}_\varepsilon=\rho\boldsymbol{f}\qquad\text{in }L^2(\Omega\times(0,T)). \tag{4.79}$$

Next, we remark that $r_\varepsilon(\boldsymbol{u}^\varepsilon)$ converges to $\boldsymbol{u}$ weakly in $L^2(0,T;V)$ and $s_\varepsilon(\boldsymbol{B}^\varepsilon)$ converges to $\boldsymbol{B}$ weakly in $L^2(0,T;W)$ (with (4.56) and (4.57)).

In order to check that we may apply the second part of Theorem 4.1, let us prove that for some $m \geq 1$ we have

$$\left|\left\langle \frac{\partial(\rho^\varepsilon \boldsymbol{u}^\varepsilon)}{\partial t}, \phi \right\rangle\right| \leq C\|\phi\|_{L^2(0,T;\mathbb{H}^m(\Omega))} \tag{4.80}$$

for all $\phi \in \mathcal{C}_0^\infty(\Omega \times (0,T))^3$ such that $\operatorname{div}\phi = 0$ on $\Omega \times (0,T)$.

First, we have

$$\begin{aligned}\langle \operatorname{div}(2\eta_\varepsilon \boldsymbol{D}(\boldsymbol{u}^\varepsilon)), \phi\rangle &= \left|\int_0^T \int_\Omega 2\eta_\varepsilon \boldsymbol{D}(\boldsymbol{u}^\varepsilon) : \boldsymbol{\nabla}\phi \, dx\, dt\right| \\ &\leq \|2\eta_\varepsilon \boldsymbol{D}(\boldsymbol{u}^\varepsilon)\|_{L^2(0,T;L^2(\Omega))}\|\phi\|_{L^2(0,T;\mathbb{H}^1(\Omega))} \\ &\leq c\|\phi\|_{L^2(0,T;\mathbb{H}^1(\Omega))},\end{aligned}$$

and

$$\langle \rho^\varepsilon \boldsymbol{f}_\varepsilon, \phi\rangle \leq \rho_2\|\boldsymbol{f}_\varepsilon\|_{L^2(0,T;\mathbb{L}^2(\Omega))}\|\phi\|_{L^2(0,T;\mathbb{L}^2(\Omega))} \leq c\|\phi\|_{L^2(0,T;\mathbb{L}^2(\Omega))}$$

where c again denotes various constants independent of ε.

Using $\|\partial_i\phi_j\|_{L^3(\Omega)} \leq c\|\phi_j\|_{H^{3/2}(\Omega)}$ and (4.56) we have

$$\begin{aligned}&|\langle \operatorname{div}(\rho^\varepsilon r_\varepsilon(\boldsymbol{u}^\varepsilon) \otimes \boldsymbol{u}^\varepsilon), \phi\rangle| \\ &= \left|\int_0^T \int_\Omega \rho^\varepsilon r_\varepsilon(\boldsymbol{u}^\varepsilon) \otimes \boldsymbol{u}^\varepsilon \cdot \boldsymbol{\nabla}\phi \, dx\, dt\right|, \\ &\leq c_1\|\rho^\varepsilon |r_\varepsilon(\boldsymbol{u}^\varepsilon)|^2\|_{L^\infty(0,T;L^1(\Omega))}\|\sqrt{\rho^\varepsilon}\boldsymbol{u}^\varepsilon\|_{L^2(0,T;\mathbb{L}^6(\Omega))}\|\phi\|_{L^2(0,T;\mathbb{H}^{3/2}(\Omega))}, \\ &\leq c_2\|\phi\|_{L^2(0,T;\mathbb{H}^{3/2}(\Omega))}.\end{aligned}$$

Finally, the inequality $\|\phi\|_{\mathbb{L}^\infty(\Omega)} \leq C\|\phi\|_{\mathbb{H}^{3/2+\alpha}(\Omega)}$ with $\alpha > 0$ and (4.57) leads to

$$\begin{aligned}&|\langle \mathbf{curl}\, \boldsymbol{B}^\varepsilon \times s_\varepsilon(\boldsymbol{B}^\varepsilon), \phi\rangle| \\ &= \left|\int_0^T \int_\Omega \mathbf{curl}\, \boldsymbol{B}^\varepsilon \times s_\varepsilon(\boldsymbol{B}^\varepsilon) \cdot \phi \, dx\, dt\right|, \\ &\leq c_1\|\mathbf{curl}\, \boldsymbol{B}^\varepsilon\|_{L^2(0,T;\mathbb{L}^2(\Omega))}\|\boldsymbol{B}^\varepsilon\|_{L^\infty(0,T;\mathbb{L}^2(\Omega))}\|\phi\|_{L^2(0,T;\mathbb{H}^{3/2+\alpha}(\Omega))}, \\ &\leq c_2\|\phi\|_{L^2(0,T;\mathbb{H}^{3/2+\alpha}(\Omega))}.\end{aligned}$$

with $\alpha > 0$. Therefore (4.80) is true for any $m > 3/2$. Part 2 of Theorem 4.1 and the convergence of ρ^ε then imply that $\rho^\varepsilon \boldsymbol{u}^\varepsilon$ converges to $\rho\boldsymbol{u}$ strongly in $L^p(0,T;\mathbb{L}^r(\Omega))$ for $2 < p < \infty$, $1 \leq r < 6p/(3p-4)$ and $\boldsymbol{u}^\varepsilon$ converges to $\boldsymbol{u}$ strongly in $L^\theta(0,T;\mathbb{L}^{3\theta}(\Omega))$ for $1 \leq \theta < 2$.

Let us prove now that $\boldsymbol{B}^\varepsilon$ converges strongly to $\boldsymbol{B}$ in $L^2(0,T;H)$. First, we check that $\partial \boldsymbol{B}^\varepsilon/\partial t$ is bounded in $L^{4/3}(0,T;W')$. Indeed, for $\phi \in L^4(0,T;W)$ we have

$$\left\langle \frac{\partial \boldsymbol{B}^\varepsilon}{\partial t}, \phi \right\rangle = \int_0^T \int_\Omega \left(-\frac{1}{\sigma_\varepsilon} \mathbf{curl}\, \boldsymbol{B}^\varepsilon + \boldsymbol{u}^\varepsilon \times s_\varepsilon(\boldsymbol{B}^\varepsilon) \right) \cdot \mathbf{curl}\, \phi \, dx\, dt$$

$$\leq \left\| \frac{1}{\sigma_\varepsilon} \mathbf{curl}\, \boldsymbol{B}^\varepsilon \right\|_{L^2(0,T;\mathbb{L}^2(\Omega))} \|\phi\|_{L^2(0,T;W)}$$

$$+ \int_0^T \|\boldsymbol{u}^\varepsilon\|_{\mathbb{L}^4(\Omega)} \|s_\varepsilon(\boldsymbol{B}^\varepsilon)\|_{\mathbb{L}^4(\Omega)} \|\phi\|_W \, dt.$$

In the last term we use (4.57) and the interpolation inequality $\|h\|_{L^4(\Omega)} \leq \|h\|_{L^6(\Omega)}^{3/4} \|h\|_{L^2(\Omega)}^{1/4}$ to obtain

$$\int_0^T \|\boldsymbol{u}^\varepsilon\|_{\mathbb{L}^4(\Omega)} \|s_\varepsilon(\boldsymbol{B}^\varepsilon)\|_{\mathbb{L}^4(\Omega)} \|\phi\|_W \, dt$$

$$\leq c \|\boldsymbol{u}^\varepsilon\|_{L^\infty(0,T;H)}^{1/4} \|\boldsymbol{u}^\varepsilon\|_{L^2(0,T;V)}^{3/4}$$

$$\|\boldsymbol{B}^\varepsilon\|_{L^\infty(0,T;H)}^{1/4} \|\boldsymbol{B}^\varepsilon\|_{L^2(0,T;W)}^{3/4} \|\phi\|_{L^4(0,T;W)}.$$

Therefore $\partial \boldsymbol{B}^\varepsilon/\partial t$ is bounded in $L^{4/3}(0,T;W')$. Moreover, we know that $\boldsymbol{B}^\varepsilon$ is bounded in $L^2(0,T;W)$. Thus, up to the extraction of a subsequence, $\boldsymbol{B}^\varepsilon$ converges strongly to $\boldsymbol{B}$ in $L^2(0,T;H)$. We deduce in particular that $s_\varepsilon(\boldsymbol{B}^\varepsilon)$ converges strongly to $\boldsymbol{B}$ in $L^2(0,T;H)$. Furthermore, in view of (4.76), note that $\boldsymbol{B}^\varepsilon$ is bounded in $L^\infty(0,T;H)$. Thus $\boldsymbol{B} \in L^\infty(0,T;H)$.

In particular $\partial \boldsymbol{B}/\partial t \in L^1(0,T;W')$, thus $\boldsymbol{B}$ is almost everywhere equal to a function continuous from $[0,T]$ into W'. Moreover, $\boldsymbol{B} \in L^\infty(0,T;H)$ and $H \subset W'$ with a continuous injection, therefore, we know that $\boldsymbol{B}$ is weakly continuous from $[0,T]$ into H (see R. Temam [227] for instance).

The weak and strong convergences obtained for $\boldsymbol{B}^\varepsilon$ and $\boldsymbol{u}^\varepsilon$ enable us to pass to the limit in the nonlinear terms

$$\iint_{\Omega\times(0,\infty)} \rho^\varepsilon r_\varepsilon(\boldsymbol{u}^\varepsilon) \otimes \boldsymbol{u}^\varepsilon \cdot \boldsymbol{\nabla} \boldsymbol{\phi} \, dxdt,$$

$$\iint_{\Omega\times(0,\infty)} (\mathbf{curl}\, \boldsymbol{B}^\varepsilon) \times s_\varepsilon(\boldsymbol{B}^\varepsilon) \cdot \boldsymbol{\phi} \, dxdt.$$

The weak convergence of $\boldsymbol{u}^\varepsilon$ in $L^2(0,T;V)$ and the strong convergence of $\boldsymbol{B}^\varepsilon$ in $L^2(0,T;H)$ enable us to pass to the limit in

$$\iint_{\Omega\times(0,\infty)} \mathbf{curl}\,(\boldsymbol{u}^\varepsilon \times s_\varepsilon(\boldsymbol{B}^\varepsilon)) \cdot \boldsymbol{\phi} \, dxdt = \iint_{\Omega\times(0,\infty)} \boldsymbol{u}^\varepsilon \times s_\varepsilon(\boldsymbol{B}^\varepsilon) \cdot \mathbf{curl}\, \boldsymbol{\phi} \, dxdt.$$

Furthermore, we have in view of (4.60) and (4.61):

$$\lim_{\varepsilon \to 0} \int_\Omega \boldsymbol{u}_0^\varepsilon \cdot \boldsymbol{\phi}(x,0)\, dx = \int_\Omega \boldsymbol{u}_0 \cdot \boldsymbol{\phi}(x,0)\, dx,$$

$$\lim_{\varepsilon \to 0} \int_\Omega \boldsymbol{B}_0^\varepsilon \cdot \boldsymbol{\phi}(x,0)\, dx = \int_\Omega \boldsymbol{B}_0 \cdot \boldsymbol{\phi}(x,0)\, dx.$$

The proof is then easy to conclude.

Remark 4.2.3 Arguing as in [155], it may be seen that any solution built as above in particular satisfies the energy inequality:

$$\int_\Omega \rho |\boldsymbol{u}|^2 + |\boldsymbol{B}|^2\, dx + \int_0^t \int_\Omega \eta |\boldsymbol{\nabla u} + \boldsymbol{\nabla u}^T|^2 + \frac{2}{\sigma} |\mathbf{curl}\, \boldsymbol{B}|^2\, dx\, ds$$
$$\leq \int_\Omega \rho_0 |\boldsymbol{u}_0|^2 + |\boldsymbol{B}_0|^2\, dx + 2 \int_0^t \int_\Omega \rho \boldsymbol{f} \cdot \boldsymbol{u}\, dx\, ds.$$

◇

Remark 4.2.4 It is useful at this stage to revisit the proof of Chapter 2, concerning the one-fluid MHD system (1.40). The present proof provides an alternative proof for the existence result of weak solutions stated in Proposition 2.19. We introduce a linear regularized problem (in the spirit of (4.39)), for which we prove existence using a Galerkin approximation. Note that the utilization of a special basis set (the eigenfunctions of the Stokes and the magnetic problem) is not required then, because the problem is linear. Next, we use a fixed point argument to obtain existence for the nonlinear, still regularized, problem (in the spirit of (4.63)), and eventually we relax the regularization. In this latter step, no sophisticated argument is needed since the one-fluid case does not require the treatment of the transport equation (4.70), which then coincides with the divergence-free constraint. ◇

4.3 On the long-time behavior

We wish to investigate in this section some issues regarding the long-time behavior of the solutions to system (1.49).

Our interest stems of course from our wish to understand as completely as possible the mathematical nature of this system, but also and as importantly, from our motivation regarding some industrial issues. Indeed, as will become clear in Chapter 6, there are some challenging practical questions concerning the *stability* of the solutions to the multifluid MHD system. One such question is the following. Suppose that, owing to adequate data (boundary conditions, say), the system has reached a steady state after a while. Suppose now that there is some perturbation of these data at some given instant: will the system return to a steady state after some transient evolution? We will see in Chapter 6 that there are many ways to address such a question. It is however clear that some theoretical understanding can do no harm in such a context. Indeed, one must then settle the following question: if some instabilities arise, are such instabilities due to the numerical approximation, or to the continuous mathematical model *per se*? In addition, knowing that the mathematical model enjoys good dissipativity properties helps in the process of designing numerical algorithms that also share the same properties. This is the purpose of the present section. The mathematical proofs will only be outlined. More details, along with extensions and comments, may be found in the original reference [92].

We will mainly concentrate, for reasons that will be clear in the next section, on a simple hydrodynamics setting. The MHD case is only briefly treated in Section 4.3.3. In the hydrodynamic case, we will however try to evaluate the impact on the question under consideration of various choices in the modeling (also valid in the MHD context). In particular, the question of the surface tension, and the question of the linearization of the system will be examined. Let us now briefly motivate our interest in these two particular settings. This owes again to our very practical purposes.

First, it is important to see, on a theoretical basis, why surface tension indeed helps for return to equilibrium. It is also important to illustrate the differences between the two situations, with and without surface tension.

On the other hand, the linearized models are quite commonly used in numerical simulations, most of the time because nonlinear models are expensive to simulate, and therefore not particularly popular among practitioners. This is all the more true for simulations in the context of stability with respect to perturbations. We will return to this in Chapter 6. It suffices to say here that linearized models are extensively used for stability analysis. They basically consist in an expansion of the solution in the neighborhood of a steady-state solution. Owing to its practical relevance, it is important to specifically address this case, and we next emphasize its difference with the nonlinear setting.

Let us also emphasize that the determination of the long-time behavior of the solutions is well known to be closely related to the existence of regular solutions for all times. This is the reason why we assume in this section that the solutions we have are sufficiently regular to give a rigorous meaning to all the manipulations we will perform. Actually, the state of the mathematical knowledge is far from providing such regularity results. In particular, the global-in-time regularity of the flow for two-fluid systems (even in the purely hydrodynamic case) is an unsolved mathematical question. A result in this direction is the study by N. Tanaka [224] that applies for some small special initial data and small forces (in a convenient functional space). Note however that the result does not cover the case we are interested in, since our fluids typically have different densities, thus the gravity force is not a pure gradient, and is not L^p in time, for $p < +\infty$.

4.3.1 The nonlinear hydrodynamics case

We return here to the Navier–Stokes equations for two incompressible immiscible fluids. The main question we consider is the following. Assume that the forces and the boundary conditions are such that for any steady state, both fluids are at rest (the velocity is zero all over the domain). Then does the viscous dissipation drive the system to such a steady state as time goes to infinity? Intuitively, if for instance the only forces are due to gravity, and if the two fluids are of different densities, it is expected that the system goes, as time goes to infinity, to the situation when the two fluids are at rest, separated by a flat interface, the heaviest fluid below this interface, the lightest above. We now investigate to what extent this simple observation is satisfied mathematically.

More explicitly, we consider $(\boldsymbol{u}, p, \rho)$ a solution to

$$\begin{cases} \dfrac{\partial(\rho\boldsymbol{u})}{\partial t} + \operatorname{div}(\rho\boldsymbol{u}\otimes\boldsymbol{u}) - \Delta\,\boldsymbol{u} = -\boldsymbol{\nabla}p + \rho\boldsymbol{f}_m + \boldsymbol{f}_v, \\ \dfrac{\partial\rho}{\partial t} + \operatorname{div}(\rho\boldsymbol{u}) = 0, \\ \operatorname{div}\boldsymbol{u} = 0, \end{cases} \tag{4.81}$$

where $\boldsymbol{f}_m$ are given forces per unit mass and $\boldsymbol{f}_v$ given forces per unit volume. We supply the system with the no-slip boundary condition, and the usual initial conditions on ρ and $\boldsymbol{u}$.

Possibly, we shall add to the right-hand side of (4.81) a term $\mathcal{T}$ modeling the effect of the *surface tension* at the interface between the two fluids. Denoting by $\boldsymbol{n}$ the normal to the interface (say from fluid 1 to fluid 2 to fix ideas), such a term is the distribution defined, for any test velocity $\boldsymbol{w}$, by

$$\langle\mathcal{T}, \boldsymbol{w}\rangle = \int_\Sigma \gamma\,\mathcal{C}\,\boldsymbol{w}\cdot\boldsymbol{n} = -\int_\Sigma \gamma(\operatorname{div}\boldsymbol{n})\boldsymbol{w}\cdot\boldsymbol{n}, \tag{4.82}$$

where the coefficient γ denotes the amplitude of the surface tension, and where $\mathcal{C} = -\operatorname{div}\boldsymbol{n}$ denotes the local *mean curvature* of the interface oriented with $\boldsymbol{n}$.

Let us assume then that the given forces per unit mass $\boldsymbol{f}_m$ and forces per unit volume $\boldsymbol{f}_v$ are such that any steady-state solution of (4.81) consists of some piecewise constant density $\rho \in \{\overline{\rho}_1, \overline{\rho}_2\}$ and of the zero velocity field $\boldsymbol{u} \equiv 0$.

For simplicity, we take $\boldsymbol{f}_m = -\boldsymbol{e}_z$ (the gravity field), and $\boldsymbol{f}_v = 0$ in the sequel.

Remark 4.3.1 For simplicity we choose for the rest of Section 4.3 a constant viscosity $\eta = 1$. There is no conceptual difficulty in dealing with a density-dependent viscosity $\eta = \eta(\rho)$ provided the solution fields are assumed sufficiently regular. On the other hand, proving such a regularity is indeed an unsolved mathematical question. ◇

Remark 4.3.2 Without aiming at being exhaustive, let us briefly point out some works in the vast literature devoted to the subject. For the case of one fluid with a free surface, local-in-time existence results can be found in G. Allain [2], V.A. Solonnikov [215], global-in-time existence results for small initial data and zero forces appeared in V.A. Solonnikov [218] (bounded case), J.T. Beale [15] (unbounded case), and also in A. Tani & N. Tanaka [226], and, for small initial data and possibly non-zero body forces, in A. Tani [225]. For the two-fluid case, local-in-time existence of strong solutions is due to I.V. Denisova [53] and I.V. Denisova, V.A. Solonnikov [54], global-in-time existence for small data is due to V.A. Solonnikov [216] and also N. Tanaka [224] (for a special initial condition).

No existence results of global weak solutions seem to have been established for the multifluid Navier–Stokes equations with surface tension. As announced above, we thus *assume* $(\boldsymbol{u}(\mathbf{x},t), p(\mathbf{x},t), \rho(\mathbf{x},t))$ are sufficiently regular. ◊

4.3.1.1 Position of the problem It is first of all to be remarked that we cannot hope to solve the question of the long-time behavior of the solution to the two-fluid Navier–Stokes equations in a very general setting. Even for the one-fluid case, this question is extremely difficult. Regarding the long-time behavior of the one-fluid Navier–Stokes equations, the main body of the theory is due to R. Temam and coworkers (see R. Temam [229,228], and P. Constantin, C. Foias, B. Nicolaenko, R. Temam [40]). In a nutshell, the long-time behavior of these equations is finite dimensional. It holds without any restriction in two dimensions and in three dimensions at least for flows that remain smooth.

In two dimensions, the solution is regular and therefore many things are known. If the force is time independent, there exists an attractor[28], and its Hausdorff dimension is finite. This attractor is all the more regular as the force is (*e.g.* $\mathcal{C}^\infty$ if the force is $\mathcal{C}^\infty$). In the space periodic case, it is even possible to show that there exists an inertial manifold. An upper bound on the finite dimension of the attractor is related to the Reynolds number of the problem (see, *e.g.* A. Miranville and X. Wang [167] and references therein). Most of these results apply to the MHD system (M. Sermange and R. Temam [208]). In three dimensions, it is only known that the functional invariant sets bounded in L^2 are of finite Hausdorff dimension, but no existence of attractor (which would exist if the solutions were regular for all time) seems to have been established to this day in the generic case.

In very particular situations, it is possible to improve these general results by proving the convergence of the flow to some stationary state. When the body force the fluid is subjected to is large (and even if it is stationary) there are some experimental situations where the flow remains turbulent and time-dependent for long times (for instance, it tends to a time-periodic solution). But when the force is small, there are many situations where the flow converges as time goes to infinity to the state where the fluid is at rest. Let us examine now the mathematical counterpart of this experimental observation.

The first result in this spirit concerns the case of one homogeneous fluid enclosed in a fixed box and goes back to Leray. In two or three dimensions with homogeneous Dirichlet boundary conditions, when there is no body force, the only steady state is the fluid at rest and the time-dependent flow converges to it in H^1 as time goes to infinity. This result has in particular been extended in the following two directions: if the body force and the data (initial velocity

[28] In case the reader is not familiar with the terminology, we refer to the bibliography for the definition of the terms *attractor, inertial manifold, Hausdorff dimension.*

and boundary conditions) are sufficiently small then the flow remains regular for all time (even in three dimensions) and the speed of convergence toward the steady state can be evaluated (see C. Guillope [117], J.G Heywood [121, 122], C. Foias, J.C. Saut [78]); if the initial velocity is large but when the force is gradient-like it is possible to show that the flow becomes smooth after a finite time, then remains smooth and converges to the steady state (see J.G Heywood [122]). Some analogous results are available under convenient hypotheses in the unbounded case (see G.P. Galdi, J.G Heywood, Y. Shibata [84], W. Borchers and T. Miyakawa [26] and references therein).

Let us also mention the work of F. Armero and J.C. Simo [8] for an enlightening presentation of the theoretical concepts of attractors and related notions, precisely in the context of MHD equations in the *one* fluid case.

We now turn to the case of one fluid with a free surface or the case of two fluids. There again, most studies deal with situations when there exist regular solutions. This is mostly the case when the data are small and the evolution is not far from equilibrium. Let us mention here the works by V.A. Solonnikov [216, 217, 219] and by J.T Beale [14, 17]. The basic result is the convergence to the steady state as time goes to infinity. Let us also mention for the sake of completeness the work by A. Tani and N. Tanaka [226], the works in the irrotational inviscid case J.T Beale, T.Y Hou, J.S Lowengrub [16], T.Y Hou, Z.H Teng, P. Zhang [126] and also the related work by H. Beirao da Veiga [18].

In the case we deal with here, it is therefore only under very restrictive assumptions that one can hope to settle this question. All the situations we shall consider below share the same following feature: there is uniqueness of the stationary velocity field (but not necessarily of the stationary interface) *and* this velocity field is zero. This is the reason why we are not able to treat the generic MHD setting (with non-zero boundary conditions), when the existence of the coupling with the magnetic field makes the steady-state velocity non-trivial.

We are going to see that the situation is the following:

- It is reasonably easy to show that the velocity field $\boldsymbol{u}$ goes to zero, at least in a weak sense, as time goes to infinity.
- As well, we can prove that ρ converges to some limit ρ_∞ (in a weak sense also), which is a solution to the Navier–Stokes equation with zero velocity field, but which, most often, cannot be explicitly identified and may thus differ from the expected solution.

4.3.1.2 Mathematical arguments First, we observe that from considerations on the equation of conservation of mass (see P.-L. Lions [155] and also earlier sections of this chapter for the details), we have

$$\|\rho(t)\|_{L^\infty(\Omega)} = \|\rho^0\|_{L^\infty(\Omega)}, \forall t \geq 0, \tag{4.83}$$

and

$$\text{meas}\{\boldsymbol{x} \in \Omega, \rho(\boldsymbol{x}) = \overline{\rho}_i\}, i = 1, 2 \text{ is independent of } t \geq 0. \tag{4.84}$$

Next, by standard manipulations we obtain

$$\frac{d}{dt}\left(\frac{1}{2}\int_\Omega \rho \boldsymbol{u}^2\,d\boldsymbol{x} + \int_\Omega \rho z\,d\boldsymbol{x}\right) + \int_\Omega |\nabla \boldsymbol{u}|^2\,d\boldsymbol{x} = \int_\Sigma \gamma \mathcal{C}\boldsymbol{u}\cdot\boldsymbol{n}. \tag{4.85}$$

For the convenience of the reader, we now reproduce these manipulations. In the absence of surface tension, we easily have

$$\int_\Omega \partial_t(\rho\boldsymbol{u})\cdot\boldsymbol{u}\,d\boldsymbol{x} + \int_\Omega \operatorname{div}(\rho\boldsymbol{u}\otimes\boldsymbol{u})\cdot\boldsymbol{u}\,d\boldsymbol{x} + \int_\Omega |\nabla\boldsymbol{u}|^2\,d\boldsymbol{x} = -\int_\Omega \rho\boldsymbol{u}\cdot\boldsymbol{e}_z\,d\boldsymbol{x}. \tag{4.86}$$

Next we write

$$\begin{aligned}
\int_\Omega \partial_t(\rho\boldsymbol{u})\cdot\boldsymbol{u}\,d\boldsymbol{x} &= \int_\Omega \frac{\partial\rho}{\partial t}\boldsymbol{u}\cdot\boldsymbol{u}\,d\boldsymbol{x} + \int_\Omega \rho\,\boldsymbol{u}\cdot\frac{\partial\boldsymbol{u}}{\partial t}\,d\boldsymbol{x}\\
&= \int_\Omega \frac{\partial\rho}{\partial t}\boldsymbol{u}\cdot\boldsymbol{u}\,d\boldsymbol{x} + \frac{1}{2}\frac{d}{dt}\int_\Omega \rho\boldsymbol{u}^2\,d\boldsymbol{x} - \frac{1}{2}\int_\Omega \frac{\partial\rho}{\partial t}\boldsymbol{u}\cdot\boldsymbol{u}\,d\boldsymbol{x}\\
&= \frac{1}{2}\int_\Omega \frac{\partial\rho}{\partial t}\boldsymbol{u}\cdot\boldsymbol{u}\,d\boldsymbol{x} + \frac{1}{2}\frac{d}{dt}\int_\Omega \rho\boldsymbol{u}^2\,d\boldsymbol{x}\\
&= \frac{1}{2}\int_\Omega \rho\,\boldsymbol{u}\cdot\boldsymbol{\nabla}(\boldsymbol{u}^2)\,d\boldsymbol{x} + \frac{1}{2}\frac{d}{dt}\int_\Omega \rho\boldsymbol{u}^2\,d\boldsymbol{x},
\end{aligned} \tag{4.87}$$

where we have used for the latter equality the equation of conservation of mass. We also have

$$\int_\Omega \operatorname{div}(\rho\boldsymbol{u}\otimes\boldsymbol{u})\cdot\boldsymbol{u}\,d\boldsymbol{x} = -\frac{1}{2}\int_\Omega \rho\boldsymbol{u}\cdot\nabla(|\boldsymbol{u}|^2)\,d\boldsymbol{x}, \tag{4.88}$$

thus, summing up (4.87) and (4.88), we obtain

$$\int_\Omega \partial_t(\rho\boldsymbol{u})\cdot\boldsymbol{u}\,d\boldsymbol{x} + \int_\Omega \operatorname{div}(\rho\boldsymbol{u}\otimes\boldsymbol{u})\cdot\boldsymbol{u}\,d\boldsymbol{x} = \frac{1}{2}\frac{d}{dt}\int_\Omega \rho\boldsymbol{u}^2\,d\boldsymbol{x}. \tag{4.89}$$

On the other hand, we have

$$\int_\Omega \rho\boldsymbol{u}\cdot\boldsymbol{e}_z\,d\boldsymbol{x} = \int_\Omega \rho\boldsymbol{u}\cdot\nabla(z)\,d\boldsymbol{x} = -\int_\Omega z\operatorname{div}(\rho\boldsymbol{u})\,d\boldsymbol{x} = \int_\Omega z\partial_t\rho\,d\boldsymbol{x} = \frac{d}{dt}\int_\Omega \rho z\,d\boldsymbol{x}. \tag{4.90}$$

Inserting (4.89) and (4.90) into (4.86) yields (4.85) when $\mathcal{C} = 0$. When a surface tension term is added to the right hand side, (4.86), and thus (4.85), are modified correspondingly. For simplicity we next assume that $\rho_2 - \rho_1 = 1$.

To compute the right-hand side of (4.85), we use the fact that:

$$\int_\Sigma \mathcal{C}\boldsymbol{u}\cdot\boldsymbol{n} = -\int_\Sigma \operatorname{div}_\Sigma(\boldsymbol{u})d\sigma_\Sigma, \tag{4.91}$$

$$= -\frac{d}{dt}\int_\Sigma d\sigma_\Sigma, \tag{4.92}$$

$$= -\frac{d}{dt}L(\Sigma), \tag{4.93}$$

where $L(\Sigma) = \int_\Sigma d\sigma$ denotes the measure of the interface Σ. To obtain (4.91), we have used the surface divergence theorem (see Section 5.1.4.2 and equation (5.59) below, with the notation $H = \mathcal{C}$ for the mean curvature). To obtain (4.92), one can use a change of variable to rewrite the integral on a non-moving interface Σ_0 and then the Liouville formula (see equation (5.18) below). For details, we refer for example to equation (4.21) of Chapter 8, Section 4.3.1 in [52].

Equation (4.85) together with (4.93) give the *first energy estimate*

$$\frac{d}{dt}\left(\frac{1}{2}\int_\Omega \rho \boldsymbol{u}^2\,d\boldsymbol{x} + \int_\Omega \rho z\,d\boldsymbol{x} + \gamma L(\Sigma)\right) + \int_\Omega |\nabla \boldsymbol{u}|^2\,d\boldsymbol{x} = 0, \tag{4.94}$$

with $\gamma = 0$ if there is no surface tension, and $\gamma = 1$ if there is.

From this energy estimate, we deduce that in particular

$$\int_0^{+\infty} \|\boldsymbol{u}\|^2_{\mathbb{H}^1(\Omega)}\,dt < +\infty, \tag{4.95}$$

and

$$\sup_{t\in[0,\infty)} \|\boldsymbol{u}\|_{\mathbb{L}^2(\Omega)} < +\infty. \tag{4.96}$$

Condition (4.95) suggests that, in a formal sense which will be made precise below, $\boldsymbol{u}$ goes to zero as time goes to infinity. We next deduce from the equation itself that, formally, $\frac{\partial(\rho\boldsymbol{u})}{\partial t}$ also goes to zero. We then recover with the Navier–Stokes equation that, in the case without surface tension, $-\nabla p - \rho \boldsymbol{e}_z \longrightarrow 0$, as t goes to infinity (see (4.100)). This means that

$$-\nabla(p + \rho z) + z\nabla\rho \longrightarrow 0,$$

which can be expressed as follows: $\mathbf{curl}\,(z\,\nabla\rho) \longrightarrow 0$, or also $\nabla\rho \times \boldsymbol{e}_z$ goes to zero. Thus ρ becomes a function of z as time goes to infinity (see (4.101)). If we admit that the two fluids do not mix in the limit $t \longrightarrow +\infty$, this implies that the interface between the two fluids consists of planes, which are parallel to the (O, x, y) plane, and which separate two consecutive layers of fluids. Notice that the interface does not necessarily consist of only one plane. There might even exist an infinite superposition of layers.

On the other hand, when there is surface tension, we obtain in the limit $t \longrightarrow +\infty$

$$-\nabla(p + \rho z) + (z - \operatorname{div}\boldsymbol{n})\nabla\rho \longrightarrow 0.$$

Taking the **curl** of the above expression, we obtain that the quantity $z - \operatorname{div}\boldsymbol{n}$ becomes constant along the connected components of the interface (assuming that $\boldsymbol{\nabla}\rho$ is normal to the interface, namely there is no homogenization in the fluids).

Of course, $z = 0$ (and thus $\boldsymbol{\nabla}\rho = \boldsymbol{e}_z$) is a solution to the equation giving position of the interface at equilibrium. However, it can be shown that there exists infinitely many steady states with zero velocity and a non-flat interface with an energy arbitrarily close to the minimal energy. A comprehensive analysis of the convergence of the density is out of reach, except for some very particular situations (such as, *e.g.* the linearized setting below). Let us make precise the above discussion.

Behavior of the velocity The results we obtain on the velocity field do not depend on the presence of surface tension. In contrast, the presence of surface tension allows us to get somewhat better insight on the behavior of the interface.

Let $(t_n)_{n\in\mathbf{N}}$ be an arbitrary sequence of positive real numbers such that $\lim_{n\to+\infty} t_n = +\infty$. We define the sequences ρ_n and $\boldsymbol{u}_n$ by $\rho_n(x,t) = \rho(x, t+t_n)$ and $\boldsymbol{u}_n(x,t) = \boldsymbol{u}(x, t+t_n)$ (in the sense of distributions).

According to estimate (4.95), we have

$$\lim_{n\to+\infty} \int_{t_n}^{+\infty} \int_\Omega |\nabla \boldsymbol{u}(x,t)|^2 \, d\boldsymbol{x} \, dt = 0,$$

therefore

$$\boldsymbol{u}_n \longrightarrow 0 \text{ in } L^2(0,\infty;\mathbb{H}^1(\Omega)) \text{ as } n \to +\infty. \tag{4.97}$$

Remark 4.3.3 [On the two-dimensional case] In the two-dimensional case, and when the viscosity is constant over the domain[29], some better information on the behavior of the velocity for large times may be obtained. By standard manipulations and using Sobolev-type inequalities specific to the two-dimensional setting (see the details in [92]), the convergence

$$\boldsymbol{u} \longrightarrow 0 \text{ in } \mathbb{H}^{1-\varepsilon}(\Omega) \text{ as } t \to +\infty, \forall \varepsilon > 0 \tag{4.98}$$

may be established. ◊

Behavior of the interface Let us first discuss the case without surface tension. In view of (4.83) the sequence (ρ_n) remains in a bounded set of $L^\infty((0,+\infty)\times\Omega)$. Therefore there exists $\rho_\infty \in L^\infty((0,+\infty)\times\Omega)$ such that

$$\rho_n \rightharpoonup \rho_\infty \qquad \text{in } L^\infty((0,+\infty)\times\Omega)\text{weak-}\star. \tag{4.99}$$

We first prove that ρ_∞ does not depend on t. Let $v \in L^2(0,\infty;\mathbb{H}^1_0(\Omega))$, we have

$$\left|\left\langle \frac{\partial \rho_n}{\partial t}, v \right\rangle\right| = |-\langle \operatorname{div}(\rho_n \boldsymbol{u}_n), v\rangle| = \left| \int_\Omega \rho_n \boldsymbol{u}_n \cdot \nabla v \, d\boldsymbol{x} \right|$$

$$\leq C \|\rho_n\|_{L^\infty((0,T)\times\Omega)} \|\boldsymbol{u}_n\|_{L^2(0,T;\mathbb{H}^1_0(\Omega))} \|v\|_{L^2(0,T;\mathbb{H}^1_0(\Omega))},$$

which proves in view of (4.97) that

$$\frac{\partial \rho_n}{\partial t} \longrightarrow 0 \text{ in } L^2(0,\infty;\mathbb{H}^{-1}(\Omega)) \text{ as } n \to +\infty.$$

Therefore, since in the sense of distributions $\partial\rho_n/\partial t \rightharpoonup \partial\rho_\infty/\partial t$, we deduce $\partial\rho_\infty/\partial t = 0$.

[29] The result might extend to the case of a slightly variable viscosity, using the results of B. Desjardins [65].

We now prove that ρ_∞ only depends on the third space variable z. We have

$$-\nabla p_n - \rho_n \boldsymbol{e}_z = \frac{\partial(\rho_n \boldsymbol{u}_n)}{\partial t} + \operatorname{div}(\rho_n \boldsymbol{u}_n \otimes \boldsymbol{u}_n) - \Delta\, \boldsymbol{u}_n$$

Let $\boldsymbol{v} \in \mathcal{C}_0^\infty((0,\infty)\times\Omega)^3$. Then

$$\left|\left\langle \frac{\partial(\rho_n \boldsymbol{u}_n)}{\partial t}, \boldsymbol{v}\right\rangle\right| \leq \|\rho_n\|_{L^\infty((0,\infty)\times\Omega)} \|\boldsymbol{u}_n\|_{L^2((0,\infty)\times\Omega)} \left\|\frac{\partial \boldsymbol{v}}{\partial t}\right\|_{L^2((0,\infty)\times\Omega)},$$

$$|\langle \operatorname{div}(\rho_n \boldsymbol{u}_n \otimes \boldsymbol{u}_n), \boldsymbol{v}\rangle| \leq C\|\rho_n\|_{L^\infty((0,\infty)\times\Omega)} \|\boldsymbol{u}_n\|^2_{L^2(0,\infty;\mathbb{H}_0^1(\Omega))} \|\boldsymbol{v}\|_{L^\infty(0,\infty;\mathbb{H}_0^1(\Omega))},$$

$$|\langle -\Delta\, \boldsymbol{u}_n, \boldsymbol{v}\rangle| \leq \|\boldsymbol{u}_n\|_{L^2(0,\infty;\mathbb{H}_0^1(\Omega))} \|\boldsymbol{v}\|_{L^2(0,\infty;\mathbb{H}_0^1(\Omega))},$$

thus the right-hand sides of the these inequalities go to zero as $n \to \infty$ (see (4.97) and (4.99)). Therefore

$$-\nabla p_n - \rho_n \boldsymbol{e}_z \longrightarrow 0, \tag{4.100}$$

in the sense of distributions. Thus, $\mathbf{curl}\,(\rho_\infty \boldsymbol{e}_z) = \nabla \rho_\infty \times \boldsymbol{e}_z = 0$, which proves that $\partial_x \rho_\infty = \partial_y \rho_\infty = 0$. Therefore

$$\rho_\infty = \rho_\infty(z). \tag{4.101}$$

Using (4.99), the global mass conservation,

$$\int_\Omega \rho_\infty(\boldsymbol{x})\, d\boldsymbol{x} = \int_\Omega \rho^0(\boldsymbol{x})\, d\boldsymbol{x},$$

is readily checked.

Notice that, according to (4.84), we know that $\text{meas}\{x \in \Omega, \rho_n(x) = \overline{\rho}_i\} = M_i$ is independent of n. Owing to the convergence of ρ_n that is only weak, we are however not able to prove that $\text{meas}\{x \in \Omega, \rho_\infty(x) = \overline{\rho}_i\} = M_i$. Homogenization may appear in the limit, *i.e.* there may exist some parts of Ω where ρ_∞ has values between ρ_1 and ρ_2. All we know is that these areas consist of horizontal layers (possibly infinitely thin).

Remark 4.3.4 If for some sequence $t_n \longrightarrow +\infty$ we have $\rho(t_n,\cdot) \longrightarrow \rho_\infty(\cdot)$ almost everywhere in Ω, then it is possible to show, using Theorem 2.4 of [155], that

$$\forall T < \infty, \forall p < \infty, \lim_{n\to\infty} \sup_{t\in[0,T]} \|\rho(t+t_n,\cdot) - \rho_\infty(\cdot)\|_{L^p} = 0,$$

which therefore prevents homogeneization. ◇

Let us now turn to the case when there is surface tension. In the sequel, $T > 0$ is fixed. We now show that the presence of surface tension allows us to improve the convergence (4.99) of ρ_n, more precisely we prove that this sequence is in a compact set of $L^p(\Omega \times (0,T))$ for any $p \geq 1$.

From estimate (4.94), we know (since $c_0 = 1$) that

$$\rho \in L^{\infty}(0, \infty; BV(\Omega)) \tag{4.102}$$

(where BV denotes the space of functions with bounded variation) and thus that our sequence ρ_n lies in a bounded set of the space $L^{\infty}(0, T; BV(\Omega) \cap L^{\infty}(\Omega))$. Using standard embeddings between functional spaces, and bootstrapping regularity on $\partial\rho_n/\partial t$ from the equation

$$\frac{\partial \rho_n}{\partial t} = -\operatorname{div}(\rho_n \boldsymbol{u}_n),$$

it follows that ρ_n is bounded in $H^{\beta}(0, T; H^{\gamma}(\Omega))$ for some $\beta > 0$ and $\gamma > 0$, thus that $(\rho_n)_{n\in\mathbf{N}}$ is a compact set of (for example) $L^1(\Omega \times (0, T))$. Since the sequence is bounded in $L^p(\Omega \times (0, T))$, $\forall p \geq 1$, we deduce that $(\rho_n)_{n\in\mathbf{N}}$ is a compact set of $L^p(\Omega \times (0, T))$. Therefore, there exists an extraction of $(\rho_n)_{n\in\mathbf{N}}$ such that

$$\rho_{n'} \longrightarrow \rho_{\infty} \text{ in } L^p(\Omega \times (0, T)), \forall p \geq 1 \text{ as } n' \to +\infty.$$

Then, $\rho_{\infty}(\boldsymbol{x}, t) = \rho_{\infty}(\boldsymbol{x})$ and the conservation of the global mass both follow.

We next show that there exists a sequence $(s_n)_{n\in\mathbf{N}}$ such that $s_n \in [0, T]$ and $\lim_{n\to+\infty} \sup_{t\in[0,T]} \|\rho_n(\cdot, t + s_n) - \rho_{\infty}(\cdot)\|_{L^p(\Omega)} = 0$, $\forall p \geq 1$.

For ease of notation, we define $X_n(t) = \|\rho_n(t) - \rho_{\infty}\|_{L^p(\Omega)}$. We recall that ρ is supposed $\mathcal{C}(0, \infty; L^p(\Omega))$, thus $X_n \in \mathcal{C}(0, \infty)$. Moreover $X_n \to 0$ as $n \to +\infty$ for the strong topology of $L^p(0, T)$. Thus, there exists a sequence $(s_n)_{n\in\mathbf{N}}$ in $[0, T]$ such that

$$\lim_{n\to+\infty} X_n(s_n) = 0. \tag{4.103}$$

Then, we denote by $(\tilde{\boldsymbol{u}}_n)_{n\in\mathbf{N}}$ and $(\tilde{\rho}_n)_{n\in\mathbf{N}}$ the sequences defined by $\tilde{\boldsymbol{u}}_n(x, t) = \boldsymbol{u}_n(x, t + s_n)$ and $\tilde{\rho}_n(x, t) = \rho_n(x, t + s_n)$. Assertion (4.103) proves the convergence of $\tilde{\rho}_n(\cdot, t = 0)$ to $\rho_{\infty}(\cdot)$ as $n \to +\infty$ for the strong topology of $L^p(\Omega)$.

Collecting the previous results, we have:

$$0 \leq \tilde{\rho}_n \leq C, \frac{\partial \tilde{\rho}_n}{\partial t} + \operatorname{div}(\tilde{\rho}_n \tilde{\boldsymbol{u}}_n) = 0,$$

$$\operatorname{div} \tilde{\boldsymbol{u}}_n = 0, \ \tilde{\rho}_n|_{t=0} \to \rho_{\infty} \text{ in } L^p(\Omega)$$

and

$$\tilde{\boldsymbol{u}}_n \to 0 \text{ in } L^2(0, T; \mathbb{H}^1(\Omega)).$$

We thus deduce from Theorem 4.1 that $\tilde{\rho}_n$ converges to ρ_{∞} in $\mathcal{C}([0, T], L^p(\Omega))$.

In other words, we have shown that, for $T > 0$, $p \geq 1$ and for any sequences $(t_n)_{n\in\mathbf{N}}$, $t_n \to +\infty$, there exists $(s_n)_{n\in\mathbf{N}}$, $s_n \in [0, T]$ such that, up to an extraction,

$$\lim_{n\to+\infty} \sup_{t\in[0,T]} \|\rho(\cdot, t + t_n + s_n) - \rho_{\infty}(\cdot)\|_{L^q(\Omega)} = 0. \tag{4.104}$$

We finally show that no homogenization appears. Indeed, for any $\beta \in \mathcal{C}^1([0,\infty),\mathbb{R})$ we have then

$$\int_\Omega \beta(\rho_n(x,t))\,d\boldsymbol{x} \longrightarrow \int_\Omega \beta(\rho_\infty(x))\,d\boldsymbol{x},$$

thus by regularization we obtain that

$$\text{meas}\{x \in \Omega, \rho_\infty(x) = \rho_i\} = \text{meas}\{x \in \Omega, \rho(x,t) = \rho_i\} \tag{4.105}$$

which is a constant of the evolution.

Summary of the results We now collect the results from [92] in the following two propositions.

Proposition 4.7 *In the nonlinear case without surface tension, a solution* $(\rho, \boldsymbol{u})$ *satisfying the estimates (4.83), (4.95), (4.96) has the following behavior as time converges to infinity:*

(i-a) The velocity field $\boldsymbol{u}$ *goes to 0 in* $\mathbb{H}^1(\Omega)$ *in the "weak" sense (4.97).*
(i-b) If we postulate that $\boldsymbol{u} \in \mathcal{C}(0,\infty;\mathbb{L}^2(\Omega))$ *and that it satisfies the energy inequality*

$$\frac{d}{dt}\left(\frac{1}{2}\int_\Omega \rho \boldsymbol{u}^2\,d\boldsymbol{x} + \int_\Omega \rho z\,d\boldsymbol{x}\right) + \int_\Omega |\nabla \boldsymbol{u}|^2\,d\boldsymbol{x} \leq 0,$$

then $\boldsymbol{u}$ *goes to 0 in* $\mathbb{L}^2(\Omega)$.
(i-c) If $\Omega \subset \mathbb{R}^2$ *then* $\boldsymbol{u} \in \mathcal{C}(0,\infty;\mathbb{H}^1(\Omega))$ *and* $\boldsymbol{u}$ *goes to 0 in* $\mathbb{H}^{1-\varepsilon}(\Omega)$, $\forall \varepsilon > 0$ *as* $t \to +\infty$.

(ii) The density ρ *goes to* ρ_∞ *in the sense of (4.99) with* $\rho_\infty = \rho_\infty(z)$. *In other words, the "interface" tends in a weak sense (and up to an extraction in time) to one or several horizontal planes. Homogenization may appear.*
(iii) There exists an infinity of steady solutions $(\boldsymbol{u} = 0, \rho_\infty)$ *whose energies are arbitrarily close to the minimal energy.*

Proposition 4.8 *In the nonlinear case with surface tension, assuming the existence of a solution sufficiently regular to give a sense to the surface tension term and satisfying the a priori estimates (4.83), (4.95), (4.96) and (4.102), the behavior of* $\boldsymbol{u}, \rho$ *as time goes to infinity is the following:*

(i) The velocity field $\boldsymbol{u}$ *goes to 0 in* $\mathbb{H}^1(\Omega)$ *in the same sense as in the case without surface tension (see (4.97)).*
(ii) The density ρ *goes to* ρ_∞ *in a stronger sense than in the case without surface tension (see (4.104)). The density* ρ_∞ *consists only of zones of densities* ρ_1 *and* ρ_2 *(see (4.105)), homogenization therefore being excluded. In addition,* ρ_∞ *is such that the quantity* $z - \text{div}\,\boldsymbol{n}$ *is constant on each connected component of the interface between zones of densities* ρ_1 *and* ρ_2.

(iii) We do not know whether the limit interface is unique or connected. Moreover, an infinity of steady solutions ($\boldsymbol{u} = 0, \rho_\infty$) *may be shown to exist, with an energy arbitrarily close to the minimal energy.*

Remark 4.3.5 Assertions (i-b) and (iii) of Proposition 4.7, together with assertion (iii) in Proposition 4.8 are proved in [92]. Assertion (i-b) of Proposition 4.7 is only mentioned in Remark 4.3.3. ♢

4.3.2 A detour by linearized models

Because this is a situation extensively addressed in the literature devoted to applications (see Chapter 6 on this matter), and because it is a case that exhibits very particular properties, we now consider the question of the long-time limit in a *linearized* setting. We will only do so for the hydrodynamic case (*i.e.* linearization of (4.81)). Some analogous considerations hold in the MHD setting. In addition, we only briefly sketch the derivation and the results. Details may be found in the bibliography.

The derivation of the linearized Navier–Stokes equations for two fluids is somewhat standard. Recall that we put ourselves in the purely gravitational case, where the forces per unit mass are $\boldsymbol{f}_m = -\boldsymbol{e}_z$ (we set the gravitational constant to unity), and the forces per unit volume $\boldsymbol{f}_v$ vanish. Then ($\boldsymbol{u} = 0, p = p_0, \rho = \rho_0$) is a steady-state solution, *i.e.*

$$0 = -\nabla p_0 - \rho_0 \boldsymbol{e}_z, \tag{4.106}$$

with

$$\rho_0(\cdot) = \begin{cases} \rho_1 > 0, & \text{constant in } \Omega_1, \\ \rho_2 > 0, & \text{constant in } \Omega_2. \end{cases} \tag{4.107}$$

The domains Ω_1 and Ω_2 are the two subdomains separated by the flat horizontal interface $\Sigma = \{z = 0\}$. The heaviest fluid is assumed to fill in the zone below the plane $z = 0$.

Consider a small constant $\varepsilon > 0$ which defines the size of the perturbation, and denote by $(\varepsilon \boldsymbol{u}_\varepsilon, \rho_0 + \varepsilon \rho_\varepsilon, p_0 + \varepsilon p_\varepsilon)$ the solution to the multifluid Navier–Stokes equations. Let us also define the function ψ such that the shape of the perturbed interface (with respect to the steady-state flat horizontal interface $z = 0$) is provided by the equation $z = \varepsilon \psi(x, y, t)$ at any time t. We assume that

$$\int_\Sigma \psi(\cdot, t = 0) = 0. \tag{4.108}$$

It is standard to show that, as ε goes to 0, the function ρ_ε converges, in the sense of distributions, to the distribution m defined for any arbitrary $\varphi \in \mathcal{D}(\Omega)$ by

$$\langle m, \varphi \rangle = (\rho_2 - \rho_1) \int_\Sigma \psi \varphi,$$

which is in fact a bounded measure on Ω, supported on the plane $z = 0$. For simplicity, we henceforth normalize the jump of densities $\rho_2 - \rho_1$ to unity.

Therefore, we obtain, letting ε go to zero, the linearized equations

$$\rho_0 \frac{\partial \boldsymbol{u}}{\partial t} - \Delta \boldsymbol{u} = -\boldsymbol{\nabla} p - \psi \delta_{z=0} \boldsymbol{e}_z,$$

$$\operatorname{div} \boldsymbol{u} = 0, \ in\ \Omega, \tag{4.109}$$

$$\frac{\partial \psi}{\partial t} - \boldsymbol{u} \cdot \boldsymbol{e}_z = 0, \ on\ \{z = 0\}.$$

The above setting may be modified in order to account for a linearized surface tension term:

$$\langle \mathcal{T}_0, \boldsymbol{w} \rangle = c_0 \int_\Sigma \Delta \psi \, \boldsymbol{w} \cdot \boldsymbol{e}_z, \tag{4.110}$$

where the coefficient c_0 has value 0 or 1 to discriminate between the case without or with surface tension, respectively. Therefore, the linearized equations read

$$\rho_0 \frac{\partial \boldsymbol{u}}{\partial t} - \Delta \boldsymbol{u} = -\boldsymbol{\nabla} p - (\psi - c_0 \Delta \psi) \delta_{z=0} \boldsymbol{e}_z,$$

$$\operatorname{div} \boldsymbol{u} = 0, \ in\ \Omega,$$

$$\frac{\partial \psi}{\partial t} - \boldsymbol{u} \cdot \boldsymbol{e}_z = 0, \ on\ \{z = 0\}. \tag{4.111}$$

We now establish the formal *a priori* estimate. First of all, we remark that

$$\int_\Sigma \psi = 0, \tag{4.112}$$

which expresses the mass conservation of each fluid.

In addition, from the above equations, the following two energy estimates can be established in a standard way:

$$\int_\Omega |\nabla \boldsymbol{u}|^2 \, d\boldsymbol{x} + \frac{1}{2} \frac{d}{dt} \left(\int_\Omega \rho_0 \boldsymbol{u}^2 \, d\boldsymbol{x} + \int_\Sigma (\psi^2 + c_0 |\nabla \psi|^2) \, d\sigma \right) = 0, \tag{4.113}$$

and

$$\int_\Omega \left| \nabla \frac{\partial \boldsymbol{u}}{\partial t} \right|^2 d\boldsymbol{x} + \frac{1}{2} \frac{d}{dt} \left(\int_\Omega \rho_0 \left| \frac{\partial \boldsymbol{u}}{\partial t} \right|^2 d\boldsymbol{x} + \int_\Sigma \left(\left| \frac{\partial \psi}{\partial t} \right|^2 + c_0 \left| \nabla \frac{\partial \psi}{\partial t} \right|^2 \right) d\sigma \right) = 0, \tag{4.114}$$

where $c_0 = 0$ in the absence of surface tension and $c_0 = 1$ when there is surface tension.

In comparison with the nonlinear setting studied above, we shall see that the convergence of the velocity fields we may prove is roughly the same (up to minor regularity issues), but, more importantly, the difference lies in the analysis of the long-time limit of the density, *i.e.* on that of the position of the interface between the fluids. Indeed, in the linearized case, it is easy to see that if $(\boldsymbol{u} = 0, \psi_\infty)$ is a steady-state solution, then $-\boldsymbol{\nabla} p + \psi_\infty \delta_{z=0} \boldsymbol{e}_z = 0$ thus ψ_∞ is a constant. Therefore, a steady-state ψ_∞ reached through an evolution subject to $\int_\Sigma \psi \, d\sigma = 0$ is necessarily $\psi_\infty = 0$. This is in contrast with the nonlinear setting where there are infinitely many steady-state solutions with $\boldsymbol{u} = 0$.

Consequently, we are able to prove that, in the case without surface-tension, the fluid does return, in some weak sense at least, to its stable steady state in this linearized setting. In the case with surface tension, if we assume a sufficient regularity for the velocity field, then we are able to completely determine the behavior of the interface as time goes to infinity.

In the case $c_0 = 0$ (no surface tension), the bounds inferred from (4.113) and (4.114) read:

$$\sup_{t \in [0,\infty)} \|\boldsymbol{u}\|_{\mathbb{L}^2(\Omega)} \leq C_1, \tag{4.115}$$

$$\sup_{t \in [0,\infty)} \|\psi\|_{L^2(\Sigma)} \leq C_1, \tag{4.116}$$

$$\int_0^{+\infty} \|\nabla \boldsymbol{u}\|^2_{\mathbb{L}^2(\Omega)} \, dt < +\infty, \tag{4.117}$$

$$\sup_{t \in [0,\infty)} \left\| \frac{\partial \boldsymbol{u}}{\partial t} \right\|_{\mathbb{L}^2(\Omega)} \leq C^{st}, \tag{4.118}$$

$$\sup_{t \in [0,\infty)} \left\| \frac{\partial \psi}{\partial t} \right\|_{L^2(\Sigma)} \leq C^{st}, \tag{4.119}$$

$$\int_0^{+\infty} \left\| \boldsymbol{\nabla} \frac{\partial \boldsymbol{u}}{\partial t} \right\|^2_{\mathbb{L}^2(\Omega)} dt < \infty. \tag{4.120}$$

Elementary manipulations then imply:

Proposition 4.9 *In the linearized case without surface tension (4.111) ($c_0 = 0$), assuming the a priori estimates and the regularity mentioned above, the behavior as time goes to infinity of a solution $\boldsymbol{u}, \psi$ is the following:*

(i) the velocity field $\boldsymbol{u}$ goes to 0 in $\mathbb{H}^1_0(\Omega)$;

(ii) the shape ψ of the interface goes to 0 in $H^{-\varepsilon}$ (for all $\varepsilon > 0$) and in weak-L^2.

In addition,

(iii) $\|\psi\|_{L^2(\Sigma)}$ has a limit as t goes to infinity;

(iv) $\displaystyle\int_0^\infty \|\boldsymbol{u}\|^2_{\mathbb{H}^1(\Omega)} + \|\partial_t \boldsymbol{u}\|^2_{\mathbb{H}^1(\Omega)} \, dt < +\infty.$

On the other hand, when $c_0 = 1$, *i.e.* when surface tension is present, we deduce from (4.113) and (4.114) the additional information:

$$\int_0^{+\infty} \|\nabla \boldsymbol{u}\|^2_{\mathbb{L}^2(\Omega)} dt < +\infty, \tag{4.121}$$

$$\sup_{t\in[0,\infty)} \|\boldsymbol{u}\|^2_{\mathbb{L}^2(\Omega)} \leq C^{st}, \tag{4.122}$$

$$\sup_{t\in[0,\infty)} \|\psi\|^2_{H^1(\Sigma)} \leq C^{st}, \tag{4.123}$$

and

$$t \longrightarrow \|\sqrt{\rho_0}\boldsymbol{u}\|^2_{\mathbb{L}^2(\Omega)} + \|\psi\|^2_{H^1(\Sigma)} \text{ is a non-increasing function of time } t. \tag{4.124}$$

This in fact allows to prove:

Proposition 4.10 *In the linearized case with surface tension (4.111) ($c_0 = 1$), assuming the a priori estimates and the regularity mentioned above, the behavior of $\boldsymbol{u}, \psi$ as time goes to infinity is the following:*

(i) the velocity field $\boldsymbol{u}$ belongs to $\mathcal{C}(0, +\infty; \mathbb{H}^1_0(\Omega))$ and goes to 0 in $\mathbb{H}^1_0(\Omega)$;

(ii) the shape ψ of the interface belongs to $\mathcal{C}(0, +\infty; H^1(\Sigma))$; there exists a sequence $t_n \to +\infty$ such that, in weak $-$ H^1, $\psi(\cdot, t_n) \to \psi_\infty$ solution to

$$\begin{cases} -\Delta\,\psi_\infty + \psi_\infty = \alpha & \text{on } \Sigma, \\ \displaystyle\int_\Sigma \psi_\infty \, d\sigma = 0, \end{cases} \tag{4.125}$$

where α is some unknown constant; in addition,

(iii) $\|\psi\|_{H^1(\Sigma)}$ has a limit as t goes to infinity;

(iv) $\displaystyle\int_0^\infty \|\boldsymbol{u}\|^2_{\mathbb{H}^1(\Omega)} + \|\partial_t \boldsymbol{u}\|^2_{\mathbb{H}^1(\Omega)} \, dt < +\infty.$

To conclude, we would like to emphasize a particular point regarding the identification of the limit ψ_∞, which is in fact related, as we will see below, to some practical techniques.

If we assume that the velocity $\boldsymbol{u}$ remains more regular, say $\mathcal{C}(0, \infty; \mathbb{H}^{1+\varepsilon}(\Omega))$, then we can improve (ii). Indeed, the function $u_z(t)|_\Sigma$ then belongs to $\mathbb{H}^{1/2+\varepsilon}(\Sigma)$ and thus has a trace on $\partial\Sigma$. Therefore, in this case

$$\frac{\partial\psi}{\partial t}(t)|_{\partial\Sigma} = 0$$

for $t \geq 0$. In particular $\psi_\infty|_{\partial\Sigma} = \psi_0|_{\partial\Sigma}$, where $\psi_0 = \psi|_{t=0}$. Then the limit ψ_∞ is the unique solution of

$$\begin{cases} -\Delta\,\psi_\infty + \psi_\infty = \alpha & \text{on } \Sigma, \\ \psi_\infty = \psi_0 & \text{on } \partial\Sigma, \\ \displaystyle\int_\Sigma \psi_\infty \, d\sigma = 0. \end{cases} \tag{4.126}$$

The possible indetermination of the limit ψ_∞ has disappeared, because the linearized system has kept a memory of the boundary value of the initial data $\psi|_{t=0}$. This is an artefact related to the question of the appropriate boundary condition to impose on the fluid. Of course, there is no reason why the arbitrary function $\psi|_{t=0}$ should agree with any steady-state ψ_∞ on the boundary. For instance, symmetry considerations on the geometry of the domain may impose some symmetry of the boundary value of all steady states. The latter is of course not necessarily satisfied by the initial condition. The difficulty regarding the boundary value of $\partial\psi/\partial t = 0$ has its numerical counterpart. The macroscopic no-slip condition $\boldsymbol{u} = 0$ on the boundary is obviously not true on the microscopic scale and one must find an adequate modelling to move the interface on the boundary of the domain, despite the no-slip condition. See Remark 5.1.6 and Section 5.1.4 for more details on this issue.

4.3.3 The MHD case

We briefly come back to the MHD case. In view of all the difficulties we have encountered above in the purely hydrodynamic case, it is easy to understand that we are not able to treat system (1.49) in its full generality.

All the studies we are aware of treat the linearized case. Many cases of magnetic and electric fields are considered, in various geometries, in two or three dimensions, under various assumptions of symmetry. The emphasis is placed on the behavior of the velocity and of the electromagnetic field, and the conclusion provided by these studies is mainly that, under convenient assumptions, the velocity goes to zero, in a more or less strong sense, while the electromagnetic field tends to some well-identified limit. Unfortunately, nothing (or almost nothing) is known about the behavior of the interface separating the two fluids (see, for instance, Remark 3.4 in [63]).

Due to the difficulty of the problem, we cannot hope to do much better by adapting the above argument to the MHD case. The only case we are able to deal with is the following simple case. We provide the system with zero boundary conditions as in (1.49). Then the presence of the magnetic coupling does not modify the fact that the only steady-state solution for the velocity is $\boldsymbol{u}_\infty \equiv 0$, together with $\boldsymbol{B}_\infty \equiv 0$ and ρ_∞ as in the purely hydrodynamic case.

The estimate (4.26) shows that

$$\frac{d}{dt}\left(\int \rho\frac{\boldsymbol{u}^2}{2} + \int \rho\, z + \int \rho\frac{\boldsymbol{B}^2}{2\overline{\mu}}\right) + \int \overline{\eta}_- \,|\boldsymbol{\nabla u}|^2 + \frac{1}{\mu\overline{\sigma}_+}\int |\mathbf{curl}\,\boldsymbol{B}|^2 \leq 0,$$

where we have used (4.28) and $\sigma(\rho) \leq \overline{\sigma}_+$. Then the arguments applied to $\boldsymbol{u}$ in the preceding sections equally apply to the field $\boldsymbol{B}$, giving rise to the same conclusions. The case with surface tension and the linearized case may be treated likewise.

5

NUMERICAL SIMULATION OF TWO-FLUID PROBLEMS

In this chapter, we present the numerical approach to deal with the multifluid feature of the MHD system (1.49).

Let us first emphasize that the problem of modeling the motion of an interface appears in many physical contexts (phase transition, flame front, bubbles, waves, jet, films, *etc.*). The designation "free surface" is used when the interface separates a liquid and a gas, since in this case, the inertia effects can usually be neglected in the gas, so that the only influence of the gas is the pressure it exerts on the liquid. The designation "interface" is used when both liquids have densities of the same order of magnitude.

The mathematical difficulties outlined in the previous chapter, which stand on the analytic side, have their companion difficulties on the numerical side. A numerical strategy is necessary to follow the heterogeneities of the density ρ. We will focus on the case when it is piecewise constant. Then, some interfaces separate the zones where ρ is constant. The numerical strategy we will adopt is to explicitly follow such interfaces in their motion (one speaks of interface *tracking* as opposed to interface *capturing*). Notice that for the numerics, we will partially adopt a Lagrangian approach, while the mathematical analysis essentially relies on an Eulerian viewpoint.

In addition, the heterogeneities on the viscosity η and the conductivity σ deteriorate the conditioning of the numerical problem. This is especially true for σ which, in the applications we have in mind, may vary from several orders of magnitudes from one zone to the other. This poor conditioning is a companion effect, on the numerical side, of the loss of regularity observed on the theoretical side, and mentioned in our mathematical discussion of the last chapter.

In Section 5.1, we present the *Arbitrary Lagrangian Eulerian* (abbreviated henceforth in ALE) approach in details. This is the strategy we will adopt in our context of the modeling of aluminum electrolysis cells. Then, in Section 5.2, we briefly review some other numerical methods to deal with moving interfaces. Finally, we give in Section 5.3 some test cases which illustrate the capabilities of the ALE method.

In this chapter, we suppose that the reader is familiar with classical Lagrangian finite elements like $\mathbb{P}_k$ finite elements on tetrahedra, and $\mathbb{Q}_k$ finite elements on hexahedra. We refer to Chapter 3 and to [39,75,192] for a presentation of these finite element spaces.

5.1 Numerical approximations in the ALE formulation

The aim of this section is to derive a finite element approximation of system (1.52) which we reproduce here for convenience:

$$\begin{cases} \text{in } \Omega_{1,t}, \begin{cases} \dfrac{\partial \boldsymbol{u}}{\partial t} + \boldsymbol{u} \cdot \boldsymbol{\nabla} \boldsymbol{u} + \boldsymbol{\nabla} p - \dfrac{2}{Re_1} \operatorname{div}(\boldsymbol{D}(\boldsymbol{u})) = \dfrac{1}{Fr} \boldsymbol{g} + S \, \mathbf{curl} \, \boldsymbol{B} \times \boldsymbol{B}, \\ \dfrac{\partial \boldsymbol{B}}{\partial t} + \dfrac{1}{Rm_1} \mathbf{curl} \, (\mathbf{curl} \, \boldsymbol{B}) = \mathbf{curl} \, (\boldsymbol{u} \times \boldsymbol{B}), \end{cases} \\ \text{in } \Omega_{2,t}, \begin{cases} M \dfrac{\partial \boldsymbol{u}}{\partial t} + M \boldsymbol{u} \cdot \boldsymbol{\nabla} \boldsymbol{u} + \boldsymbol{\nabla} p - \dfrac{2M}{Re_2} \operatorname{div}(\boldsymbol{D}(\boldsymbol{u})) = \dfrac{M}{Fr} \boldsymbol{g} + S \, \mathbf{curl} \, \boldsymbol{B} \times \boldsymbol{B}, \\ \dfrac{\partial \boldsymbol{B}}{\partial t} + \dfrac{1}{Rm_2} \mathbf{curl} \, (\mathbf{curl} \, \boldsymbol{B}) = \mathbf{curl} \, (\boldsymbol{u} \times \boldsymbol{B}), \end{cases} \\ \operatorname{div} \boldsymbol{u} = 0, \\ \operatorname{div} \boldsymbol{B} = 0, \\ \dfrac{\partial \rho}{\partial t} + \operatorname{div}(\rho \boldsymbol{u}) = 0. \end{cases}$$

We recall that this system is complemented with initial conditions (1.54), boundary conditions (1.53) and transmission conditions at the interface between the two fluids (1.55).

In order to deal with the motion of the interface separating the two fluids, we use an ALE approach. The bottom line for this approach is to let the nodes on the interface move with the same normal displacement as the fluid, thus the "Lagrangian" denomination. In contrast, the motion of the other nodes of the mesh ("internal" to each fluid) is not related to any real kinematics of the fluid, thus the "Eulerian" denomination. In summary, the mesh follows the displacement of the interface, but not the trajectories of the fluid particles.

The ALE formulation, first introduced by C.W. Hirt, A.A. Amsden and J.L. Cook [123], has been used by a huge number of authors in the case of *one* fluid with a free surface: see for example A. Soulaïmani et al. [220, 221], B. Maury [160] A. Huerta [128] T.A. Baer et al. [12] R.A. Cairncross et al. [33], and L.W. Ho [125]. In the case of two fluids, see for example H.K. Rasmussen et al. [196]. For the specific MHD context, see the early works of J.U. Brackbill et al. [29, 27], using finite difference methods in 2D.

The ALE method is particularly well adapted to the industrial application we have in mind (see Chapter 6). We need a rather inexpensive and accurate method to capture some small displacements of the interface, which do not imply any topological changes. In addition, one of the advantage of the ALE method is that its implementation in the context of finite element methods is rather simple and does not require many tailored implementation details, as is the case for many alternative methods.

We first present the weak ALE formulation at the continuous level in Section 5.1.1, and then the time and space discretization in Section 5.1.2. In Section 5.1.3, we analyze the stability properties of the numerical scheme. Finally, Section 5.1.4 is devoted to surface tension effects.

5.1.1 *Weak ALE formulation*

We concentrate on system (1.52)–(1.55). The domain Ω is typically a rectangle if $d = 2$ or a parallelepiped if $d = 3$. We recall that, for the sake of simplicity, we assume no-slip boundary conditions, and homogeneous boundary conditions on the normal component of the magnetic field (1.53) (see Section 3.8 for other possible magnetic boundary conditions and Remark 5.1.6 about the boundary conditions used in practice). In addition, we impose initial conditions $\boldsymbol{u}_0$ on the velocity, $\boldsymbol{B}_0$ on the magnetic field, and ρ_0 for the density, thereby fixing the domains initially occupied by each fluid (see (1.54)).

The three main ingredients of the ALE formulation we use are the following:

(1) *An application $\hat{\mathcal{A}}_t$ that maps a reference domain $\hat{\Omega}$ to the current domain Ω_t*. The *arbitrary* feature of the algorithm arises from the fact that this application does not follow trajectories of the fluid particles, but only of some parts of the fluid (like the interface in our context, or the free surface for free-surface flow simulations). This is detailed in Section 5.1.1.1.

(2) *A Galerkin formulation with test functions that follow the deformation given by $\hat{\mathcal{A}}_t$*. The test functions $\boldsymbol{v}(t, \mathbf{x})$ depend on time but $\boldsymbol{v}(t, \hat{\mathcal{A}}_t(\hat{\mathbf{x}}))$ do not. Equivalently,

$$\frac{\partial \boldsymbol{v}}{\partial t} + \boldsymbol{w} \cdot \boldsymbol{\nabla} \boldsymbol{v} = 0,$$

in terms of the velocity of the domain $\boldsymbol{w}$ (defined by (5.5) and (5.7)). This is presented in Section 5.1.1.2.

(3) *A special formulation of the terms involving a time-derivative in the Galerkin formulation.* This is made precise in Section 5.1.1.3 (see Formula (5.21)).

In these first three Sections 5.1.1.1, 5.1.1.2 and 5.1.1.3 we suppose that we know *a priori* the history of the velocity and the domains occupied by each fluid: $(\boldsymbol{u})_{t \geq 0}$ and $(\rho)_{t \geq 0}$. After having presented these three ingredients, we finally derive the coupled weak ALE formulation of (1.52) in Section 5.1.1.4.

5.1.1.1 The geometric setting and the function $\hat{\mathcal{A}}_t$ For times $t \geq 0$, $\Omega_{i,t}$ ($i = 1$ or 2) denotes the domain occupied by fluid i[30]. We typically have in mind a tank filled with two fluids, one lighter than the other. By convention, the heavier and lower fluid is numbered 1, and the lighter and upper fluid is numbered 2. We suppose that for all $t \geq 0$, $\overline{\Omega} = \overline{\Omega}_{1,t} \cup \overline{\Omega}_{2,t}$ and $\Omega_{1,t} \cap \Omega_{2,t} = \emptyset$. We denote by $\Sigma_t = \partial\Omega_{1,t} \cap \partial\Omega_{2,t}$ the interface between the two domains $\Omega_{1,t}$ and $\Omega_{2,t}$ (see Figure 5.1). The normal to Σ_t, oriented from Ω_1 to Ω_2, is denoted by $\boldsymbol{n}_1$ and we set $\boldsymbol{n}_2 = -\boldsymbol{n}_1$.

[30]With the notation of system (1.52), the domain $\Omega_{1,t}$ (resp. $\Omega_{2,t}$) is defined as the points in Ω with density 1 (resp. M).

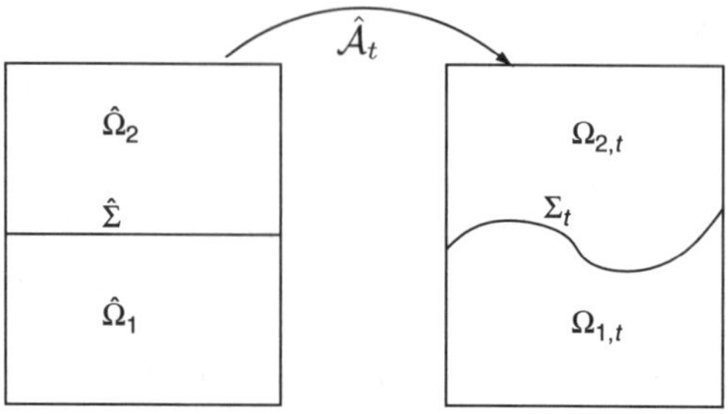

Fig. 5.1 The partition of the domain Ω.

We assume that for any time $t \geq 0$, there exists a mapping $\hat{\mathcal{A}}_t$ from a reference domain $\overline{\hat{\Omega}} = \overline{\hat{\Omega}_1} \cup \overline{\hat{\Omega}_2}$ to the current domain, such that (see Figure 5.1)[31]:

$$\hat{\mathcal{A}}_t \text{ is a } \mathcal{C}^1\text{-diffeomorphism with respect to } \mathbf{x}, \text{ differentiable with respect to } t, \tag{5.1}$$

$$\hat{\mathcal{A}}_t \text{ maps } \hat{\Omega}_i \text{ to } \Omega_{i,t} \ (i = 1 \text{ or } 2). \tag{5.2}$$

One can think of the reference domain $\overline{\hat{\Omega}} = \overline{\hat{\Omega}_1} \cup \overline{\hat{\Omega}_2}$ as the position of the domain at a fixed time t_0 (in which case one can choose $\hat{\mathcal{A}}_{t_0} = \mathrm{Id}$). Therefore, $\hat{\Omega}_1 \cap \hat{\Omega}_2 = \emptyset$ and let $\hat{\Sigma} = \partial\hat{\Omega}_1 \cap \partial\hat{\Omega}_2$ be the interface between the two domains $\hat{\Omega}_1$ and $\hat{\Omega}_2$. The inverse function (with respect to the space variable) of $\hat{\mathcal{A}}_t$ is denoted $\hat{\mathcal{A}}_t^{-1}$. Let $\hat{\mathbf{J}}_t$ be the Jacobian matrix $\hat{\mathbf{J}}_t$ associated with $\hat{\mathcal{A}}_t$ defined by: for $1 \leq i, j \leq d$,

$$[\hat{\mathbf{J}}_t(\hat{\mathbf{x}})]_{i,j} = \frac{\partial(\hat{\mathcal{A}}_t(\hat{\mathbf{x}}))_i}{\partial \hat{\mathbf{x}}_j}. \tag{5.3}$$

The determinant of $\hat{\mathbf{J}}_t(\hat{\mathbf{x}})$ is denoted $\hat{J}_t(\hat{\mathbf{x}})$. Notice that $\hat{\mathbf{J}}_t(\hat{\mathbf{x}})$ is an invertible matrix and that we can assume without loss of generality that:

$$\forall t \geq 0, \quad \forall \hat{\mathbf{x}} \in \hat{\Omega}, \quad \hat{J}_t(\hat{\mathbf{x}}) > 0. \tag{5.4}$$

The velocity of the domain $\hat{\boldsymbol{w}}$ is defined by:

$$\hat{\boldsymbol{w}}(t, \hat{\mathbf{x}}) = \frac{\partial \hat{\mathcal{A}}_t}{\partial t}(\hat{\mathbf{x}}). \tag{5.5}$$

For any function $\psi(t, .)$ defined on Ω, we denote by $\hat{\psi}(t, .)$ the corresponding function defined on the reference domain $\hat{\Omega}$ by

$$\hat{\psi}(t, \hat{\mathbf{x}}) = \psi(t, \hat{\mathcal{A}}_t(\hat{\mathbf{x}})). \tag{5.6}$$

[31] We recall that a function is a $\mathcal{C}^1$-diffeomorphism if this function is of class $\mathcal{C}^1$, bijective, and with inverse function of class $\mathcal{C}^1$.

For example, the velocity of the domain $\boldsymbol{w}$ on the current frame is defined by

$$\boldsymbol{w}(t, \mathbf{x}) = \hat{\boldsymbol{w}}(t, \hat{\mathcal{A}}_t^{-1}(\mathbf{x})). \tag{5.7}$$

Notice that the functions ψ and $\hat{\psi}$ are such that:

$$\frac{\partial \hat{\psi}}{\partial t}(t, \hat{\mathbf{x}}) = \frac{\partial \psi}{\partial t}(t, \hat{\mathcal{A}}_t(\hat{\mathbf{x}})) + \boldsymbol{w}(t, \hat{\mathcal{A}}_t(\hat{\mathbf{x}})) \cdot \boldsymbol{\nabla}\psi(t, \hat{\mathcal{A}}_t(\hat{\mathbf{x}})). \tag{5.8}$$

The density ρ of the fluid is such that:

$$\rho(t, \mathbf{x}) = \hat{\rho}(\hat{\mathcal{A}}_t^{-1}(\mathbf{x})), \tag{5.9}$$

where $\hat{\rho}$ is equal to $\overline{\rho}_1$ on $\hat{\Omega}_1$ and $\overline{\rho}_2$ on $\hat{\Omega}_2$. Moreover, we have the following differentiation formula: in a distributional sense,

$$\boldsymbol{\nabla}\rho = \delta\rho\, \boldsymbol{n}_1\, \delta_{\Sigma_t}, \tag{5.10}$$

where $\delta\rho = \overline{\rho}_2 - \overline{\rho}_1$, and the distribution δ_{Σ_t} is defined by, for any smooth function ψ,

$$\int \psi \delta_{\Sigma_t} = \int_{\Sigma_t} \psi(\mathbf{x}) d\sigma_{\Sigma_t}(\mathbf{x}), \tag{5.11}$$

where σ_{Σ_t} denotes the Lebesgue measure on Σ_t.

We would now like to translate assumption (5.2) on $\hat{\mathcal{A}}_t$ onto some assumptions on the domain velocity $\boldsymbol{w}$, namely:

$$\boldsymbol{w} \cdot \boldsymbol{n} = 0 \text{ on } \partial\Omega, \tag{5.12}$$

$$\boldsymbol{w} \cdot \boldsymbol{n}_1 = \boldsymbol{u} \cdot \boldsymbol{n}_1 \text{ on } \Sigma_t. \tag{5.13}$$

Let us rigorously state this:

Lemma 5.1 *Suppose that the velocity field $\mathbf{u}(t, .)$ and the density $\rho(t, .)$ satisfy the conservation of mass*

$$\frac{\partial \rho}{\partial t} + \operatorname{div}(\rho \boldsymbol{u}) = 0$$

and the incompressibility constraint $\operatorname{div}\boldsymbol{u} = 0$ and $\boldsymbol{u} \cdot \boldsymbol{n} = 0$ on $\partial\Omega$. We suppose moreover that

$$\rho(t = 0, .) = \begin{cases} 1 & \text{on } \Omega_{1,0} \\ M & \text{on } \Omega_{2,0} \end{cases} \text{(with } \overline{\Omega}_{1,0} \cup \overline{\Omega}_{2,0} = \overline{\Omega} \text{ and meas } (\Omega_{i,0}) > 0,\ i = 1, 2).$$

We then define $\Omega_{1,t} = \{\mathbf{x} \in \Omega, \rho(t, \mathbf{x}) = 1\}$ and $\Omega_{2,t} = \{\mathbf{x} \in \Omega, \rho(t, \mathbf{x}) = M\}$[32].

Let $\hat{\mathcal{A}}_t : \hat{\Omega} \to \Omega$ be an application which satisfies (5.1) and $\boldsymbol{w}$ be the associated domain velocity defined by (5.5) and (5.7).

Then, assumption (5.2) on $\hat{\mathcal{A}}_t$ implies assumptions (5.12)–(5.13) on $\boldsymbol{w}$. Conversely if $\boldsymbol{w}$ satisfies (5.12)–(5.13) and $\hat{\mathcal{A}}_0$ maps $\hat{\Omega}_i$ to $\Omega_{i,0}$ ($i = 1$ or 2), then $\hat{\mathcal{A}}_t$ satisfies (5.2).

[32] We recall that the fact that ρ is piecewise constant equal to 1 or M propagates in time. Moreover, $\overline{\Omega} = \overline{\Omega}_{1,t} \cup \overline{\Omega}_{2,t}$ and $\operatorname{meas}(\Omega_{i,t}) = \operatorname{meas}(\Omega_{i,0}) > 0$ for $i = 1, 2$ (see Theorem 4.4).

Proof It is easy to check that (5.12) is equivalent to the fact that the domain $\hat{\mathcal{A}}_t(\hat{\Omega})$ does not depend on time.

From the conservation of mass

$$\frac{\partial \rho}{\partial t} + \mathrm{div}\,(\rho \boldsymbol{u}) = 0$$

and $\mathrm{div}\, \boldsymbol{u} = 0$, we obtain

$$\frac{\partial \rho}{\partial t} + \boldsymbol{u} \cdot \boldsymbol{\nabla} \rho = 0.$$

Let $\hat{r}(t, \hat{\mathbf{x}}) = \rho(t, \hat{\mathcal{A}}_t(\hat{\mathbf{x}}))$. We have, using (5.8):

$$\begin{aligned}
\frac{\partial \hat{r}}{\partial t}(t, \hat{\mathcal{A}}_t^{-1}(\mathbf{x})) &= \left(\frac{\partial \rho}{\partial t} + \boldsymbol{w} \cdot \boldsymbol{\nabla} \rho\right)(t, \mathbf{x}), \\
&= ((\boldsymbol{w} - \boldsymbol{u}) \cdot \boldsymbol{\nabla} \rho)(t, \mathbf{x}), \\
&= \delta\rho\, (\boldsymbol{w} - \boldsymbol{u}) \cdot \boldsymbol{n}_1(t, \mathbf{x}) \delta_{\Sigma_t}(\mathbf{x}).
\end{aligned}$$

Now, if we assume that $\hat{\mathcal{A}}_t$ satisfies (5.2), then $\hat{r}$ does not depend on time which implies (5.13).

Conversely, if $\boldsymbol{w}$ satisfies (5.13), then $\hat{r}$ does not depend on time, and if $\hat{\mathcal{A}}_0$ maps $\hat{\Omega}_i$ to $\Omega_{i,0}$ ($i = 1$ or 2), this implies that $\hat{\mathcal{A}}_t$ satisfies (5.2). □

Concerning the assumption (5.1) on $\hat{\mathcal{A}}_t$, if the domain velocity $\boldsymbol{w}$ is such that:

$$\begin{aligned}
&\boldsymbol{w} \text{ is } \mathcal{C}^0 \text{ with respect to time, } \mathcal{C}^2 \text{ with respect to } \mathbf{x} \\
&\qquad \text{and globally Lipschitz with respect to } \mathbf{x}
\end{aligned} \tag{5.14}$$

then, if we choose $\hat{\Omega}_i = \Omega_{i,t=0}$ and define $\hat{\mathcal{A}}_t$ by:

$$\begin{cases}
\dfrac{\partial \hat{\mathcal{A}}_t}{\partial t}(\hat{\mathbf{x}}) = \boldsymbol{w}(t, \hat{\mathcal{A}}_t(\hat{\mathbf{x}})), \\
\hat{\mathcal{A}}_{t=0} = \mathrm{Id},
\end{cases} \tag{5.15}$$

then $\hat{\mathcal{A}}_t$ satisfies (5.1). Therefore, we can also translate the assumption (5.1) on $\hat{\mathcal{A}}_t$ to the assumption (5.14) on $\boldsymbol{w}$. This will be useful to set the ALE problem in terms of $\boldsymbol{w}$ rather than in terms of $\hat{\mathcal{A}}_t$ (see Remark 5.1.5 below).

5.1.1.2 Functional spaces and test functions The following functional spaces will be needed, respectively, for the velocity $\boldsymbol{u}$, the magnetic field $\boldsymbol{B}$, and the pressure p:

$$V = L^2(0, T; \mathbb{H}_0^1(\Omega)), \quad W = L^2(0, T; \mathbb{H}_{\boldsymbol{n}}^1(\Omega)), \quad M = L^2(0, T; L_0^2(\Omega)).$$

The principle of the derivation of the weak formulation is to use a variational formulation (Galerkin method) with test functions which do not depend on time when transported in the reference domain $\hat{\Omega}$ by formula (5.6), and which therefore depend on time in the moving frame.

We introduce the test function spaces on the reference domain

$$\hat{V} = \mathbb{H}_0^1(\hat{\Omega}), \qquad \hat{W} = \mathbb{H}_{\boldsymbol{n}}^1(\hat{\Omega}), \qquad \hat{M} = L_0^2(\hat{\Omega}).$$

In the moving frame, the test function spaces are defined by

$$V_T = \{\boldsymbol{v} : [0,T] \times \Omega \to \mathbb{R}^d,\ \boldsymbol{v}(t,\mathbf{x}) = \hat{\boldsymbol{v}}(\hat{\mathcal{A}}_t^{-1}(\mathbf{x})), \hat{\boldsymbol{v}} \in \hat{V}\},$$

$$W_T = \{\boldsymbol{C} : [0,T] \times \Omega \to \mathbb{R}^d,\ \boldsymbol{C}(t,\mathbf{x}) = \hat{\boldsymbol{C}}(\hat{\mathcal{A}}_t^{-1}(\mathbf{x})), \hat{\boldsymbol{C}} \in \hat{W}\},$$

$$M_T = \{q : [0,T] \times \Omega \to \mathbb{R},\ q(t,\mathbf{x}) = \hat{q}(\hat{\mathcal{A}}_t^{-1}(\mathbf{x})), \hat{q} \in \hat{M}\}.$$

Thus, the test functions do not depend on time in the reference frame $\hat{\Omega}$ whereas they do on the current one. More precisely, let $\boldsymbol{v}$ be in V_T. Then for a fixed $\hat{\mathbf{x}} \in \hat{\Omega}$, $\boldsymbol{v}(t, \hat{\mathcal{A}}_t(\hat{\mathbf{x}}))$ does not depend on time while for a fixed $\mathbf{x} \in \Omega$, $\boldsymbol{v}(t,\mathbf{x})$ does. This is equivalent to the fact that the test functions satisfy (using (5.8))

$$\frac{\partial \boldsymbol{v}}{\partial t} + \boldsymbol{w} \cdot \boldsymbol{\nabla} \boldsymbol{v} = 0. \tag{5.16}$$

Remark 5.1.1 [On the inclusion $W_T \subset W$] In the following, we suppose that the space W_T is included in W. This is satisfied for example if $\boldsymbol{w} = 0$ on the boundary of Ω, or more generally if, for any $\hat{\mathbf{x}} \in \partial\hat{\Omega}$, $\boldsymbol{n}(\hat{\mathcal{A}}_t(\hat{\mathbf{x}})) = \boldsymbol{n}(\hat{\mathbf{x}})$. In our practical cases, this is satisfied, since the points on the boundary move along a line (in 2D) or in a plane (in 3D), or along a line at a fixed polar angle in the case when Ω is a cylinder in 3D. Notice that the same remark applies for the functional space for the velocity in the case of pure slip boundary conditions on the side walls. ♢

5.1.1.3 A fundamental transport formula One crucial formula for the derivation of the weak formulation is the following identity known as the *transport formula* or the *Reynolds transport formula*[33]:

Lemma 5.2 *For any smooth function ψ depending on time t and space $\mathbf{x}$, and any function ϕ such that $\hat{\phi}$ (defined by $\hat{\phi}(t,\hat{\mathbf{x}}) = \phi(t, \hat{\mathcal{A}}_t(\hat{\mathbf{x}}))$) is time-independent, we have:*

$$\frac{d}{dt} \int_\Omega \psi(t,\mathbf{x})\, \phi(t,\mathbf{x})\, d\mathbf{x} \tag{5.17}$$

$$= \int_\Omega \phi(t,\mathbf{x}) \frac{\partial \psi}{\partial t}(t,\mathbf{x}) + \phi(t,\mathbf{x}) \boldsymbol{w}(t,\mathbf{x}) \cdot \boldsymbol{\nabla}\psi(t,\mathbf{x}) + \phi(t,\mathbf{x})\, div\,(\boldsymbol{w}(t,\mathbf{x}))\psi(t,\mathbf{x})\, d\mathbf{x}.$$

Notation In Lemma 5.2 and in the sequel, the spatial differential operators are taken with respect to the Eulerian variable $\mathbf{x}$. We omit to explicitly denote this for conciseness.

[33] Actually, in the literature, the Reynolds transport formula usually refers to equation (5.19) below, which is a straightforward consequence of (5.17).

Proof The proof is based on the well-known *Liouville formula*:

$$\frac{\partial \hat{J}_t(\hat{\mathbf{x}})}{\partial t} = \operatorname{div} \boldsymbol{w}(t, \hat{\mathcal{A}}_t(\hat{\mathbf{x}}))\hat{J}_t(\hat{\mathbf{x}}). \tag{5.18}$$

We have (using (5.4)):

$$\begin{aligned}
\frac{d}{dt}\int_{\Omega} \phi(t,\mathbf{x})\psi(t,\mathbf{x})\,d\mathbf{x} &= \frac{d}{dt}\int_{\hat{\Omega}} \hat{\phi}(\hat{\mathbf{x}})\psi(t,\hat{\mathcal{A}}_t(\hat{\mathbf{x}}))\,\hat{J}_t(\hat{\mathbf{x}})d\hat{\mathbf{x}} \\
&= \int_{\hat{\Omega}} \hat{\phi}(\hat{\mathbf{x}})\frac{\partial \psi}{\partial t}(t,\hat{\mathcal{A}}_t(\hat{\mathbf{x}}))\,\hat{J}_t(\hat{\mathbf{x}})\,d\hat{\mathbf{x}} + \int_{\hat{\Omega}} \hat{\phi}(\hat{\mathbf{x}})\frac{\partial \hat{\mathcal{A}}_t}{\partial t}(\hat{\mathbf{x}})\cdot\boldsymbol{\nabla}\psi(t,\hat{\mathcal{A}}_t(\hat{\mathbf{x}}))\,\hat{J}_t(\hat{\mathbf{x}})d\hat{\mathbf{x}} \\
&\quad + \int_{\hat{\Omega}} \hat{\phi}(\hat{\mathbf{x}})\psi(t,\hat{\mathcal{A}}_t(\hat{\mathbf{x}}))\,\frac{\partial \hat{J}_t}{\partial t}(\hat{\mathbf{x}})d\hat{\mathbf{x}} \\
&= \int_{\hat{\Omega}} \hat{\phi}(\hat{\mathbf{x}})\frac{\partial \psi}{\partial t}(t,\hat{\mathcal{A}}_t(\hat{\mathbf{x}}))\,\hat{J}_t(\hat{\mathbf{x}})\,d\hat{\mathbf{x}} + \int_{\hat{\Omega}} \hat{\phi}(\hat{\mathbf{x}})\hat{\boldsymbol{w}}(t,\hat{\mathbf{x}})\cdot\boldsymbol{\nabla}\psi(t,\hat{\mathcal{A}}_t(\hat{\mathbf{x}}))\,\hat{J}_t(\hat{\mathbf{x}})\,d\hat{\mathbf{x}} \\
&\quad + \int_{\hat{\Omega}} \hat{\phi}(\hat{\mathbf{x}})\psi(t,\hat{\mathcal{A}}_t(\hat{\mathbf{x}}))\operatorname{div}\boldsymbol{w}(t,\hat{\mathcal{A}}_t(\hat{\mathbf{x}}))\hat{J}_t(\hat{\mathbf{x}})\,d\hat{\mathbf{x}} \\
&= \int_{\Omega} \phi(t,\mathbf{x})\frac{\partial \psi}{\partial t}(t,\mathbf{x})\,d\mathbf{x} + \int_{\Omega} \phi(t,\mathbf{x})\boldsymbol{w}(t,\mathbf{x})\cdot\boldsymbol{\nabla}\psi(t,\mathbf{x})\,d\mathbf{x} \\
&\quad + \int_{\Omega} \phi(t,\mathbf{x})\psi(t,\mathbf{x})\operatorname{div}\boldsymbol{w}(t,\mathbf{x})d\mathbf{x}.
\end{aligned}$$

Let us give two simple applications of Lemma 5.2. Taking $\phi(t,\mathbf{x}) = 1_{\hat{\Omega}_1}(\hat{\mathcal{A}}_t^{-1}(\mathbf{x}))$, we obtain, for any smooth function ψ depending on time t and space $\mathbf{x}$,

$$\frac{d}{dt}\int_{\Omega_{1,t}} \psi(t,\mathbf{x})\,d\mathbf{x} = \int_{\Omega_{1,t}} \frac{\partial \psi}{\partial t}(t,\mathbf{x}) + \boldsymbol{w}(t,\mathbf{x})\cdot\boldsymbol{\nabla}\psi(t,\mathbf{x}) + \operatorname{div}(\boldsymbol{w}(t,\mathbf{x}))\psi(t,\mathbf{x})\,d\mathbf{x}, \tag{5.19}$$

which is the classical formula for differentiating an integral set on a time-dependent domain.

On the other hand, taking $\phi(t,\mathbf{x}) = \rho(t,\mathbf{x})$, we have, for any smooth function ψ depending on time t and space $\mathbf{x}$,

$$\begin{aligned}
&\frac{d}{dt}\int_{\Omega} \rho(t,\mathbf{x})\psi(t,\mathbf{x})\,d\mathbf{x} \qquad (5.20)\\
&= \int_{\Omega}\Big(\rho(t,\mathbf{x})\frac{\partial \psi}{\partial t}(t,\mathbf{x}) + \rho(t,\mathbf{x})\boldsymbol{w}(t,\mathbf{x})\cdot\boldsymbol{\nabla}\psi(t,\mathbf{x}) \\
&\quad + \rho(t,\mathbf{x})\operatorname{div}(\boldsymbol{w}(t,\mathbf{x}))\psi(t,\mathbf{x})\Big)\,d\mathbf{x}.
\end{aligned}$$

As expected, the right-hand side of equation (5.20) actually depends on $\boldsymbol{w}$ only through the values of $\boldsymbol{w}\cdot\boldsymbol{n}_1$ on Σ_t. Indeed, by integration by parts and using (5.12) and (5.10), we have:

$$\int_\Omega \rho(t,\mathbf{x})\boldsymbol{w}(t,\mathbf{x})\cdot\boldsymbol{\nabla}\psi(t,\mathbf{x}) + \rho(t,\mathbf{x})\mathrm{div}\,(\boldsymbol{w}(t,\mathbf{x}))\psi(t,\mathbf{x})\,d\mathbf{x}$$

$$= \int_\Omega \rho(t,\mathbf{x})\mathrm{div}\,(\boldsymbol{w}(t,\mathbf{x})\psi(t,\mathbf{x}))\,d\mathbf{x}$$

$$= \int_{\Sigma_t} \delta\rho\,\boldsymbol{w}(t,\mathbf{x})\cdot\boldsymbol{n}_1(\mathbf{x})\,\psi(t,\mathbf{x})\,d\sigma_{\Sigma_t}(\mathbf{x}).$$

Therefore, by (5.13), the right-hand side of (5.20) depends only on the derivative of ψ with respect to time, on the value of ψ on the interface Σ_t, and on the normal velocity of the fluid on the interface. The result is thus independent of the specific arbitrary choice of $\boldsymbol{w}$ inside the domain $\Omega\backslash\Sigma$.

In the following, we use Lemma 5.2 in the weak ALE formulation to transform the terms involving a time derivative: the derivative with respect to time is taken out of the integration sign over the domain Ω. Let us detail this on the example of the term $\rho(\partial\boldsymbol{u}/\partial t)$ in (1.52). If $\boldsymbol{v}(t,\mathbf{x})\in V_T$ denotes a velocity test function[34], we multiply $\rho(\partial\boldsymbol{u}/\partial t)$ by $\boldsymbol{v}$, integrate over Ω, and then use the following identity:

$$\int_\Omega \rho\frac{\partial\boldsymbol{u}}{\partial t}\cdot\boldsymbol{v} = \frac{d}{dt}\int_\Omega \rho\boldsymbol{u}\cdot\boldsymbol{v} - \int_\Omega \rho\boldsymbol{w}\cdot\boldsymbol{\nabla}\boldsymbol{u}\cdot\boldsymbol{v} - \int_\Omega \rho\,\mathrm{div}\,(\boldsymbol{w})\boldsymbol{u}\cdot\boldsymbol{v}. \tag{5.21}$$

Formula (5.21) is obtained from (5.17) with $\phi=\rho\boldsymbol{v}_i$ and $\psi=\boldsymbol{u}_i$ and summing over $i\in\{1,\ldots,d\}$. This will also be used for the time derivative of the magnetic field (namely with $\boldsymbol{u}$ replaced by $\boldsymbol{B}$, and $\boldsymbol{v}$ replaced by a test function $\boldsymbol{C}\in W_T$). □

5.1.1.4 The weak ALE formulation of (1.52) We are now in position to derive the weak ALE formulation of (1.52). As explained in Section 5.1.1.2, we use a variational formulation (Galerkin method) with test functions which do not depend on time when transported in the reference domain $\hat{\Omega}$ by formula (5.6). To obtain the weak formulation, we multiply the equations on $\boldsymbol{u}$ and $\boldsymbol{B}$ in (1.52) by test functions respectively in V_T and W_T, integrate over Ω, and then get the derivative with respect to time out of the integration sign, using (5.21). Thus, two corrective terms involving $\boldsymbol{w}$ appear (see the right-hand side of (5.21), and the first three terms in (5.25) and (5.27)).

We will denote by $(\cdot,\cdot)$ the $(L^2(\Omega))^d$ scalar product

$$(\boldsymbol{u},\boldsymbol{v}) = \int_\Omega \boldsymbol{u}\cdot\boldsymbol{v}\,d\mathbf{x}.$$

We introduce the following bilinear and trilinear forms: $\forall\boldsymbol{v},\boldsymbol{v}_1,\boldsymbol{v}_2,\boldsymbol{v}_3\in\mathbb{H}^1_0(\Omega)$, $\forall q\in L^2_0(\Omega)$, $\forall\boldsymbol{C}_1,\boldsymbol{C}_2\in\mathbb{H}^1_{\boldsymbol{n}}(\Omega)$,

$$a_1(\boldsymbol{v}_1,\boldsymbol{v}_2) = \int_\Omega 2\eta\boldsymbol{D}(\boldsymbol{v}_1):\boldsymbol{D}(\boldsymbol{v}_2)\,d\mathbf{x},$$

[34] We recall that, by definition of V_T, $\boldsymbol{v}$ is such that $\hat{\boldsymbol{v}}(t,\hat{\mathbf{x}}) = \boldsymbol{v}(t,\hat{\mathcal{A}}_t(\hat{\mathbf{x}}))$ is time-independent.

$$a_2(\boldsymbol{C}_1, \boldsymbol{C}_2) = \int_\Omega \left(\frac{1}{\overline{\mu}\sigma} \mathbf{curl}\, \boldsymbol{C}_1 \cdot \mathbf{curl}\, \boldsymbol{C}_2 + \alpha \operatorname{div} \boldsymbol{C}_1 \operatorname{div} \boldsymbol{C}_2 \right) d\mathbf{x},$$

α being a given positive constant,

$$b(\boldsymbol{v}, q) = \int_\Omega q \operatorname{div} \boldsymbol{v}\, d\mathbf{x},$$

$$c(\boldsymbol{v}_1, \boldsymbol{v}_2, \boldsymbol{v}_3) = \int_\Omega \boldsymbol{v}_1 \cdot \boldsymbol{\nabla} \boldsymbol{v}_2 \cdot \boldsymbol{v}_3\, d\mathbf{x},$$

$$c_{\boldsymbol{w}}(\boldsymbol{v}_1, \boldsymbol{v}_2, \boldsymbol{v}_3) = \int_\Omega \rho\, (\boldsymbol{v}_1 - \boldsymbol{w}) \cdot \boldsymbol{\nabla} \boldsymbol{v}_2 \cdot \boldsymbol{v}_3\, d\mathbf{x}, \tag{5.22}$$

$$d(\boldsymbol{v}_1, \boldsymbol{v}_2, \boldsymbol{v}_3) = \int_\Omega \boldsymbol{v}_2 \cdot \boldsymbol{v}_3 \operatorname{div} \boldsymbol{v}_1\, d\mathbf{x}, \tag{5.23}$$

$$l(\boldsymbol{v}, \boldsymbol{C}_1, \boldsymbol{C}_2) = \int_\Omega \boldsymbol{v} \times \boldsymbol{C}_1 \cdot \mathbf{curl}\, \boldsymbol{C}_2\, d\mathbf{x}.$$

The weak ALE formulation is the following coupled problem: we look for a function $\hat{\mathcal{A}}_t : \hat{\Omega} \to \Omega$ and $(\boldsymbol{u}, \boldsymbol{B}, p)$ in $V \times W \times M$ such that $\boldsymbol{u}(t = 0, .) = \boldsymbol{u}_0$, $\boldsymbol{B}(t = 0, .) = \boldsymbol{B}_0$ and:

- The function $\hat{\mathcal{A}}_t$ satisfies the assumptions (5.1)–(5.2) of Section 5.1.1.1, and thus defines the domains $\Omega_{i,t}$ occupied by each fluid. Then, the density of the fluid ρ is *defined* by:

$$\rho(t, \mathbf{x}) = \hat{\rho}(\hat{\mathcal{A}}_t^{-1}(\mathbf{x})) = \overline{\rho}_i \qquad \text{for } \mathbf{x} \in \Omega_{i,t}, \tag{5.24}$$

  and the velocity of the domain $\boldsymbol{w}$ defined by (5.5) and (5.7) satisfies (5.12)–(5.13) (see Lemma 5.1).
- For all $(\boldsymbol{v}, \boldsymbol{C}, q)$ in $V_T \times W_T \times M_T$,

$$\frac{d}{dt}(\rho \boldsymbol{u}, \boldsymbol{v}) + c_{\boldsymbol{w}}(\boldsymbol{u}, \boldsymbol{u}, \boldsymbol{v}) - d(\boldsymbol{w}, \rho \boldsymbol{u}, \boldsymbol{v}) + a_1(\boldsymbol{u}, \boldsymbol{v})$$

$$-b(\boldsymbol{v}, p) + \frac{1}{\overline{\mu}} l(\boldsymbol{v}, \boldsymbol{B}, \boldsymbol{B}) = (\rho \boldsymbol{f}, \boldsymbol{v}), \tag{5.25}$$

$$b(\boldsymbol{u}, q) = 0, \tag{5.26}$$

$$\frac{d}{dt}(\boldsymbol{B}, \boldsymbol{C}) - c(\boldsymbol{w}, \boldsymbol{B}, \boldsymbol{C}) - d(\boldsymbol{w}, \boldsymbol{B}, \boldsymbol{C}) + a_2(\boldsymbol{B}, \boldsymbol{C}) = l(\boldsymbol{u}, \boldsymbol{B}, \boldsymbol{C}). \tag{5.27}$$

Remark 5.1.2 With the notation of the non-dimensionalized system (1.52), we have:

$$\boldsymbol{f} = \frac{\boldsymbol{g}}{Fr} \quad \overline{\mu} = \frac{1}{S} \quad \rho = \begin{cases} \overline{\rho}_1 = 1 \text{ on } \Omega_{1,t} \\ \overline{\rho}_2 = M \text{ on } \Omega_{2,t} \end{cases}$$

$$\eta = \begin{cases} \overline{\eta}_1 = \dfrac{1}{Re_1} \text{ on } \Omega_{1,t} \\[2ex] \overline{\eta}_2 = \dfrac{M}{Re_2} \text{ on } \Omega_{2,t} \end{cases} \qquad \sigma = \begin{cases} \overline{\sigma}_1 = S\, Rm_1 \text{ on } \Omega_{1,t} \\ \overline{\sigma}_2 = S\, Rm_2 \text{ on } \Omega_{2,t}. \end{cases}$$

◇

Let us now explain why system (1.52)–(1.55) is at least formally equivalent to the weak ALE formulation (5.24)–(5.27).

Let us start with a solution $(\rho, \boldsymbol{u}, \boldsymbol{B}, p)$ to (1.52). We suppose that the velocity field is regular enough so that there exists a reference domain $\overline{\hat{\Omega}} = \overline{\hat{\Omega}}_1 \cup \overline{\hat{\Omega}}_2$ and a $\mathcal{C}^1$-diffeomorphism $\hat{\mathcal{A}}_t$, differentiable with respect to time t, which maps for all $t \geq 0$ the domains $\hat{\Omega}_i$ on $\Omega_{i,t}$ (for $i = 1$ and $i = 2$). The velocity of the domain $\boldsymbol{w}$ is defined by (5.5) and (5.7). Then, ρ indeed satisfies (5.24). Moreover, from Lemma 5.1, $\boldsymbol{w}$ satisfies (5.12)–(5.13).

As for the momentum equation (5.25), we have from the equation on $\boldsymbol{u}$ in (1.52), multiplying by $\boldsymbol{v} \in V_T$ and integrating on Ω:

$$\int_\Omega \rho \frac{\partial \boldsymbol{u}}{\partial t} \cdot \boldsymbol{v} \, d\mathbf{x} + \int_\Omega \rho \boldsymbol{u} \cdot \boldsymbol{\nabla u} \cdot \boldsymbol{v} \, d\mathbf{x} = \int_\Omega \mathbf{h} \cdot \boldsymbol{v} \, d\mathbf{x},$$

with $\mathbf{h} = \rho \boldsymbol{f} - \boldsymbol{\nabla} p + \operatorname{div}(2\eta \boldsymbol{D}(\boldsymbol{u})) + \frac{1}{\mu} \mathbf{curl}\, \boldsymbol{B} \times \boldsymbol{B}$. Using (5.21), we then obtain:

$$\frac{d}{dt} \int_\Omega \rho \boldsymbol{u} \cdot \boldsymbol{v} \, d\mathbf{x} + \int_\Omega \rho (\boldsymbol{u} - \boldsymbol{w}) \cdot \boldsymbol{\nabla u} \cdot \boldsymbol{v} \, d\mathbf{x} - \int_\Omega \rho \, \boldsymbol{u} \cdot \boldsymbol{v} \operatorname{div} \boldsymbol{w} \, d\mathbf{x} = \int_\Omega \mathbf{h} \cdot \boldsymbol{v} \, d\mathbf{x}.$$

Then, standard integrations by parts on the right-hand side of this equation yield (5.25). Analogous computations give the left-hand side of equation (5.27) and its right-hand side is a straightforward consequence of the formula

$$\int_\Omega \mathbf{curl}\, \boldsymbol{B} \cdot \boldsymbol{C} \, d\mathbf{x} = \int_\Omega \boldsymbol{B} \cdot \mathbf{curl}\, \boldsymbol{C} \, d\mathbf{x} + \int_{\partial\Omega} \boldsymbol{n} \times \boldsymbol{B} \cdot \boldsymbol{C} \, d\mathbf{x}$$

and the zero boundary conditions (1.53). To perform the integrations by parts, since the integrals contain some discontinuous coefficients (viscosity or conductivity), it is convenient to split the integrals into two integrals on each fluid ($\Omega_{1,t}$ and $\Omega_{2,t}$) and then perform integrations by parts. Jump terms across the interface appear. Using this weak formulation amounts to imposing that these terms on the interface vanish, in a weak sense, which are precisely the transmission conditions at the interface (1.55).

Suppose conversely that $(\hat{\mathcal{A}}_t, \boldsymbol{u}, \boldsymbol{B}, p)$ is a solution to the weak ALE formulation (5.24)–(5.27). Concerning the mass conservation, using (5.8) with $\psi = \rho$, $\operatorname{div} \boldsymbol{u} = 0$ (see (5.26)), (5.10) and (5.13), we have:

$$\begin{aligned}
\frac{\partial \rho}{\partial t} + \operatorname{div}(\rho \boldsymbol{u}) &= -\boldsymbol{w} \cdot \boldsymbol{\nabla} \rho + \boldsymbol{u} \cdot \boldsymbol{\nabla} \rho + \operatorname{div}(\boldsymbol{u}) \rho \\
&= \delta\rho \, (\boldsymbol{u} - \boldsymbol{w}) \cdot \boldsymbol{n}_1 \delta_{\Sigma_t} \\
&= 0.
\end{aligned}$$

The same manipulations as above show that the equations on $\boldsymbol{u}$ and $\boldsymbol{B}$ in (1.52) hold.

We now turn to the divergence-free constraint on $\boldsymbol{B}$. We first notice that the weak formulation (5.27) can equivalently be rewritten in the following way (using (5.17)):

$$\forall \boldsymbol{C} \in W_T, \left(\frac{\partial \boldsymbol{B}}{\partial t}, \boldsymbol{C}\right) + a_2(\boldsymbol{B}, \boldsymbol{C}) = l(\boldsymbol{u}, \boldsymbol{B}, \boldsymbol{C}). \tag{5.28}$$

By a density argument, this equality holds for any $\boldsymbol{C} \in L^2(0, T; \mathbb{H}^1_{\boldsymbol{n}}(\Omega))$.

Let us now introduce a function ϕ, a solution to the following backward-in-time equation:

$$\begin{cases} \dfrac{\partial \phi}{\partial t} + \alpha \Delta \phi = \operatorname{div} \boldsymbol{B} & \text{in } \Omega, \\ \dfrac{\partial \phi}{\partial \boldsymbol{n}} = 0 & \text{on } \partial\Omega, \\ \phi(t = T) = 0 & \text{in } \Omega. \end{cases}$$

Taking $\boldsymbol{C} = \boldsymbol{\nabla}\phi$ as a test function in (5.28), we obtain:

$$\int_\Omega \frac{\partial \boldsymbol{B}}{\partial t} \cdot \boldsymbol{\nabla}\phi \, d\mathbf{x} + \alpha \int_\Omega \operatorname{div}(\boldsymbol{\nabla}\phi) \operatorname{div} \boldsymbol{B} = 0.$$

Integrating with respect to $t \in [0, T]$ and performing two integrations by parts in the first term, with respect to time and space, respectively, one obtains:

$$\int_0^T \int_\Omega \operatorname{div} \boldsymbol{B} \left(\frac{\partial \phi}{\partial t} + \alpha \Delta \phi\right) = 0,$$

using $\partial\phi/\partial\boldsymbol{n} = 0$, and $\phi(t = T) = \operatorname{div} \boldsymbol{B}(t = 0) = 0$. This yields $\int_0^T\int_\Omega |\operatorname{div}\boldsymbol{B}|^2 = 0$. Therefore, we see that the constraint $\operatorname{div} \boldsymbol{B}(t = 0) = 0$ is propagated forward in time.

Remark 5.1.3 [On free surface flows] We have described the ALE method for modeling the motion of an interface between two fluids in a fixed domain Ω. All the derivations we have performed are also valid if $\boldsymbol{w} \cdot \boldsymbol{n} \neq 0$ on $\partial\Omega$, *i.e.* if the boundary of the domain Ω[35] is moving.

Of particular interest is the case of a free surface (see, for example, the test cases of Sections 5.3.3 and 5.3.1). In this case, one imposes the following boundary conditions on the free surface $\Gamma_t \subset \partial\Omega_t$:

- for the velocity of the domain $\boldsymbol{w}$: $\boldsymbol{w} \cdot \boldsymbol{n} = \boldsymbol{u} \cdot \boldsymbol{n}$ on Γ_t;
- for the velocity $\boldsymbol{u}$ and the pressure p: $\left(-p\mathrm{Id} + \dfrac{2}{Re}\boldsymbol{D}(\boldsymbol{u})\right) \cdot \boldsymbol{n} = 0$ on Γ_t;
- for the magnetic field $\boldsymbol{B}$: $\left(\dfrac{1}{Rm}\mathbf{curl}(\boldsymbol{B}) - \boldsymbol{u} \times \boldsymbol{B}\right) \times \boldsymbol{n} = 0$ on Γ_t.

◇

[35] which thus becomes a time-dependent domain Ω_t.

Remark 5.1.4 [On the treatment of the time derivative] In the weak ALE formulation, we have used (5.21) to obtain time derivatives of integrals over Ω rather than integrals of time derivatives. This may seem useless at the continuous level, since we would have obtained an equivalent formulation by leaving the time derivative inside the integral, but it has important consequences on the stability of the time discretized version. This will be discussed in Section 5.1.3.3. ♢

Remark 5.1.5 [On the choice of the set of unknowns] We have presented the ALE formulation with the set of unknowns $(\hat{\mathcal{A}}_t, \boldsymbol{u}, \boldsymbol{B}, p)$. Notice that one can equivalently consider the set of unknowns $(\boldsymbol{w}, \boldsymbol{u}, \boldsymbol{B}, p)$, and require that the velocity field $\boldsymbol{w}$ is sufficiently smooth (see (5.14) for a sufficient condition) and satisfies (5.12)–(5.13) in order that $\hat{\mathcal{A}}_t$, defined by (5.15), satisfy (5.1)–(5.2) (see Section 5.1.1.1). At the discrete level, we will actually define the problem in terms of $(\boldsymbol{w}, \boldsymbol{u}, \boldsymbol{B}, p)$ rather than in terms of $(\hat{\mathcal{A}}_t, \boldsymbol{u}, \boldsymbol{B}, p)$. ♢

Remark 5.1.6 [On the boundary conditions used in practice] In practice (see in particular the test cases of Section 5.3), we impose a pure slip boundary condition ($\boldsymbol{u}\cdot\boldsymbol{n} = 0$) on the side walls, and no-slip on the other boundaries. In addition, the magnetic boundary conditions are, of course, non-zero. We consider in Section 5.1 simple homogeneous boundary conditions for clarity, but the extension to the practical cases is straightforward. In particular, we refer to Section 3.8 for some details on the implementation of the magnetic boundary conditions.

Notice that it is important to assume a pure slip boundary condition on the side walls (or at least a slip boundary condition with respect to the vertical direction) to obtain a physically admissible solution, since the nodes of the mesh on the interface follow the interface motion. A no-slip boundary condition on the side walls would keep fixed the nodes of the interface on the side walls, which would lead to large distortions of the mesh. This is the so-called kinematic paradox (see, for example, T.A. Baer et al. [12]). This is actually a question of modeling: the boundary condition models the dynamics of the *contact line*[36], and more general slip boundary conditions can be used, like the *Navier slip condition*:

$$\begin{cases} \boldsymbol{u}\cdot\boldsymbol{n} = 0 \\ \alpha\boldsymbol{u}\cdot\boldsymbol{t} + (1-\alpha)\,(\boldsymbol{\tau}\cdot\boldsymbol{n})\cdot\boldsymbol{t} = 0 \end{cases},$$

where $\boldsymbol{t}$ denotes any tangent vector on the boundary, and $\alpha \in [0,1]$ is a slip coefficient. We recall that $\boldsymbol{\tau} = -pId + 2\eta\boldsymbol{D}(\boldsymbol{u})$ is the stress tensor. The case $\alpha = 0$ corresponds to pure slip boundary conditions, and the case $\alpha = 1$ to no-slip boundary conditions (which is not acceptable in our case). The case $\alpha \in (0,1)$ corresponds to the case the fluid slips on the wall with friction. In [190, 191], T.Z. Qian et al. propose a generalization of the Navier boundary condition to model the dynamics of the contact line and a very interesting review about the modeling of a moving contact line. We come back on this issue in Section 5.1.4 where the modeling of surface tension effects is discussed. ♢

[36] The contact line is the boundary of the interface Σ_t, where three materials are in contact: the two liquids, and the solid boundary.

5.1.2 *Time and space discretization*

This section presents the space and time discretizations of the weak ALE formulation (5.24)–(5.27). We recall the ALE formulation amounts to finding both the displacement of the domain $\hat{\mathcal{A}}_t$ and the unknowns $(\boldsymbol{u}, \boldsymbol{B}, p)$ (see Section 5.1.1.4 above).

In the following, we need to modify the trilinear form $c_{\boldsymbol{w}}$ (see (5.22) above) in order to prove the stability of the space and time discretized scheme. We define $\tilde{c}_{\boldsymbol{w}}$ by:

$$\begin{aligned}\tilde{c}_{\boldsymbol{w}}(\boldsymbol{v}_1, \boldsymbol{v}_2, \boldsymbol{v}_3) =& c_{\boldsymbol{w}}(\boldsymbol{v}_1, \boldsymbol{v}_2, \boldsymbol{v}_3) + \frac{1}{2}\int_\Omega \rho \operatorname{div} \boldsymbol{v}_1\, \boldsymbol{v}_2 \cdot \boldsymbol{v}_3 \, d\mathbf{x} \\ &+ \frac{\delta\rho}{2}\int_{\Sigma_t} (\boldsymbol{v}_1 - \boldsymbol{w}) \cdot \boldsymbol{n}_1\, \boldsymbol{v}_2 \cdot \boldsymbol{v}_3 \, d\sigma_{\Sigma_t}.\end{aligned} \tag{5.29}$$

Notice that at the continuous level, $\operatorname{div}(\boldsymbol{u}) = 0$ and relation (5.13) yield

$$c_{\boldsymbol{w}}(\boldsymbol{u}, \boldsymbol{u}, \boldsymbol{v}) = \tilde{c}_{\boldsymbol{w}}(\boldsymbol{u}, \boldsymbol{u}, \boldsymbol{v}).$$

The term $1/2\int_\Omega \rho \operatorname{div} \boldsymbol{v}_1\, \boldsymbol{v}_2 \cdot \boldsymbol{v}_3 \, d\mathbf{x}$ in (5.29) is standard. It is analogous to the well-known modification introduced by Temam of the convective term (see [227] Section III.5) which allows us to recover at the discrete level the skew-symmetry property of the advection term. The second integral is more specific. It will be motivated in Section 5.1.3.4 (see equation (5.53) below).

The discretization uses a finite element method in space, and a semi-implicit Euler time-discretization. Here, the domain $\overline{\Omega}^n = \overline{\Omega}_1^n \cup \overline{\Omega}_2^n$ at the beginning of the n-th timestep, where Ω_i^n is the domain occupied by the fluid i at time t_n, plays the role of the reference domain $\overline{\hat{\Omega}} = \overline{\hat{\Omega}}_1 \cup \overline{\hat{\Omega}}_2$. Notice that the superscript n (in Ω^n) emphasizes that we consider the domain at time t^n, even if the boundary of the domain is *not* moving. For example, when we integrate over Ω^n, this is to indicate that the test functions and the functions ρ, η and σ (whose values are deduced from the domains occupied by each fluid) are taken at time t_n.

Given a mesh $\mathcal{M}^n = \mathcal{M}_1^n \cup \mathcal{M}_2^n$ of the domain[37] $\overline{\Omega}^n = \overline{\Omega}_1^n \cup \overline{\Omega}_2^n$, velocity $\boldsymbol{u}^n$ and magnetic field $\boldsymbol{B}^n$ discretized in finite element spaces at time t_n, we aim to propagate these three items to time t_{n+1}, using the weak ALE formulation (5.24)–(5.27).

In addition to $(\mathcal{M}^n, \boldsymbol{u}^n, \boldsymbol{B}^n)$, let us give ourselves a space discretization of the domain velocity $\boldsymbol{w}^n$ at time t_n. We will come back to its computation in Section 5.1.2.3. We introduce the application

$$\mathcal{A}_{n,n+1} : \begin{cases} (\Omega_i^n)_{i=1,2} \to & (\Omega_i^{n+1})_{i=1,2} \\ \quad\mathbf{y} & \mapsto \mathbf{x} = \mathbf{y} + \delta t\, \boldsymbol{w}^n(\mathbf{y}) \end{cases} \tag{5.30}$$

which might be seen as an approximation of $\hat{\mathcal{A}}_{t_{n+1}} \circ \hat{\mathcal{A}}_{t_n}^{-1}$. This application defines the domain occupied by each fluid[38] at time t_{n+1}: $\Omega_i^{n+1} = \mathcal{A}_{n,n+1}(\Omega_i^n)$, for

[37] and therefore the domains occupied by each fluid.

[38] $\mathcal{A}_{n,n+1}$ also defines the mesh at time t_{n+1}: for $i = 1, 2$, each node of $\mathcal{M}_i^n$ is transported from Ω_i^n to Ω_i^{n+1} by $\mathcal{A}_{n,n+1}$, thus defining the mesh $\mathcal{M}_i^{n+1}$ of Ω_i^{n+1} at time t_{n+1}.

$i = 1, 2$. Without loss of generality, the timestep $\delta t = t_{n+1} - t_n$ is supposed to be constant. In the sequel, our convention is that $\mathbf{y}$ denotes a point in $(\Omega_i^n)_{i=1,2}$ and $\mathbf{x}$ a point in $(\Omega_i^{n+1})_{i=1,2}$. Notice that the position of the domains $(\Omega_i^n)_{i=1,2}$ gives the value of the fluid-dependent quantities ρ, η and σ at each point of Ω. The Jacobian matrix

$$[\mathbf{J}_{n,n+1}]_{i,j} = \frac{\partial(\mathcal{A}_{n,n+1})_i}{\partial \mathbf{y}_j}$$

of $\mathcal{A}_{n,n+1}$ is assumed to have a positive determinant $J_{n,n+1}$ (see Remark 5.1.7 below).

5.1.2.1 Space discretization We consider a finite element discretization of the domain $(\Omega_i^n)_{i=1,2}$. It is transported by the application $\mathcal{A}_{n,n+1}$ to a finite element discretization of the domain $(\Omega_i^{n+1})_{i=1,2}$. The finite element spaces at time t_n for the velocity, the magnetic field and the pressure are, respectively, denoted by

$$V_{h,n} \subset \mathbb{H}_0^1(\Omega), \qquad W_{h,n} \subset \mathbb{H}_{\boldsymbol{n}}^1(\Omega), \qquad M_{h,n} \subset L_0^2(\Omega).$$

Notice that these finite element spaces depend on n, since the mesh is moving. In the numerical tests we present in Sections 5.3 and 6.3, $V_{h,n}$ and $W_{h,n}$ are always based on the same Lagrangian finite element. In addition, as explained in Chapter 3, one can use either stable or stabilized spaces. We recall that by "stable spaces" we mean a pair $(V_{h,n}, M_{h,n})$ of spaces satisfying the following standard inf–sup condition (see Section 3.1.7)

$$\inf_{q_h \in M_{h,n}} \sup_{\boldsymbol{v}_h \in V_{h,n}} \frac{b(\boldsymbol{v}_h, q_h)}{\|\boldsymbol{v}_h\|_{\mathbb{H}^1} \|q_h\|_{L^2}} \geq \beta > 0.$$

Examples of stable spaces are $\mathbb{Q}_2$ elements for the velocity and the magnetic field and either discontinuous piecewise $\mathbb{P}_1$ or continuous piecewise $\mathbb{Q}_1$ elements for the pressure (with possible discontinuities on the interface). On the other hand, one can also use a stabilized formulation as explained in Section 3.5. In this case, one can use equal-order finite elements for the three unknown fields (typically Lagrangian $\mathbb{Q}_1$ finite elements) and this formulation improves the stability at high Reynolds numbers (see Section 3.2).

As explained in Section 5.1.1.2, we use test functions which follow the deformation of the domain given by $\mathcal{A}_{n,n+1}$: the test functions at time t_{n+1} belong to the following spaces:

$$\begin{aligned} V_{h,n+1} = \{&\boldsymbol{v}(t_{n+1}, .) : \Omega \to \mathbb{R}^d, \\ &\boldsymbol{v}(t_{n+1}, \mathbf{x}) = \boldsymbol{v}(t_n, \mathcal{A}_{n,n+1}^{-1}(\mathbf{x})), \boldsymbol{v}(t_n, .) \in V_{h,n}\}, \\ W_{h,n+1} = \{&\boldsymbol{C}(t_{n+1}, .) : \Omega \to \mathbb{R}^d, \\ &\boldsymbol{C}(t_{n+1}, \mathbf{x}) = \boldsymbol{C}(t_n, \mathcal{A}_{n,n+1}^{-1}(\mathbf{x})), \boldsymbol{C}(t_n, .) \in W_{h,n}\}, \\ M_{h,n+1} = \{&q(t_{n+1}, .) : \Omega \to \mathbb{R}, \\ &q(t_{n+1}, \mathbf{x}) = q(t_n, \mathcal{A}_{n,n+1}^{-1}(\mathbf{x})), q(t_n, .) \in M_{h,n}\}. \end{aligned}$$

Unless there is a risk of confusion, we omit the index h for the functions belonging to the finite element spaces $V_{h,n}$, $W_{h,n}$ or $M_{h,n}$.

5.1.2.2 Time discretization and linearization We use the following semi-implicit Euler discretization of (5.25)–(5.27): for a given $(\boldsymbol{u}^n, \boldsymbol{B}^n) \in V_{h,n} \times W_{h,n}$, $(\Omega_i^n)_{i=1,2}$, $\boldsymbol{w}^n$ and $(\Omega_i^{n+1})_{i=1,2}$, compute $(\boldsymbol{u}^{n+1}, \boldsymbol{B}^{n+1}, p^{n+1}) \in V_{h,n+1} \times W_{h,n+1} \times M_{h,n+1}$ such that, for all $(\boldsymbol{v}(t_n,.), \boldsymbol{C}(t_n,.), q(t_n,.)) \in V_{h,n} \times W_{h,n} \times M_{h,n}$,

$$\left\{\begin{aligned} &\frac{1}{\delta t}(\rho \boldsymbol{u}^{n+1}, \boldsymbol{v})^{n+1} + \tilde{c}_{\boldsymbol{w}^n}^{n+1}(\boldsymbol{u}^n, \boldsymbol{u}^{n+1}, \boldsymbol{v}) - d^{n+1}(\boldsymbol{w}^n, \rho \boldsymbol{u}^{n+1}, \boldsymbol{v}) \\ &\quad + a_1^{n+1}(\boldsymbol{u}^{n+1}, \boldsymbol{v}) + b^{n+1}(\boldsymbol{v}, p^{n+1}) + \frac{1}{\overline{\mu}} l^{n+1}(\boldsymbol{v}, \boldsymbol{B}^n, \boldsymbol{B}^{n+1}) = (\rho \boldsymbol{f}^{n+1}, \boldsymbol{v})^{n+1} \\ &\qquad\qquad + \frac{1}{\delta t}(\rho \boldsymbol{u}^n, \boldsymbol{v})^n, \\ &b^{n+1}(\boldsymbol{u}^{n+1}, q) = 0, \\ &\frac{1}{\delta t}(\boldsymbol{B}^{n+1}, \boldsymbol{C})^{n+1} - c^{n+1}(\boldsymbol{w}^n, \boldsymbol{B}^{n+1}, \boldsymbol{C}) - d^{n+1}(\boldsymbol{w}^n, \boldsymbol{B}^{n+1}, \boldsymbol{C}) \\ &\quad + a_2^{n+1}(\boldsymbol{B}^{n+1}, \boldsymbol{C}) - l^{n+1}(\boldsymbol{u}^{n+1}, \boldsymbol{B}^n, \boldsymbol{C}) = \frac{1}{\delta t}(\boldsymbol{B}^n, \boldsymbol{C})^n. \end{aligned}\right. \tag{5.31}$$

The superscripts n (resp. $n+1$) on the forms $(\cdot,\cdot)$, a_1, a_2, b, c, $\tilde{c}$, d and l indicate that the test functions $(\boldsymbol{v}, \boldsymbol{C}, q)$ and the functions ρ, η and σ are taken at time t_n (resp. t_{n+1}). For example:

$$(\rho \boldsymbol{u}^{n+1}, \boldsymbol{v})^{n+1} = \int_{\Omega^{n+1}} \rho \boldsymbol{u}^{n+1} \cdot \boldsymbol{v} \, d\mathbf{x} = \sum_{i=1,2} \overline{\rho}_i \int_{\Omega_i^{n+1}} \boldsymbol{u}^{n+1} \cdot \boldsymbol{v}(t_{n+1}) \, d\mathbf{x},$$

$$(\rho \boldsymbol{u}^n, \boldsymbol{v})^n = \int_{\Omega^n} \rho \boldsymbol{u}^n \cdot \boldsymbol{v} \, d\mathbf{y} = \sum_{i=1,2} \overline{\rho}_i \int_{\Omega_i^n} \boldsymbol{u}^n \cdot \boldsymbol{v}(t_n) \, d\mathbf{y},$$

$$\begin{aligned} a_1^{n+1}(\boldsymbol{u}^{n+1}, \boldsymbol{v}) &= \int_{\Omega^{n+1}} 2\eta \, \boldsymbol{D}(\boldsymbol{u}^{n+1}) : \boldsymbol{D}(\boldsymbol{v}) \, d\mathbf{x} \\ &= \sum_{i=1,2} 2\overline{\eta}_i \int_{\Omega_i^{n+1}} \boldsymbol{D}(\boldsymbol{u}^{n+1}) : \boldsymbol{D}(\boldsymbol{v}(t_{n+1})) \, d\mathbf{x}. \end{aligned}$$

When a form with the superscript $n+1$ is applied to a function defined on Ω^n, this means that this function is transported on Ω^{n+1} by $\mathcal{A}_{n,n+1}$. For example:

$$\begin{aligned} d^{n+1}(\boldsymbol{w}^n, \rho \boldsymbol{u}^{n+1}, \boldsymbol{v}) &= \int_{\Omega^{n+1}} \rho \operatorname{div}(\boldsymbol{w}^n \circ \mathcal{A}_{n,n+1}^{-1}) \boldsymbol{u}^{n+1} \cdot \boldsymbol{v} \, d\mathbf{x} \\ &= \sum_{i=1,2} \overline{\rho}_i \int_{\Omega_i^{n+1}} \operatorname{div}(\boldsymbol{w}^n \circ \mathcal{A}_{n,n+1}^{-1}) \boldsymbol{u}^{n+1} \cdot \boldsymbol{v}(t_{n+1}) \, d\mathbf{x}, \end{aligned}$$

$$\begin{aligned} \tilde{c}_{\boldsymbol{w}^n}^{n+1}(\boldsymbol{u}^n, \boldsymbol{u}^{n+1}, \boldsymbol{v}) = \sum_{i=1,2} \overline{\rho}_i \int_{\Omega_i^{n+1}} \Big(&(\boldsymbol{u}^n - \boldsymbol{w}^n) \circ \mathcal{A}_{n,n+1}^{-1} \cdot \boldsymbol{\nabla} \boldsymbol{u}^{n+1} \cdot \boldsymbol{v}(t_{n+1}) \\ &+ \frac{1}{2} \operatorname{div}(\boldsymbol{u}^n \circ \mathcal{A}_{n,n+1}^{-1}) \boldsymbol{u}^{n+1} \cdot \boldsymbol{v}(t_{n+1}) \Big) \, d\mathbf{x} \\ + \frac{\delta\rho}{2} \int_{\Sigma^{n+1}} &(\boldsymbol{u}^n - \boldsymbol{w}^n) \circ \mathcal{A}_{n,n+1}^{-1} \cdot \boldsymbol{n}_1 \, \boldsymbol{u}^{n+1} \cdot \boldsymbol{v}(t_{n+1}) \, d\sigma_{\Sigma^{n+1}}. \end{aligned} \tag{5.32}$$

The definitions for l, b, c and a_2 are similar.

In practice, all these integrals are easy to compute since they only involve functions which are considered at the same time (t_n or t_{n+1}) and therefore, functions that are discretized on the same mesh. For example, in formula (5.32), $\boldsymbol{u}^n \circ \mathcal{A}_{n,n+1}^{-1}(\mathbf{x})$ (or $\boldsymbol{w}^n \circ \mathcal{A}_{n,n+1}^{-1}(\mathbf{x})$) is defined on Ω^{n+1}. No interpolation (*e.g.* of a function defined on Ω_n on the mesh at t_{n+1}) is required in this scheme.

Indeed, functions such as $\boldsymbol{u}^n \circ \mathcal{A}_{n,n+1}^{-1}(\mathbf{x})$ (or $\boldsymbol{w}^n \circ \mathcal{A}_{n,n+1}^{-1}(\mathbf{x})$) are naturally defined on the mesh at t_{n+1}. For example, if the decomposition of $\boldsymbol{u}^n$ at time t_n was $\boldsymbol{u}^n(\mathbf{y}) = \sum_i \boldsymbol{U}_i^n \boldsymbol{v}_i(t_n, \mathbf{y})$, then the decomposition of $\boldsymbol{u}^n \circ \mathcal{A}_{n,n+1}^{-1}$ at time t_{n+1} is $\boldsymbol{u}^n \circ \mathcal{A}_{n,n+1}^{-1}(\mathbf{x}) = \sum_i \boldsymbol{U}_i^n \boldsymbol{v}_i(t_{n+1}, \mathbf{x})$ (where $(\boldsymbol{v}_i)_{1 \le i \le I}$ here denote the finite element test functions for the velocity: $V_{h,n} = \mathrm{span}(\boldsymbol{v}_i(t_n),\, 1 \le i \le I)$). This follows from the fact that the value $\boldsymbol{v}(t_n, \mathbf{y}_i)$ of the test function $\boldsymbol{v} \in V_{h,n}$ at the node $\mathbf{y}_i$ on Ω^n is the same as the value $\boldsymbol{v}(t_{n+1}, \mathbf{x}_i)$ of this test function $\boldsymbol{v}$ at the node $\mathbf{x}_i = \mathcal{A}_{n,n+1}(\mathbf{y}_i)$ on Ω^{n+1}.

We emphasize that the system to be solved at each timestep is linear but is a coupled system. The hydrodynamic and magnetic equations are coupled through the terms $l^{n+1}(\boldsymbol{v}, \boldsymbol{B}^n, \boldsymbol{B}^{n+1})$ and $l^{n+1}(\boldsymbol{u}^{n+1}, \boldsymbol{B}^n, \boldsymbol{C})$. As will be shown in Section 5.1.3.4, the specific implicit-in-time discretization of these terms is one of the ingredients for the overall stability of the scheme. This has been already noticed and employed for the one-fluid problem, in Sections 3.6.1 and 3.6.3.

5.1.2.3 Complete algorithm To complete the presentation of the numerical scheme, we now describe the computation of the domain velocity $\boldsymbol{w}^n$. The basic requirements are the kinematic conditions (5.12) and (5.13). The first requirement (5.12) ensures that the boundary of the domain $\partial\Omega$ remains fixed while the second requirement (5.13) ensures that the nodes of the mesh which are initially on the interface Σ_0 remain on the interface Σ_{t_n} separating the two fluids, for all the following timesteps t_n.

In addition, $\mathcal{A}_{n,n+1}$ defined from $\boldsymbol{w}^n$ by (5.30) must satisfy the constraint $J_{n,n+1} > 0$. More generally, this application which is used to define the mesh $\mathcal{M}^{n+1}$ from the mesh $\mathcal{M}^n$, should be such that the mesh remains sufficiently regular for finite element computations. This can be seen as the numerical counterpart of the regularity requirement (5.1) on $\hat{\mathcal{A}}_t$ at the continuous level.

Notice that for a given admissible displacement of the interface (in the sense $\Sigma^{n+1} \subset \Omega$), it is always possible to define an extension of the interface velocity in order to ensure $J_{n,n+1} > 0$. A deformation of the mesh based on models for elastic materials may for instance be used (see C. Farhat et al. [77] or L. Gastaldi [85]). See also R.A. Cairncross et al. [33] for a presentation of a method adapted to the cases when the mesh experiences large strains. In the practical problem we are interested in, it seems sufficient to adopt the very standard method that consists in solving a simple Poisson problem to compute the velocity of the mesh (see *e.g.* A. Soulaimani and Y. Saad [221]). Moreover, we choose the displacement purely vertical, so that we actually solve a *scalar* Poisson problem (see (5.33) below). This choice, which is definitely reasonable in the physical situation that we consider, has important favorable consequences on the quality of the algorithm. This will be made precise in next section. Moreover, we discretize the velocity of the domain $\boldsymbol{w}^n$ in space using the same finite element space as for the components of $\boldsymbol{u}^n$ (typically $\mathbb{P}_2$ or $\mathbb{P}_1$ on tetrahedra, and $\mathbb{Q}_2$ or $\mathbb{Q}_1$ on hexahedra). Notice that what is actually needed in practice is the value of $\boldsymbol{w}^n$ at each node of the mesh, so that the interpolation used for $\boldsymbol{w}^n$ is only required to solve (5.33) below.

We may now write the full algorithm. Let us be given $(\Omega_i^n)_{i=1,2}$ and $(\boldsymbol{u}^n, \boldsymbol{B}^n, p^n)$. Then $\boldsymbol{w}^n$, $(\Omega_i^{n+1})_{i=1,2}$ and $(\boldsymbol{u}^{n+1}, \boldsymbol{B}^{n+1}, p^{n+1})$ are computed as follows:

(i) Compute the terms (defined on Ω^n) $\dfrac{1}{\delta t}\displaystyle\int_{\Omega^n} \rho^n \boldsymbol{u}^n \cdot \boldsymbol{v}\, d\mathbf{x}$ and $\dfrac{1}{\delta t}\displaystyle\int_{\Omega^n} \boldsymbol{B}^n \cdot \boldsymbol{C}\, d\mathbf{x}$ in the system (5.31).

(ii) Compute $\boldsymbol{w}^n = (0, 0, w)$ with w such that

$$\begin{cases} -\Delta w = 0, & \text{on } \Omega_i^n, i = 1, 2, \\ w = \dfrac{\boldsymbol{u}^n \cdot \boldsymbol{n}_1}{\boldsymbol{n}_1^{(3)}}, & \text{on } \Sigma^n, \\ \dfrac{\partial w}{\partial \boldsymbol{n}} = 0, & \text{on } \partial\Omega, \end{cases} \tag{5.33}$$

where $\boldsymbol{n}_1^{(3)}$ denotes the vertical component of the normal $\boldsymbol{n}_1$.

(iii) Move the nodes of the mesh according to:

$$\mathcal{A}_{n,n+1} : \begin{cases} (\Omega_i^n)_{i=1,2} \to & (\Omega_i^{n+1})_{i=1,2}, \\ \quad\mathbf{y} & \mapsto \mathbf{x} = \mathbf{y} + \delta t\, \boldsymbol{w}^n(\mathbf{y}). \end{cases}$$

(iv) Compute the remaining terms (defined on Ω^{n+1}) in the system (5.31) (which in particular means assembling the matrix).

(v) Solve (5.31) to determine $(\boldsymbol{u}^{n+1}, \boldsymbol{B}^{n+1}, p^{n+1})$. The resolution is typically performed by a GMRES iterative procedure with an ILU preconditioner and $(\boldsymbol{u}^n, \boldsymbol{B}^n, p^n)$ as the initial guess.

In step (ii), the implementation of the Dirichlet boundary condition on w would be made easier by defining the normals $\boldsymbol{n}_1$ at each node of the discretized surface Σ^n. However, such a definition is delicate, since Σ^n is piecewise smooth, and the nodes are typically singular points of Σ^n. We come back to this question in Section 5.1.3.2 below where we define approximate normals $\boldsymbol{n}_h$ at each node of the interface, by requiring that the Stokes integration by parts formula holds at the discrete level. A precise definition of the Dirichlet boundary condition on $\boldsymbol{w}^n$ is given there (see equation (5.41)).

Remark 5.1.7 [On the choice of the timestep] In practice, a sufficiently small timestep δt is used. This avoids some undesirable deformation of the elements of the mesh and ensures that $\mathcal{A}_{n,n+1}$ defines a bijective orientation preserving map (as a perturbation of the identical map), which translates mathematically into $J_{n,n+1} > 0$. For a general domain velocity, $J_{n,n+1} > 0$ is proved to be automatically satisfied if δt is sufficiently small with respect to $\left\| \dfrac{\partial \boldsymbol{w}^n}{\partial \mathbf{y}} \right\|_{L^\infty(\Omega)}^{-1}$.

With our particular choice of vertical displacement, $J_{n,n+1} > 0$ as soon as the displacement of the points on the interface $\left(i.e. \text{ the product of } \delta t \text{ with } \dfrac{\boldsymbol{u}^n \cdot \boldsymbol{n}_1}{\boldsymbol{n}_1^{(3)}} \right)$ is sufficiently small so that the interface remains inside Ω[39]. Therefore, in our framework, this theoretical limitation on the timestep is not too demanding, since it is satisfied for an admissible motion of the interface.

More quantitatively, we can prove that in our framework, $J_{n,n+1} > 0$ if and only if:

$$\delta t < \left\| \frac{\partial w}{\partial \mathbf{y}_3} \right\|_{L^\infty(\Omega)}^{-1} = \left\| \frac{h(\mathbf{y})}{w(\mathbf{y})} \right\|_{L^\infty(\Sigma^n)}, \tag{5.34}$$

where, for any $\mathbf{y} \in \Sigma^n$, $h(\mathbf{y})$ denotes the height of the upper fluid if $w(\mathbf{y}) > 0$ and the height of the lower fluid if $w(\mathbf{y}) < 0$. We of course assume here that $w(\mathbf{y}) \neq 0$ ♢

Remark 5.1.8 [On the segregated algorithm to compute $\boldsymbol{w}$] It is also possible to consider a fully implicit problem coupling the velocity of the domain and the unknowns $(\boldsymbol{u}, \boldsymbol{B}, p)$, by writing the kinematic constraint (5.13) at time t_{n+1}: given $(\Omega_i^n)_{i=1,2}$ and $(\boldsymbol{u}^n, \boldsymbol{B}^n, p^n)$, compute a domain velocity $\boldsymbol{w}^{n+1}$ (defined on $(\Omega_i^{n+1})_{i=1,2}$) along with $(\boldsymbol{u}^{n+1}, \boldsymbol{B}^{n+1}, p^{n+1})$ such that:

- the following kinematic constraint holds:
$$\boldsymbol{w}^{n+1} \cdot \boldsymbol{n}_1 = \boldsymbol{u}^{n+1} \cdot \boldsymbol{n}_1 \text{ on } \Sigma^{n+1};$$
- system (5.31) (with $\boldsymbol{w}^n$ replaced by $\boldsymbol{w}^{n+1}$) is satisfied;
- and $(\Omega_i^{n+1})_{i=1,2} = \mathcal{A}_{n+1,n}^{-1}((\Omega_i^n)_{i=1,2})$, where
$$\mathcal{A}_{n+1,n} \begin{cases} (\Omega_i^{n+1})_{i=1,2} \to & (\Omega_i^n)_{i=1,2}, \\ \quad \mathbf{x} & \mapsto \mathbf{y} = \mathbf{x} - \delta t\, \boldsymbol{w}^{n+1}(\mathbf{x}). \end{cases}$$

The algorithm we have presented above is the first iteration of a fixed point algorithm to solve this nonlinear problem, iterating separately on $\boldsymbol{w}$ and $(\boldsymbol{u}, \boldsymbol{B}, p)$. This strategy seems to us satisfactory:

- Our numerical observations show that performing more than one iteration of the fixed point algorithm does not change significantly the results.
- Our algorithm enjoys very good stability and conservation properties (see Section 5.1.3).

However, in some specific problems (for example in fluid–structure interaction problems, see [37]), the motion of the mesh and the velocity field are so entangled that the segregated fixed point algorithm may not converge. In this case, a fully coupled algorithm on the set of unknowns $(\boldsymbol{w}^{n+1}, \boldsymbol{u}^{n+1}, \boldsymbol{B}^{n+1}, p^{n+1})$ may be useful (see for example [221]). ♢

[39] Indeed, even if the interface remains inside Ω for the continuous-in-time problem, the timestepping scheme may lead to the unphysical situation that the interface leaves the domain Ω, if the timestep is not small enough.

Remark 5.1.9 [ALE method and space-time finite element method] One can re-interpret the ALE method in the context of *space-time finite elements*. Indeed, the deformation of the mesh amounts to using time-varying finite elements: the space-time frame is divided into time slabs, which links the mesh at one timestep to the mesh at the next timestep. This viewpoint has been used, together with a least-square stabilization method, by T.E. Tezduyar et al. in [231, 232, 187]. ◇

Remark 5.1.10 [ALE method and moving finite element method] The ALE method we have presented is also related to *moving finite elements methods*. In such methods, the problem is to find the best velocity of the domain, in order to reduce the errors. For example, in advection–diffusion problems, one would ideally move the mesh using the advective part of the equation (Lagrangian method). In hyperbolic problems with shocks, one would like to move the mesh in order to have many points in the shock region. These requirements must also be made compatible with the fact that the mesh remains sufficiently regular to obtain meaningful solutions. We refer to M.J. Baines [13], W. Huang [127], K. Miller et al. [166, 165] or G.J. Liao et al. [151]. ◇

Remark 5.1.11 [Algebraic parallelization and ALE method] Since the mesh is moving, the mass matrix and the right-hand side of the linear system need to be recomputed at each timestep. This is the most time-consuming part of the algorithm (nearly 90 % of the computational time). However, the assembling procedure can be parallelized, since it only requires local computations on each finite element. Therefore, it is possible to improve the efficiency of the algorithm using an algebraic parallelization, which consists in:

- distributing in an initial step the nodes of the mesh (and their related degrees of freedom) between the processors;
- assembling at each timestep in parallel on each processor only the lines of the matrix (and of the right-hand side) which correspond to the degrees of freedom related to the nodes associated to the processor;
- using then a parallel iterative solver of a linear system (see the libraries Aztec[40] or PETSc[41]). ◇

5.1.3 Geometric conservation law, stability and conservation properties

In this section, two important properties of the ALE algorithm described in Section 5.1.2.3 are proved: the mass conservation, and the energy stability.

5.1.3.1 Geometric conservation law The geometric conservation law (hereafter abbreviated as GCL) refers to properties of some timestepping schemes related to the evolution of the metric on a moving mesh. There are various definitions of the GCL. The simplest one is the following: a numerical scheme is said to satisfy the GCL if it preserves the constant solution on

[40]see `http://www.cs.sandia.gov/CRF/aztec1.html`

[41]see `http://www-unix.mcs.anl.gov/petsc/petsc-2/`

a moving mesh (of course, when the constants are solution to the continuous problem[42]).

The notion of GCL has been primarily investigated in the framework of the finite volume method. See in particular H. Guillard and C. Farhat [116], M. Lesoinne and C. Farhat [150], B. Nkonga and H. Guillard [180]. H. Guillard and C. Farhat prove in [116] that the GCL is a sufficient condition for a numerical scheme to be first-order time-accurate on a moving grid, independently of the grid motion. More generally, these authors show that a higher accuracy is obtained with schemes satisfying the GCL compared to the schemes that violate it and that the schemes satisfying the GCL generally allow for a larger timestep. In the framework of finite element methods, F. Nobile and L. Formaggia prove in [80] that the GCL is a sufficient condition to ensure the unconditional stability of a backward Euler scheme applied to an advection–diffusion equation on a moving domain (as shown in Section 5.1.3.3). Let us also cite the work by P. Le Tallec and J. Mouro [143] where implications of the GCL in the framework of fluid structure interaction problems are discussed.

If we restrict ourselves to vertical displacements, the ALE scheme of Section 5.1.2.3 satisfies the GCL in the following sense.

Lemma 5.3 *Suppose that the domain velocity $\boldsymbol{w}^n$ has the form $(0, 0, w)$. Let ϕ be a function defined on Ω_i^{n+1}, for $i = 1$ or 2. Then the ALE scheme satisfies the GCL in the following sense:*

$$\int_{\Omega_i^{n+1}} \phi(\mathbf{x})\, d\mathbf{x} - \int_{\Omega_i^n} \phi \circ \mathcal{A}_{n,n+1}(\mathbf{y})\, d\mathbf{y}$$

$$= \delta t \int_{\Omega_i^n} \phi \circ \mathcal{A}_{n,n+1}(\mathbf{y})\, div_{\mathbf{y}} \boldsymbol{w}^n(\mathbf{y})\, d\mathbf{y} \tag{5.35}$$

$$= \delta t \int_{\Omega_i^{n+1}} \phi(\mathbf{x})\, div_{\mathbf{x}} \left(\boldsymbol{w}^n \circ \mathcal{A}_{n,n+1}^{-1}(\mathbf{x})\right)\, d\mathbf{x}. \tag{5.36}$$

Proof The change of variable defined by $\mathbf{x} = \mathcal{A}_{n,n+1}(\mathbf{y})$ gives in the first integral:

$$\int_{\Omega_i^{n+1}} \phi(\mathbf{x})\, d\mathbf{x} = \int_{\Omega_i^n} \phi \circ \mathcal{A}_{n,n+1}(\mathbf{y}) J_{n,n+1}(\mathbf{y})\, d\mathbf{y}.$$

Considering the mesh velocity is purely vertical, the Jacobian matrix has the following form

$$\mathbf{J}_{n,n+1} = \begin{bmatrix} 1 & 0 & 0 \\ 0 & 1 & 0 \\ \delta t \dfrac{\partial w}{\partial \mathbf{y}_1} & \delta t \dfrac{\partial w}{\partial \mathbf{y}_2} & 1 + \delta t \dfrac{\partial w}{\partial \mathbf{y}_3} \end{bmatrix},$$

and therefore

$$J_{n,n+1} = 1 + \delta t \ \mathrm{div}_{\mathbf{y}} \boldsymbol{w}^n, \tag{5.37}$$

which concludes the proof of (5.35).

[42]This is, for example, the case for a simple advection–diffusion equation on a moving mesh, see Section 5.1.3.3.

For (5.36), we perform in the second integral of the left-hand side the change of variable $\mathbf{y} = \mathcal{A}_{n,n+1}^{-1}(\mathbf{x})$. Noticing that $\mathbf{y} = \mathbf{x} - \delta t \boldsymbol{w}^n \circ \mathcal{A}_{n,n+1}^{-1}(\mathbf{x})$, analogous computations as in the proof of (5.35) give:

$$\int_{\Omega_i^n} \phi(\mathcal{A}_{n,n+1}(\mathbf{y}))\, d\mathbf{y} = \int_{\Omega_i^{n+1}} \phi(\mathbf{x}) \left(1 - \delta t \operatorname{div}_{\mathbf{x}} \left(\boldsymbol{w}^n \circ \mathcal{A}_{n,n+1}^{-1}(\mathbf{x})\right)\right) d\mathbf{x},$$

which is (5.36). □

Remark 5.1.12 [Interpretation of GCL] The GCL (5.35) and (5.36) can be seen as discrete counterparts of the following formula: for any smooth function ϕ such that $\hat{\phi}$ (defined by $\hat{\phi}(t, \hat{\mathbf{x}}) = \phi(t, \hat{\mathcal{A}}_t(\hat{\mathbf{x}}))$) is time-independent,

$$\frac{d}{dt} \int_{\Omega} \phi(t, \mathbf{x})\, d\mathbf{x} = \int_{\Omega} \phi(t, \mathbf{x}) \operatorname{div} \boldsymbol{w}(t, \mathbf{x})\, d\mathbf{x},$$

which is obtained from (5.17) by taking $\psi = 1$. ◊

Remark 5.1.13 [GCL and vertical displacement] Considering a purely vertical displacement of the nodes ensures the GCL. The question arises as to know whether it is necessary that the mesh moves in only one direction to have the GCL. Let us denote by $(\lambda_1, \lambda_2, \lambda_3)$ the (possibly complex) eigenvalues of $\partial \boldsymbol{w}^n / \partial \mathbf{y}$. One can check that

$$J_{n,n+1} = 1 + \delta t \operatorname{div}_{\mathbf{y}} \boldsymbol{w}^n + (\delta t)^2 \sum_{1 \leq i \neq j \leq 3} \lambda_i \lambda_j + (\delta t)^3 \prod_{i=1}^{3} \lambda_i. \tag{5.38}$$

The GCL holds if and only if the last two terms in (5.38) vanish, which means the Hessian $\partial \boldsymbol{w}^n / \partial \mathbf{y}$ has at least two zero eigenvalues. The displacement must therefore be essentially one-dimensional. For example, a velocity of the form $\boldsymbol{w}^n = w(\theta) \boldsymbol{e_r}$ in cylindrical coordinate satisfies the GCL. ◊

In the next three sections, we show that (5.35) and (5.36) are key properties to ensure discrete mass conservation and energy inequality.

5.1.3.2 Discrete mass conservation We now present the ingredients that allow us to obtain an exact mass conservation of each fluid after time and space discretizations.

The *first ingredient* is an appropriate computation of the normal velocities on the surfaces (boundary of the domain or interface between the two fluids), in such a way that the Stokes integration by parts formula holds at the discrete level (see (5.42) below). It is convenient, with a view to enforcing Dirichlet boundary conditions in (5.33), to compute approximate discrete normals at the *nodes* of the interface. But this approximation must be done carefully. We now explain how to perform this approximation. Let i be a node on the interface Σ^n and let φ_i be the basis function associated to this node. Following M.S. Engelman, R.L. Sani and P.M. Gresho [74], we define $\boldsymbol{n}_{h,i}$ by

$$\boldsymbol{n}_{h,i} = \frac{1}{\int_{\Sigma^n} \varphi_i \, d\sigma} \int_{\Omega_1^n} \nabla \varphi_i \, d\mathbf{x}. \tag{5.39}$$

The approximate normal $\boldsymbol{n}_{h,i}$ on node i is oriented from Ω_1^n to Ω_2^n. The components of $\boldsymbol{n}_{h,i}$ are denoted by $(\boldsymbol{n}_{h,i}^{(1)}, \boldsymbol{n}_{h,i}^{(2)}, \boldsymbol{n}_{h,i}^{(3)})$. Let $\boldsymbol{v}_h = (\boldsymbol{v}_h^{(1)}, \boldsymbol{v}_h^{(2)}, \boldsymbol{v}_h^{(3)})$ be an element of $V_{h,n}$. By convention, the value of $\boldsymbol{v}_h$ along the normal $\boldsymbol{n}_h$ to Σ_n is defined by:

$$\boldsymbol{v}_h \cdot \boldsymbol{n}_h = \sum_i (\boldsymbol{v}_{h,i}^{(1)} \boldsymbol{n}_{h,i}^{(1)} + \boldsymbol{v}_{h,i}^{(2)} \boldsymbol{n}_{h,i}^{(2)} + \boldsymbol{v}_{h,i}^{(3)} \boldsymbol{n}_{h,i}^{(3)}) \varphi_i, \tag{5.40}$$

where $(\boldsymbol{v}_{h,i}^{(1)}, \boldsymbol{v}_{h,i}^{(2)}, \boldsymbol{v}_{h,i}^{(3)})$ denotes the value of $\boldsymbol{v}_h$ at node i. This is the convention used to define the Dirichlet boundary conditions in (5.33). Therefore, the velocity $\boldsymbol{w}^n$ of the mesh is such that the following discrete version of (5.13) is satisfied: $\forall n \geq 0$,

$$\boldsymbol{w}^n \cdot \boldsymbol{n}_h = \boldsymbol{u}^n \cdot \boldsymbol{n}_h \text{ on } \Sigma^n, \tag{5.41}$$

where the normal velocities are defined by (5.40).

The definition of the value of $\boldsymbol{v}_h$ along the outward normal on $\partial\Omega$ is extended likewise. Then, formula (5.39) yields the following key property. Let $\boldsymbol{v}_h$ be a function of $V_{h,n}$. Then

$$\begin{cases} \displaystyle\int_{\Omega_1^n} \operatorname{div} \boldsymbol{v}_h \, dx = \int_{\Sigma^n} \boldsymbol{v}_h \cdot \boldsymbol{n}_h \, d\sigma + \int_{\partial\Omega \cap \partial\Omega_1^n} \boldsymbol{v}_h \cdot \boldsymbol{n}_h \, d\sigma, \\ \displaystyle\int_{\Omega_2^n} \operatorname{div} \boldsymbol{v}_h \, dx = -\int_{\Sigma^n} \boldsymbol{v}_h \cdot \boldsymbol{n}_h \, d\sigma + \int_{\partial\Omega \cap \partial\Omega_2^n} \boldsymbol{v}_h \cdot \boldsymbol{n}_h \, d\sigma. \end{cases} \tag{5.42}$$

Remark 5.1.14 [On $\boldsymbol{n}_h$ and $\boldsymbol{n}_1$] Notice that the normal vectors $\boldsymbol{n}_h$ introduced here are only used in practice to define the Dirichlet boundary conditions for the equation satisfied by $\boldsymbol{w}$ in (5.33). In particular, the normal $\boldsymbol{n}_1$ which appears in the modified trilinear form $\tilde{c}_{\boldsymbol{w}}$ defined by (5.29) is not the approximate node normal $\boldsymbol{n}_h$ but rather the "real" normal defined almost everywhere on the interface, and in particular on the Gauss–Legendre integration points located inside the boundary elements. ◊

The *second ingredient* to ensure mass conservation is the following property: $\forall n \geq 0$,

$$\int_{\Omega_i^n} \operatorname{div} \boldsymbol{u}^n \, dx = 0, \qquad \text{for } i = 1, 2. \tag{5.43}$$

This can be achieved in many ways. First, a space of *discontinuous* finite elements may be used for the discretization of the pressure (for example the mixed element

$\mathbb{Q}_2/\mathbb{P}_1$). Second, a Lagrange multiplier may be introduced to impose that the numerical flux through the interface vanishes:

$$\int_{\Sigma^n} \boldsymbol{u}^n \cdot \boldsymbol{n}_h \, d\sigma = 0 \tag{5.44}$$

(thus (5.43) by integration by parts). The latter strategy amounts to including the characteristic function $1_{\Omega_1^n}$ in the functional space for the pressure $M_{h,n}$. For more details, we refer the interested reader to L. Formaggia *et al.* [79]. Third, finite element spaces for the pressure that are continuous on each fluid, but discontinuous on the interface, may be used. We have tested the three approaches, and they all give satisfactory results.

On a fixed domain, (5.42) and (5.43) would be sufficient to ensure mass conservation. But the GCL property of Lemma 5.3 is needed to extend it to the case of moving domains. We use (5.35) as the *third ingredient* to obtain mass conservation. This is stated in the following proposition.

Proposition 5.4 *Suppose that:*

- *the discrete normal velocities are such that (5.42) holds;*
- *(5.43) is satisfied;*
- *the motion of the mesh is such that the geometric conservation law (5.35) is satisfied.*

Then the mass of each fluid is preserved:

$$\overline{\rho}_i |\Omega_i^n| = \overline{\rho}_i |\Omega_i^{n+1}|, \ \textit{for } i = 1, 2, \tag{5.45}$$

where $|\Omega_i^n|$ denotes the measure of Ω_i^n.

Proof Relation (5.35) gives, with $\phi = \overline{\rho}_1$ and $i = 1$,

$$\overline{\rho}_1 |\Omega_1^{n+1}| - \overline{\rho}_1 |\Omega_1^n| = \delta t \overline{\rho}_1 \int_{\Omega_1^n} \operatorname{div}_{\mathbf{y}} \boldsymbol{w}^n \, d\mathbf{y}.$$

Thus, using successively (5.42), (5.41) and (5.43), we have

$$\begin{aligned} \overline{\rho}_1 |\Omega_1^{n+1}| - \overline{\rho}_1 |\Omega_1^n| &= \delta t \overline{\rho}_1 \int_{\Sigma^n} \boldsymbol{w}^n \cdot \boldsymbol{n}_h \, d\sigma \\ &= \delta t \overline{\rho}_1 \int_{\Sigma^n} \boldsymbol{u}^n \cdot \boldsymbol{n}_h \, d\sigma \\ &= \delta t \overline{\rho}_1 \int_{\Omega_1^n} \operatorname{div}_{\mathbf{y}} \boldsymbol{u}^n \, d\mathbf{y} = 0. \end{aligned}$$

□

We refer to Section 5.3.2 for some numerical experiments that confirm that if either the geometric conservation law (5.35) or relation (5.43) are not satisfied, then the mass of each fluid is not conserved. See also Remark 5.1.17 below for another discussion of the case when the GCL is not satisfied.

5.1.3.3 Stability of the time-discretized ALE weak formulation Anticipating the next section, we would like to discuss here the stability of our numerical scheme on a simple toy model, comparing it with another ALE scheme. In this section, we explain why we rather consider time derivatives of integrals over Ω than integrals of time derivatives in our ALE formulation (see (5.21) and Remark 5.1.4 above). Our discussion follows L. Formaggia and F. Nobile [80,81].

Let us simply consider a convection–diffusion equation:

$$\begin{cases} \dfrac{\partial \phi}{\partial t} + \boldsymbol{u} \cdot \boldsymbol{\nabla} \phi - \Delta\, \phi = 0 \quad \text{on } \Omega, \\ \qquad\qquad\qquad\quad \phi = 0 \quad \text{on } \partial\Omega, \end{cases}$$

where $\boldsymbol{u}$ is a given velocity field assumed steady and divergence-free. The equation is supplied with an initial condition ϕ_0.

Let us in addition introduce the internal motion of the domain Ω, through a bijective application $\hat{\mathcal{A}}_t : \hat{\Omega} \to \Omega$, which satisfies the regularity assumptions (5.1). The domain velocity $\boldsymbol{w}$ is then defined by (5.5) and (5.7) and we suppose that it is such that (5.12) is satisfied, so that $\hat{\mathcal{A}}_t(\hat{\Omega}) = \Omega$ (the domain Ω is thus fixed, and the motion $\hat{\mathcal{A}}_t$ is artificial: it is not related to any physical motion). Notice that we suppose here that the motion of the domain is defined *a priori*, which is not the case in the other sections, where $\boldsymbol{w}$ depends on the velocity of the fluids, which is part of the unknowns.

Our approach considers the following weak formulation of the problem:

$$\frac{d}{dt}(\phi, \psi) + c_{\boldsymbol{w}}(\boldsymbol{u}, \phi, \psi) - d(\boldsymbol{w}, \phi, \psi) + a(\phi, \psi) = 0, \tag{5.46}$$

where $c_{\boldsymbol{w}}$ is defined by (5.22) (with here $\rho = 1$), d is defined by (5.23), and $a(\phi, \psi) = \int_\Omega \boldsymbol{\nabla}\phi \cdot \boldsymbol{\nabla}\psi \, d\mathbf{x}$. We recall that the test function ψ is time-dependent in the current frame, but does not depend on time when transported in the reference frame: $\exists \hat{\psi}$, $\psi(t, \mathbf{x}) = \hat{\psi}(\hat{\mathcal{A}}_t^{-1}(\mathbf{x}))$. The first three terms in (5.46) come from the special treatment of the time derivative (see (5.21)). Following Section 5.1.2.2, (5.46) is discretized into:

$$\begin{aligned} &\frac{1}{\delta t}(\phi^{n+1}, \psi)^{n+1} + \tilde{c}_{\boldsymbol{w}^n}(\boldsymbol{u}, \phi^{n+1}, \psi) \\ &-d^{n+1}(\boldsymbol{w}^n, \phi^{n+1}, \psi) + a^{n+1}(\phi^{n+1}, \psi) = \frac{1}{\delta t}(\phi^n, \psi)^n \end{aligned} \tag{5.47}$$

where $a^{n+1}(\phi^{n+1}, \psi) = \int_{\Omega^{n+1}} \boldsymbol{\nabla}\phi^{n+1} \cdot \boldsymbol{\nabla}\psi \, d\mathbf{x}$ and $\tilde{c}_{\boldsymbol{w}^n}$ is defined by (5.29) (with here $\rho = 1$ and $\delta\rho = 0$). We recall that the superscript indicates the time of the test function: for example, $(\phi^{n+1}, \psi)^{n+1} = \int_{\Omega^{n+1}} \phi^{n+1}\psi(t_{n+1})$. We recall that the change of $c_{\boldsymbol{w}}$ to $\tilde{c}_{\boldsymbol{w}}$ is motivated by the fact that, once the problem is discretized in space by a finite element method, numerical integrations may result into a not pointwise divergence-free velocity field $\boldsymbol{u}$.

For comparison, we consider another approach, where the derivative with respect to time is kept inside the space integration sign:

$$\int_{\Omega} \frac{\partial((\phi\psi)\circ\hat{\mathcal{A}}_t)}{\partial t}(\hat{\mathcal{A}}_t^{-1}(\mathbf{x}))\, d\mathbf{x} + c_{\boldsymbol{w}}(\boldsymbol{u},\phi,\psi) + a(\phi,\psi) = 0. \tag{5.48}$$

Notice that (5.48) is equivalent to (5.46) at the continuous level since:

$$\begin{aligned}
\frac{d}{dt}(\phi,\psi) &= \frac{d}{dt}\int_{\Omega}\phi\psi\, d\mathbf{x}\\
&= \frac{d}{dt}\int_{\hat{\Omega}}\phi\psi(\hat{\mathcal{A}}_t(\hat{\mathbf{x}}))\,\hat{J}_t(\hat{\mathbf{x}})\, d\hat{\mathbf{x}}\\
&= \int_{\hat{\Omega}}\frac{\partial(\phi\psi(\hat{\mathcal{A}}_t(\hat{\mathbf{x}})))}{\partial t}\,\hat{J}_t(\hat{\mathbf{x}})\, d\hat{\mathbf{x}} + \int_{\hat{\Omega}}\phi\psi(\hat{\mathcal{A}}_t(\hat{\mathbf{x}}))\,\frac{\partial\hat{J}_t(\mathbf{x})}{\partial t}\, d\hat{\mathbf{x}}\\
&= \int_{\hat{\Omega}}\frac{\partial(\phi\psi(\hat{\mathcal{A}}_t(\hat{\mathbf{x}})))}{\partial t}\,\hat{J}_t(\hat{\mathbf{x}})\, d\hat{\mathbf{x}} + \int_{\hat{\Omega}}\phi\psi(\hat{\mathcal{A}}_t(\hat{\mathbf{x}}))\,\mathrm{div}\,(\boldsymbol{w}(t,\hat{\mathcal{A}}_t(\hat{\mathbf{x}})))\hat{J}_t(\hat{\mathbf{x}})\, d\hat{\mathbf{x}}\\
&= \int_{\Omega}\frac{\partial((\phi\psi)\circ\hat{\mathcal{A}}_t)}{\partial t}(\hat{\mathcal{A}}_t^{-1}(\mathbf{x}))\, d\mathbf{x} + \int_{\Omega}\phi\psi(\mathbf{x})\,\mathrm{div}\,(\boldsymbol{w}(t,\mathbf{x}))\, d\mathbf{x}.
\end{aligned}$$

As the discrete level, starting from (5.48), only integrations on Ω^{n+1} are involved:

$$\frac{1}{\delta t}(\phi^{n+1},\psi)^{n+1} + \tilde{c}_{\boldsymbol{w}^n}(\boldsymbol{u},\phi^{n+1},\psi) + a^{n+1}(\phi^{n+1},\psi) = \frac{1}{\delta t}(\phi^n,\psi)^{n+1}. \tag{5.49}$$

More explicitly, the last integral is $(\phi^n,\psi)^{n+1} = \int_{\Omega^{n+1}}\phi^n\circ\mathcal{A}_{n,n+1}^{-1}\psi(t_{n+1})$. Notice that (5.47) contains a correction term involving $\mathrm{div}\,\boldsymbol{w}^n$, while (5.49) does not.

From a practical point of view, the computation of each term of the formulation (5.49) is no more complicated than in (5.47). For example, if the decomposition of ϕ^n at time t_n was $\phi^n(\mathbf{y}) = \sum_i \Phi_i^n\psi_i(t_n,\mathbf{y})$, then the decomposition of $\phi^n\circ\mathcal{A}_{n,n+1}^{-1}$ at time t_{n+1} is $\phi^n\circ\mathcal{A}_{n,n+1}^{-1}(\mathbf{x}) = \sum_i\Phi_i^n\psi_i(t_{n+1},\mathbf{x})$. The difference in the discrete time derivative between the two formulations is that, if we denote by $M_{i,j}^n = \int_{\Omega^n}\psi_i(t_n)\psi_j(t_n)$ the mass matrix at time t_n, the formulation (5.47) contains the term:

$$\frac{1}{\delta t}\left(M^{n+1}\Phi^{n+1} - M^n\Phi^n\right)$$

while the formulation (5.49) contains the term:

$$\frac{1}{\delta t}M^{n+1}\left(\Phi^{n+1} - \Phi^n\right).$$

Let us now compare the stability properties of the schemes (5.47) and (5.49). Choosing $\psi = \phi^{n+1}$ in (5.47), we readily obtain:

$$\frac{1}{\delta t}(\phi^{n+1},\phi^{n+1})^{n+1} + a^{n+1}(\phi^{n+1},\phi^{n+1}) = \frac{1}{2}d^{n+1}(\boldsymbol{w}^n,\phi^{n+1},\phi^{n+1}) + \frac{1}{\delta t}(\phi^{n+1},\phi^n)^n$$

and therefore, using $\phi^{n+1}\,\phi^n = -\frac{1}{2}|\phi^{n+1} - \phi^n|^2 + \frac{1}{2}|\phi^{n+1}|^2 + \frac{1}{2}|\phi^n|^2$,

$$\frac{1}{2\delta t}\left((\phi^{n+1}, \phi^{n+1})^{n+1} - (\phi^n, \phi^n)^n\right) + a^{n+1}(\phi^{n+1}, \phi^{n+1})$$

$$= -\frac{1}{2\delta t}(\phi^{n+1} - \phi^n, \phi^{n+1} - \phi^n)^n$$

$$-\frac{1}{2\delta t}(\phi^{n+1}, \phi^{n+1})^{n+1} + \frac{1}{2\delta t}(\phi^{n+1}, \phi^{n+1})^n + \frac{1}{2}d^{n+1}(\boldsymbol{w}^n, \phi^{n+1}, \phi^{n+1})$$

$$= -\frac{1}{2\delta t}(\phi^{n+1} - \phi^n, \phi^{n+1} - \phi^n)^n,$$

using the geometric conservation law (see Lemma 5.3). This yields the stability of the numerical scheme (5.47) in the energy norm.

On the other hand, similar computations on (5.49) show:

$$\frac{1}{2\delta t}\left((\phi^{n+1}, \phi^{n+1})^{n+1} - (\phi^n, \phi^n)^n\right) + a^{n+1}(\phi^{n+1}, \phi^{n+1})$$

$$= -\frac{1}{2\delta t}(\phi^{n+1} - \phi^n, \phi^{n+1} - \phi^n)^n$$

$$-\frac{1}{2}d^{n+1}(\boldsymbol{w}^n, \phi^{n+1} - \phi^n, \phi^{n+1} + \phi^n).$$

Since the last term is not *a priori* non-positive, it is not clear how to prove the stability of the numerical scheme (5.49) in the energy norm. On the other hand, there is no numerical evidence showing a definite instability of (5.49), at least using reasonable numerical parameters.

In the particular framework of this section (convection–diffusion equation with a given domain velocity field), it is however possible to prove stability of the scheme (5.49) if the timestep is smaller than a maximum timestep that depends on the domain velocity (see F. Nobile [181]). However, this result is not conclusive if the domain velocity field is not known *a priori*, but depends on the solution ϕ itself (as is the case in the two-fluid MHD problem).

5.1.3.4 Energy stability It is assumed throughout this section that stable finite element spaces are used (for example $\mathbb{P}_2/\mathbb{P}_1$ or $\mathbb{P}_1$-bubble/$\mathbb{P}_1$ on tetrahedra, or $\mathbb{Q}_2/\mathbb{Q}_1$ on hexahedra, see Chapter 3).

Proposition 5.5 *We denote by*

$$E^n = \int_{\Omega^n} \frac{\rho|\boldsymbol{u}^n|^2}{2}\,d\mathbf{y} + \int_{\Omega^n} \frac{|\boldsymbol{B}^n|^2}{2\overline{\mu}}\,d\mathbf{y} \tag{5.50}$$

the total energy of the system at time t_n. If the body force $\boldsymbol{f}$ vanishes, the solution computed by the algorithm of Section 5.1.2.3 satisfies the energy inequality

$$\frac{E^{n+1} - E^n}{\delta t} + \int_{\Omega^{n+1}} 2\eta|\boldsymbol{D}(\boldsymbol{u}^{n+1})|^2\,d\mathbf{x} + \int_{\Omega^{n+1}} \frac{1}{\overline{\mu}\sigma}|\mathbf{curl}\,\boldsymbol{B}^{n+1}|^2\,d\mathbf{x} \le 0. \tag{5.51}$$

Proof We take $\boldsymbol{v}(t_{n+1}) = \boldsymbol{u}^{n+1}$ in the hydrodynamic equation and $\boldsymbol{C}(t_{n+1}) = \frac{1}{\overline{\mu}}\boldsymbol{B}^{n+1}$ in the magnetic equation, in the system of equations (5.31). The proof of (5.51) falls in three steps.

The *first step* of the proof is once again based on the GCL. Elementary manipulations give

$$\begin{aligned}
&\frac{1}{\delta t}\int_{\Omega^{n+1}} \rho|\boldsymbol{u}^{n+1}|^2\,d\mathbf{x} - \frac{1}{\delta t}\int_{\Omega^n} \rho\boldsymbol{u}^n\cdot(\boldsymbol{u}^{n+1}\circ\mathcal{A}_{n,n+1})\,d\mathbf{y}\\
&= \frac{1}{\delta t}\int_{\Omega^{n+1}} \rho|\boldsymbol{u}^{n+1}|^2\,d\mathbf{x}\\
&\quad -\frac{1}{2\delta t}\int_{\Omega^n} \rho|\boldsymbol{u}^{n+1}|^2\circ\mathcal{A}_{n,n+1}\,d\mathbf{y} - \frac{1}{2\delta t}\int_{\Omega^n}\rho|\boldsymbol{u}^n|^2\,d\mathbf{y}\\
&\quad +\frac{1}{2\delta t}\int_{\Omega^n}\rho|\boldsymbol{u}^{n+1}\circ\mathcal{A}_{n,n+1}-\boldsymbol{u}^n|^2\,d\mathbf{y}.
\end{aligned}$$

Then, using Lemma 5.3, we obtain

$$\begin{aligned}
&\frac{1}{\delta t}\int_{\Omega^{n+1}} \rho|\boldsymbol{u}^{n+1}|^2\,d\mathbf{x} - \frac{1}{\delta t}\int_{\Omega^n} \rho|\boldsymbol{u}^{n+1}|^2\circ\mathcal{A}_{n,n+1}\,d\mathbf{y}\\
&\qquad = \int_{\Omega^{n+1}}\rho|\boldsymbol{u}^{n+1}|^2\mathrm{div}\,(\boldsymbol{w}^n\circ\mathcal{A}_{n,n+1}^{-1})\,d\mathbf{x},
\end{aligned}$$

and therefore

$$\begin{aligned}
&\frac{1}{\delta t}\int_{\Omega^{n+1}} \rho\;|\boldsymbol{u}^{n+1}|^2\,d\mathbf{x} - \frac{1}{\delta t}\int_{\Omega^n} \rho\boldsymbol{u}^n\cdot(\boldsymbol{u}^{n+1}\circ\mathcal{A}_{n,n+1})\,d\mathbf{y}\\
&\qquad = \frac{1}{2\delta t}\int_{\Omega^{n+1}} \rho|\boldsymbol{u}^{n+1}|^2\,d\mathbf{x} - \frac{1}{2\delta t}\int_{\Omega^n}\rho|\boldsymbol{u}^n|^2\,d\mathbf{y}\\
&\qquad\quad +\frac{1}{2\delta t}\int_{\Omega^n}\rho|\boldsymbol{u}^{n+1}\circ\mathcal{A}_{n,n+1}-\boldsymbol{u}^n|^2\,d\mathbf{y}\\
&\qquad\quad +\frac{1}{2}\int_{\Omega^{n+1}}\rho|\boldsymbol{u}^{n+1}|^2\mathrm{div}\,(\boldsymbol{w}^n\circ\mathcal{A}_{n,n+1}^{-1})\,d\mathbf{x}.
\end{aligned}\tag{5.52}$$

Following the same lines, a similar relation is readily obtained for the magnetic field.

The *second step* uses the corrected expression (5.32) of the discrete advection term. By (5.32), we have

$$\begin{aligned}
\tilde{c}_{\boldsymbol{w}^n}^{n+1}(\boldsymbol{u}^n,\boldsymbol{u}^{n+1},\boldsymbol{u}^{n+1}) &= \int_{\Omega^{n+1}}\rho\left((\boldsymbol{u}^n-\boldsymbol{w}^n)\circ\mathcal{A}_{n,n+1}^{-1}\right)\cdot\boldsymbol{\nabla}\left(\frac{|\boldsymbol{u}^{n+1}|^2}{2}\right)\,d\mathbf{x}\\
&\quad +\frac{1}{2}\int_{\Omega^{n+1}}\rho|\boldsymbol{u}^{n+1}|^2\mathrm{div}\,(\boldsymbol{u}^n\circ\mathcal{A}_{n,n+1}^{-1})\,d\mathbf{x}\\
&\quad +\frac{\delta\rho}{2}\int_{\Sigma^{n+1}}\left((\boldsymbol{u}^n-\boldsymbol{w}^n)\circ\mathcal{A}_{n,n+1}^{-1}\right)\cdot\boldsymbol{n}_1|\boldsymbol{u}^{n+1}|^2\,d\sigma.
\end{aligned}$$

The first integral reads

$$\int_{\Omega^{n+1}} \rho\left((\boldsymbol{u}^n - \boldsymbol{w}^n)\circ \mathcal{A}_{n,n+1}^{-1}\right)\cdot \boldsymbol{\nabla}\left(\frac{|\boldsymbol{u}^{n+1}|^2}{2}\right)\, d\mathbf{x}$$
$$= -\frac{1}{2}\int_{\Omega^{n+1}} |\boldsymbol{u}^{n+1}|^2 \mathrm{div}\,\left(\rho((\boldsymbol{u}^n - \boldsymbol{w}^n)\circ \mathcal{A}_{n,n+1}^{-1})\right)\, d\mathbf{x}$$
$$= -\frac{1}{2}\int_{\Omega^{n+1}} \rho|\boldsymbol{u}^{n+1}|^2 \mathrm{div}\,\left((\boldsymbol{u}^n - \boldsymbol{w}^n)\circ \mathcal{A}_{n,n+1}^{-1}\right)\, d\mathbf{x}$$
$$- \frac{\delta\rho}{2}\int_{\Sigma^{n+1}} \left((\boldsymbol{u}^n - \boldsymbol{w}^n)\circ \mathcal{A}_{n,n+1}^{-1}\right)\cdot \boldsymbol{n}_1 |\boldsymbol{u}^{n+1}|^2\, d\sigma.$$

Thus,

$$\tilde{c}_{\boldsymbol{w}^n}^{n+1}(\boldsymbol{u}^n, \boldsymbol{u}^{n+1}, \boldsymbol{u}^{n+1}) = \frac{1}{2}\int_{\Omega^{n+1}} \rho|\boldsymbol{u}^{n+1}|^2 \mathrm{div}\,(\boldsymbol{w}^n \circ \mathcal{A}_{n,n+1}^{-1})\, d\mathbf{x}. \tag{5.53}$$

The sum of this term and the last term of (5.52) exactly compensates for the quantity $-d^{n+1}(\boldsymbol{w}^n, \rho\boldsymbol{u}^{n+1}, \boldsymbol{u}^{n+1})$. For the magnetic field, we also have $-c^{n+1}(\boldsymbol{w}^n, \boldsymbol{B}^{n+1}, \boldsymbol{B}^{n+1}) = \frac{1}{2}\int_{\Omega^{n+1}} |\boldsymbol{B}^{n+1}|^2 \mathrm{div}\,(\boldsymbol{w}^n \circ \mathcal{A}_{n,n+1}^{-1})\, d\mathbf{x}$. And similarly, the sum of this term and the last term of (5.52) (where $\boldsymbol{u}^{n+1}$ is replaced by $\boldsymbol{B}^{n+1}$) exactly compensates for the quantity $-d^{n+1}(\boldsymbol{w}^n, \boldsymbol{B}^{n+1}, \boldsymbol{B}^{n+1})$.

Finally, the *third step* of the proof consists in noticing that the term coming from the Lorentz force $\frac{1}{\mu}l^{n+1}(\boldsymbol{u}^{n+1}, \boldsymbol{B}^n, \boldsymbol{B}^{n+1})$ exactly balances the coupling term in the magnetic equation. This fact comes from the way we linearize the equation in the Euler scheme (5.31).

After some standard integrations by parts, we obtain (5.51). □

Remark 5.1.15 [On the energy conservation up to order $O(\delta t)$] The left-hand side of (5.51) is actually equal to:

$$-\alpha\int_{\Omega^{n+1}} \left|\mathrm{div}\,\boldsymbol{B}^{n+1}\right|^2\, d\mathbf{x}$$
$$-\frac{1}{2\delta t}\int_{\Omega^n} \rho|\boldsymbol{u}^{n+1}\circ \mathcal{A}_{n,n+1} - \boldsymbol{u}^n|^2\, d\mathbf{y} - \frac{1}{2\delta t}\int_{\Omega^n} \frac{1}{\mu}|\boldsymbol{B}^{n+1}\circ \mathcal{A}_{n,n+1} - \boldsymbol{B}^n|^2\, d\mathbf{y}.$$

In practice, the first term turns out to be very small compared to the other two terms. Therefore, in practice, the right-hand side is not only non-positive, but also of order $O(\delta t)$. ◊

Remark 5.1.16 [Physical interpretation of the energy inequality] From a physical point of view, E^n is the sum of the kinetic energy and the magnetic energy, and the cancellation which occurs between the Lorentz forces in the momentum equation and the induced currents in the Maxwell equation expresses a possible conversion of magnetic energy to kinetic energy or of kinetic energy to magnetic energy. The two terms $\int_{\Omega^{n+1}} 2\eta|\boldsymbol{D}(\boldsymbol{u}^{n+1})|^2\, d\mathbf{x} + \int_{\Omega^{n+1}} \frac{1}{\mu\sigma}|\mathbf{curl}\,\boldsymbol{B}^{n+1}|^2 d\mathbf{x}$ in the left-hand side of (5.51) are successively the viscous dissipation and the magnetic dissipation. Thus, the numerical scheme obeys some energy conservation and dissipation properties which are very natural from a physical point of view. Proposition 5.5 shows that the scheme does not bring some spurious energy into the system[43] and is the counterpart in a moving domain of the energy inequality derived in Section 3.6.3 for a fixed domain. Notice that such a property can be obtained by solving a *linear* problem at each timestep. ◇

Remark 5.1.17 [Mass conservation and energy stability without GCL] We would like to discuss in this remark what remains of the mass conservation and energy stability results if the GCL is not satisfied. One can check that for a general domain velocity $\boldsymbol{w}^n$, the GCL properties (5.35)–(5.36) become:

$$\int_{\Omega_i^{n+1}} \phi(\mathbf{x})\, d\mathbf{x} - \int_{\Omega_i^n} \phi \circ \mathcal{A}_{n,n+1}(\mathbf{y})\, d\mathbf{y}$$

$$= \delta t \int_{\Omega_i^n} \phi \circ \mathcal{A}_{n,n+1}(\mathbf{y}) \mathrm{div}_{\mathbf{y}} \boldsymbol{w}^n(\mathbf{y})\, d\mathbf{y} + O\left((\delta t)^2\right) \int_{\Omega_i^n} \phi \circ \mathcal{A}_{n,n+1}(\mathbf{y})\, d\mathbf{y},$$

$$= \delta t \int_{\Omega_i^{n+1}} \phi(\mathbf{x}) \mathrm{div}_{\mathbf{x}} \left(\boldsymbol{w}^n \circ \mathcal{A}_{n,n+1}^{-1}(\mathbf{x})\right) d\mathbf{x} + O\left((\delta t)^2\right) \int_{\Omega_i^{n+1}} \phi(\mathbf{x})\, d\mathbf{x},$$

where $O\left((\delta t)^2\right)$ denotes a function bounded from above by $C\,(\delta t)^2 \|\partial \boldsymbol{w}^n / \partial \mathbf{y}\|^3_{L^\infty(\Omega)}$, where C is a universal constant. Therefore, one can prove that, in the framework of Proposition 5.4,

$$\left|\overline{\rho}_i |\Omega_i^n| - \overline{\rho}_i |\Omega_i^{n+1}|\right| \leq C'(\delta t)^2 \left\| \frac{\partial \boldsymbol{w}^n}{\partial \mathbf{y}} \right\|^3_{L^\infty(\Omega)}, \quad \text{for } i = 1, 2,$$

and, in the framework of Proposition 5.5,

$$\frac{E^{n+1} - E^n}{\delta t} + \int_{\Omega^{n+1}} 2\eta |\boldsymbol{D}(\boldsymbol{u}^{n+1})|^2\, d\mathbf{x} + \int_{\Omega^{n+1}} \frac{1}{\overline{\mu}\sigma} |\mathbf{curl}\, \boldsymbol{B}^{n+1}|^2\, d\mathbf{x}$$

$$\leq C''(\delta t)^2 \left\| \frac{\partial \boldsymbol{w}^n}{\partial \mathbf{y}} \right\|^3_{L^\infty(\Omega)} \frac{E^{n+1}}{\delta t}.$$

Therefore, if δt is sufficiently small with respect to the inverse of the normal velocity at the interface, it can be proven that, on a finite time interval, the mass is conserved up to an error proportional to δt, and that the scheme is stable (using a discrete Gronwall lemma). ◇

Remark 5.1.18 [About the precision of the numerical scheme] We have only discussed so far the stability of the timestepping scheme, and not its consistency. Actually, it is straightforward to check that it is a first-order scheme in time, and this is confirmed by the numerical experiments.

In case the displacement of the domain is known *a priori*, D. Boffi and L. Gastaldi propose in [23] a second-order time-advancing scheme which satisfies the GCL, for sufficiently small timesteps (see also L. Formaggia et al. [81]

[43] This is particularly important when studying the stability of the solutions, see Chapter 6.

and Ph. Geuzaine et al. [103] on this subject). However, in our coupled framework where the displacement of the domain depends on the solution itself, we are not aware of a second-order time-advancing scheme which satisfies discrete mass conservation (5.45) and energy stability (5.51).

◇

5.1.4 *Surface tension effects*

Surface tension is a force appearing at the interface between two liquids or a liquid and a gas that tends to reduce the area of the interface. In our context, we would like to explain how to take into account surface tension at the interface Σ_t between the two liquids. Even in problems where surface tension effects do not play a major role (*e.g.* for the modeling of aluminum electrolysis cells), it might become non negligible for the study of instabilities.

5.1.4.1 Modeling of surface tension effects The classical Laplace formulation of the surface tension correlates the normal force with the main curvature of the surface (see L. Landau and E. Lifchitz [139] for example). The first equation in (1.55) giving the continuity of the normal stress on Σ_t is replaced by: $\forall \mathbf{x} \in \Sigma_t$,

$$(\boldsymbol{\tau}|_{\Omega_1}(\mathbf{x})\boldsymbol{n}_1(\mathbf{x}) + \boldsymbol{\tau}|_{\Omega_2}(\mathbf{x})\boldsymbol{n}_2(\mathbf{x})) = \gamma H(\mathbf{x})\boldsymbol{n}_1(\mathbf{x}), \tag{5.54}$$

where we recall that $\boldsymbol{\tau} = -pId + 2\eta \boldsymbol{D}(\boldsymbol{u})$ is the stress tensor. The parameter γ is the surface tension coefficient (with dimension N/m), which is henceforth assumed to be constant, and H is the *mean curvature* of Σ_t positively counted with respect to the normal $\boldsymbol{n}_1$. For example, the surface tension coefficient is equal to $0.07\,N/m$ for the water/air interface, and to $0.5\,N/m$ for the aluminum/cryolite interface.

Let us give a definition of the mean curvature in terms of radius of curvature. For a "1D surface" (*i.e.* a curve) $H = 1/R$ is simply the *curvature* of the curve, with R the radius of curvature positively counted along the normal. For a 2D surface, $H = \frac{1}{R_1} + \frac{1}{R_2}$, with R_1 and R_2 the principal radii of curvature positively counted along the normal. Depending on the textbooks, other definitions can be found, such as $\frac{1}{2}\left(\frac{1}{R_1} + \frac{1}{R_2}\right)$ or $-\left(\frac{1}{R_1} + \frac{1}{R_2}\right)$, which modify the definition of the surface tension term (5.54) accordingly.

Let us make precise the surface tension term in the equations we have considered so far. In the formulation (1.49), the following quantity must be added to the right-hand side of the equations on the velocity $\boldsymbol{u}$:

$$\gamma H \boldsymbol{n}_1 \delta_{\Sigma_t},$$

where δ_{Σ_t} is defined by (5.11). The non-dimensional form of this term is $\dfrac{1}{We}$ $H\boldsymbol{n}_1\delta_{\Sigma_t}$, where H is the mean curvature of the non-dimensionalized surface Σ_t and the non-dimensional number

$$We = \rho_1 U^2 L/\gamma \tag{5.55}$$

is the Weber number. The weak formulation (5.25) is modified correspondingly, adding the term

$$\int_{\Sigma_t} \gamma H \boldsymbol{v} \cdot \boldsymbol{n}_1 d\sigma_{\Sigma_t} \tag{5.56}$$

in the right-hand side. One recognizes here the term (4.82) we considered in the mathematical analysis of the system with surface tension (since $\operatorname{div}(\boldsymbol{n}) = -H$, see equation (5.57) below).

Remark 5.1.19 [Definition of H in terms of $\boldsymbol{n}_1$] In this remark, we would like to give another expression for the mean curvature, which explains the link between the two expressions (5.56) and (4.82) of the surface tension term. The*Gauss formula* states that the mean curvature H is the opposite of the trace of the derivative of the *Gauss application* which maps each point of the surface to the normal of the surface at this point: $\forall \mathbf{x} \in \Sigma_t$

$$H(\mathbf{x}) = -\operatorname{tr}(\boldsymbol{\nabla}_s \boldsymbol{n}_1)(\mathbf{x}) \tag{5.57}$$

where $\boldsymbol{\nabla}_s$ is the gradient along the surface (with value in the tangent space to Σ_t). If we extend the definition of the normal $\boldsymbol{n}_1$ on a neighborhood of Σ_t, for example by

$$\boldsymbol{n}_1(\mathbf{x}) = \frac{\boldsymbol{\nabla} d(\mathbf{x}, \Sigma_t)}{|\boldsymbol{\nabla} d(\mathbf{x}, \Sigma_t)|},$$

where $d(\mathbf{x}, \Sigma_t)$ denotes the signed distance of $\mathbf{x}$ to Σ_t, then we have $\boldsymbol{\nabla}_s \boldsymbol{n}_1(\mathbf{x}) = P_s(\mathbf{x}) \boldsymbol{\nabla} \boldsymbol{n}_1(\mathbf{x})$, where $P_s(\mathbf{x}) = (\mathrm{Id} - \boldsymbol{n}_1(\mathbf{x}) \otimes \boldsymbol{n}_1(\mathbf{x}))$ is the orthogonal projection operator onto the tangent space to Σ_t at point $\mathbf{x}$. Notice that for any smooth vector field $\boldsymbol{X}$ defined on Σ_t,

$$\boldsymbol{\nabla}_s \boldsymbol{X} = P_s(\mathbf{x}) \boldsymbol{\nabla} \boldsymbol{X} \tag{5.58}$$

does not depend on the way we expand the definition of $\boldsymbol{X}$ on a neighborhood of Σ_t. The operator $\operatorname{tr}(\boldsymbol{\nabla}_s .)$ is also called *the surface divergence.* ◇

5.1.4.2 The surface divergence theorem The main difficulty to compute the surface tension term (5.56) is that it requires, at least in principle, to evaluate the mean curvature H of the surface, which is not easy, especially for a discretized surface. This is a challenging problem, especially in 3D, since it requires to compute normal and tangent vectors of a discrete surface (typically a Lipschitz 2D manifold)[44].

Instead of computing directly the quantity H in (5.56), it is more convenient to use an integration by parts formula over Σ_t to find another expression of (5.56), which is better suited for computations. This requires the *surface divergence theorem*, which is a generalization of the divergence theorem to surfaces

[44]We already discussed a similar problem in Section 5.1.3.2.

with non-zero curvature. For any smooth hypersurface Σ in $\mathbb{R}^d$ (*i.e.* a submanifold of $\mathbb{R}^d$ with codimension 1) with a smooth boundary $\partial\Sigma$ and normal $\boldsymbol{n}_1(\mathbf{x})$ at point $\mathbf{x}$, one has: for any smooth function $\boldsymbol{\Phi} : \Sigma \to \mathbb{R}^d$,

$$-\int_{\Sigma} H\boldsymbol{\Phi} \cdot \boldsymbol{n}_1 \, d\sigma_{\Sigma} = \int_{\Sigma} \operatorname{tr}(\boldsymbol{\nabla}_s \boldsymbol{\Phi}) \, d\sigma_{\Sigma} - \int_{\partial\Sigma} \boldsymbol{\Phi} \cdot \boldsymbol{m} \, dl_{\partial\Sigma}, \tag{5.59}$$

where the surface gradient $\boldsymbol{\nabla}_s$ is defined by (5.58). The vector $\boldsymbol{m}$ is the normal vector to $\partial\Sigma$ (in the tangent space of Σ) pointing outwards from Σ. The measure l_{∂_Σ} is the Lebesgue measure on $\partial\Sigma$. We refer to C.E. Weatherburn [238, equation (24) p. 239] or to L. Ambrosio and H.M. Soner [3, equation (3.8)].

Using formula (5.59), the weak formulation of the surface tension term (5.56) is then

$$\int_{\Sigma_t} \gamma H \boldsymbol{v} \cdot \boldsymbol{n}_1 d\sigma_{\Sigma_t} = -\gamma \int_{\Sigma_t} \operatorname{tr}(\boldsymbol{\nabla}_s \boldsymbol{v}) \, d\sigma_{\Sigma_t} + \gamma \int_{\partial\Sigma_t} \boldsymbol{v} \cdot \boldsymbol{m} \, dl_{\partial\Sigma_t}. \tag{5.60}$$

Physically, the vector $\boldsymbol{m}$ defines the *contact angle* between the interface and the wall, which is a parameter of the problem.

Remark 5.1.20 [On the contact angle] As we already discussed in Remark 5.1.6 concerning the dynamics of the contact line, the determination of the contact angle is a question of modeling. Actually, it turns out to be a difficult problem in the case of dynamic contact lines, and it is still a matter of current research. In our specific framework, this angle does not influence very much the whole dynamics, since it only perturbs the geometry of the surface in a small neighborhood of the side walls. This is of course not the case for flows at smaller scales, which we do not discuss here. We refer, for example, to [12] where T.A. Baer et al. study the motion of a droplet of fluid sliding on an inclined plane at constant velocity. We also refer to W. Dettmer et al. [67] for some applications of a similar method to flows driven by surface tension, like the stretching of a liquid bridge. Finally, we would like to mention the very interesting review [191] by T.Z. Qian et al. where an appropriate boundary condition at the moving contact line is proposed. This so-called generalized Navier boundary condition is particularly well-suited to an implementation in the ALE formulation of surface tension presented in this section (see [97]). ◇

5.1.4.3 Computation of (5.60) We now turn to implementation details. As explained in Remark 5.1.20, we focus on the first term in the right-hand side of (5.60), since for the applications we have in mind, the last term can be neglected.

The problem is to compute the integral of the surface divergence $\operatorname{tr}(\boldsymbol{\nabla}_s \boldsymbol{v})$ of $\boldsymbol{v}$ over Σ_t, when Σ_t is a discretized curve or surface and $\boldsymbol{v}$ is a test function belonging to the velocity finite element space. By the finite element methods, the curve Σ_t, in case $d = 2$ (or the surface Σ_t, in case $d = 3$), is discretized into a collection of surface elements (also called boundary elements). We decompose the integral over Σ_t as the sum of the integrals on each surface element. Therefore, it suffices to compute

$$\int_K \mathrm{tr}(\boldsymbol{\nabla}_s \boldsymbol{v})\, d\sigma_K \tag{5.61}$$

where K denotes a surface element.

Let us start with the case where Σ_t is one dimensional ($d = 2$). In this case K is a segment which is naturally parameterized by the application $\boldsymbol{f} : [0,1] \rightarrow K$ which maps the reference element $[0,1]$ onto the "surface" element K. This application is generally available in a finite element code, since the computations of the integrals are usually done on the reference element, after a change of variable. It is then easy to evaluate $\mathrm{tr}(\boldsymbol{\nabla}_s \boldsymbol{v})$ using the formula

$$\mathrm{tr}(\boldsymbol{\nabla}_s \boldsymbol{v}) = \frac{d\boldsymbol{v}(\boldsymbol{e_u}) \cdot \boldsymbol{e_u}}{|\boldsymbol{e_u}|^2}$$

where $d\boldsymbol{v}(\boldsymbol{e_u})$ denotes the derivative of $\boldsymbol{v}$ along $\boldsymbol{e_u}$, and $\boldsymbol{e_u}$ is the vector in the tangent space defined by $\boldsymbol{e_u} = \dfrac{d\boldsymbol{f}}{du}$. More explicitly, $d\boldsymbol{v}(\boldsymbol{e_u}) = \dfrac{d}{du}(\boldsymbol{v} \circ \boldsymbol{f})$. This derivative is computed at each Gauss point of the finite element to then obtain (5.61).

This approach can be generalized in the case where Σ_t is two dimensional ($d=3$). Let us introduce again the parameterization $\boldsymbol{f}$ given by the finite element method. For example, for isoparametric Lagrangian hexahedral finite elements, the parametrization is

$$\boldsymbol{f} : \begin{cases} [0,1]^2 \rightarrow & K \\ (u,v) \mapsto & \displaystyle\sum_{P \in \mathcal{P}(K)} \phi_P(u,v) \boldsymbol{X_P} \end{cases}$$

where $\phi_P : [0,1]^2 \rightarrow \mathbb{R}$ is the shape function related to the node P, $\boldsymbol{X_P}$ is the coordinate vector of P and $\mathcal{P}(K)$ is the set of nodes of the element K. Notice that for hexahedral finite elements, K is not in general included in a plane, while it is the case for tetrahedral finite elements. Since we compute the integral over K using a Gauss integration rule, we need to evaluate $\mathrm{tr}(\boldsymbol{\nabla}_s \boldsymbol{v})$ at each Gauss point of K. We consider in the following that the Gauss point $\mathbf{x}$ is fixed. We introduce the basis $(\boldsymbol{e_u}, \boldsymbol{e_v})$ of the tangential space to K at $\mathbf{x}$, with $\boldsymbol{e_u} = \partial \boldsymbol{f}/\partial u$ and $\boldsymbol{e_v} = \partial \boldsymbol{f}/\partial v$ (see Figure 5.2). The metric tensor at point $\mathbf{x}$ is $\begin{bmatrix} E & F \\ F & G \end{bmatrix}$ with $E = |\boldsymbol{e_u}|^2$, $F = (\boldsymbol{e_u} \cdot \boldsymbol{e_v})$ and $G = |\boldsymbol{e_v}|^2$. The element of area is $d\sigma = \sqrt{EG - F^2}\, du\, dv$. We want to evaluate the trace of the surface gradient $\boldsymbol{\nabla}_s \boldsymbol{v}$, at point $\mathbf{x}$. We define $e = (d\boldsymbol{v}(\boldsymbol{e_u}) \cdot \boldsymbol{e_u})$, $f_1 = (d\boldsymbol{v}(\boldsymbol{e_u}) \cdot \boldsymbol{e_v})$, $f_2 = (d\boldsymbol{v}(\boldsymbol{e_v}) \cdot \boldsymbol{e_u})$ and $g = (d\boldsymbol{v}(\boldsymbol{e_v}) \cdot \boldsymbol{e_v})$. By definition, we have (at point $\mathbf{x}$) $d\boldsymbol{v}(\boldsymbol{e_u}) = \dfrac{\partial}{\partial u}(\boldsymbol{v} \circ \boldsymbol{f})$ and $d\boldsymbol{v}(\boldsymbol{e_v}) = \dfrac{\partial}{\partial v}(\boldsymbol{v} \circ \boldsymbol{f})$. Notice that in the finite element

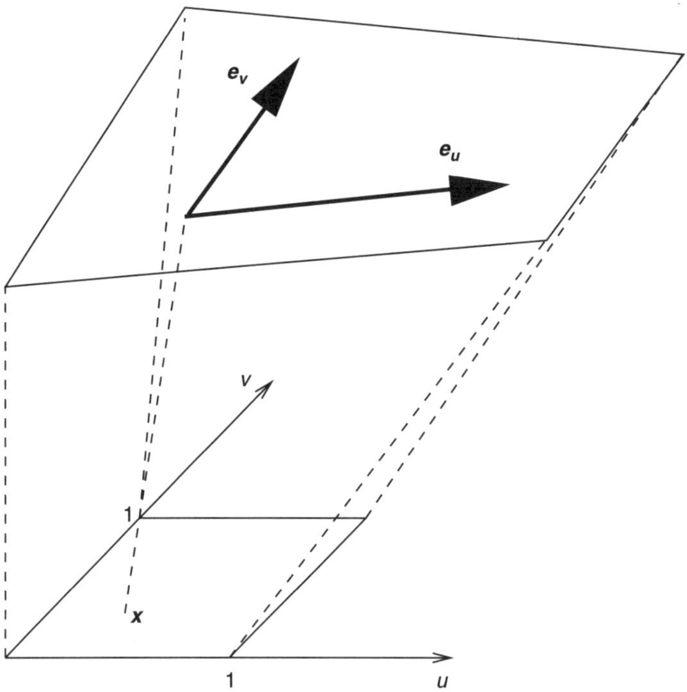

Fig. 5.2 One face of a hexahedral finite element parameterized by $[0,1]^2$, and the tangent space $(\boldsymbol{e_u}, \boldsymbol{e_v})$ at a Gauss point of the element.

code, these derivatives of the test function $\boldsymbol{v} \circ \boldsymbol{f}$ defined on the reference element are available since they are also needed for the computation of many operators. We then have the following equality:

$$\begin{bmatrix} e & f_1 \\ f_2 & g \end{bmatrix} = \begin{bmatrix} a_{1,1} & a_{1,2} \\ a_{2,1} & a_{2,2} \end{bmatrix} \begin{bmatrix} E & F \\ F & G \end{bmatrix}$$

where $(a_{i,j})_{1 \le i,j \le 2}$ is the matrix expressing $\boldsymbol{\nabla}_s \boldsymbol{v}$ in the basis $(\boldsymbol{e_u}, \boldsymbol{e_v})$. A simple computation then gives the following expression for $\mathrm{tr}(\boldsymbol{\nabla}_s \boldsymbol{v}) = (a_{1,1} + a_{2,2})$ at point $\mathbf{x}$ (we recall that the trace of a linear operator does not depend on the basis in which it is represented as a matrix):

$$\mathrm{tr}(\boldsymbol{\nabla}_s \boldsymbol{v}) = \frac{eG + Eg - F(f_1 + f_2)}{EG - F^2}.$$

The weighted sum of this quantity at each Gauss point of the element then yields (5.61).

The introduction of the surface tension term in the finite element ALE framework is thus rather straightforward.

5.2 Other approaches

For comparison, we give here a short overview of some numerical methods that have been proposed over the past 30 years for simulating the motion of surfaces, and which are different from the ALE approach.

In the following, we discuss various methods in the particular context of an interface which moves without breakup or coalescence of fluids. The topology of the volume occupied by each fluid is conserved along the simulation. Our discussion is specific to this context. The reader should bear in mind that the method of choice depends on the physical problem.

Let us further specify the problem. The problem we have in mind is the modeling of a moving interface between two incompressible non-miscible fluids. The natural boundary condition at the interface is the continuity of the normal stress (in the case of zero surface tension, see equation (1.55)), while the motion of the interface is completely defined by the motion of the fluids (see equation (5.13) above). In this context any numerical scheme needs at some stage to solve the following problem: for a given velocity field $\boldsymbol{u}(t, \mathbf{x})$ and a given position of the interface Σ_t at time t, compute the position of the interface $\Sigma_{t+\delta t}$ at time $t + \delta t$. This is the question we want to discuss, in dimension 2 or 3.

We distinguish between *fixed-mesh methods* (in Section 5.2.1) and *moving-mesh methods* (in Section 5.2.2). There are many other ways proposed in the literature to classify the various methods. For instance *interface tracking* methods (in which the exact position of the interface is known) *versus interface capturing* methods (in which the position of the interface is known through the values of an additional parameter), or *Lagrangian versus Eulerian* methods. The classification is always schematic in nature. Actually, many methods combine various strategies: both a fixed and a moving mesh, both interface tracking or capturing, both a Lagrangian and a Eulerian viewpoint are used. To cite only one example, the latter is the case for the ALE method. Our classification based on moving and fixed-mesh methods is only used for clarity of exposition. We will not go deep into the details of the implementation. Our aim is to present the main ideas underlying the methods for comparison purposes.

Likewise, we do not claim to be exhaustive on such a broad and very lively field. Applied mathematicians, mechanical engineers, and fluid mechanicists continuously develop new numerical methods to treat this problem, and no approach seems to supplant all others. Actually, each method has advantages and drawbacks, which may make it more or less appealing, depending on the targeted application. For example, we will not discuss the following methods to simulate the motion of an interface:

- Smooth particle hydrodynamics (see R.A. Gingold and J.J. Monaghan [105, 170]). This is a mesh-free method. The whole fluid is modeled by a set of particles and the field variables are computed by averaging over the region of interest. One of the difficulties is to correctly model the interactions between the particles.

- Boltzmann lattice-gas method (see D.H. Rothman *et al.* [199]). This is a probabilistic method. It consists in introducing a lattice with some sites where particles stochastically evolve, jumping from one site to another. The fluid is thus represented by microscopic interacting particles.

Before going into the heart of the matter, we would like to mention a few examples of the methods which have been used in our particular context of MHD flows with free interfaces: J.R. Mooney et al. [171] use an ALE method for a 3D finite element computation of free-surface flow. J.U. Brackbill et al. [29] propose an ALE method with a 2D finite difference method. R. Samulyak et al. [201] use a VOF method in two dimensions to simulate instability in liquid jets in strong magnetic fields. D. Munger [179] uses a level set method within a 3D finite volume discretization for the simulation of aluminum electrolysis cells.

5.2.1 *Fixed-mesh methods*

Our first class of method keeps the mesh fixed, so that the position of the interface is usually known from the values of some additional field on the mesh.

5.2.1.1 Lagrangian methods Let us first consider fixed-mesh methods within a Lagrangian viewpoint. These methods are usually *marker methods*. The first method which ever appeared in the literature for simulating time-dependent free-surface flows is of this type: it is the marker-and-cell (MAC) method (see F.H. Harlow and J.E. Welch [119]). The idea is to throw some massless particles in one fluid (say fluid 1), and to let them evolve along the flow. The basic equation to solve is:

$$\frac{d\boldsymbol{X}(t)}{dt} = \boldsymbol{u}(t, \boldsymbol{X}(t)). \tag{5.62}$$

The velocity $\boldsymbol{u}$ within the cell is computed by interpolation. Loosely speaking, the cells that contain some particles are supposed to be filled with fluid 1, while the others are supposed to be filled with fluid 2 (which is a vacuum in the case of a free-surface). A special treatment is made for cells called surface cells, that contain some particles, but have some neighboring cells that are empty.

The main difficulty with this method is the repartition of the markers: how to ensure that the distribution of markers remains sufficiently homogeneous so as to accurately follow the interface. This is especially true if there are some regions with recirculations or stagnation points. This usually leads to the introduction of an enormous number of particles, which require large memory and CPU time.

In order to improve the method, one can notice that only the particles on the interface are actually needed to define the interface. This idea has been used to develop the so-called *surface marker method* (see S. Chen et al. [38], S.O. Unverdi et al. [235], C.S. Peskin [183] J. Glimm et al. [108]). In this method, the markers are kept only on the surface. The markers are linked by segments in 2D, or triangles in 3D, which define explicitly the position of the interface. This method is therefore an interface tracking method. The breakup or coalescence around the surface is more complicated to treat than in the original MAC method. The method is, however, less CPU and memory demanding. The implementation of the method requires some know-how for maintaining a fine repartition of the particles at the surface, and to handle the possible topological changes, especially in 3D.

5.2.1.2 Eulerian methods The idea of Eulerian methods is to replace the sum of Dirac functions at each particle position of the Lagrangian methods above by a smoother marker function χ. This implicitly defines the position of the interface. For example, χ is a function positive in one fluid and negative in the other, or is constant equal to one in one fluid, and constant equal to zero in the other. The difficulty related to the repartition of discrete particles is thus solved by the continuous nature of χ.

The basic equation underlying these methods is:

$$\frac{\partial \chi(t, \mathbf{x})}{\partial t} + \boldsymbol{u}(t, \mathbf{x}) \cdot \boldsymbol{\nabla} \chi(t, \mathbf{x}) = 0, \tag{5.63}$$

which may also be written in the conservative form:

$$\frac{\partial \chi(t, \mathbf{x})}{\partial t} + \operatorname{div}\left(\boldsymbol{u}(t, \mathbf{x}) \chi(t, \mathbf{x})\right) = 0,$$

using the fact that $\operatorname{div} \boldsymbol{u} = 0$.

This equation is of hyperbolic type, so that the following difficulties are expected:

- some wild oscillations may appear near discontinuities of χ (this is a stability issue);
- the numerical scheme may bring spurious diffusion, which implies in our framework an imprecise determination of the position of the interface.

There are many numerical schemes which have been specifically developed to address these difficulties, especially finite difference schemes. This is why most of the methods presented in this section use a regular structured mesh.

The first method we would like to introduce is the *volume of fluid* (VOF) method (see C.W. Hirt and B.D. Nichols [124], B. Lafaurie et al. [138] or D. Gueyffier et al. [115]). The function χ here represents the volume fraction of fluid 1 (for example) in each cell. Therefore, the function χ is meant to be the characteristic function of the volume occupied by fluid 1. In practice, ensuring the conservation of the volume of each fluid with this method is easy, since the advection equation on χ can be interpreted as a conservation law. However, the numerical approximation of (5.63) usually brings in some numerical diffusion and dispersion, so that χ does not remain a step-function. What characterizes the VOF method is a reconstruction step, which transforms back χ into a step function. A precise definition of the position of the interface is also needed to define the properties (density, viscosity) of the fluid at different positions in the mesh and to compute surface tension terms, for example. Various methods of reconstruction have been proposed in the literature, and this step is certainly the most complicated in the algorithm, especially in 3D. The drawback of this method is the technicalities involved in this reconstruction step. The VOF method is implemented in many commercial software packages.

A second method in the same vein is the *level set method* (see M. Sussman et al. [223] or J.A. Sethian et al. [209]). The function χ is now meant to be the signed distance function to the interface, which is implicitly defined as the zero set of χ. The advantage of this choice is that χ is smoother (it has no discontinuities), which helps the numerical resolution of (5.63) with high accuracy. Moreover, it seems intuitively clear that the signed distance function to Σ gives more information on Σ than the step function used in VOF approaches. Again, the difficulty lies in the fact that the solution to (5.63) does not remain the signed distance function to Σ (*i.e.* $|\nabla\chi| \neq 1$): the solution becomes very steep or flat in some regions, which makes difficult the precise determination of the position of the interface. To solve this difficulty, a first idea is to modify the velocity field $\boldsymbol{u}$ in (5.63) away from Σ, so that the solution remains closer to the signed distance function. Another idea in the spirit of the reconstruction step of the VOF method is to use a reinitialization step, in order to recover the signed distance function. This can be performed for example using the following fictitious dynamics:

$$\frac{\partial\chi}{\partial s} = \operatorname{sgn}(\chi)(1 - |\nabla\chi|) \tag{5.64}$$

whose solution in the long-time limit is the signed distance function to the zero set of the initial condition $\chi(s = 0)$. Of course, the practical implementation of this reinitialization step requires some know-how, like the reconstruction step in the VOF method. The interest of this method is that no explicit reconstruction of the interface is needed. For example, the normal to the surface or the curvature can be computed by differentiation of the function χ. It is also possible to take into account surface tension without explicitly reconstructing the interface, using both an artificial thickening of the interface (using the signed distance function), and a computation of the curvature in the vicinity of the interface (see the continuum surface force model in J.U. Brackbill et al. [28]). Therefore, this method is an interface capturing method. One difficulty of the level set method is to ensure the conservation of the volume of each fluid, since (5.63) is not a conservation law. The level-set method is used in many applications in very different fields: imaging, vision, graphics, and computational mechanics (see, *e.g.* [44]).

To summarize, the main advantage of fixed-mesh methods is their ability to capture topological changes. Their common difficulty is the maintenance of a sharp boundary between the fluids.

5.2.2 Moving-mesh methods

In moving-mesh methods, the idea is to explicitly follow the motion of the interface with some elements of the mesh. The idea is conceptually very simple: a Lagrangian grid, embedded within the fluid, moves with the fluid. However, the implementation of this idea directly may lead to large distortions of the mesh. Actually it is not necessary that the mesh everywhere follows exactly the trajectory of the fluid particles. Only the normal velocity of the mesh at the interface needs to coincide with the normal velocity of the fluids at the interface (see (5.13)). This is the bottom line of the Arbitrary Lagrangian Eulerian method described in much detail in Section 5.1 above. The ALE method is a prototypical example of *interface tracking methods*.

In contrast to fixed-mesh methods, moving-mesh methods cannot handle easily some topological changes of the fluids. Moreover, large displacements of the interface may lead to important distortions of the mesh, so that a remeshing may be necessary. This remeshing can be seen as the counterpart of the reconstruction steps presented above for VOF or level set methods.

The approach is however appealing for many applications. It is usually considered that moving-mesh methods are potentially more accurate than fixed-mesh methods. In addition, the fact that the mesh moves only using the normal velocities at the interface may lead to more stable schemes than fixed-mesh methods if the tangential velocities are large (which may require some small timesteps for solving (5.62) or (5.63) due to some CFL stability conditions). One interest of this approach is also theoretical: we have seen in Section 5.1 that the ALE method can be mathematically analyzed. We can prove that the numerical scheme conserves the mass of each fluid, and does not bring spurious energy into the system (*i.e.* that the scheme is stable in the energy norm). We are not aware of such analysis of the complete algorithm for other methods.

5.3 Example of test cases and simulations

In this section, we illustrate the ALE formulation presented above by numerical experiments. The first two cases are purely hydrodynamic, the third one deals with the MHD equations. We also refer to Sections 6.3.2 and 6.3.3 for other MHD test cases related to aluminum electrolysis.

5.3.1 A benchmark problem with a free surface

In order to test the capability of the ALE solver to address complex free-surface flows, we consider the so-called *dam-break problem* and we compare the solution given by the Saint-Venant (nonlinear shallow water) system (see, for example, [102]) and that obtained by direct simulations of the ALE Navier–Stokes equations with free surface[45]. The water depth before the dam is $h_l = 2$ and $h_r = 1$ after the dam. Gravity is $g = 2$, density is $\rho = 1$, viscosity is $\mu = 0.01$, and the length of the computational domain is $L = 100$.

The solution of the Saint-Venant equation is analytically known: it consists in a rarefaction wave and a shock wave. The shock velocity and the intermediate water depth are the important parameters of this benchmark, the solution being fully defined by these two quantities.

Figure 5.3 shows the free-surface elevations given by the Saint-Venant system and by the Navier–Stokes equations for various times. Note that, although the

[45]Notice that in this section, there is only one fluid.

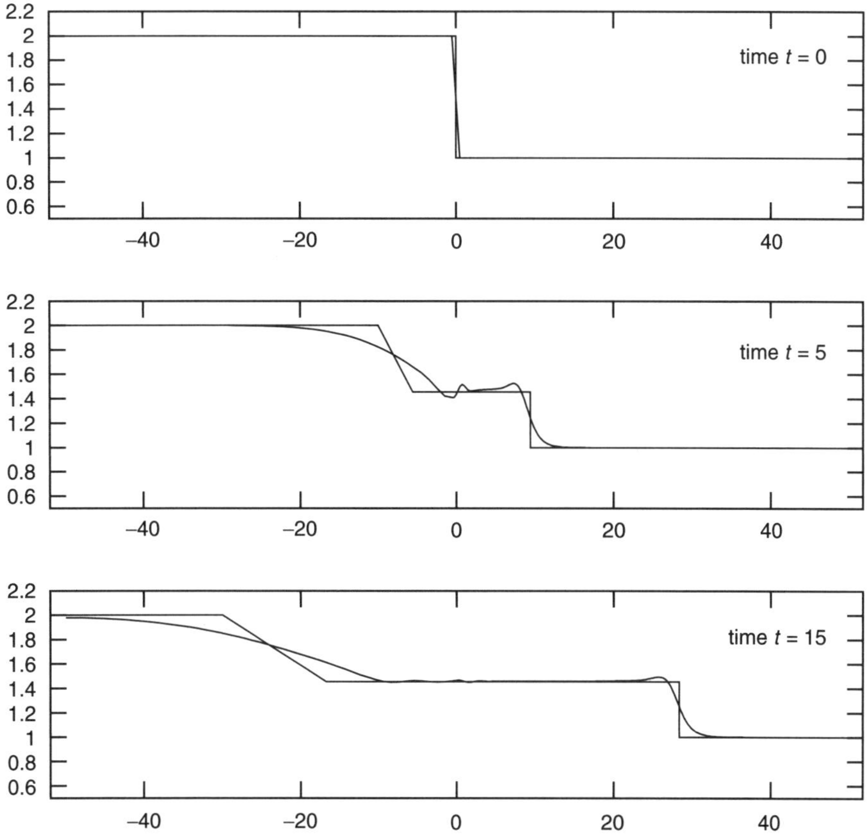

Fig. 5.3 Free surface in the dam-break problem. The shock velocity and the intermediate water depths given by the Saint-Venant equations and the direct simulation of the ALE–Navier–Stokes equations are in good agreement.

solutions do not match perfectly, the shock speed and the intermediate water depth are in very good agreement.

Shallow water equations will also be addressed in Section 6.2.2.2 in the context of two-fluid MHD flows.

5.3.2 On the discrete mass conservation

We have described in Section 5.1.3.2 the three ingredients that allow for mass conservation in two-fluid problems, which we recall here for clarity:

1. the discrete normal velocities are such that (5.42) holds;
2. (5.43) is satisfied;
3. the motion of the mesh is such that the geometric conservation law (5.35) is satisfied.

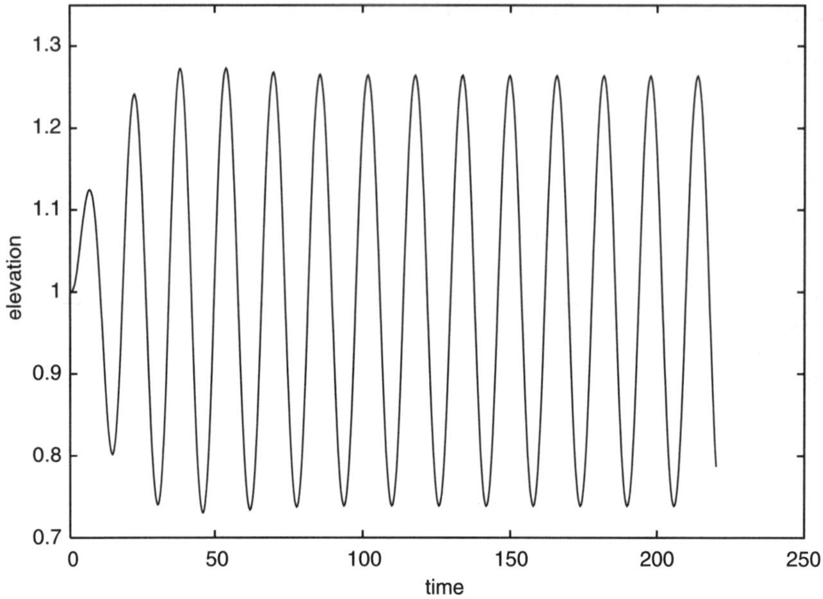

Fig. 5.4 The time history of the elevation of a point of the interface.

The aim of this section is to illustrate by numerical experiments that the second and third ingredients are indeed necessary in order to ensure mass conservation, at least for general cases.

For this purpose, we consider a two-dimensional case when the two fluids are only subjected to oscillating gravity, *i.e.* equations (1.52) with the magnetic field set to zero. The computational domain is $\Omega = [-2, 2] \times [0, 2]$, the lower (resp. upper) fluid occupies at $t = 0$ the subdomain $\Omega_{1,t=0} = [-2, 2] \times [0, 1]$ (resp. $\Omega_{2,t=0} = [-2, 2] \times [1, 2]$). The gravity vector $\boldsymbol{g}$ in (1.52) is set to:

$$\boldsymbol{g} = A \sin(2\pi\nu t)\boldsymbol{e_x} - \boldsymbol{e_y},$$

where A and ν are two positive constants, and $(\boldsymbol{e_x}, \boldsymbol{e_y})$ is the orthonormal basis of the computational domain. We choose the following non-dimensional parameters: $Re_1 = Re_2 = 100$, $M = 0.91$, $\nu = 0.0625$, $A = 0.05$. Figure 5.4 shows the evolution in time of the elevation of a point on the boundary of the interface.

Let us first show that the second ingredient is necessary to conserve the mass. In Figure 5.5, we show the evolution in time of the mass of fluid 1 with three finite element spaces: $\mathbb{Q}_2/\mathbb{Q}_1$ (with continuous pressure), $\mathbb{Q}_2/\mathbb{P}_1$ (with discontinuous pressure), and $\mathbb{Q}_2/\mathbb{Q}_1$ with the flux constraint (5.44). In each case, we use a vertical displacement of the nodes so that the GCL (5.35) is satisfied (see Lemma 5.3). Contrary to the other finite element spaces, the pair $\mathbb{Q}_2/\mathbb{Q}_1$ does not satisfy (5.43) and the mass is indeed not conserved.

We now illustrate the role of the third ingredient (the GCL) in mass conservation. We use the $\mathbb{Q}_1/\mathbb{Q}_1$ stabilized finite element with flux constraint (5.44)

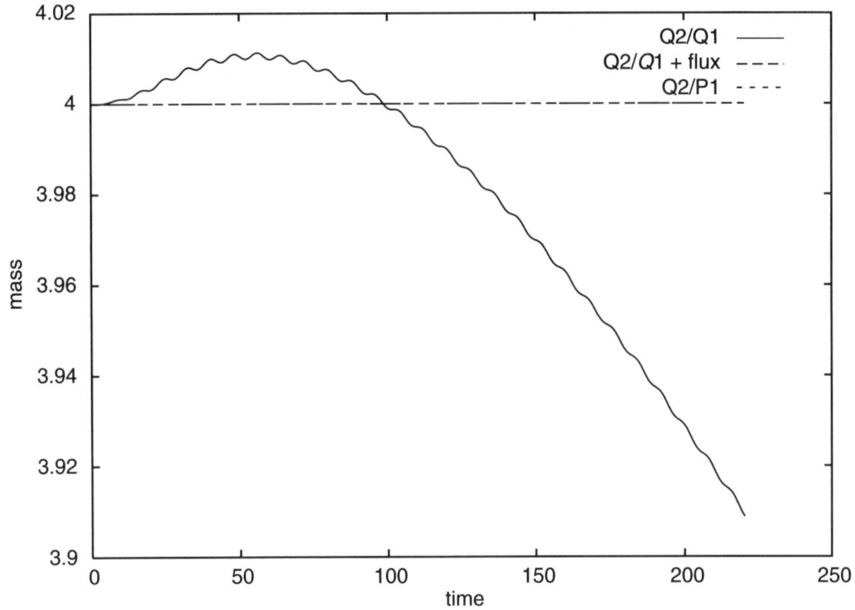

Fig. 5.5 Mass conservation is ensured with $\mathbb{Q}_2/\mathbb{P}_1$ or $\mathbb{Q}_2/\mathbb{Q}_1$ with a flux constraint (the two curves coincide). It is not with $\mathbb{Q}_2/\mathbb{Q}_1$ elements.

so that (5.42) is satisfied. If the displacement of the mesh is purely vertical, the GCL property (5.35) is satisfied, and the mass is preserved (see Lemma 5.3 and Figure 5.7). But, if we replace system (5.33) by

$$\begin{cases} -\Delta\, \boldsymbol{w} = 0, & \text{on } \Omega_i^n, i = 1, 2, \\ \boldsymbol{w} = \boldsymbol{u}, & \text{on } \Sigma^n, \\ \dfrac{\partial \boldsymbol{w}}{\partial \boldsymbol{n}} = 0, & \text{on } \partial\Omega, \end{cases} \tag{5.65}$$

the displacement of the mesh is now arbitrary, and relation (5.35) is no longer true (an additional term appears in the development of the Jacobian determinant). Figure 5.6 shows on the left-hand side the mesh obtained with (5.33) and on the right-hand side, that obtained with (5.65). Although a mesh displacement along the x axis is not natural for this test case, our purpose is illustrative. Figure 5.7 shows the corresponding evolution of the mass of fluid 1. The fact that relation (5.35) is necessary on a moving mesh to ensure mass conservation is striking.

5.3.3 An MHD experiment with a free surface and a free interface

We present here the numerical simulation of a laboratory MHD experiment (described by R. Moreau [172, 173]). A uniform vertical electric current flows in a cylindrical tank through two layers of fluid subjected to gravity. The interface between the fluids and the upper surface are both moving (see Figure 5.8). The ALE

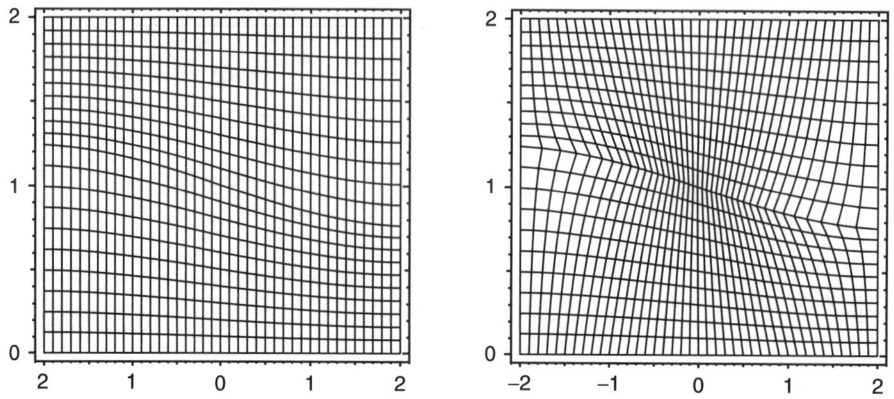

Fig. 5.6 The velocity of the mesh is computed with (5.33) on the left-hand side, and with (5.65) on the right-hand side.

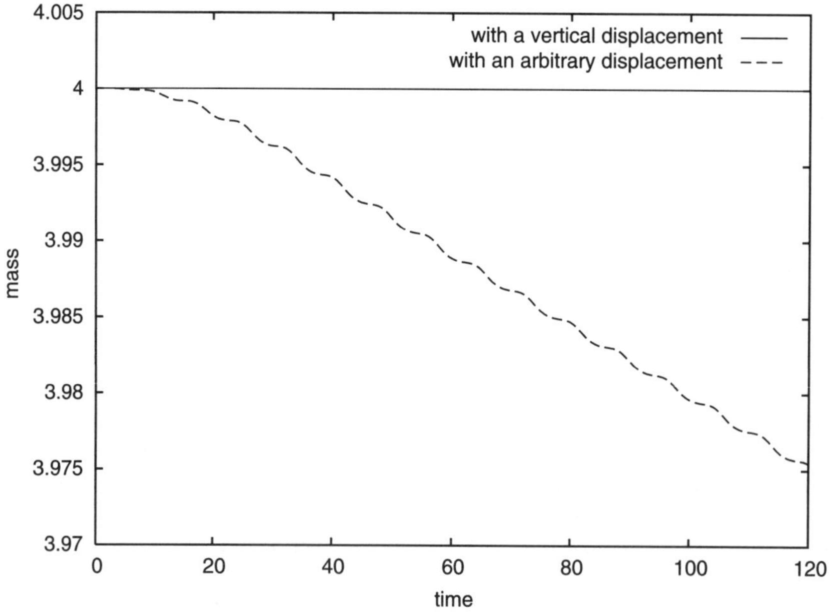

Fig. 5.7 Both curves are obtained with $\mathbb{Q}_1/\mathbb{Q}_1$ finite element with flux constraint (5.44). When the displacement of the mesh is arbitrary, Lemma 5.3 does not apply and we observe a mass loss, contrary to the case when the displacement is purely vertical.

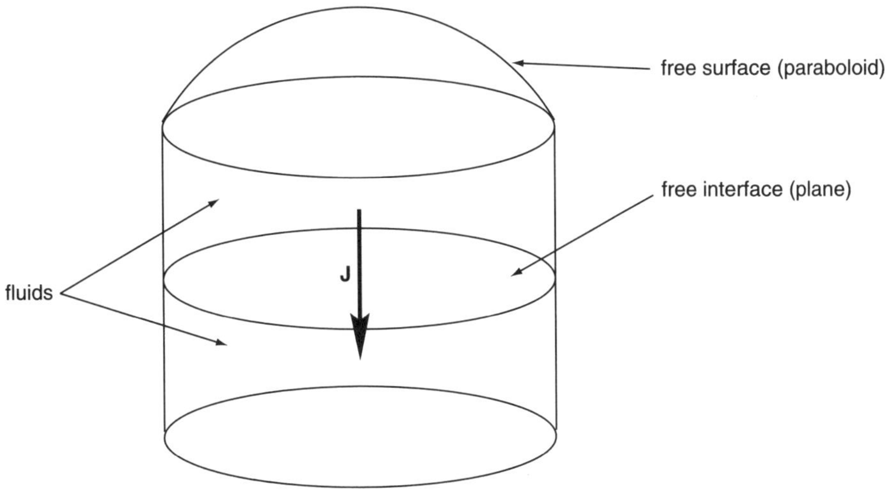

Fig. 5.8 Schematic representation of the experiment of Section 5.3.3.

formulation described in Section 5.1 must be slightly modified to treat this case due to the presence of two free surfaces, but the modifications are straightforward and we do not detail them here (see Remark 5.1.3). This test case is interesting since an analytical steady-state solution is known.

We denote by J_0 the intensity of the non-dimensional homogeneous density of current and by R the radius of the cylinder. Working with the natural cylindrical coordinates $(\boldsymbol{e}_r, \boldsymbol{e}_\theta, \boldsymbol{e}_z)$ associated with the cylindrical tank, we have:

$$\boldsymbol{B} = B(r)\boldsymbol{e}_\theta, \qquad \text{with } B(r) = -\frac{J_0}{2}r \quad (\text{for } r \leq R),$$

and the magnetic force is

$$S\,\mathbf{curl}\,\boldsymbol{B} \times \boldsymbol{B} = -\frac{SJ_0^2}{2} r\,\boldsymbol{e}_r.$$

Gravity is directed along $\boldsymbol{g} = -\boldsymbol{e}_z$. Looking for a solution with $\boldsymbol{u} = 0$, we obtain

$$\boldsymbol{\nabla} p = -\frac{m}{Fr}\boldsymbol{e}_z - \frac{SJ_0^2}{2} r\,\boldsymbol{e}_r, \tag{5.66}$$

with $m = M$ in the upper fluid and $m = 1$ in the lower one. Thus

$$p(r, z) = -\frac{m}{Fr} z - \frac{SJ_0^2}{4} r^2 + C,$$

where C is a constant. We denote by h_1^0 (resp. h_2^0) the elevation of the free interface (resp. surface) before the application of the electric current, and by $h_1(r)$ (resp. $h_2(r)$) their steady-state elevation, in the presence of the electric

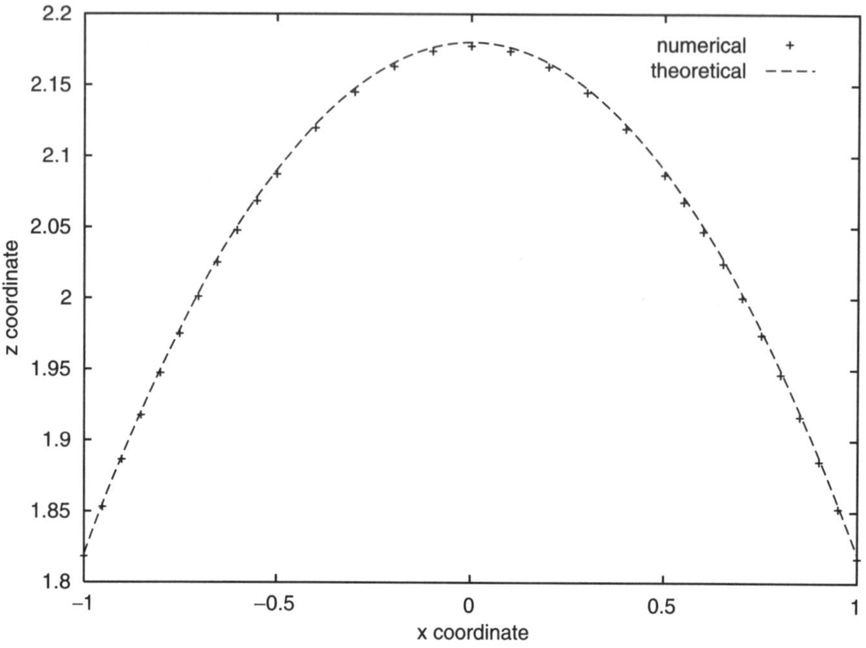

Fig. 5.9 Elevation of the steady-state top surface in the simulation of Section 5.3.3. Comparison between theoretical and numerical results.

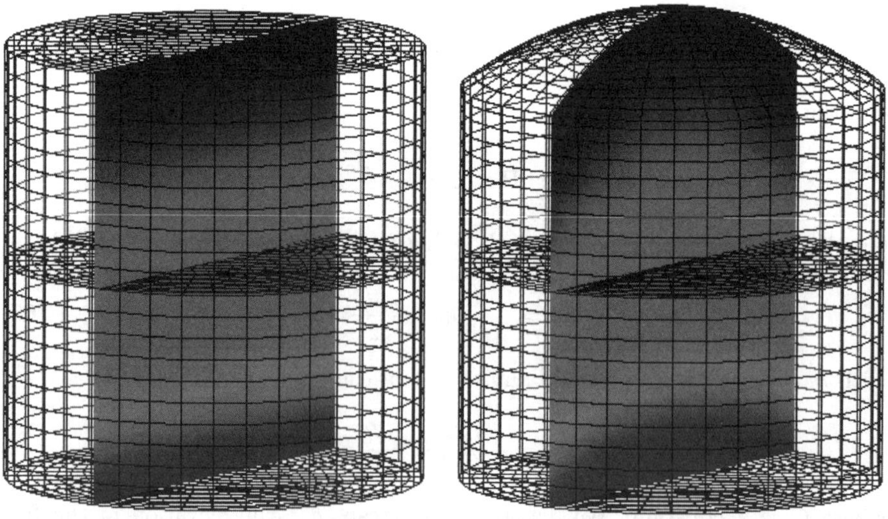

Fig. 5.10 Mesh and isovalues of pressure on a portion of the mesh, in the presence only of gravity (left), and after the application of an electric current (right). **See plate 2.**

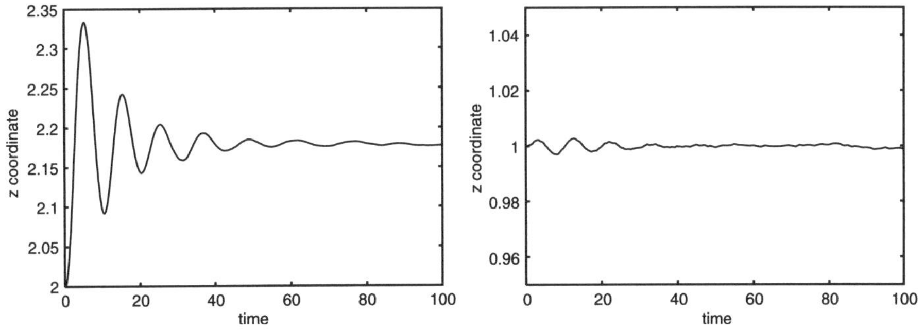

Fig. 5.11 Evolution in time of the elevation of centers of the top surface (left) and of the interface (right) in the simulation of Section 5.3.3.

current. The pressure above the free surface is supposed to be constant and equal to zero. Writing

$$0 = p(r, h_2(r)) = -\frac{M}{Fr} h_2(r) - \frac{SJ_0^2}{4} r^2 + C,$$

the constant C is determined by the conservation of the volume of the fluid, and we finally obtain:

$$h_2(r) = h_2^0 + \frac{S\,Fr\,J_0^2 R^2}{8M}\left(1 - \frac{2r^2}{R^2}\right). \tag{5.67}$$

Next, taking the **curl** of equation (5.66), we have (using (5.10)):

$$0 = \mathbf{curl}\left(\frac{m}{Fr}\boldsymbol{e}_z\right) = \boldsymbol{\nabla}\left(\frac{m}{Fr}\right) \times \boldsymbol{e}_z = \frac{M-1}{Fr}\boldsymbol{n}_1 \times \boldsymbol{e}_z \delta_{\Sigma_t}, \tag{5.68}$$

where $\boldsymbol{n}_1$ is the normal to the interface Σ_t directed from Ω_1 to Ω_2. Therefore, $\boldsymbol{n}_1 = \boldsymbol{e}_z$, which proves that the steady-state interface is horizontal:

$$h_1(r) = h_1^0.$$

We use the following numerical values: $Re_1 = Re_2 = 200$, $S = 1$, $Rm_1 = Rm_2 = 1$, $Fr = 8$, $J_0 = 0.3$, $M = 0.5$, $We = 10^2$. Figure 5.9 shows a comparison between the elevation of the upper surface and the analytical solution (5.67) along the plane $y = 0$. Figure 5.10 shows the steady-state upper surface and the interface. Figure 5.11 shows the time evolution of the elevation of the center of the upper surface (left-hand side) and of the interface (right-hand side).

6

MHD MODELS FOR ONE INDUSTRIAL APPLICATION

The aim of this chapter is to give some details of the industrial application which has motivated the work of the previous chapters. We focus on the modeling of aluminum electrolysis cells. However, the theory and the numerical methods developed in the previous chapters can of course be employed to simulate other industrial processes, especially in the metallurgical industry where magnetic fields are used to heat, pump, stir and levitate liquid metals (see Chapter 1.4 in [47]).

In Section 6.1, we briefly present aluminum electrolysis, and the industrial challenges raised by the modeling of aluminum electrolysis cells. We conclude this section by a qualitative approach of such modeling, namely the Sele criterion. Then, we turn to more quantitative approaches, starting with linearized models in Section 6.2. We next present in Section 6.3 the results we have obtained using the nonlinear model and the numerical methods presented in the previous chapters. Section 6.4 is devoted to other nonlinear approaches and general conclusions.

Throughout this chapter, the numerical values of the physical and geometric parameters used for the simulations are realistic but are not the exact values of industrial cells.

6.1 Presentation of aluminum electrolysis

Before going into the details of the MHD modeling of an aluminum electrolysis cell, we present in Section 6.1.1 some of the main features of the industrial process. Our aim is to investigate what kind of understanding of the process can be obtained from MHD simulations, and also what is left out by such models. Another objective of this presentation is to introduce in Section 6.1.2 the notions of stability and efficiency of the cell, and the industrial challenge raised by the design of stable cells working with higher current intensities. Section 6.1.3 is then devoted to some basics on MHD models for aluminum electrolysis cells. Finally, Section 6.1.4 gives a first glance at what kind of results regarding cells stability can be obtained by MHD models, using very qualitative reasoning.

6.1.1 The electrolysis process

For a more detailed introduction to aluminum electrolysis, we refer to K. Grjotheim and H. Kvande [113] or to J.-F. Gerbeau [88].

6.1.1.1 Generalities Aluminum is the most abundant metal element in the Earth's crust. However, since it is a reactive metal, it does not occur in the metallic state in nature: it is mainly contained in rocks composed of hydrated aluminum oxides, such as bauxite. Extracting the metal from most minerals is very energy-intensive, and therefore expensive. The first step consists in processing the bauxite in order to extract alumina (alumina is an aluminum oxide, Al_2O_3). Depending on the quality of the bauxite, two to three tonnes of bauxite yield one tonne of alumina. The second step is then to reduce alumina to aluminum through an electrolytic process.

The electrolytic process was discovered independently by the American Charles Hall and the Frenchman Paul L.T. Héroult in 1886, and it is still today the only method used to isolate aluminum on a commercial scale. In this process, alumina is dissolved in molten cryolite (cryolite is an aluminum fluoride mineral, Na_3AlF_6) to form the bath. Other fluor products are added in order to get an adequate fusion temperature of the bath. Alumina represents actually only 2 to 3% of the mass of the bath. Alumina is then reduced to aluminum by electrolysis, and the bath is periodically fed with fresh alumina from the top of the cell. This oxydoreduction reaction requires a very intense electric current. State-of-the-art smelters operate with about 350 kA. This process is carried out in electrolysis cells (or pots), at a high temperature (around 950 °C), due to the Joule effect. This is why both the bath and the produced metallic aluminum are liquid. Liquid aluminum, being heavier than the bath, sinks to the bottom of the cell. Periodically, it is siphoned out of the cell into a vacuum crucible (approximately every two days from each cell). The production of one tonne of aluminum typically requires 1.9 tonnes of alumina, 0.42 tonnes of carbon products and 13 500 kW h.

6.1.1.2 The aluminum electrolysis cell In Figure 6.1, we present a schematic vertical cut of an aluminum electrolysis cell. An industrial cell is typically 3 meters wide, 13 meters long and 1 meter high. The cell itself is a steel shell protected from the hot liquids by refractory materials. A layer of frozen electrolyte forms on the sides (ledge) and on the top (crust) of the cell: this protects the steel shell and contributes to the thermal equilibrium of the cell. The (DC) current enters the cell through the anodes which are immersed in the bath (typically 40 anodes per cell). The anodes consist of carbon and are periodically moved down because they are burnt away to produce carbon monoxide and carbon dioxide in the electrolysis reaction (see equation (6.2) below). They are replaced every 20 days. The current runs through the two layers of non-miscible, conducting and incompressible fluids (the bath and below the aluminum). The height of the bath is about 20 cm, but the distance between the anodes immersed therein and the aluminum (also called *anode to cathode distance* or ACD) is about 5 cm. The height of the liquid metal layer (also called metal pad) ranges from 20 cm to 30 cm.

The oxydoreduction reaction takes place both at the interface between the two fluids:

$$\mathrm{Al_2O_3} + 6\,\mathrm{e}^- \longrightarrow 2\,\mathrm{Al}^{3+} + 3\,\mathrm{O}^{2-} + 6\,\mathrm{e}^- \longrightarrow 2\,\mathrm{Al} + 3\,\mathrm{O}^{2-} \qquad (6.1)$$

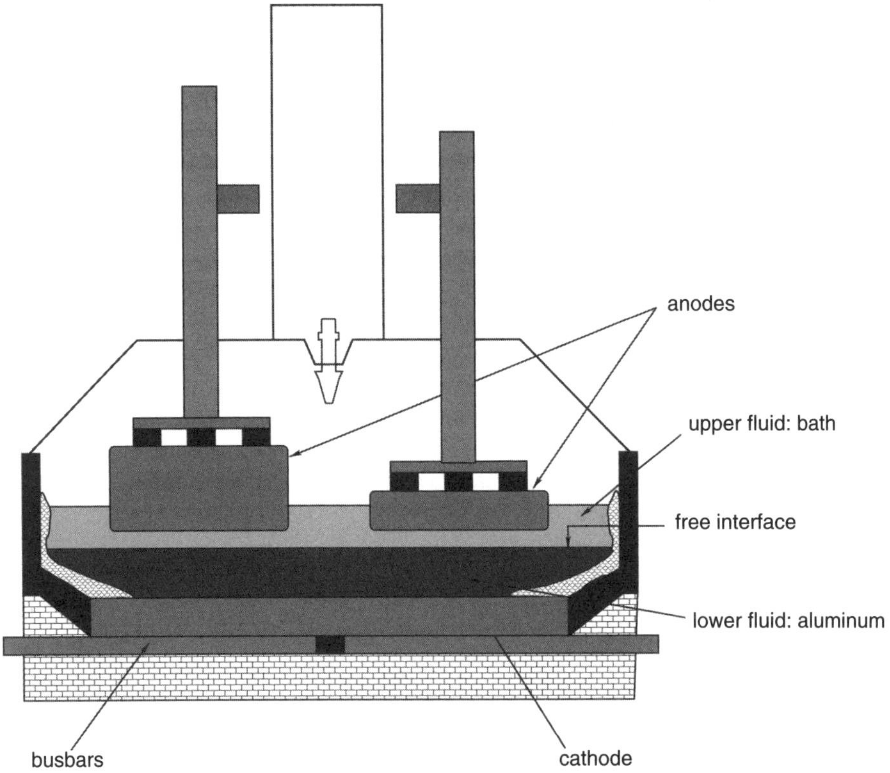

Fig. 6.1 Schematic vertical cut of an aluminum cell. **See plate 3.**

and at the surface of anodes that provide the carbon needed for the reaction:

$$3\,O^{2-} + 3/2\,C \longrightarrow 3/2\,CO_2(\text{gas}) + 6\,e^- . \tag{6.2}$$

The global balance reads:

$$2\,Al_2\,O_3 + 3\,C \longrightarrow 4\,Al + 3\,CO_2. \tag{6.3}$$

The difference of electric potential needed for the oxydoreduction is small ($1.85\,V$). The aluminum is a highly conducting metal compared to the bath where an important Joule effect heats the liquids. This difference of conductivity between the two liquids also has very important consequences on the repartition of the electric currents within the cell (see, for example, Section 6.3.3).

A very important feature of an electrolysis cell is not represented on Figure 6.1: the array of conductors (or busbars) which carries the current from one cell to another. Due to the large current intensity, this array of conductors create some relatively large magnetic fields (about $10^{-2}\,T = 100\,G$, compared with $5.10^{-5}\,T = 0.5\,G$ for the Earth's magnetic field). Since the two fluids in the cells are conducting, this results in the motion of the fluids due to

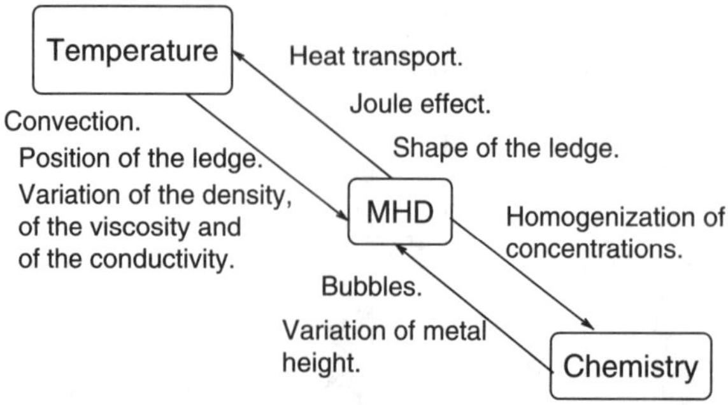

Fig. 6.2 Coupling of physical phenomena in the cell.

Laplace-Lorentz forces, which usually implies some motion of the interface. The typical flow velocity in the cell is $0.2\,m/s$. These dynamical phenomena are one of the important features to understand the whole behavior of the cell.

6.1.2 Questions of stability and efficiency of the cell

6.1.2.1 The stability of the cell The stability of the cells results from three coupled equilibria (see Figure 6.2):

- magnetic and *magnetohydrodynamic equilibrium* (current repartition within the liquids, induced currents, magnetic field created by the conductors around the cell);
- *chemical equilibrium* (alumina concentration, alumina dissolution, composition of the bath);
- *thermal equilibrium* (Joule effect, heat loss, solidification at the boundary, dissolution of alumina).

If the cell deviates from one of these equilibria, the cell is said to be unstable.

For illustration, let us further describe one feature of the chemical equilibrium, namely the control of alumina concentration. When this concentration reaches a critical level the cell enters what is called an anode effect: this is an abrupt decrease of the current flow going through the anode, due to formation of an insulating layer on the anode surface. This layer is composed of some gas resulting from the electrolysis of other components than alumina, occurring when the alumina concentration is too small. This results in an increase of the cell voltage (up to 60 V in the worst cases). Thus, schematically, the chemical equilibrium of the cell can be reached by monitoring alumina feeding *versus* cell voltage, and working at the edge of the anode effect.

We do not discuss here the thermal equilibrium.

In relation with the focus of this book, we concentrate throughout this chapter on the magnetohydrodynamic equilibrium. Loosely speaking, a *magnetohydrodynamic instability* is characterized by so large motions of the interface that the efficiency of the cell decreases (see Section 6.1.2.2 below). Of course, this is a very loose definition and more precise mathematical definitions will be given in Section 6.2 for linearized approaches, and in Section 6.3 for nonlinear approaches respectively. We also refer to Section 6.3.1 for a discussion of the relations between these different notions of stability[46]. More precisely, some waves of periods ranging from a few seconds to one minute are observed at the interface (see, for example, K. Mori et al. [178]). These waves are monitored through the variations of the resistance (either locally, measuring the current through the anodes, or globally, measuring the cell voltage). In some cases, the amplitude of these waves grows, which affects the global equilibrium and the efficiency of the cell (see Section 6.1.2.2); these amplified waves are magnetohydrodynamic instabilities.

In practice, the equilibrium of the cell is perturbed by many factors: anode change, aluminum siphoning, disequilibrium of neighboring cells, presence of some sludge at the bottom of the cell (typically due to undissolved alumina), bad positioning of an anode, disturbance in feeding, *etc.*

6.1.2.2 The efficiency of the cell The efficiency of the cell is commonly measured by the *Faraday number*, which is the ratio of the mass of aluminum actually produced versus the theoretical production if all the electric intensity had been used only to perform the electrolytic reaction (6.3). For modern cells, this number typically ranges between 93% and 97%. The loss of Faraday efficiency is mainly caused by reduced species which are deoxidized by carbon dioxide forming carbon monoxide and metal oxide. Any disequilibrium in one of the above three factors has direct or indirect consequences on the Faraday number. The main direct factor which lowers the Faraday number is deoxidization, namely the inverse reaction of (6.3). Thus, the hydrodynamics in the cell influences the Faraday number since deoxidization is typically enhanced by large motions of the fluids and of the interface between the aluminum and the electrolytic bath. On the other hand, large circulations in the cell are also useful since they allow for a uniform temperature field and a uniform concentration of the chemical species (and therefore a good dissolution of alumina, for example).

Another aspect to consider is *energy efficiency*, which can be measured for example by the energy needed to produce 1 tonne of aluminum (about 13 500 kW h for modern cells). Though the difference of electric potential between the anodes and the cathode is of the order of 4 V, the intensity of the current is so large that the electric power spent in the cell is huge. The reason for such a large intensity is of course that the quantity of aluminum produced in the cell is proportional to the intensity (accordingly to (6.1)), and that larger intensities thus lead to economies of scale (in particular by reducing labor costs and heat loss). Notice that in order to keep the current density small enough in order not to create too strong motions of the fluids, cells with larger intensity are also bigger. For an intensity of 100 kA, the production is typically of a few tonnes of aluminum per cell and per day. An aluminum plant consists of one (or many) line(s) of, say, 250 cells connected in series. The electric power needed for the whole plant is

[46] Actually, one of the difficulties of the modeling is to find a common definition of stability, suitable for both mathematical analysts and practitioners.

typically of the order of magnitude of that produced by one unit of a contemporary nuclear plant. Only half of the electric power is actually used to perform the chemical reaction. A lot of energy is wasted in heat loss.

The latter is mainly due to the potential drop (about 2 V) which occurs within the badly conducting bath. Thus, a diminution of the anode–metal distance reduces heat loss. On the other hand, since the fluids (and the interface between the fluids) are moving within the cell, this distance, which is rather small compared to the horizontal sizes of the cell, should be kept sufficiently large for two reasons. First, short-circuits (which stop the electrolytic reaction) should be avoided. Second, it has been observed that a too small anode–metal distance causes large motions of the liquids in the cell, and therefore enhances deoxidization, which lowers the efficiency of the cell, namely here the Faraday number (see Section 6.1.2.1). Moreover, the gas bubbles generated at the anode also impose a lower bound on the anode–metal distance. Therefore, there exists an optimal anode–metal distance. This is difficult to reach since it depends on many factors (the configuration of the busbars around the cell, the history of the cell) and in practice, many perturbations affect the equilibrium of the cell (see Section 6.1.2.1).

6.1.3 The magnetohydrodynamic modeling of the cell

The previous sections have demonstrated the importance of understanding the behavior of the cell in order to maintain its stability (control), and possibly enhance its efficiency (optimization). Experiments are the method of choice for this purpose. They are, however, difficult to conduct. The high temperature and the corrosive nature of the media make measurements difficult. Modeling and numerical simulation provide a useful complement to experiments.

The variety of physical phenomena occurring in the cell make a complete modeling very complicated: thermal, magnetic, dynamic and chemical effects (see Figure 6.2) would have to be accounted for. However, to model the behavior of the cell on a time interval ranging from a few seconds to a few minutes, some coupling phenomena mentioned in Figure 6.2 *can* be neglected. For example, the chemical reaction itself does not affect the behavior of the cell on such short periods of time. Likewise, temperature changes (which can lead for example to solidification or fusion of some cryolite on the ledge, and therefore affect the geometry of the domain where the liquids move) are not sufficiently fast to influence the cell behavior on this time scale. It can also be shown that the characteristic times associated with thermal convection are larger than those associated with magnetic phenomena (see D. Munger [179] for a discussion of these aspects). For comparison, magnetohydrodynamic instabilities linked to a metal pad roll (see Section 6.3.3) are typically associated with periodic oscillations of the cell voltage with a period ranging between a few seconds to one minute. Therefore, it seems reasonable to try to understand the dynamics of the fluids in the cell by considering magnetic and dynamic effects, decoupled from thermal and chemical effects. This approach has been followed by many authors (see for example [50, 56, 58, 175, 177, 206, 214]). To summarize, we will now consider that an aluminum reduction cell is adequately modeled by two shallow layers of conducting, incompressible and non-miscible fluids confined in a box, subjected to a magnetic field due to the busbar network and to the current which runs through the cell. This is exactly the framework of the previous chapters of this book. One important feature of this model is the difference of conductivity of the two fluids (the heaviest one, namely aluminum, being the most conducting).

Notice that we will not take into account in dynamic MHD models the influence of the ferromagnetic shell of the cell. More precisely, this effect can be somewhat accounted for in a magnetostatic pre-computation (see for example J. Descloux et al. [60]), in order to determine the background magnetic field, which can subsequently be used to give boundary conditions to the transient MHD system, or to perform a linearization around a stationary state. We are not aware of any approach taking this effect into account in a genuine transient manner.

Many works have been devoted to the magnetohydrodynamic modeling of Hall–Héroult aluminum electrolysis cells since the early paper of J.P. Givry [107]. Most of them deal with the stability of the cells. The variety of results can be explained by the variety of concerns of the authors. Schematically, three approaches have been followed:

(i) the derivation of very simple models to qualitatively identify some basic mechanisms leading to instability;
(ii) the derivation of simple models to quantitatively analyze the linear stability of the system, sometimes through numerical computations;
(iii) the use of complicated nonlinear models and appropriate numerical methods to numerically test the stability of the cells.

Of course, all these approaches complement one another. Improvements of the stability of the cells require a deep understanding of the physical phenomena at the origin of instabilities, thus approximations and simplifications like in approaches (i). Complete nonlinear models (iii) contribute to consolidate the assumptions made to derive simple models, and ensure that these over-simplified models indeed keep the essence of the cell behavior. On the other hand, complete nonlinear models (iii) usually require large computational times, so that numerical tests on simpler models (ii) are needed to perform real-time computations. Historically, many studies of type (i) or (ii) were conducted in the 1970s and 1980s. Since the 1990s, the trend is to use in a more systematic way the increasing capabilities of numerical investigations through complex nonlinear models (iii).

In the following, we focus on *time-dependent models*, since it does not seem possible to infer the stability of the cell from some simple characteristics of the stationary magnetic field or shape of the interface. For example, it is not clear that there is a correlation between the flatness of the interface and the fact that the cell is stable (see A.F. LaCamera et al. [137] and N. Urata [236]), neither that reducing the maximum value of the vertical component of the magnetic field will reduce the instabilities (see [236]). However, we will see in the following that some linear approaches require a delicate computation of a stationary state. We shall come back to this in Sections 6.2.1.1 and 6.2.5.2.

Before describing quantitative approaches based on various systems of equations in Sections 6.2, 6.3 and 6.4, we first present in the next section a qualitative study of the stability of aluminum electrolysis cells.

6.1.4 Qualitative interpretation of MHD instability and the Sele criterion

It has been observed in practice that either strong *vertical magnetic fields* or strong *horizontal components of the electric current* in the aluminum favor instabilities. The cell design (busbar arrangement, positioning of the cells lines) is the practical way to control the first factor, in a permanent manner. On the other hand, some horizontal components of the electric current may also occur during the process (position of the ledge, sludge at the bottom of the cell). On the basis of these observations, theoretical efforts for understanding instability mechanisms have focused on either the vertical component of the background magnetic field or the horizontal components of the background current.

6.1.4.1 Effects of the horizontal components of the background electric current

Some analyses of the effects of the horizontal components of the background electric current can be found in A.D. Sneyd [213], R. Moreau et al. [176] and S. Pigny et al. [185]. The instability mechanisms involve some horizontal components of the background current and therefore some spatial variations of the background magnetic field. We will present these studies in Section 6.2.3 below. These mechanisms have, however, not received as much attention as the Sele mechanism we now present.

6.1.4.2 Sele mechanism and Sele criterion In [207], T. Sele identified a basic mechanism for the interface motions. It relies on the presence of a vertical component of the background magnetic field, and is completely different from the mechanisms mentioned in the previous section.

The basic phenomenon is described on Figure 6.3. A crest of the interface at the edge of the cell leads to a redistribution of the currents within the cell: the current flows through the peak rather than through the trough. Since the aluminum is much more conducting than the bath, this induces horizontal currents in the aluminum, while the current remains approximately vertical in the bath. Then, another important parameter to sustain the perturbation of the interface is the vertical component of the magnetic field, *which is supposed to be of constant sign.* The horizontal currents interact with the vertical component of the magnetic field to produce a horizontal electromagnetic force in the metal, perpendicular to the current. This induces a motion of the crest along the edge.

The stability of the cell depends on whether the perturbation of the interface subjected to the Lorentz force grows or vanishes. A simple way to quantify this phenomenon, described in [207], is to consider a particular deformation, namely

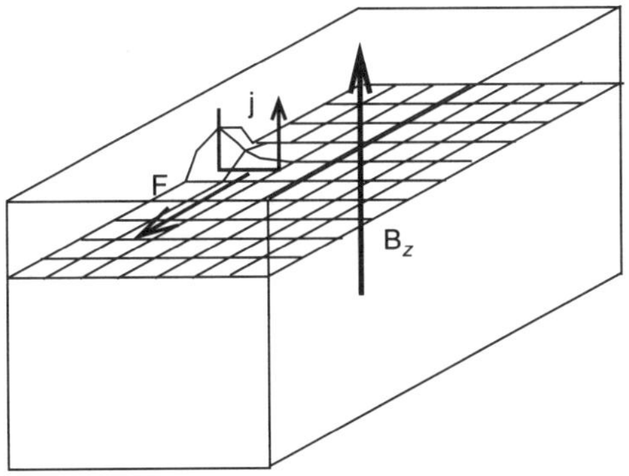

Fig. 6.3 The basic mechanism creating waves at the interface. A perturbation of the position of the interface creates a perturbed current ($\boldsymbol{j}$) which interacts with the vertical component of the magnetic field ($\boldsymbol{B}_z$) to create a horizontal magnetic force ($\boldsymbol{F}$) in the metal. This in turn moves the crest.

a tilting of the flat interface, with a uniform constant vertical component of the magnetic field. Based on qualitative arguments that we sketch below, the famous Sele criterion for the stability of the cell is obtained. The cell is stable if

$$\beta = \frac{I_0 \boldsymbol{B}_z}{h_1\, h_2\, g\,(\rho_1 - \rho_2)} < \beta_{\text{critical}}. \tag{6.4}$$

Here, I_0 denotes the current intensity in the cell, $\boldsymbol{B}_z$ denotes a typical value of the vertical component of the magnetic field, h_1 (resp. h_2) denotes the metal (resp. bath) height, $g = 9.81\, m\, s^{-2}$ denotes the acceleration of gravity, ρ_1 (resp. ρ_2) denotes the metal (resp. bath) density, and β_{critical} is an empirical parameter.

This formula is obtained by comparing the magnetic force due to the perturbation of current with the gravity force exerted on the metal pad for the same displacement. The former is of order

$$\frac{I_0\, \eta}{h_1\, h_2}\, \boldsymbol{B}_z, \tag{6.5}$$

where η denotes a typical value of the vertical displacement of the interface from its equilibrium state, while the latter can be estimated as

$$g\,(\rho_1 - \rho_2)\, \eta. \tag{6.6}$$

Equations (6.5) and (6.6) actually define pressures[47] associated with magnetic effects and gravity, respectively. The ratio of these two pressures yield the non-dimensional parameter β in (6.4).

Let us now explain how formulae (6.5) and (6.6) can be obtained. Equation (6.5) derives from the formula for Lorentz forces: current times magnetic field. The formula for the perturbed current $I_0\, \eta/(h_1\, h_2)$ relies on the fact that

[47] *i.e.* quantities measured in $N\, m^{-2}$.

the perturbed current is vertical in the bath, and spreads horizontally in the whole metal pad. The perturbed current in the bath is of the order $I_0\,\eta/h_2$ (think for example of the formula $R = \rho\, L/S$ giving the resistance of a wire of length L and cross section S to derive that $J_0/h_2 \propto j/\eta$ where J_0 is the equilibrium current density, and j is the perturbed current) and the term $1/h_1$ then appears from the conservation of current, when it runs through the metal pad of height h_1 (see also the derivation of equation (6.19) below for a more rigorous justification). On the other hand, equation (6.6) derives from classical considerations for two-liquid flow, and can be seen as an Archimedes force.

6.1.4.3 Discussion of Sele criterion The stability criterion (6.4) is considered important since it illustrates the following practical observations. First, the practical way to reduce an instability in a cell is to increase the anode to cathode distance, *i.e.* h_2. This fact is illustrated by (6.4) since increasing h_2 indeed decreases β. Second, it has been observed that the vertical component of the magnetic field is of crucial importance for the stability of the cell (see, for example, A.F. LaCamera et al. [137] p. 1185). This is contained in (6.4) since increasing $\boldsymbol{B}_z$ indeed increases β. Third, designing a stable cell is more and more difficult with increasing intensity. The Sele criterion (6.4) indeed shows that β increases with growing I_0. These agreements explain the success of the Sele criterion.

Moreover, the elementary mechanism underlying the Sele criterion has been thoroughly confirmed by subsequent works. Indeed, the model behind the Sele mechanism has been successfully used to improve the stability of some industrial cells (see [236]). In addition, the Sele mechanism has been confirmed by complete 3D MHD simulations (see [99, 101, 179] and Section 6.3.3).

Actually, many stability criteria analogous to the form (6.4) have been obtained by many authors following various analyses. Those sometimes give some explicit values for the parameter β_{critical} in the function of the geometry of the cell, the mechanical dissipation or some frequencies of the gravitational modes: see for example formulae (4.20), (4.21) and (4.22) in V. Bojarevics et al. [25], or formula (7.3) in P.A. Davidson et al. [50], or formulae (4.11) and (5.6) in A.D. Sneyd et al. [214]. We also refer to Section 6.2.4 where some details on these approaches are given. In all these works, the basic idea is again to compare the gravity force with the Lorentz force due to the interaction of the vertical component of the magnetic field with the horizontal currents created by the tilting of the interface. The differences between these works essentially come from the initial perturbation of the equilibrium interface the authors consider.

For other qualitative analyses of the instability mechanisms, in the spirit of the Sele mechanism, we refer to P.A. Davidson et al. [50] where a simple mechanical system which exhibits similar behavior is described, and to S. Molokov et al. [156, 169], where the authors put particular emphasis on the reflection mechanisms at the boundary of the cell.

However, it is not clear whether the instability mechanism described by T. Sele in [207] and yielding formula (6.4) is responsible for the practical observations. For example, in many industrial cells, the vertical component of the magnetic field is not of constant sign, so that the classical interpretation (see Figure 6.3) cannot apply[48]. In particular, it is not clear what is a good value to consider for $\boldsymbol{B}_z$ in (6.4): T. Sele in [207] considers the mean value over the cell (without any real justification), V. Bojarevics and M.V. Romerio in [25] consider some Fourier transforms of $\boldsymbol{B}_z$ (see (6.38) below).

More generally, we would like to stress that even if the mechanisms described by various theories may be different (see below), the design criteria they yield often look quite similar. The qualitative agreement of a global criterion with the experimental recipes to reduce instabilities does not guarantee that the underlying instability mechanisms are correctly described.

6.2 Linearized approaches

In this section, we present the main approaches which use linear stability analysis to study the stability of aluminum electrolysis cells. We also refer to the review paper by A.F. LaCamera et al. [137].

Since the complete MHD system with a free interface is complicated, many approaches simplify the original problem, to obtain simpler linear equations to solve and derive analytical expressions for stability criteria. These simplifications also aim at a better insight into the physical origin of the MHD instabilities. A classical method to simplify nonlinear equations is to consider their linearization.

Linear stability analysis consists in a three-step procedure: first determine a stationary state, second perform a linearization of the nonlinear equations at the vicinity of this stationary state, and, third, compute the eigenmodes and the eigenvalues of this linearized system. More explicitly, the third step consists in searching for the unknown fields under the generic form:

$$\boldsymbol{A}(t,\boldsymbol{x}) = \boldsymbol{A}_0(\boldsymbol{x}) + \mathrm{Re}\big(\exp(-i\omega t)\widetilde{\boldsymbol{A}}(\boldsymbol{x})\big). \tag{6.7}$$

Here $\boldsymbol{A}_0$ is the stationary value of the field under consideration and

$$\mathrm{Re}\big(\exp(-i\omega t)\widetilde{\boldsymbol{A}}(\boldsymbol{x})\big)$$

is the perturbation. This amounts to replacing the time derivative by a multiplication by $-i\omega$ in the linearized equations.

The field $\widetilde{\boldsymbol{A}}(\boldsymbol{x})$, called an eigenmode, is determined by its approximation on a finite dimensional vector space so that the problem reduces to determining the eigenvalues of a matrix. In our framework, two finite-dimensional spaces are classically used: finite element spaces (see, for example, Section 6.2.5.2), or spaces spanned by the gravitational modes[49] (see, for example, Section 6.2.4.2).

[48] Many works we have mentioned so far actually suppose that the vertical component of the background magnetic field is uniform.

[49] See (6.16) for a precise definition of gravitational modes.

Notice that the "frequency" ω is a complex number. The sign of its imaginary part discriminates between stable and unstable modes. The system is stable if all the eigenvalues are such that the imaginary part of ω is non-positive (or smaller than a positive value called a stability threshold)[50]. Under some special assumptions on the geometry, the eigenmodes of the linearized MHD system are traveling plane waves (the so-called gravity waves solutions), so that an explicit dispersion relation is obtained, relating ω to the wave number of the plane wave (see Section 6.2.3). This can be used to derive explicit stability criteria. But in a general framework, only approximations of the eigenvalues can be obtained (see Sections 6.2.4 and 6.2.5).

Before going into the details, let us mention the common qualitative conclusions of all these analyses:

- Instability comes from a complex interaction involving the magnetic fields created by the external conductors, and the perturbation of currents and magnetic field due to the motion of the interface.
- Increasing the height of aluminum or bath contributes to stabilizing the cell.
- The configuration of the ambient magnetic field, and the presence of horizontal currents in the aluminum, are important in instability mechanisms.

These conclusions agree with observations on real industrial cells.

In the following, we go into the details of these linearized approaches, going from qualitative ones, to more quantitative ones. In Section 6.2.1, we list the assumptions used in linearized approaches. We then use these assumptions to derive in Section 6.2.2 a prototypical linear system, which has been used by many authors. We then distinguish between three kinds of results obtained from the analysis of such a system:

- analysis of single Fourier modes on cells without boundaries, around a simple stationary state (see Section 6.2.3);
- analysis of the coupling between Fourier modes on cells with boundaries, again around a simple stationary state (see Section 6.2.4);
- numerical computations of the eigenvalues, around a precomputed complicated stationary state (see Section 6.2.5).

6.2.1 Basic assumptions

In this section, we present the main assumptions used in these approaches in order to simplify and linearize the MHD equations. In Section 6.2.1.1, we present the basic hypothesis of any linearized analysis study. Then, we list the assumptions related to the derivation of the simplified equations. We first examine in Section 6.2.1.2 the common employed assumptions concerning the equations

[50]This is of course a matter of convention. Depending on the authors, the sign convention on ω may be the opposite: $\boldsymbol{A}(t,\boldsymbol{x}) = \boldsymbol{A}_0(\boldsymbol{x}) + \mathrm{Re}(\exp(i\omega t)\widetilde{\boldsymbol{A}}(\boldsymbol{x}))$, so that a stable system is associated with the non-negative imaginary part of ω.

themselves. Finally, we draw up the list of hypotheses on the geometry of the computational domain in Section 6.2.1.3.

6.2.1.1 Assumptions related to linearized approaches The linear stability analysis approach we sketched above is essentially based upon the following two assumptions:

(L1) There exists a stationary state.

(L2) The fluctuations around this stationary state are sufficiently small so that the stability of the nonlinear system can be studied on the linearized system.

Let us now comment on these assumptions.

Concerning assumption (L2), let us simply indicate that its validity is questionable in our context. This is related to a discussion about the validity of the linearized approach, which we postpone until Section 6.3.1. Concerning this assumption, let us also mention that all these studies suppose that the fluctuations only affect the physical quantities (velocity field, magnetic field, current distribution) within the cell. For example, the current external to the cell is assumed time-independent. This seems to be a fair assumption in practice.

Let us now question assumption (L1) both from a theoretical and a practical point of view. Actually, the existence of a stationary state itself may be questionable. As reported by O. Zikanov et al. in [241], some observations on industrial cells show that the actual flow may exhibit large variations in time, including possibly complete reversals of the circulation. Evidently, such flows are unlikely to be small perturbations around a stationary state.

Having said that, let us however *postulate* the existence of this steady state. In the following we will refer to this stationary state as the "background" velocity field, interface deformation, current repartition or magnetic field. The question is now to compute an approximation of the stationary state.

Actually, the stationary state is very often assumed to have a flat interface with zero velocity field in the literature. This implies that $\partial \boldsymbol{B}_0/\partial z = 0$, where $\boldsymbol{B}_0$ is the stationary magnetic field (see equation (2.1) in A.D. Sneyd [212]). For some discussions about this assumption, we refer to D.P. Ziegler [240] for work on the effect of the steady velocity, in a purely hydrodynamic framework or to O. Zikanov et al. [241] and to A. Kurenkov et al. [136] for a study based on some MHD numerical experiments.

Notice that it may also be possible to directly measure the stationary state. For example, in [6], J. Antille et al. propose a method to infer the velocity field from the measurement of the anodic current fluctuations.

However, except for very specific models such as the Moreau–Evans model (see [175] for some analytical derivations further investigated in [177] with numerical computations), elaborate numerical computations are necessary. It is generally believed that an accurate description of the stationary state should at least take into account the geometry of the cell, the distribution of currents around the cell, the presence of ferromagnetic materials and turbulence effects on the flow (see M.V. Romerio et al. [197]). For this purpose, involved numerical precomputations are necessary (see for example M.V. Romerio et al. [198] or J. Descloux et al. [62]).

Notice that the difficulties associated with the computation of a stationary state disappear if the stability analysis is performed with a purely nonlinear time-dependent approach such as the one we describe in Sections 6.3 and 6.4 below.

6.2.1.2 Assumptions to simplify the equations Let us now list some common assumptions used to simplify the equations:

- (E1) The induced currents $\boldsymbol{u} \times \boldsymbol{B}$ in Ohm's law (1.36) are neglected.
- (E2) The time derivative $\partial \boldsymbol{B}/\partial t$ in the partial differential equation on $\boldsymbol{B}$ in (1.49) is neglected.
- (E3) The electric problem is simplified by passing to the limit $\sigma_2 \ll \sigma_C \ll \sigma_1$ where σ_C denotes the carbon[51] conductivity.
- (E4) The vertical dimensions (height of bath and aluminum) are considered small compared to the horizontal dimensions (wavelength of the perturbations, horizontal dimensions of the cell).
- (E5) Mechanical dissipations (viscosity, surface tension) are neglected or replaced by a linear damping term.

Let us now comment on these assumptions.

Assumption (E1) is often justified by the fact that since induced currents always have a stabilizing effect, omitting them corresponds to a worst-case scenario. However, A.F. LaCamera et al. have shown in [137] that induced currents could represent as large as 40% of the total current density in aluminum.

Assumption (E2) is a very important assumption which reduces the magnetic problem to a simple static problem on the electric potential in each layer. It is commonly justified by the fact that the magnetic Reynolds number is small ($Rm \ll 1$), so that the characteristic time associated with magnetic diffusion is very small compared to the other characteristic times. This is, however, questionable in the aluminum, where this characteristic time becomes comparable with the characteristic time associated with gravity forces (see D. Munger [179, p. 44]). As noticed in J.-F. Gerbeau et al. [91] and recalled in Section 2.2.5, the MHD system simplified using assumption (E2) but not (E1) may be ill-posed. Actually, these assumptions are usually used jointly: in the following, we call the pair of hypotheses (E1)–(E2) *the magnetostatic approximation.*

Assumption (E3) is used in order to simplify the electric problem. In particular, it enables us to derive some approximate boundary conditions on the electric potential at each interface, and contributes (together with (E4)) to justify that

[51] We recall that the anodes are made of carbon.

the horizontal component of the current can be neglected in the bath (see *e.g.* Section 6.2.2.4 below).

Assumption (E4) underlies the shallow water approximation (see, *e.g.* Section 6.2.2.2 below).

We mention in assumption (E5) the introduction of a damping term to model the mechanical dissipation. Usually, this damping term is a linear term $-\kappa \boldsymbol{u}$ in the momentum equation. In the present context, this damping term was first introduced by R. Moreau and J.W. Evans in [175] (see formula (8)) and then used by many authors (see for example A.D. Sneyd [213], S. Pigny et al. [185] or H. Sun et al. [222]). This simplification can be justified by integrating across the layer the Navier–Stokes equations for a nearly horizontal flow (see S.D. Lympany et al. [157]). In [24], V. Bojarevics shows that in the shallow-water approximation, this amounts to introducing a damping term of the form $-\kappa\,(\partial\eta/\partial t)$ in the classical wave equation ruling the evolution of the interface elevation η (gravity waves) (see the end of Section 6.2.4.2 below for some details). This is also used in this form by S. Molokov et al. in [169]. The main difficulty with such a damping term is to estimate the value of the damping parameter κ. The instability thresholds are indeed sensitive to this parameter (see, for example, V. Bojarevics [24, p. 838] J. Descloux et al. [59, p. 130–131] or the end of Section 6.2.4.2 below). We however mention that the same kind of difficulty occurs, to a lesser extent, in models using an effective viscosity. These difficulties are somehow related to the absence of a satisfactory model for turbulence in our framework (see Remark 1.1.1).

To conclude this section, we recall that we are not aware of any study taking into account the ferromagnetic material in a time-dependent model. A common assumption of these works is that these magnetic nonlinearities do not influence the fluctuations of the unknowns around the stationary state.

Likewise, the effects related to the generation and release of gas bubbles at the anodes surface are ignored (see [135] for a model of the detachment of bubbles). This effect has been so far considered as secondary regarding the stability of the cell.

6.2.1.3 Assumptions to simplify the geometry As announced above, a third set of assumptions concerns the geometry of the computational domain:

- (G1) The anode is supposed to be flat, and to entirely cover the bath.
- (G2) The cell is approximated by a parallelepiped, not taking into account the geometry of the ledge.
- (G3) The cell has no boundaries (infinite horizontal dimensions).
- (G4) The horizontal dimension of the cell is infinite in one direction.

Let us now comment on these assumptions.

Concerning assumption (G1), according to R. Moreau and J.W. Evans [175], the anode channels (namely the channels around the anode blocks) may affect the stationary state. It is however generally admitted that these effects are secondary (see O. Zikanov et al. [241]).

Assumption (G2) is more questionable. Rather than influencing the fluid dynamics, the ledge is overall important for the distribution of the currents. Indeed, the geometry of the side walls may create some horizontal currents in the metal pad, which are considered as an important source of instability. One way to implicitly account for the ledge is to use a parallelepiped for the computational domain, and to assume that the background current distribution is not purely vertical, but contains some horizontal components in the metal pad. Notice that this trick can also be used to model the presence of sludge at the bottom of the cell under the molten aluminum pad, or heterogeneities in the conductivity of the cathode.

Assumptions (G3) and (G4) are used to simplify the resolution of the system of equations.

Assumption (G3) appears in a series of papers [49, 176, 185, 212, 213] we discuss below. This simplification is used to study some traveling wave solutions and derive explicit dispersion relations.

Assumption (G4) may be found in a few papers [50, 63, 159, 198], as an intermediate assumption between (G3) and a cell with boundaries. It particularly simplifies the computation of stationary states (see M.V. Romerio et al. [198]).

6.2.2 A prototypical derivation of a linearized system

We present in this section a prototypical derivation of a simplified and linearized version of the MHD system (1.49), which has been used by many authors (see for example [25, 50, 156, 178, 214, 236, 237]). The motivation for the derivation of this specific model is the quantitative understanding of Sele mechanism presented in Section 6.1.4. Therefore, the focus of this model is the influence of the background vertical magnetic field. In contrast, it does not take into account some horizontal components for the background current.

In Sections 6.2.2.1–6.2.2.5, we propose a detailed five-step derivation of the simplified system (6.25) starting from the MHD system (1.49). We think that it is interesting to understand the derivation for two reasons. First, it illustrates the links between stability studies performed on the MHD system (1.49) and on the simplified system (6.25). Second, it shows the inherent limitations of a stability analysis on the simplified system (6.25). The reader already familiar with the derivation of (6.25) may directly proceed to Section 6.2.2.6.

6.2.2.1 The magnetostatic approximation We start from the MHD system (1.49). We assume (G1)–(G2) for the geometry. In the first step, we use assumptions (E1), (E2) and (E5) to obtain the following system:

$$\begin{cases} \begin{cases} \dfrac{\partial}{\partial t}(\rho \boldsymbol{u}) + \operatorname{div}(\rho \boldsymbol{u} \otimes \boldsymbol{u}) = -\boldsymbol{\nabla} p + \rho \boldsymbol{g} + \dfrac{1}{\mu} \mathbf{curl}\, \boldsymbol{B} \times \boldsymbol{B}, \\ \operatorname{div} \boldsymbol{u} = 0, \end{cases} \\ \dfrac{\partial \rho}{\partial t} + \operatorname{div}(\rho \boldsymbol{u}) = 0, \\ \begin{cases} \dfrac{1}{\mu} \mathbf{curl} \left(\dfrac{1}{\sigma(\rho)} \mathbf{curl}\, \boldsymbol{B} \right) = 0, \\ \operatorname{div} \boldsymbol{B} = 0. \end{cases} \end{cases} \tag{6.8}$$

This system of equations is supplied with initial conditions as in (1.49). In addition, we consider here a pure slip boundary condition ($\boldsymbol{u} \cdot \boldsymbol{n} = 0$) on the sides of the cell, and we will make precise magnetic boundary conditions below.

We can rewrite the magnetic part of the equation in terms of the electric potential Φ. We denote $\boldsymbol{J} = \frac{1}{\mu}\, \mathbf{curl}\, \boldsymbol{B}$ the current, and we have $\mathbf{curl} \left(\frac{1}{\sigma(\rho)} \boldsymbol{J} \right) = 0$ so that there exists a scalar function Φ (the electric potential) such that $\boldsymbol{J} = -\sigma(\rho) \boldsymbol{\nabla} \Phi$. Since $\boldsymbol{J} = \frac{1}{\mu}\, \mathbf{curl}\, \boldsymbol{B}$, we have $\operatorname{div}(\boldsymbol{J}) = 0$. Thus, the last two equations of (6.8) are rewritten in the following way:

$$\begin{cases} \boldsymbol{J} = -\sigma(\rho) \boldsymbol{\nabla} \Phi, \\ \operatorname{div} \boldsymbol{J} = 0. \end{cases} \tag{6.9}$$

The magnetic field $\boldsymbol{B}$ appearing in the right-hand side of the momentum equation (6.8) is produced both by the electric current $\boldsymbol{J}$ running through the two liquids in the cell, and by the electric current running through the busbars and the other neighboring cells. The boundary conditions on the electric potential Φ at each interface are derived from the continuity of the normal component of the current $\boldsymbol{J}$ and of the tangential component of the electric field $\boldsymbol{\nabla} \Phi$ (see, for example, the appendix of A.D. Sneyd [213]).

6.2.2.2 The shallow water approximation In a second step, we use a *shallow water* approximation (see (E4)). There are many more or less formal ways to derive the shallow water equations. The most rigorous derivations use integration of the equations along a vertical slice and expansion in term of a small parameter associated with assumption (E4) (this is sometimes called the *Saint-Venant* approach, see H. Sun et al. [222], J.-F. Gerbeau et al. [102], and Section 5.3.1 for some numerical experiments). In a more intuitive way, the shallow water approximation consists in neglecting the vertical dependency so that the velocity only depends on the horizontal components (x, y), approximating the pressure by the hydrostatic pressure:

$$p = p_{\text{int}} + \rho\, g\, (h - z)$$

where $p_{\text{int}}(t,x,y)$ denotes the pressure at the interface and $h(t,x,y)$ the height of aluminum (*i.e.* the position of the interface), and neglecting the velocity in the vertical direction in both fluids. Using these approximations, we obtain the two-fluid shallow water equations[52] (corresponding to the hydrodynamic part of the system (6.8)):

$$\begin{cases} \begin{cases} \rho_1 \dfrac{\partial}{\partial t}(\boldsymbol{u}_H^a) + \rho_1 \boldsymbol{u}_H^a \cdot \boldsymbol{\nabla}_H \boldsymbol{u}_H^a = -\boldsymbol{\nabla}_H p_{\text{int}} - \rho_1 g\, \boldsymbol{\nabla}_H h + (\boldsymbol{J} \times \boldsymbol{B})_H^a, \\ \\ \rho_2 \dfrac{\partial}{\partial t}(\boldsymbol{u}_H^c) + \rho_2 \boldsymbol{u}_H^c \cdot \boldsymbol{\nabla}_H \boldsymbol{u}_H^c = -\boldsymbol{\nabla}_H p_{\text{int}} - \rho_2 g\, \boldsymbol{\nabla}_H h + (\boldsymbol{J} \times \boldsymbol{B})_H^c, \end{cases} \\ \\ \dfrac{\partial h}{\partial t} = -\text{div}_H(h\boldsymbol{u}_H^a) = \text{div}_H(h'\boldsymbol{u}_H^c), \end{cases} \tag{6.10}$$

where $h' = h_1 + h_2 - h$ denotes the height of bath. The subscript H indicates that only horizontal components are taken into account (for example, $\boldsymbol{\nabla}_H = (\partial_x, \partial_y)$) and the superscript a (resp. c) denotes quantities in the aluminum (resp. in the bath) (for example, $\boldsymbol{u}_H^a$ denotes the horizontal components of the velocity field in the aluminum, and $(\boldsymbol{J} \times \boldsymbol{B})_H^a$ denotes a vertical average of the horizontal components of the Lorentz force in the aluminum). These equations are defined on a domain $\Omega_H \subset \mathbb{R}^2$ which is a rectangle:

$$\Omega_H = (0, L_x) \times (0, L_y).$$

6.2.2.3 Linearization The third step consists of a linearization: we consider only perturbations around a stationary state for which the velocity field is zero and the interface is horizontal and flat (see assumptions (L1)–(L2)). To be consistent, this means that the stationary magnetic field and current are such that the associated Lorentz force is irrotational. Moreover, we suppose that the stationary current is uniform and vertical (see (6.12) below).

We introduce the perturbed quantities η, $\boldsymbol{j}$ and $\boldsymbol{b}$ defined by:

$$h = h_1 + \eta, \qquad \boldsymbol{J} = \boldsymbol{J}_0 + \boldsymbol{j}, \qquad \boldsymbol{B} = \boldsymbol{B}_0 + \boldsymbol{b}, \tag{6.11}$$

with the following assumptions on the stationary state $(h_1, \boldsymbol{J}_0, \boldsymbol{B}_0)$:

$$h_1 \text{ is constant}, \qquad \boldsymbol{J}_0 = -J_0 \boldsymbol{e}_z, \qquad \textbf{curl}\,(\boldsymbol{J}_0 \times \boldsymbol{B}_0) = 0, \tag{6.12}$$

where J_0 is a positive constant. It may be checked that this last property is equivalent to

$$\frac{\partial \boldsymbol{B}_0}{\partial z} = 0$$

[52] We give here the non-conservative form of the two-fluid shallow water equations, since in our setting, we only consider regular solutions. In the presence of shocks, it is necessary to use the conservative form (see [10, 36]).

by using the fact that $\operatorname{div} \boldsymbol{B}_0 = 0$ (see A.D. Sneyd [212]). The fact that the interface is flat and horizontal if $\mathbf{curl}\,(\boldsymbol{J}_0 \times \boldsymbol{B}_0) = 0$ is explained in Section 5.3.3 (see equation (5.68)).

By inserting (6.11) in (6.10), we obtain by a linearization procedure (we do not linearize the Lorentz force at this stage, see Section 6.2.2.5 below):

$$\begin{cases} \begin{cases} \rho_1 \dfrac{\partial}{\partial t}(\boldsymbol{u}_H^a) = -\boldsymbol{\nabla}_H p_{\text{int}} - \rho_1\, g\, \boldsymbol{\nabla}_H \eta + (\boldsymbol{J} \times \boldsymbol{B})_H^a, \\ \\ \rho_2 \dfrac{\partial}{\partial t}(\boldsymbol{u}_H^c) = -\boldsymbol{\nabla}_H p_{\text{int}} - \rho_2\, g\, \boldsymbol{\nabla}_H \eta + (\boldsymbol{J} \times \boldsymbol{B})_H^c, \end{cases} \\ \\ \dfrac{\partial \eta}{\partial t} = -h_1 \operatorname{div}_H(\boldsymbol{u}_H^a) = h_2 \operatorname{div}_H(\boldsymbol{u}_H^c), \end{cases} \tag{6.13}$$

which reduces to an equation on the displacement of the interface η by simple algebra to eliminate the unknown pressure at the interface p_{int}:

$$\left(\frac{\rho_1}{h_1} + \frac{\rho_2}{h_2}\right) \frac{\partial^2 \eta}{\partial t^2} - (\rho_1 - \rho_2)\, g \Delta_H \eta = \operatorname{div}_H \left((\boldsymbol{J} \times \boldsymbol{B})_H^c - (\boldsymbol{J} \times \boldsymbol{B})_H^a \right). \tag{6.14}$$

Equation (6.14) may be more rigorously obtained using *a priori* expansions in terms of two small parameters, related to assumptions (E4) and (L2) (see equation (2.24) in V. Bojarevics et al. [25]). For a box with finite horizontal dimensions $L_x \times L_y$, the pure slip boundary condition on the velocity translates into the following boundary condition on η: on $\partial\Omega_H$

$$\Big((\rho_1 - \rho_2)\, g \boldsymbol{\nabla}_H \eta + \big((\boldsymbol{J} \times \boldsymbol{B})_H^c - (\boldsymbol{J} \times \boldsymbol{B})_H^a \big) \Big) \cdot \boldsymbol{n} = 0, \tag{6.15}$$

which can be obtained from (6.13) and $\boldsymbol{u}_H^a \cdot \boldsymbol{n} = \boldsymbol{u}_H^c \cdot \boldsymbol{n} = 0$ (where $\boldsymbol{n}$ denotes the outward normal to Ω_H).

If the right-hand side of (6.14) is set to zero, one can obtain the usual *gravity waves* solutions (traveling plane waves), which are, for a box with infinite horizontal dimensions: $\forall k, l \in \mathbb{R}$,

$$\eta(t, x, y) = \cos(kx + ly - \omega_{k,l} t) \tag{6.16}$$

with[53]

$$\omega_{k,l} = \sqrt{(k^2 + l^2) g \frac{\rho_1 - \rho_2}{\frac{\rho_1}{h_1} + \frac{\rho_2}{h_2}}}. \tag{6.17}$$

In the case of a box with finite horizontal dimensions, the gravity wave solutions are standing waves. The wave number (k, l) is restricted to the particular form

[53] A similar formula can be derived by assuming that the flow is potential, but without the shallow water hypothesis, see formula (6.43) below.

$(m\pi/L_x, n\pi/L_y)$ where m and n are two integers and the gravity wave solutions are standing plane waves:

$$\eta(t,x,y) = A\cos\left(\omega_{\frac{m\pi}{L_x},\frac{n\pi}{L_y}}(t-t_0)\right)\cos\left(\frac{m\pi}{L_x}x\right)\cos\left(\frac{n\pi}{L_y}y\right). \tag{6.18}$$

6.2.2.4 Approximation of the currents in the cell The fourth step is to deal with the magnetic part (6.9) of the system of equations (6.8). This is actually the most delicate step.

Using assumptions (E3)–(E4) and (L2), it can be shown that in the bath, the perturbed current can be approximated by

$$\boldsymbol{j}^c = -\frac{J_0\eta}{h_2}\boldsymbol{e}_z, \tag{6.19}$$

while in the aluminum, the perturbed current can be approximated by:

$$\boldsymbol{j}^a = \boldsymbol{j}^a_H(x,y) - \frac{J_0\eta}{h_2}\frac{z}{h_1}\boldsymbol{e}_z. \tag{6.20}$$

The horizontal component $\boldsymbol{j}^a_H$ of the perturbed current in the aluminum can be expressed in the following form:

$$\boldsymbol{j}^a_H = -\sigma_1\boldsymbol{\nabla}_H\phi, \tag{6.21}$$

where ϕ is a real-valued function which depends on (x,y) and is a solution to:

$$\begin{cases} -\Delta_H\phi = \dfrac{J_0\eta}{h_1h_2\sigma_1}, & \text{in } \Omega_H, \\ \dfrac{\partial\phi}{\partial\boldsymbol{n}} = 0, & \text{on } \partial\Omega_H. \end{cases} \tag{6.22}$$

Let us now explain how these approximations are justified.

The expression (6.19) for $\boldsymbol{j}^c$ comes from the fact that due to the big difference in conductivity between the bath and the aluminum, the current is nearly vertical in the bath. Since $\operatorname{div}\boldsymbol{j}^c = 0$, the current thus depends on (x,y) only, through the height of the interface. Using the fact that the difference between the electric potential Φ at the anode and at the interface (the cell voltage) is approximately constant, one then gets (6.19).

More precisely, let us denote by $\delta\Phi$ the cell voltage. From $\Phi(h_1+h_2)-\Phi(h_1+\eta) = \delta\Phi$ and the fact that Φ is approximately an affine function of z, we obtain

$$\boldsymbol{J}^c = -\frac{\sigma_2\,\delta\Phi}{h_2-\eta}\boldsymbol{e}_z.$$

Using now the fact that, for $\eta = 0$,

$$\boldsymbol{J}_0 = -\frac{\sigma_2\,\delta\Phi}{h_2}\boldsymbol{e}_z,$$

one easily gets the expression (6.19) for $\boldsymbol{j}^c = \boldsymbol{J}^c - \boldsymbol{J}_0$.

The expression (6.20) for $\boldsymbol{j}^a$ can be understood by the following reasoning. We first assume, in a "shallow water"-like approximation, that the horizontal component $\boldsymbol{j}_H^a$ of $\boldsymbol{j}^a$ only depends on (x, y). Then, using $\operatorname{div} \boldsymbol{j}^a = 0$, one obtains that the vertical component $\boldsymbol{j}_z^a$ of $\boldsymbol{j}^a$ is such that $\boldsymbol{j}_z^a = C - \operatorname{div}_H(\boldsymbol{j}_H^a)\, z$. Now, for small perturbations of the interface, the continuity of the normal component of $\boldsymbol{j}$ at the interface shows that $\boldsymbol{j}_z^a(h) = \boldsymbol{j}_z^c(h) = -J_0\eta/h_2$. At the bottom of the cell, $\boldsymbol{j}_z^a(0) = 0$ can be shown by using the fact that the conductivity of the cathode is small compared to the conductivity of the aluminum (see (E3) and Appendix A.2 in A.D. Sneyd [213]). From these two boundary conditions, one easily gets

$$\boldsymbol{j}_z^a = -\operatorname{div}_H(\boldsymbol{j}_H^a)\, z = -\frac{J_0\eta}{h_2}\frac{z}{h_1}.$$

The function ϕ introduced in (6.21) is the perturbation of the electric potential in the aluminum. The equation (6.22) on ϕ is derived from

$$\operatorname{div}_H(\boldsymbol{j}_H^a) = \frac{J_0\eta}{h_1 h_2},$$

and the fact that the sides of the cell are insulating.

6.2.2.5 Linearization of the Lorentz forces The fifth and last step is the linearization of the Lorentz forces in (6.14): $((\boldsymbol{J} \times \boldsymbol{B})_H^c - (\boldsymbol{J} \times \boldsymbol{B})_H^a)$. Using (6.12), we have by linearization:

$$\begin{aligned} &((\boldsymbol{J} \times \boldsymbol{B})_H^c - (\boldsymbol{J} \times \boldsymbol{B})_H^a) \\ &\approx ((\boldsymbol{j} \times \boldsymbol{B}_0)_H^c - (\boldsymbol{j} \times \boldsymbol{B}_0)_H^a) + ((\boldsymbol{J}_0 \times \boldsymbol{b})_H^c - (\boldsymbol{J}_0 \times \boldsymbol{b})_H^a). \end{aligned} \tag{6.23}$$

We have used the fact that, since $\mathbf{curl}\,(\boldsymbol{J}_0 \times \boldsymbol{B}_0) = 0$ (see (6.12) above), the tangential components of $\boldsymbol{J}_0 \times \boldsymbol{B}_0$ are continuous across the interface: $(\boldsymbol{J}_0 \times \boldsymbol{B}_0)_H^c = (\boldsymbol{J}_0 \times \boldsymbol{B}_0)_H^a$. By a simple scaling argument, one can see that $\boldsymbol{j}^c \ll \boldsymbol{j}_H^a$ and $\boldsymbol{j}_z^a \ll \boldsymbol{j}_H^a$ (using (E4)). The first term in the right-hand side of (6.23) can thus be approximated as follows: $((\boldsymbol{j} \times \boldsymbol{B}_0)_H^c - (\boldsymbol{j} \times \boldsymbol{B}_0)_H^a) \approx -\boldsymbol{j}_H^a \times (\boldsymbol{B}_{0,z}\boldsymbol{e}_z)$.

Another scaling argument shows that the second term in the right-hand side of (6.23) is of order $\mu h_1 |\boldsymbol{j}||\boldsymbol{J}_0|$, and thus can be neglected compared to the first term which is of order $\mu \min(L_x, L_y)|\boldsymbol{j}||\boldsymbol{J}_0|$ (see P.A. Davidson et al. [50, p. 278] and A.D. Sneyd et al. [214, equation (4.15)]). This approximation is actually questioned in the literature: we refer to M. Segatz et al. [206] and H. Sun et al. [222] for a discussion of the relative importance of these two forces.

The right-hand side of (6.14) can therefore be approximated as:

$$((\boldsymbol{J} \times \boldsymbol{B})_H^c - (\boldsymbol{J} \times \boldsymbol{B})_H^a) \approx -\boldsymbol{j}_H^a \times (\boldsymbol{B}_{0,z}\boldsymbol{e}_z). \tag{6.24}$$

6.2.2.6 The simplified system Collecting (6.14), (6.15), (6.19), (6.20), (6.21), (6.22) and (6.24), we obtain the following simplified system:

$$\left\{\begin{array}{ll}\left\{\begin{array}{l}\left(\dfrac{\rho_1}{h_1}+\dfrac{\rho_2}{h_2}\right)\dfrac{\partial^2\eta}{\partial t^2}-(\rho_1-\rho_2)\,g\Delta_H\eta=\sigma_1\left(\dfrac{\partial\phi}{\partial y}\dfrac{\partial \boldsymbol{B}_{0,z}}{\partial x}-\dfrac{\partial\phi}{\partial x}\dfrac{\partial \boldsymbol{B}_{0,z}}{\partial y}\right),\\ -\Delta_H\phi=\dfrac{J_0\eta}{h_1h_2\sigma_1},\end{array}\right. & \text{in } \Omega_H,\\ \left\{\begin{array}{l}(\rho_1-\rho_2)\,g\dfrac{\partial\eta}{\partial \boldsymbol{n}}=\sigma_1\boldsymbol{B}_{0,z}\left(\dfrac{\partial\phi}{\partial x}\boldsymbol{n}_y-\dfrac{\partial\phi}{\partial y}\boldsymbol{n}_x\right),\\ \dfrac{\partial\phi}{\partial \boldsymbol{n}}=0.\end{array}\right. & \text{on } \partial\Omega_H.\end{array}\right. \tag{6.25}$$

A non-dimensionalized version of this system can be obtained by introducing the non-dimensional quantities:

$$(\widetilde{x},\widetilde{y})=\left(\frac{x}{L},\frac{y}{L}\right),\ \widetilde{t}=\frac{t}{T},\ \widetilde{\eta}=\frac{\eta}{\eta_0},\ \widetilde{\boldsymbol{B}_{0,z}}=\frac{\boldsymbol{B}_{0,z}}{B_0},\ \widetilde{\phi}=\frac{\sigma_1\,B_0}{(\rho_1-\rho_2)\,g\,\eta_0}\phi, \tag{6.26}$$

where B_0 is a typical value of the vertical magnetic field, η_0 is a typical value of the displacement of the interface, T is a characteristic time and $L\simeq\min(L_x,L_y)$ is a characteristic horizontal dimension. The non-dimensionalized form of (6.25) is then, omitting the *tilde* for clarity:

$$\left\{\begin{array}{ll}\left\{\begin{array}{l}\dfrac{\partial^2\eta}{\partial t^2}-c^2\Delta_H\eta=c^2\left(\dfrac{\partial\phi}{\partial y}\dfrac{\partial \boldsymbol{B}_{0,z}}{\partial x}-\dfrac{\partial\phi}{\partial x}\dfrac{\partial \boldsymbol{B}_{0,z}}{\partial y}\right),\\ -\Delta_H\phi=\beta\,\eta,\end{array}\right. & \text{in } \Omega_H,\\ \left\{\begin{array}{l}\dfrac{\partial\eta}{\partial \boldsymbol{n}}=\boldsymbol{B}_{0,z}\left(\dfrac{\partial\phi}{\partial x}\boldsymbol{n}_y-\dfrac{\partial\phi}{\partial y}\boldsymbol{n}_x\right),\\ \dfrac{\partial\phi}{\partial \boldsymbol{n}}=0,\end{array}\right. & \text{on } \partial\Omega_H,\end{array}\right. \tag{6.27}$$

with the following non-dimensional numbers

$$c^2=\frac{(\rho_1-\rho_2)g}{\frac{\rho_1}{h_1}+\frac{\rho_2}{h_2}}\frac{T^2}{L^2},\ \beta=\frac{J_0B_0L^2}{h_1h_2(\rho_1-\rho_2)g}. \tag{6.28}$$

In our framework it is reasonable to consider a natural characteristic time T such that $c^2=1$ (this corresponds to $T\simeq 6\,\mathrm{s}$ for $h_1=0.2\,\mathrm{m}$, $h_2=0.05\,\mathrm{m}$ and $L=1\,\mathrm{m}$). Notice that the non-dimensional parameter β which appears in the Poisson equation is the same as that introduced by Sele (see equation (6.4)). This parameter can be seen as the coupling parameter between the two equations. For a typical value of the magnetic field B_0 between[54] 10 G and 100 G, a typical intensity J_0L^2 between 100 000 A and 500 000 A, $h_1=0.2\,\mathrm{m}$ and $h_2=0.05\,\mathrm{m}$, we observe that β varies between 6 and 340.

[54]We recall that $1\,\mathrm{G}=10^{-4}\,\mathrm{T}$.

The original system is thus approximated by a wave equation on the displacement of the interface η coupled with a Poisson equation on the perturbation of the electric field ϕ in the aluminum. The coupling comes from the right-hand sides of each equations, and also from the boundary conditions. A particular case which is often considered is the case of a constant vertical magnetic field $\boldsymbol{B}_{0,z}$, which reduces the first equation to a simple wave equation with zero right-hand side. In any case, we see that the vertical component of the background magnetic field plays an important role in this model.

A more rigorous derivation of this system (in particular of the magnetic part) can be found in the article [25] by V. Bojarevics and M.V. Romerio. It is based on an asymptotic analysis using the small parameters

$$\epsilon = \frac{\max(h_1, h_2)}{\min(L_x, L_y)} \text{ (see assumption (E4))}$$

and

$$\delta = \frac{\sup|\eta|}{h_1} \text{ (see assumption (L2)).}$$

Many assumptions need to be made for performing the analysis more rigorously (*a priori* expansions, $\sigma_2/\sigma_1 \sim \epsilon^2$). We also refer to A. Lukyanov et al. [156] for a similar approach. An analysis essentially based on scaling arguments to derive (6.27) can be found in A.D. Sneyd et al. [214] and P.A. Davidson et al. [50].

It is interesting to note that such a model has been used to successfully improve the stability of a 200 kA cell (see N. Urata [236]) which provides some *a posteriori* justification of the validity of the model, at least in a certain range of the parameter values.

6.2.2.7 A first discussion on the stability of (6.25) As already mentioned, the system of equation (6.25) (or a very similar one) has been used by many authors as a starting point to give a quantitative basis to Sele's argument. Let us discuss how instabilities can occur for the solution to (6.25). In a very qualitative way, we observe that instabilities for the solution to (6.25) can only come from the coupling between the equations, which requires:

(i) a non-uniform vertical component of the magnetic field in the case of an infinite geometry ($\Omega_H = \mathbb{R}^2$);
(ii) a finite geometry ($L_x < \infty$ or $L_y < \infty$) in the case of a uniform vertical component of the magnetic field.

Case (i) is related to a horizontal current in the cell, and will be studied in Section 6.2.3 and case (ii) is linked to the value of the vertical magnetic field and not its variation and will be studied in Section 6.2.4.

The Sele mechanism described in Section 6.1.4 above is necessarily linked to (ii), since it is based on the value of $\boldsymbol{B}_{0,z}$, and not on its variation, and this value only appears in the boundary condition.

6.2.2.8 Higher order approximations We have described one of the simplest ways to obtain a linear set of equations. We now briefly give a few examples of improvements of this simplified approach.

Concerning assumption (E5), it is possible to better take into account the mechanical dissipation using friction terms (see J.-F. Gerbeau et al. [102], A.D. Sneyd [213], S. Pigny et al. [185], A.D. Sneyd et al. [222], or S. Molokov et al. [169]) or a viscous shallow water model (see J.-F. Gerbeau et al. [102]).

Concerning assumption (L2), higher order terms with respect to the height of the waves at the interface can be introduced (see V. Bojarevics [24]). For a discussion of nonlinear wave theory in hydrodynamics, we refer, for example, to B. Le Méhauté [142].

Concerning the assumptions on the stationary state, a more complicated stationary state can be used, with a non-zero velocity field and deformation of the interface. We refer to H. Sun et al. [222].

We will now described how simplified systems such as (6.25) can be used to study the stability of the cell.

6.2.3 Analysis of a single Fourier mode and dispersion relations

In this section, we briefly overview some works analyzing the stability of traveling plane wave solution of a linearized system (in the spirit of system (6.25) above): [49,176,185,212,213]. In particular, assumptions (L1)–(L2), (E1)–(E2)–(E3) and (G1)–(G2)–(G3) are used. Assumption (E5) is used to model the mechanical dissipation by a linear damping term. Assumption (E4) does not play an essential role: in particular, these analyses do not rely on a shallow water approximation.

The noticeable differences between (6.25) and the system considered in these works are the following. In order to obtain traveling plane wave solutions, the authors consider a cell with infinite horizontal dimensions (see (G3)). In addition, in some papers, the stationary state is more complicated than the one we have considered to obtain (6.25): some horizontal components of the background current or a non-zero velocity field are taken into account.

As in the simplified system (6.25), the divergence of the perturbed Lorentz force $\boldsymbol{f} = \boldsymbol{j} \times \boldsymbol{B}_0 + \boldsymbol{J}_0 \times \boldsymbol{b}$ appears in the equations. An important remark made in [212] is that $\operatorname{div}(\boldsymbol{f}) = \boldsymbol{J}_0 \cdot \boldsymbol{j}$ (using $\mathbf{curl}\, \boldsymbol{b} = \boldsymbol{j}$ and $\mathbf{curl}\, \boldsymbol{B}_0 = \boldsymbol{J}_0$). Compared to the previous analysis made to obtain (6.25), both perturbations $\boldsymbol{j} \times \boldsymbol{B}_0$ and $\boldsymbol{J}_0 \times \boldsymbol{b}$ are used (see Section 6.2.2.5 above). A typical example of the system used in these works is written below in (6.41).

We first illustrate in Section 6.2.3.1 the kinds of result obtained in these works considering the simplified system (6.27) on a cell with infinite horizontal dimensions. We then briefly overview the conclusions obtained in the literature in Section 6.2.3.2 before discussing these results in Section 6.2.3.3.

6.2.3.1 A simple illustration based on (6.27) For illustration, the results typically obtained may be understood on the simplified system (6.27). We assume that $\Omega_H = \mathbb{R}^2$, and we search for traveling plane wave solutions of the form (see (6.16) above): $\eta = \operatorname{Re}(\eta_0 \exp(i(k_x x - \omega t)))$, $\phi = \operatorname{Re}(\phi_0 \exp(i(k_x x - \omega t)))$, where η_0 and ϕ_0 are two real constants. In addition, for simplicity, we suppose that $\partial \boldsymbol{B}_{0,z}/\partial y$ is constant. The following dispersion relation is readily obtained, inserting these solutions in (6.27):

$$\omega^2 = c^2 \left(k_x^2 + i \frac{\beta}{k_x} \frac{\partial \boldsymbol{B}_{0,z}}{\partial y} \right). \tag{6.29}$$

If $\beta = 0$, there is no current in the two liquids and we recover the non-dimensionalized version of the classical dispersion relation (6.17) for two-fluid shallow water waves. If $\beta \neq 0$, we see from (6.29) that traveling waves are likely to be unstable, since the imaginary part of ω becomes non-zero. This unstable behavior is enhanced for large β (accordingly with the Sele criterion (6.4)) or for small wave numbers. Notice that in order to obtain instabilities, we need some gradients of the vertical magnetic field. Moreover, only the variation of the vertical magnetic field appears in the dispersion relation. This very simple result follows the spirit of the works we now present in more details.

6.2.3.2 A review of the bibliography Basically, two types of instabilities are described in these approaches: some originate from horizontal components in the stationary currents in the liquid metal, and the others from gradients in the stationary magnetic field. In particular, the Sele mechanism is not identified as a source of instability in these approaches, since the vertical component of the magnetic field does not appear in the stability criteria (we recall that the Sele mechanism relies on the value of the vertical component of the magnetic field, and not on its variations). A uniform magnetic field has no effects in the derived dispersion relations. One common qualitative conclusion of these analyses is that the most unstable modes have large wavelengths. This seems physically correct since, as the wavelength increases, the stabilizing effects of surface tension and gravity are weaker. Moreover, these analyses show that both the augmentation of the anode to cathode distance and the aluminum pool depth are stabilizing.

Let us now detail the main assumptions and conclusions of each work:

- A.D. Sneyd [212]: *Geometry:* $h_1 = \infty$, the anode is part of the computational domain. *Stationary state:* flat horizontal interface, fluid at rest, uniform vertical current, magnetic field is a linear function of position. *Main conclusions:* The dispersion relation (see (3.11), p. 229 of [212]) shows that the irrotational part of the background magnetic field may be destabilizing. More precisely, the horizontal gradient of the horizontal components of this field plays the essential role, while, for example, the constant part of the background magnetic field has no effect on the stability of the cell. One conclusion is that the perturbation of the magnetic field due to the perturbed current has a stabilizing effect. A stability criterion, very similar to the Sele criterion (6.4) is derived (see (5.1), p. 234 of [212]).
- R. Moreau and D. Ziegler [176]: *Geometry:* The computational domain is the two fluids. *Stationary state:* flat horizontal interface, horizontal uniform velocities, non-uniform current with some horizontal components. *Main conclusion:* The analysis takes into account mechanical dissipation through a linear friction term. Only the interaction between

the background current and the perturbation of the magnetic field plays a role in instabilities. Here again, a uniform external magnetic field has no influence on instabilities. A dispersion relation is derived (see (21), p. 362 of [176]). It appears that the value of the friction term (see (E5)) is critical for the stability of the cell. Instabilities are generated by sufficiently large horizontal electric currents and propagate in their direction.

- A.D. Sneyd [213]. This article is a generalization of [212] with, in particular, mechanical dissipation, a finite aluminum pool depth and some horizontal component for the background current. *Geometry:* the computational domain is only the two fluids. *Stationary state:* flat horizontal interface, horizontal uniform velocities, uniform current with some horizontal components, magnetic field is a linear function of position. *Main conclusions:* A dispersion relation is derived (see equation (4.11), p. 119 of [213]). It is shown that spatial variations in the background magnetic field could destabilize the interface. The friction has a stabilizing effect. The most important destabilizing effects come from vertical gradients of the horizontal magnetic field components (or horizontal gradients of the vertical component of the magnetic field), and only depends on the vertical component of the background current. The analysis also detects less important instability mechanisms related to the horizontal gradient of the horizontal components of the magnetic field (as a generalization of [212]) and to the horizontal components of the background current (accordingly to [176]). The paper [49] presents similar results and proposes a physical interpretation, which relates such instabilities with those described in [176] (see also [46]).
- S. Pigny and R. Moreau [185]: This article is a generalization of [176] in order to take into account some spatial variation of the background magnetic field and current. The assumptions are similar to the ones made in [213], but with a more general repartition of the background current density. *Geometry:* the anode and the cathode are part of the computational domain. *Stationary state:* flat horizontal interface, horizontal uniform velocities, non-uniform current with some horizontal components. *Main conclusion:* A dispersion relation is derived (see equation (2.35), p. 10). The analysis concludes that large-wavelength disturbance may be unstable, due to the horizontal background current and that magnetohydrodynamics effects are predominant compared with classical Kelvin–Helmholtz instabilities. This article also gives qualitative results with two basic instability mechanisms: the differential pinch effect and levitation phenomena.

6.2.3.3 Discussion of this approach All these analyses neglect the effects of lateral boundaries (this was already mentioned in [212, p. 236]). These boundaries however play an essential role in the cell behavior, in particular in the Sele mechanism. This has been particularly emphasized by S. Molokov et al. in [156,169]. Loosely speaking, the boundaries imply some interactions between the traveling plane waves, and the coupling between different gravitational modes may lead to instabilities. Contrary to the results presented above, the uniform component of the magnetic field will play an important role. This is presented in next Section 6.2.4.

In order to distinguish the instability mechanisms we have described here, from those we introduce in the next section, notice that the vertical component of the background magnetic field cannot play a role if one considers only traveling plane waves perturbations (our discussion follows A.F. LaCamera et al. [137, p. 1182]). Indeed, if the traveling plane wave propagates in direction $\boldsymbol{k}$, the horizontal perturbed current in the aluminum is also along $\boldsymbol{k}$. Thus the Lorentz force $\boldsymbol{F}$ created by the interaction of this perturbed current with the vertical magnetic field is directed at right angles to the wave propagation direction, and thus cannot bring energy into the system since $\boldsymbol{F} \cdot \boldsymbol{k} = 0$.

6.2.4 Taking into account the boundaries: analysis of coupling of standing plane waves

In this section, we present some works which take into account the actual geometry of the cell. This additional difficulty compared to the works presented in the last section is compensated for by a simplification of the stationary state. In particular, the background velocity field is supposed to be zero, and the background current to be purely vertical. All these articles rely on the simplified system (6.25) we have presented above, or on a very similar one. They aim at providing a quantitative basis to Sele mechanism.

As in Section 6.2.3, we first illustrate some of the main conclusions of these works, starting from the simplified system (6.27). We show in Section 6.2.4.1 how explicit solutions in cylindrical domains may be obtained, and then we explain in Section 6.2.4.2 how unstable modes appear from the interaction between stable modes, in rectangular domains. Then we briefly overview the most representative works of this type in Section 6.2.4.3. We finally conclude in Section 6.2.4.4 with a discussion of these results and a comparison with the works presented in Section 6.2.3 above.

6.2.4.1 Explicit solutions in cylindrical domains In this section, we illustrate why (6.27) is linked to the Sele mechanism, by giving explicit solutions to this system in cylindrical domains. Our discussion follows [50, 156].

In this case, Ω_H is a disk. We denote by R its radius, and by (r, θ) the polar coordinates. We suppose that the background magnetic field is such that $\boldsymbol{B}_{0,z}$ is uniform. We choose the characteristic time T such that $c = 1$ (namely $T \simeq 6\,s$), the characteristic magnetic field value B_0 such that $\boldsymbol{B}_{0,z} = 1$, and the characteristic length L such that $R = 1$. The system (6.27) thus becomes

$$\begin{cases} \begin{cases} \dfrac{\partial^2 \eta}{\partial t^2} - \Delta_H \eta = 0, \\ -\Delta_H \phi = \beta\,\eta, \end{cases} & \text{in } \Omega_H \\ \begin{cases} \dfrac{\partial \eta}{\partial \boldsymbol{n}} = \left(\dfrac{\partial \phi}{\partial x}\boldsymbol{n}_y - \dfrac{\partial \phi}{\partial y}\boldsymbol{n}_x \right), \\ \dfrac{\partial \phi}{\partial \boldsymbol{n}} = 0, \end{cases} & \text{on } \partial\Omega_H \end{cases} \tag{6.30}$$

where

$$\beta = \frac{J_0 \boldsymbol{B}_{0,z} R^2}{h_1 h_2 (\rho_1 - \rho_2) g}. \tag{6.31}$$

This system admits explicit solutions of the form:

$$\begin{cases} \eta(t,r,\theta) = \hat{\eta}(r)\exp(i(\nu\theta - \omega t)), \\ \phi(t,r,\theta) = \hat{\phi}(r)\exp(i(\nu\theta - \omega t)), \end{cases} \tag{6.32}$$

where ν is an integer, and the functions $\hat{\eta}$ and $\hat{\phi}$ satisfy:

$$\begin{cases} \begin{cases} r^2\hat{\eta}''(r) + r\hat{\eta}'(r) + \left(\omega^2 r^2 - \nu^2\right)\hat{\eta}(r) = 0, \\ r^2\hat{\phi}''(r) + r\hat{\phi}'(r) - \nu^2\hat{\phi}(r) = -\beta r^2 \hat{\eta}(r), \end{cases} \\ \begin{cases} \hat{\eta}'(1) = -i\nu\hat{\phi}(1), \\ \hat{\phi}'(1) = 0. \end{cases} \end{cases} \tag{6.33}$$

Using the fact that the solutions are finite at $r = 0$, explicit solutions are given by:

$$\begin{cases} \hat{\eta}(r) = C\, J_\nu(\omega r), \\ \hat{\phi}(r) = C\left(\dfrac{\beta}{\omega^2} J_\nu(\omega r) - \dfrac{\beta}{\omega\nu}\dfrac{r^\nu}{R^\nu} J_\nu'(\omega r)\right), \end{cases} \tag{6.34}$$

where C is a real constant, J_ν is the Bessel function of the first kind (see M. Abramowitz and I.A. Stegun [1, p. 358]) and ω is linked to ν by the following dispersion relation:

$$-\nu\omega\frac{J_\nu(\omega)}{J_{\nu+1}(\omega)} + \omega^2 = i\beta. \tag{6.35}$$

To derive the dispersion relation, we have used the following property of Bessel functions of the first kind: $J_\nu'(z) = nz^{-1}J_\nu(z) - J_{\nu+1}(z)$ (see [1, p. 361]). As we have already mentioned, this derivation is justified for small values of ν[55], since higher modes will be damped by some mechanical dissipation phenomena (viscosity, surface tension) that we have neglected.

From this derivation, we can observe that:

- All these solutions will be unstable as long as $\beta > 0$.
- For small $|\omega|$ and for large $|\omega|$, the imaginary part of ω grows with β, in agreement with the Sele criterion (6.4) (see the asymptotic relations in [1]).
- The interface for these unstable modes is indeed in rotation due to the term $\exp(i(\nu\theta - \omega t))$, in agreement with the Sele mechanism.

We refer to S. Molokov et al. [156, 169] for a detailed analysis of these solutions, in particular in the case of large β.

[55] *i.e.* for the first modes.

6.2.4.2 Numerical computations in parallelepipedic domains: the phenomenon of collision of eigenvalues We have seen that in the framework of the model (6.27), cylindrical cells are unstable, whatever β is. The situation is quite different in parallelepipedic domains for which unstable modes appear with growing β by collision of eigenvalues. Our discussion follows [25].

In this case, $\Omega_H = (0, L_x) \times (0, L_y)$. Here again, we choose the characteristic time T such that $c = 1$, which corresponds to $T \simeq 6\,\mathrm{s}$. Let us rewrite the system (6.27):

$$\left\{\begin{array}{ll}\left\{\begin{array}{l}\dfrac{\partial^2\eta}{\partial t^2} - \Delta_H \eta = \left(\dfrac{\partial\phi}{\partial y}\dfrac{\partial \boldsymbol{B}_{0,z}}{\partial x} - \dfrac{\partial\phi}{\partial x}\dfrac{\partial \boldsymbol{B}_{0,z}}{\partial y}\right), \\ -\Delta_H \phi = \beta\,\eta,\end{array}\right. & \text{in } \Omega_H \\ \left\{\begin{array}{l}\dfrac{\partial\eta}{\partial \boldsymbol{n}} = \boldsymbol{B}_{0,z}\left(\dfrac{\partial\phi}{\partial x}\boldsymbol{n}_y - \dfrac{\partial\phi}{\partial y}\boldsymbol{n}_x\right), \\ \dfrac{\partial\phi}{\partial \boldsymbol{n}} = 0,\end{array}\right. & \text{on } \partial\Omega_H\end{array}\right. \tag{6.36}$$

where β is defined in (6.28). We recall that $\boldsymbol{B}_{0,z}$ is non-dimensionalized by a characteristic value of the magnetic field B_0. By classical integration by parts and using the boundary conditions on $\partial\Omega_H$, we obtain the following variational formulation of (6.36): search for η and ϕ in $H^1(\Omega_H)$ such that, for all ζ and ψ in $H^1(\Omega_H)$,

$$\left\{\begin{array}{l}\displaystyle\int_{\Omega_H}\frac{\partial^2\eta}{\partial t^2}\zeta + \int_{\Omega_H}\boldsymbol{\nabla}_H\eta\cdot\boldsymbol{\nabla}_H\zeta = \int_{\Omega_H}\boldsymbol{B}_{0,z}\left(\frac{\partial\phi}{\partial x}\frac{\partial\zeta}{\partial y} - \frac{\partial\phi}{\partial y}\frac{\partial\zeta}{\partial x}\right), \\ \displaystyle\int_{\Omega_H}\boldsymbol{\nabla}_H\phi\cdot\boldsymbol{\nabla}_H\psi = \beta\int_{\Omega_H}\eta\psi.\end{array}\right. \tag{6.37}$$

Let us introduce the functions

$$f_{m,n}(x,y) = \frac{2}{\sqrt{L_x L_y}}\epsilon_{m,n}\cos\left(\frac{m\pi}{L_x}x\right)\cos\left(\frac{n\pi}{L_y}y\right),$$

where

$$\epsilon_{m,n} = \left\{\begin{array}{ll}1 & \text{if } m \neq 0 \text{ and } n \neq 0, \\ \frac{\sqrt{2}}{2} & \text{if } m = 0 \text{ or } n = 0, \text{ and } (m,n) \neq (0,0), \\ \frac{1}{2} & \text{if } (m,n) = (0,0).\end{array}\right.$$

The set of functions $\{f_{m,n},\ m, n \in \mathbb{N}\}$ defines an orthogonal basis of $H^1(\Omega_H)$, orthonormal on $L^2(\Omega_H)$. These functions $f_{m,n}$ are the gravitational modes (see equation (6.18) above). They are the eigenvectors of the Neumann operator $-\Delta_H$, associated with the eigenvalues

$$k_{m,n}^2 = \left(\frac{m\pi}{L_x}\right)^2 + \left(\frac{n\pi}{L_y}\right)^2.$$

We express η and ϕ on this basis (spectral method):

$$\eta(t,x,y) = \sum_{m,n} \eta_{m,n}(t) f_{m,n}(x,y), \qquad \phi(t,x,y) = \sum_{m,n} \phi_{m,n}(t) f_{m,n}(x,y),$$

and by taking test functions in the set $\{f_{m,n},\ m,n \in \mathbb{N}\}$, we first easily obtain that, $\forall (m,n) \in \mathbb{N}^2$, $\phi_{m,n} = \beta \eta_{m,n}/k_{m,n}^2$ and thus that $\forall (m,n) \in \mathbb{N}^2$,

$$\begin{aligned}
&\frac{d^2\eta_{m,n}}{dt^2} + k_{m,n}^2 \eta_{m,n} \\
&\quad = \beta \frac{4\pi^2}{(L_x L_y)^2} \sum_{(m',n') \in \mathbb{N}^2} \frac{\eta_{m',n'}}{k_{m',n'}^2} \epsilon_{m,n}\, \epsilon_{m',n'} \\
&\quad \int_{\Omega_H} \boldsymbol{B}_{0,z} \Bigg(m'n \sin\left(\frac{m'\pi}{L_x}x\right) \cos\left(\frac{m\pi}{L_x}x\right) \cos\left(\frac{n'\pi}{L_y}y\right) \sin\left(\frac{n\pi}{L_y}y\right) \\
&\qquad - n'm \cos\left(\frac{m'\pi}{L_x}x\right) \sin\left(\frac{m\pi}{L_x}x\right) \sin\left(\frac{n'\pi}{L_y}y\right) \cos\left(\frac{n\pi}{L_y}y\right) \Bigg).
\end{aligned} \tag{6.38}$$

The $(0,0)$ mode evolves independently from the other modes and its evolution is linear. Therefore, we focus on the other modes in the following. For $(m,n) \neq (0,0)$, let us now introduce $\tilde{\eta}_{m,n} = \eta_{m,n}/k_{m,n}$. The system of equations (6.38) can be rewritten in the following form: $\forall (m,n) \in \mathbb{N}^2 \setminus (0,0)$,

$$\frac{d^2\tilde{\eta}_{m,n}}{dt^2} + k_{m,n}^2 \tilde{\eta}_{m,n} = \beta \sum_{(m',n') \in \mathbb{N}^2} G_{(m,n),(m',n')} \tilde{\eta}_{m',n'} \tag{6.39}$$

where $\forall (m,n),(m',n') \in (\mathbb{N}^2 \setminus (0,0)) \times (\mathbb{N}^2 \setminus (0,0))$,

$$\begin{aligned}
G_{(m,n),(m',n')} = {} & \frac{\epsilon_{m,n}\, \epsilon_{m',n'}}{L_x\, L_y\, k_{m,n}\, k_{m',n'}} \\
& \big(m'n (b_{m'+m,n'+n} - b_{m'-m,n'-n} + b_{m'-m,n'+n} - b_{m+m',n'-n}) \\
& + n'm(-b_{m'+m,n'+n} + b_{m'-m,n'-n} + b_{m'-m,n'+n} - b_{m'+m,n'-n}) \big),
\end{aligned}$$

with $\forall (m,n) \in \mathbb{Z}^2$,

$$b_{m,n} = \frac{\pi^2}{L_x L_y} \int_{\Omega_H} \boldsymbol{B}_{0,z} \sin\left(\frac{m\pi}{L_x}x\right) \sin\left(\frac{n\pi}{L_y}y\right).$$

Notice that since for all ϵ_m, $\epsilon_n \in \{-1,1\}$, $b_{\epsilon_m m, \epsilon_n n} = \epsilon_m \epsilon_n b_{m,n}$, G is an anti-symmetric matrix: $\forall (m,n),(m',n') \in (\mathbb{N}^2 \setminus (0,0)) \times (\mathbb{N}^2 \setminus (0,0))$, $G_{(m,n),(m',n')} = -G_{(m',n'),(m,n)}$.

Thus, the linear stability of the system (6.36) amounts to the spectral analysis of the matrix $K - \beta G$, where the matrices are indexed by the double indices $(m, n) \in \mathbb{N}^2 \setminus (0, 0)$, and K is the diagonal matrix defined by: $K_{(m,n),(m',n')} = k^2_{m,n}\delta_{(m,n),(m',n')}$, where $\delta_{(m,n),(m',n')}$ is the identity matrix. Indeed, if (λ, η) is an eigenmode of $K - \beta G$, a solution to (6.36) is $\exp(-i\omega t)\eta$ with $\omega^2 = \lambda$. It is easy to check that the real part of λ is positive (using the fact that K is a real symmetric positive definite matrix and G is a real antisymmetric matrix). Therefore, either all the eigenvalues of $K - \beta G$ are real and the system is stable, or there exists an eigenvalue with a non zero imaginary part and the system is unstable. With growing β, some unstable modes (namely some eigenvalues with non-zero imaginary parts) appear.

To understand this phenomenon, one can simply consider the toy problem $K = \begin{bmatrix} k_1^2 & 0 \\ 0 & k_2^2 \end{bmatrix}$ and $G = \begin{bmatrix} 0 & 1 \\ -1 & 0 \end{bmatrix}$. Then the eigenvalues of $K - \beta G = \begin{bmatrix} k_1^2 & -\beta \\ \beta & k_2^2 \end{bmatrix}$ solve $\lambda^2 - (k_1^2 + k_2^2)\lambda + (k_1^2 k_2^2 + \beta^2) = 0$ which is equivalent to

$$\left(\lambda - \frac{k_1^2 + k_2^2}{2}\right)^2 = \left(\frac{k_1^2 - k_2^2}{2}\right)^2 - \beta^2.$$

Thus, with growing β, the two eigenvalues starting from k_1^2 and k_2^2 for $\beta = 0$, get closer and closer until they meet when $\beta = \frac{1}{2}\left|k_1^2 - k_2^2\right|$. Beyond this threshold $(\beta > \frac{1}{2}\left|k_1^2 - k_2^2\right|)$, the two eigenvalues are complex conjugate: the absolute value of their imaginary part is $\left(\beta^2 - \left(\frac{k_1^2 - k_2^2}{2}\right)^2\right)^{1/2}$. We have here illustrated the fact that an instability occurs with growing β by "collision" of two eigenvalues. Notice that the critical value $\beta_{\text{critical}} = \frac{1}{2}\left|k_1^2 - k_2^2\right|$ is zero if the two eigenvalues are initially the same: $k_1^2 = k_2^2$. This situation occurs in practice in square geometries $(L_x = L_y)$, for example, since $k_{1,0}^2 = k_{0,1}^2$. In this case, like for cylindrical cells (see the section above), the model predicts that the cell is unstable as soon as $\beta > 0$.

We conclude this section with an experiment on a "realistic" geometry. We choose $L_x = 3\,\text{m}$, $L_y = 10\,\text{m}$ and a constant vertical magnetic field $\boldsymbol{B}_{0,z} \equiv 1$. In this case, $b_{m,n} = b_m\, b_n$, where

$$b_m = \int_0^\pi \sin(mx)\, dx = \begin{cases} 0 & \text{if } m \text{ is even} \\ \frac{2}{m} & \text{if } m \text{ is odd} \end{cases}$$

The eigenvalues of $K - \beta G$ for growing β can thus be numerically computed. We restrict ourselves to the modes (m, n) with $0 \leq m, n \leq 3$ (and $(m, n) \neq (0, 0)$). The results are plotted on Figure 6.4. As β grows the first eigenvalue with imaginary part appears for $\beta \in (0.244, 0.245)$, from the collision of modes $(0, 3)$ and $(1, 0)$. Then, this unstable mode restabilizes for $\beta \in (0.502, 0.503)$. The system is then stable for $\beta \in (0.503, 0.514)$. Finally, the system becomes definitely unstable for β larger than a critical value in the interval $(0.514, 0.515)$.

This numerical experiment illustrates the surprising fact that the system may be unstable for some values of β, and stable for larger values of β: unstable modes may "restabilize". It also appears that the critical β for which genuine unstable modes appears is rather small compared to realistic values (we recall that β typically varies between 6 and 340, see Section 6.2.2.6). This illustrates the fact that it is necessary to determine a stability threshold for the imaginary parts of the eigenvalues, in order to obtain a realistic value for β_{critical}.

Let us now discuss how this stability threshold is related to the friction term mentioned in (E5). In [24], it is shown that introducing a friction law as in [175] amounts to changing the left-hand side of the equation on η in (6.25) to:

$$\left(\frac{\rho_1}{h_1}+\frac{\rho_2}{h_2}\right)\frac{\partial^2\eta}{\partial t^2}+\left(\frac{\rho_1\gamma_1}{h_1}+\frac{\rho_2\gamma_2}{h_2}\right)\frac{\partial\eta}{\partial t}-(\rho_1-\rho_2)\,g\Delta_H\eta,$$

where γ_1 and γ_2 are the so-called friction coefficients, respectively, in the aluminum and in the bath, which have the dimension of the inverse of a time. Therefore, in the non-dimensionalized version, the left-hand side of the equation on η in (6.27) becomes:

$$\frac{\partial^2\eta}{\partial t^2}+\gamma\frac{\partial\eta}{\partial t}-c^2\Delta_H\eta, \tag{6.40}$$

where γ is a dimensionless friction coefficient:

$$\gamma = T\frac{\dfrac{\rho_1\gamma_1}{h_1}+\dfrac{\rho_2\gamma_2}{h_2}}{\dfrac{\rho_1}{h_1}+\dfrac{\rho_2}{h_2}}.$$

We recall that in this section, we have chosen a characteristic time $T \simeq 6\,\mathrm{s}$ so that $c = 1$. Typical values for the friction coefficients are $0.01\,\mathrm{s}^{-1} \le \gamma_1 = \gamma_2 \le 0.1\,\mathrm{s}^{-1}$ (see Table 1 in [241] where these coefficients are fitted by for example comparing computed velocities with experiments, or [176]), and thus $\gamma \in (0.06, 0.6)$. If (λ, η) is an eigenmode of $K - \beta G$, a solution to the linear model (6.27) with a damping term (6.40) is given by $\exp(-i\omega t)\eta$, with $-\omega^2 - i\gamma\omega + \lambda = 0$. As $\mathrm{Re}(\lambda) > 0$, one can check that for $\gamma \ge |\mathrm{Im}(\lambda)|/\sqrt{\mathrm{Re}\lambda}$, $\mathrm{Im}(\omega) \le 0$ and the system is therefore stable. The damping coefficient γ is thus related to the stability threshold (that we mentioned above) for the imaginary parts of the eigenvalues of the linear model (6.27).

On our numerical example, we observe that for $\gamma = 0.06$, as β grows, the first unstable mode appears for some $\beta \in (0.289, 0.290)$, then the system "restabilizes" for some $\beta \in (0.492, 0.493)$, and finally becomes definitely unstable for some $\beta \in (0.521, 0.522)$. For $\gamma = 0.6$, the system becomes definitely unstable for β larger than a critical value in the interval (0.800, 0.801). These are still very pessimistic values of β_{critical} compared to the typical values of β in real cells: $6 \le \beta \le 340$. Conversely, for $\beta = 6$, the system is stable for γ larger than 6.1. For $\beta = 340$, the system is stable for γ larger than 295. These values of γ correspond to unrealistic friction coefficients γ_1 and γ_2. In conclusion, the linear model seems too pessimistic with respect to the stability of industrial cells.

6.2.4.3 A review of the bibliography Let us now detail the assumptions and conclusions of some articles using this approach:

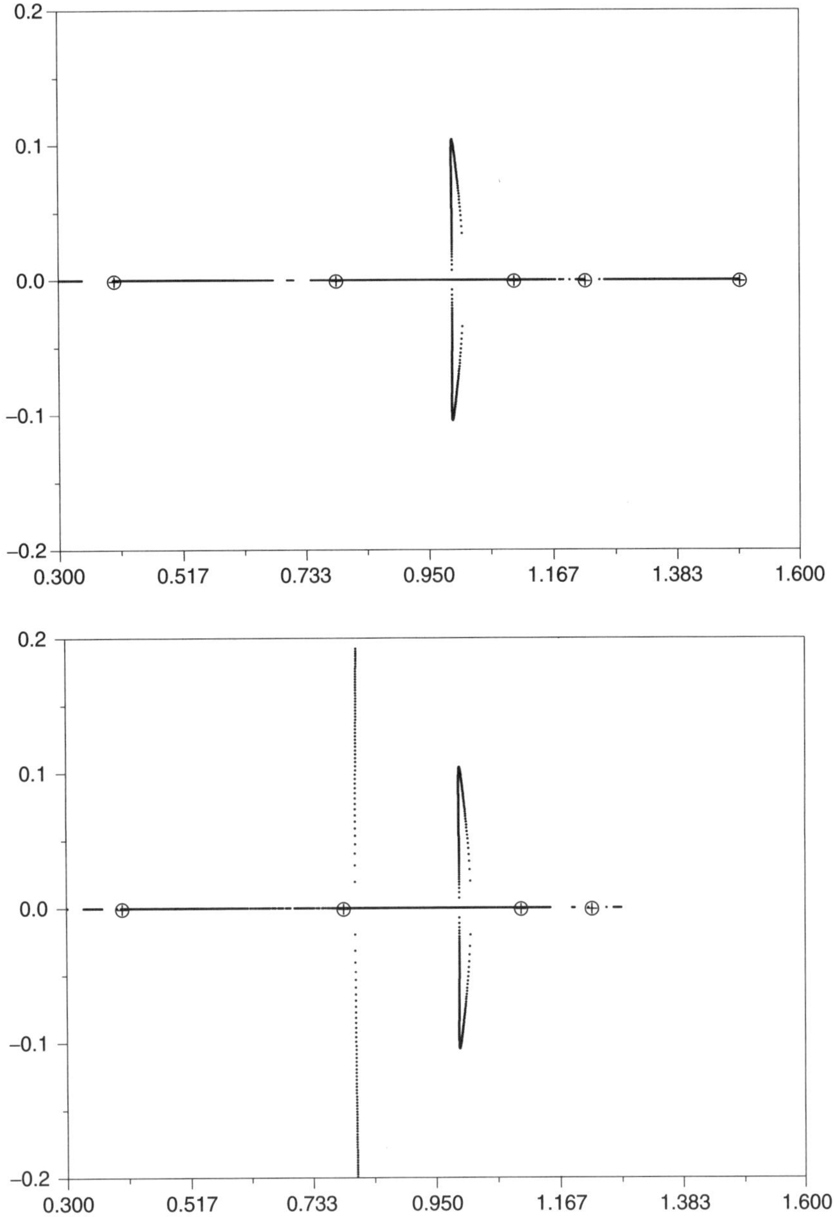

Fig. 6.4 Trajectories of the eigenvalues of $K - \beta G$ in the complex plane as β evolves (between two points, the increment on β is 0.001): $\beta \in [0, 0.5]$ at the top, and $\beta \in [0, 0.59]$ at the bottom. The initial positions of the eigenvalues when $\beta = 0$ are indicated by circles.

- N. Urata, K. Mori and H. Ikeuchi [237,236]: In these early articles, the authors propose a system similar to (6.25) and compare theoretical predictions with experiments on industrial cells. Using this model, they are able to enhance the stability of the cell, which shows the validity of the approach. An important observation is that the period of the interface oscillations are close to the periods of gravity waves, which can be analytically derived (see also J. Descloux et al. [64] on this subject). This explains partly why in many following works, the gravity waves will be used as a Galerkin basis to decompose the solutions of the MHD problem, or that some methods consider the MHD system as a perturbation of the pure gravity case to compute the eigenvalues (typically, a continuation method is used, from the pure gravity case to the MHD case, see for example Section 6.2.5 below). The authors suggest that the slight discrepancy between the observed periods and the theoretical gravity wave periods are due to MHD effects. The main conclusion of their rather qualitative analysis concerning instabilities is that the major roles are played by the interactions between the vertical magnetic field (and particularly its horizontal gradient) and the horizontal currents.
- A.D. Sneyd and A. Wang [214]: The authors analyze a simplified MHD system by projecting it onto the gravity waves. In particular, instabilities coming from the interaction of two gravity modes with close eigenvalues are studied. A stability parameter similar to β (see (6.4)) is derived and an estimate of the stability threshold β_{critical} is given. It is shown that the cell becomes more unstable with increasing external field, and the influence of the vertical component is predominant. They also study the stability in function of the angle between the symmetry axis of the external field and the longer cell wall. The general conclusion is that stability theories such as those presented in Section 6.2.3 above are incomplete, since interaction of gravity waves may induce instabilities with a very low current. In particular, the uniform part of the vertical component of the background magnetic field is a crucial parameter, while it did not appear in the analyses not taking into account the boundaries.
- V. Bojarevics and M.V. Romerio [25]: To our knowledge, this paper presents the most complete and most rigorous approach to derive the system of equations (6.25). The authors project this system onto a finite basis of gravity waves to analyze the linear stability of the cell numerically. Some stability criteria similar to (6.4) are derived by considering only two particular gravity modes. This analysis is further improved in [24] by taking into account some additional nonlinear terms in the asymptotic expansion (nonlinear waves) and a linear damping term to model mechanic dissipation (see (E5)). The influence of a non-zero background velocity field is taken into account. The eigenvalue problem is then numerically solved.
- P.A. Davidson and R.I. Lindsay [50]: This paper summarizes the two previous articles [214,25], and provides a mechanical analog, namely a compound pendulum. In cylindrical domains, the authors show that unstable modes correspond to a rotating tilted interface. The behavior of the eigenmodes with growing current is studied, showing how unstable modes arise from

interaction of two stable modes. A stability criterion of the form (6.4) is derived. These are some conclusions about the way unstable eigenvalues appear. First, it is not the gravitational modes with the closest eigenvalues which necessarily interact first to go unstable. In addition, the growth rate of the imaginary part of the unstable modes in function of the parameter β depends on the modes. Finally, it is important to consider more than two gravitational modes to completely describe instabilities.

Basically, the mechanism of instability behind all these studies is the Sele mechanism, namely instabilities created by the interaction of the horizontal component of the perturbed current in aluminum with the vertical component of the background magnetic field.

6.2.4.4 Comparison with previous works and discussion To conclude this section, we compare the results presented here [25, 50, 214, 237, 236] (where boundaries are taken into account) with the works presented in Section 6.2.3 [49, 176, 185, 212, 213] (with dispersion relations on traveling plane waves). One can notice that we do not recover here the instability mechanisms described in the last section since, according to formula (6.24):

- the vertical component of the perturbed current is neglected with respect to the horizontal component, and the basic mechanism of instability in [49, 212, 213] relies on the interaction of the vertical current with the horizontal field;
- the perturbation of the magnetic field is neglected, whereas it plays an important role in [176];
- the horizontal components of the background current are supposed to be zero, whereas they play an essential role in [176, 185, 213].

In other words, these works suppose that the instability mechanisms described in [49, 176, 185, 212, 213] can be neglected compared to the Sele mechanism.

The main criticism to the approaches presented in this Section 6.2.4 is the approximation of a uniform vertical background current, since it is known from practical knowledge that the horizontal components play an important role in the cell stability. These horizontal currents are due to the ledge, to the heterogeneities in conductivity of the cathode, and sometimes to some sludge at the bottom of the cell (some undissolved alumina).

Another criticism is that many dissipation phenomena are neglected (mechanic dissipation due to viscosity, surface tension, induced currents) which leads to rather severe instability criteria when they are directly applied to industrial cells: some stability threshold needs then to be determined, and this is not an easy task. This has been illustrated in Section 6.2.4.2.

6.2.5 Numerical computations of the eigenvalues around a precomputed stationary state

In this section, we present some works using more elaborate stationary states, and a finite element (or finite difference) method (rather than a discretization on gravitational modes) to discretize the spectral problem in space.

6.2.5.1 A first approach, based on [212] In [206], M. Segatz and C. Droste notice that the ledge should play an important role in the stability of the cell. They adapt the approach of [212] for finite horizontal dimensions. In particular, the hydrodynamics equations are reduced to a Poisson equation on the pressure (computed in the aluminum and the bath), while the electric potential is computed in the aluminum, the bath and the anode. The linearized system is thus slightly different from (6.25), but the assumptions used to derive it are similar. The linearization is performed around an horizontal and flat interface. The global system reduces to a system of coupled Poisson equations:

$$\left\{\begin{array}{l}
\left\{\begin{array}{l}
\Delta p^a = -2\mu\sigma_1 \boldsymbol{J}_0 \cdot \boldsymbol{j}, \\
\Delta p^c = -2\mu\sigma_2 \boldsymbol{J}_0 \cdot \boldsymbol{j}, \\
\Delta\phi^a = \Delta\phi^c = 0,
\end{array}\right. \\[2ex]
\left\{\begin{array}{ll}
\dfrac{\partial p^a}{\partial \boldsymbol{n}} = \dfrac{\partial p^c}{\partial \boldsymbol{n}} = \boldsymbol{f} \cdot \boldsymbol{n} & \text{on the boundary of the cell,} \\
\dfrac{\partial \phi^a}{\partial \boldsymbol{n}} = \dfrac{\partial \phi^c}{\partial \boldsymbol{n}} = 0 & \text{on the lateral sides of the cell,} \\
\dfrac{\partial \phi^a}{\partial z} = 0 & \text{at the bottom of the cell,} \\
\phi^c = 0 & \text{at the top of the anode,} \\
\dfrac{\partial p^a}{\partial z} = \boldsymbol{f} \cdot \boldsymbol{e}_z + \omega^2 \dfrac{\rho_1}{(\rho_1 - \rho_2) g}(p^a - p^c) & \text{at the interface,} \\
\dfrac{\partial p^c}{\partial z} = \boldsymbol{f} \cdot \boldsymbol{e}_z + \omega^2 \dfrac{\rho_2}{(\rho_1 - \rho_2) g}(p^a - p^c) & \text{at the interface,} \\
\sigma_1 \dfrac{\partial \phi^a}{\partial z} = \sigma_2 \dfrac{\partial \phi^c}{\partial z} & \text{at the interface,} \\
\phi^a - \dfrac{\eta}{\sigma_1} \dfrac{\partial \phi_0}{\partial z} = \phi^c - \dfrac{\eta}{\sigma_2} \dfrac{\partial \phi_0}{\partial z} & \text{at the interface,}
\end{array}\right. \\[2ex]
\text{where} \left\{\begin{array}{l}
\boldsymbol{j} = -\sigma \boldsymbol{\nabla}\phi \text{ is the perturbed current,} \\
\boldsymbol{f} = \boldsymbol{j} \times \boldsymbol{B}_0 + \boldsymbol{J}_0 \times \boldsymbol{b} \text{ is the perturbed Lorentz force,} \\
\eta = \dfrac{p^a - p^c}{(\rho_1 - \rho_2) g} \text{ is the displacement of the interface.}
\end{array}\right.
\end{array}\right. \tag{6.41}$$

In (6.41), ϕ^a (resp. ϕ^c) denotes the perturbation of the electric potential in the aluminum (resp. in the cryolitic bath and the anode) and p^a (resp. p^c) denotes the perturbation of the pressure in the aluminum (resp. in the bath). The complex number ω is the oscillation parameter in time according to (6.7). The perturbed magnetic field $\boldsymbol{b}$ (which is required to compute the force $\boldsymbol{f}$ at the boundary and at the interface) is computed by the Biot–Savart formula from the perturbed current $\boldsymbol{j}$. Notice that the stationary state only appears through the background current $\boldsymbol{J}_0$, the background magnetic field $\boldsymbol{B}_0$ and the background potential ϕ_0, which are all related. The background flow does not appear. The authors then compute numerically the eigenmodes (using a finite element like discretization in space) and show that with increasing current, the number of unstable modes grows rapidly. They also show that increasing the metal height, increasing the anode to cathode distance or decreasing the bath density all decrease the growth rate of unstable modes. Finally, they analyze changes of the behavior due to an anode change (cold anode), or to some sludge on the cathode block (cathodic current disturbance).

6.2.5.2 A second approach, using a more complete MHD model In a first step [64], J. Descloux and M.V. Romerio propose an approximation of the spectrum of the linearized system through a perturbation method assuming that the Lorentz forces are much smaller than the gravitational forces. The aim of this work is to explain why the observed oscillation frequencies are close to the gravitational ones. A basic assumption is that the frequency spectrum consists of simple (non-degenerate) eigenvalues.

In a second preliminary step, J. Descloux, Y. Jaccard, P. Maillard, M.V. Romerio and M.-A. Secrétan study in [63, 158, 159, 198] the case of a cell infinitely long in one direction (see assumption (G4)). In [63], it is proven that if the fields are invariant in one horizontal direction (say x), then instabilities cannot develop. More precisely, the y and z components of the velocity field vanish as time goes to infinity. It is important to mention that this result requires neither any linearization nor simplification of the MHD equations: the system the authors consider is that considered in this book. In the same situation, a linearization performed in [159] proves that the system is unstable if the background longitudinal current is too large. This paper relies on [198] to compute the stationary state around which the linearization procedure is performed.

In a third step, J.P. Antille, J. Descloux, M. Flueck and M.V. Romerio use a computational approach to analyze the linear stability of the cells in the series of papers [5, 55–59, 61, 197]. Here, the Lorentz forces are not supposed to be small compared to gravity forces, and the geometry of the cell is finite. Compared to the works presented so far, the main differences are:

- the background stationary state is not zero (the velocity field is not zero, and the interface is not supposed to be flat and horizontal);
- no assumptions on the current direction in the bath are needed;
- induced currents are taken into account;
- a precise description of the geometry of the cell is used (with anode channels and ledges);
- the discretization of the linearized system is performed with a finite element method, so that the computed eigenmodes are not (by construction) a linear combination of the gravitational modes.

The only assumptions used to derive the linearized system of equation are (E2) and (E5): the system of equations is thus much more complicated than the one mentioned earlier. Compared to the methods described above, this is definitely more a numerical approach.

The method consists of two steps. First, a stationary state (including the shape of interface) is computed using a precise description of the cell and of the busbars (see for example [62, 198]). The assumptions we have mentioned above do not concern this preliminary step, for which ferromagnetic material or viscous mechanical dissipation, for example, are taken into account. In other words, the stationary state is computed from a system of equations which is different from that used to derive the linearized equations (see Remark 3.1 in [55] about this apparent incoherence). Then, after a linearization of the MHD system around the stationary state using assumptions (E2) and (E5), the eigenmodes are computed using a continuation procedure with the intensity of the electric current running through the cell as the continuation parameter. Only the first twelve eigenmodes are computed, since higher modes are meaningless due to the fact that viscous effects have been neglected. An empirical stability threshold on the imaginary part of the eigenvalue is determined (as a function of the damping parameter, see (E5)). The continuation procedure is justified by the fact that the observed frequencies of dominant oscillating modes are close to the gravitational modes. For an overview of the method, we refer to [59].

Some theoretical results Let us first show how the system is derived and the theoretical results the authors have obtained. Let us mention that, while the derivation of the linearized system is formal and based on a classical *ansatz* procedure, these studies rely on a rigorous analysis of the spectral properties of the linearized system (see in particular [55, 57, 61]). In particular, theoretical and numerical problems related to multiple eigenvalues are extensively discussed.

Let us indicate the spectral problem the authors obtain:

$$\left\{\begin{array}{l} \left\{\begin{array}{l} -i\omega\rho\widetilde{\boldsymbol{u}} + \boldsymbol{\nabla}\widetilde{p} = F(\widetilde{\boldsymbol{u}}, \widetilde{H}), \\ \operatorname{div}(\widetilde{\boldsymbol{u}}) = 0, \\ -i\omega\widetilde{H} - \widetilde{\boldsymbol{u}}\cdot\boldsymbol{\nabla}(z - H_0) = G(\widetilde{\boldsymbol{u}}, \widetilde{H}), \end{array}\right. \\ \left\{\begin{array}{ll} \widetilde{\boldsymbol{u}} = 0, & \text{on the boundary of the cell,} \\ \widetilde{\boldsymbol{u}}\cdot\boldsymbol{n} = 0, & \text{on the interface,} \\ \widetilde{p}^a + \widetilde{H}\dfrac{\partial p_0^a}{\partial z} = \widetilde{p}^c + \widetilde{H}\dfrac{\partial p_0^c}{\partial z}, & \text{on the interface,} \end{array}\right. \\ \text{where} \\ \left\{\begin{array}{l} F(\widetilde{\boldsymbol{u}}, \widetilde{H}) = \frac{1}{\mu}\big(\mathbf{curl}\,(\boldsymbol{B}_0)\times\widetilde{\boldsymbol{B}}(\widetilde{\boldsymbol{u}}, \widetilde{H}) + \mathbf{curl}\,\big(\widetilde{\boldsymbol{B}}(\widetilde{\boldsymbol{u}}, \widetilde{H})\big)\times\boldsymbol{B}_0\big) \\ \qquad\qquad -\rho\boldsymbol{u}_0\cdot\boldsymbol{\nabla}\widetilde{\boldsymbol{u}} - \rho\widetilde{\boldsymbol{u}}\cdot\boldsymbol{\nabla}\boldsymbol{u}_0, \\ G(\widetilde{\boldsymbol{u}}, \widetilde{H}) = \dfrac{\partial\boldsymbol{u}_0}{\partial z}\cdot\boldsymbol{\nabla}(z - H_0)\,\widetilde{H} - \boldsymbol{u}_0\cdot\boldsymbol{\nabla}\widetilde{H}, \end{array}\right. \end{array}\right. \tag{6.42}$$

where $\boldsymbol{u} = \boldsymbol{u}_0 + \operatorname{Re}(\widetilde{\boldsymbol{u}}\exp(-i\omega t))$ (resp. $H = H_0 + \operatorname{Re}\big(\widetilde{H}\exp(-i\omega t)\big)$) denotes the velocity field, and its expansion in terms of a stationary field plus an oscillating fluctuation. The equation on $\widetilde{H}$ is posed on the stationary interface: for example, the velocity $\widetilde{\boldsymbol{u}}$ is taken at the point $(x, y, H_0(x, y))$. The fluctuation of the magnetic field $\widetilde{\boldsymbol{B}}(\widetilde{\boldsymbol{u}}, \widetilde{H})$ which appears in F is linear with respect to $(\widetilde{\boldsymbol{u}}, \widetilde{H})$ and is obtained through a separate magnetostatic problem, where induced currents are taken into account (see, for example, Section 5 in [55]). Biot–Savart law is used to solve this magnetostatic problem. The magnetic fluctuations are of course computed in the fluids, but also in the anode and in the cathode. In case $F = G = 0$ and the stationary state is such that the interface is flat and horizontal and the velocity field is zero, one recovers the usual gravitational modes when solving the spectral problem (6.42).

For computational and theoretical purposes, a variational formulation of the spectral problem (6.42) is derived.

In Section 6 of [57], it is shown that this spectral problem is well-posed under the following assumptions:

- the advective terms $-\rho \boldsymbol{u}_0 \cdot \boldsymbol{\nabla} \widetilde{\boldsymbol{u}} - \rho \widetilde{\boldsymbol{u}} \cdot \boldsymbol{\nabla} \boldsymbol{u}_0$ in the definition of F are neglected;
- the functions F and G in the right-hand sides are small compared to the left-hand side in (6.42) (see Propositions 6.5 and 6.6 in [57] for a more precise statement).

These assumptions are only used for the theoretical study, and are omitted in the numerical practice.

The numerical methods The algorithms used to compute the eigenvalues are based on the fact that the effects of the convective and electromagnetic terms are small in comparison with the gravity effects.

The algorithm is based on two principles. First, the authors use a continuation procedure with the intensity of the electric current running through the cell as the continuation parameter, to obtain an approximation (say Y) of the eigenspace. Then, the spectral problem at each step of the continuation procedure is solved using another iterative algorithm consisting in (let $Y_0 = Y$): for $k = 0, 1, 2, \ldots$, until a stopping criterion is fulfilled,

- in the first step, solve the spectral problem on the approximate Riesz–Galerkin space Y_k;
- in the second step, use one iteration of the inverse power method with shift on each eigenvector in Y_k to determine a new Riesz–Galerkin space Y_{k+1} (this is done in a finite element space).

The function spaces Y_k are subspaces of finite element spaces. The interest of this algorithm is that it can handle situations corresponding to a degenerate spectrum.

In this approach, the linear damping term used to model mechanical dissipation in the linearized hydrodynamics equations directly determines the stability threshold on the imaginary part of ω (see p. 130–131 in [59], and also Section 6.2.4 above). This is why this linear damping term is not included in the linearized system (6.42): it is rather used in a final step to determine the stability of the cell. The value for this damping factor may be determined from an analysis in infinitely long cell (see [158]).

Some numerical results and comparison with other works Let us now mention a few conclusions of some numerical studies (see [56,58,197]) based on this method. As expected, the authors observe in [58] that an increase of the anode to cathode distance decreases the imaginary part of the eigenvalues, and thus the unstable character of the cell. In [197], it is mentioned that the imaginary part of the eigenvalues strongly depends on the background current and flow. The induced currents are also responsible for significant contributions.

Let us now underline the qualitatively different conclusions these studies lead to, compared to the results obtained with the simpler methods we have presented above:

- As noticed in Section 4 of [57], if the stationary state is supposed to be the simplest one (zero velocity field, flat horizontal interface) and without induced currents, the only phenomenon of instability is the collision of two real eigenvalues (when increasing the current) to create two complex conjugate eigenvalues, as observed in [25, 50], and which can be related to the Sele mechanism (see Section 6.2.4.2 above). On the other hand, for a non-zero stationary state and induced currents, instabilities may occur as long as the current is non-zero, without such a collision phenomenon: this is an important difference with the works in the spirit of [25, 50].
- In [56], it is observed that instabilities disappear for large values of the vertical component of the background magnetic field, which is an opposite effect to many other works. The authors suggest that this discrepancy is due to induced currents.

As a conclusion, we would like to mention that some other groups have derived similar approaches but have published only a few papers on their results (see, for example, the ESTER/PHOENICS code: V. Potocnik et al. [188, 189] and Ch. Droste [71], or the numerical experiments reported by M. Dupuis et al. and L. Leboucher et al. in [72, 144]).

6.3 A nonlinear approach

This section is devoted to the application of the numerical approach described in the former chapters to the modeling of aluminum electrolysis cells. We first describe in Section 6.3.2 some experiments on realistic cells which enable us to compare numerical results with experimental data. Then we present in Section 6.3.3 how such an approach can give an insight into a well-known physical phenomenon for industrial cells: metal pad rolling.

6.3.1 Generalities about linear versus nonlinear approaches

The nonlinear approach models the problem with system (1.52). Therefore, the model at the continuous level contains only a few simplifications: Navier–Stokes equations and Maxwell equations are simply coupled by Ohm's law and the Lorentz forces. This is a genuinely nonlinear approach. In this section, we give some results obtained with this model.

We first need to give a definition of the cell stability in the nonlinear framework. We propose to use a rather "practical" definition of stability as follows: the system is stable if the interface remains inside the domain of computation[56] and is unstable if not.

We would like to explain why a classical stability analysis based on the study of the sign of the imaginary part of the eigenvalues (see Section 6.2) may not be relevant in the framework of strongly nonlinear partial differential equations. These are a few facts we would like to stress:

- For nonlinear PDEs, linear stability analysis may erroneously conclude that a stationary state is stable. The most famous example is turbulence in shear flows which may develop for Reynolds numbers such that the laminar shear flow is still linearly stable (see for example R.O. Grigoriev and A. Handel [112]). This is related to the fact that for infinite-dimensional non-normal operators, spectral analysis does not guarantee stability.
- Linear stability analysis cannot help to conclude if a perturbation of a stationary state will lead to unbounded response (*i.e.* if the stationary state is "unstable"). Indeed, the linearized equations around a stationary state are valid for a small perturbation. In the case of large deviations from the stationary state, this assumption is questionable. An example of such a discrepancy between a linear stability analysis and a nonlinear stability analysis in the framework of aluminum electrolysis is given in the two papers of J. Descloux et al. [63] and P. Maillard et al. [159]. The authors consider the same configuration (a cell infinitely long in one horizontal direction, and fields which are transversionaly invariant in this direction). The nonlinear analysis performed in [63] concludes that the system is stable. However, in the second paper [159], the linearization around a stationary state, for which a constant longitudinal electric current runs in the cell, leads to unstable situations.
- Careless linearization may induce instabilities. For an example in the fluid–structure interaction, we refer to the work of D. Errate et al. [76], where it is shown that neglecting the Navier term in the Navier–Stokes equations for the fluid may prevent establishing the energy inequality, and therefore lead to unstable numerical results.
- Linear stability analysis is based on the analysis of the spectrum of a linearized operator. One important item of information which may be lost through the linearization procedure is the time needed for unstable modes to develop. In our particular framework, it may happen that a disturbance occurs on a finite time range (like for example during an anode change), and the period of time perturbation is an important parameter to take into account. Actually, the inverse of the imaginary part of the eigenvalues can

[56] More precisely, in the continuous-in-time problem, the interface remains inside Ω but can get very close to the anode if the interface motion is too large. At the discrete level, this translates into the unphysical situation that the interface leaves the domain Ω. It corresponds to the case that the motion of the interface is so large that the aluminum pad touches the anode, which clearly corresponds to an unstable industrial situation.

	Aluminum	Bath
Density	$\rho_1 = 2300\,\mathrm{kg\,m}^{-3}$	$\rho_2 = 2150\,\mathrm{kg\,m}^{-3}$
Viscosity	$\eta_1 = 1.196.10^{-3}\,\mathrm{kg\,m}^{-1}.\mathrm{s}^{-1}$	$\eta_2 = 2.558.10^{-3}\,\mathrm{kg\,m}^{-1}.\mathrm{s}^{-1}$
Conductivity	$\sigma_1 = 3.5.10^6\,\Omega^{-1}\,\mathrm{m}^{-1}$	$\sigma_2 = 2.5.10^2\,\Omega^{-1}\,\mathrm{m}^{-1}$
Surface tension	$\gamma = 0.5\,N.m^{-1}$	

Table 6.1 Physical parameters of the fluids.

	Aluminum	Bath
Velocity	$U = 0.1\,\mathrm{m\,s}^{-1}$	
Magnetic field	$B = 0.01\,\mathrm{T}$	
Distance	$L = 1\,\mathrm{m}$	
Reynolds number	$Re_1 = 1.92.10^5$	$Re_2 = 8.41.10^4$
Magnetic Reynolds number	$Rm_1 = 0.44$	$Rm_2 = 3.14.10^{-5}$
Coupling parameter	$S = 3.46$	
Froude number	$Fr = 10^{-3}$	
Density ratio	$M = 0.935$	
Weber number	$We = 46$	

Table 6.2 Typical values for the characteristic parameters and non-dimensional numbers in the cell (see system (1.52) and equation (5.55)).

be used as a typical time scale for the instability to develop, but an accurate quantitative result is difficult to obtain.

6.3.2 Some experiments on realistic cells

We now describe some results on the behavior of realistic aluminum electrolysis cells, obtained with the MHD model and the numerical schemes described in the preceding chapters. What we demonstrate is the ability of the modeling to reproduce some well-known phenomenological parameters (*e.g.* the number of vortices; see also Section 3.7.2) as well as to provide some evaluation of parameters that are difficult to measure (*e.g.* the shape of the interface between the bath and the metal).

The specific data needed by the nonlinear model are:

- the geometric characteristics of the cell to define and to mesh the domain Ω;
- the physical parameters of the fluids (see Tables 6.1 and 6.2);
- some components of the magnetic field on the boundary (see (1.53)).

These boundary conditions contain for example information about the intensity of the current running through the cell, or about the design of the array of conductors around the cell. Some remarks regarding these boundary conditions are in order. In a first approximation, the magnetic field on the boundary can

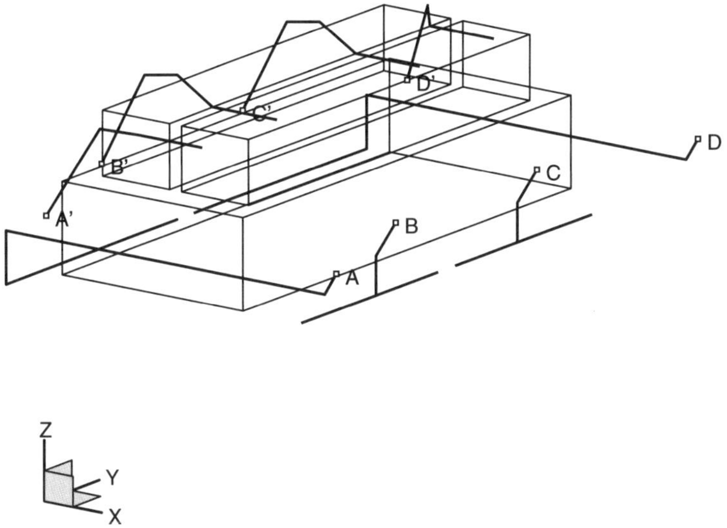

Fig. 6.5 A schematic circuit of conductors of an idealized cell. In factories, cells function in series, so that the points A (resp. B, C and D) are connected to the points A′ (resp. B′, C′ and D′) of the next cell. The cell is represented by the lower parallelepiped, while the two smaller parallelepipeds above represent the anodes.

be obtained either from some experimental measurements, or from a precomputation. In the second case, some magnetostatic models are employed, using the current repartition within the cell, the configuration of the network of conductors around the cell and accounting for the ferromagnetic characteristics of the metallic shell of the cell. In principle, the magnetic field on the boundary should be updated, considering that the flow inside the cell influences the magnetic field on the boundary. It is, however, a fair assumption to neglect these perturbations. To give a basis to this strategy, we notice that numerical experiments in simple cases have shown that the results obtained by a fully coupled computation between a MHD model in Ω and a magnetostatic model in $\mathbb{R}^3$ are similar to those obtained with fixed boundary conditions. An additional favorable feature is that only some of the components of the magnetic fields are needed, and sometimes, the boundary conditions do not depend on the magnetic field inside the cell (see for example the boundary condition $\boldsymbol{E} \times \boldsymbol{n} = 0$ on conducting boundaries, namely on the anodes).

In this section, we only give results on a simplified geometry, with a schematic network of conductors (see Figure 6.5). However, these results are representative of that obtained on more realistic cells. Here, the electric current in the cell is $90\,kA$ and the dimensions of the cell are about $3\,m \times 9\,m \times 1.5\,m$. We compute the boundary conditions on the magnetic field by simple Biot–Savart laws using either linear or parallelepipedic conductors (see Figure 6.5).

The numerical method we have described in Chapter 3 (and more precisely in Sections 3.5 and 3.6.1) can be used to compute the stationary state of an

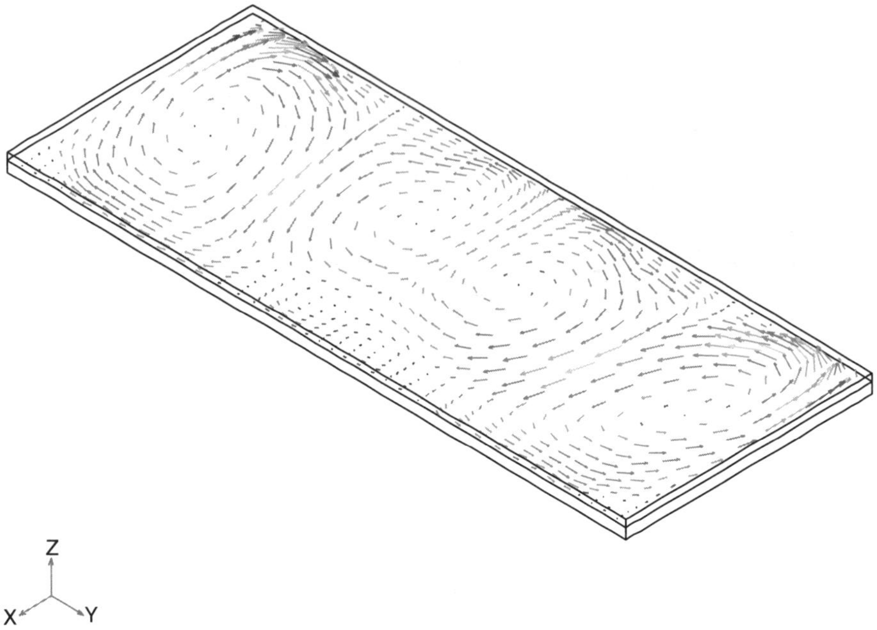

Fig. 6.6 Velocity profile in a plane situated 5 cm under the interface. Velocity magnitude is between 1 $\text{cm}\,\text{s}^{-1}$ and 9 $\text{cm}\,\text{s}^{-1}$. **See plate 4**.

aluminum electrolysis cell, for a given fixed position of the interface. The aim here is thus not to study the stability of the cell, but the velocity and the magnetic field in the cell, for a fixed position of the interface that we supposed to be *a priori* known. This stationary position can be obtained through transient ALE simulations (see below) or from measurements.

The geometry of the computational domain is that of the industrial cell. The boundary conditions on the magnetic fields come either from experimental measurements in the factory or from independent numerical simulation of the magnetic field created by the conductors around the cell. To obtain the presented results, we have used a continuation method on two parameters (the Reynolds number and a multiplicative coefficient of the boundary conditions on the magnetic field).

In Figures 6.6 and 6.7, we show the result of a computation for a schematic cell represented on Figure 6.5. On the magnetic field profile, one can see the influence of the two positive conductors.

We can use the ALE method presented in Chapter 5 to compute the deformation of the interface. In Figure 6.8, we show the shape of the interface between the two fluids, obtained after a long-time computation. Here, we compute the stationary state of the cell without *a priori* knowing the position of the interface.

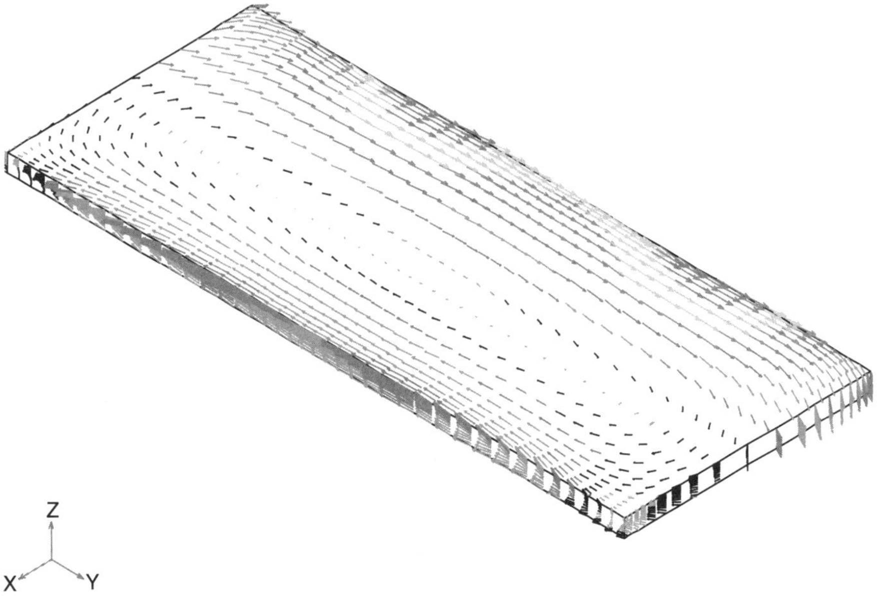

Fig. 6.7 Magnetic field on the boundary of the cell. *See plate 5.*

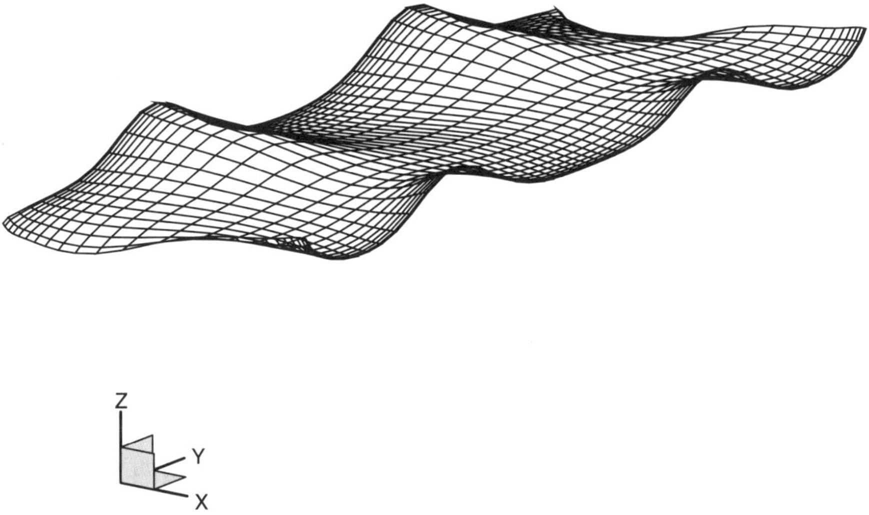

Fig. 6.8 The shape of the interface computed with a long-time simulation, with the simplified circuit of conductors represented in Figure 6.5.

6.3.3 Metal pad roll instabilities

In this section, we present a test case, inspired from a phenomenon observed in aluminum electrolysis cells and thoroughly investigated over the past few years: *metal pad rolling*. It is an oscillation of the bath/aluminum interface with a period ranging from five seconds up to more than one minute (see, for example, K. Mori et al. [178]). The aim of most of the theoretical and experimental studies of MHD cells has been to understand, forecast and avoid this phenomenon (see Section 6.2 above).

6.3.3.1 The physical phenomenon One of the explanations of metal pad rolling is the presence of a vertical component in the magnetic field. The Sele criterion follows this line (see T. Sele [207] and Section 6.1.4 above). T. Sele was the first to propose a physical grounding for the rotation by the interaction of the vertical component of the magnetic field with horizontal perturbed currents. More recently, V. Bojarevics and M.V. Romerio [25] and then P.A. Davidson and R.I. Lindsay [50] have used a more general linearized system to study this phenomenon (see Section 6.2.4 above). Their analysis leads to quantitative results for the instability of standing and traveling waves in rectangular and circular cells. In [50], they also give a mechanical analog which provides an interesting physical insight into the phenomenon.

The physical phenomenon creating the rolling of the metal is generally explained in the physics literature as follows (see Figure 6.9). An initial tilting (or a long-wavelength disturbance) creates a perturbed current flow $\boldsymbol{j} = \boldsymbol{J} - \boldsymbol{J_0}$ ($\boldsymbol{J_0}$ denotes the unperturbed – or background – current and $\boldsymbol{J}$ the total current in the cell) which is mainly vertical in the bath and horizontal in the aluminum (because of the strong difference of electric conductivity). The interaction of this current with the vertical magnetic field results in a horizontal Lorentz force $\boldsymbol{F} = \boldsymbol{j} \times \boldsymbol{B}$, in the direction perpendicular to $\boldsymbol{j}$. It consequently induces a rotating motion of the interface. This is in the vein of the basic mechanism we explained in Section 6.1.4 (see Figure 6.3).

In some cases, this phenomenon may lead to an instability: when the vertical field is too large, the amplitude of the oscillation may grow with time. Reportedly, the metal can even escape from the cell!

6.3.3.2 Numerical simulations In [25, 50], the authors propose that the cause of unstable rolling waves is the interaction between gravitational modes (see Section 6.2.4.2). In particular, they show that a cylindrical cell becomes instable whenever the vertical component $\boldsymbol{B}_z$ of the magnetic field is not zero. We gave details about this case in Section 6.2.4.1. Their approach based on many assumptions and linearization may well reproduce qualitatively metal pad rolling, but a few of the assumptions are quite questionable as far as quantitative results are concerned (see Section 6.2 and Section 6.3.1). This simple experiment of a cylindrical cell is reproduced here, with a view to checking this strong result of instability, which is obtained after many assumptions and linearization

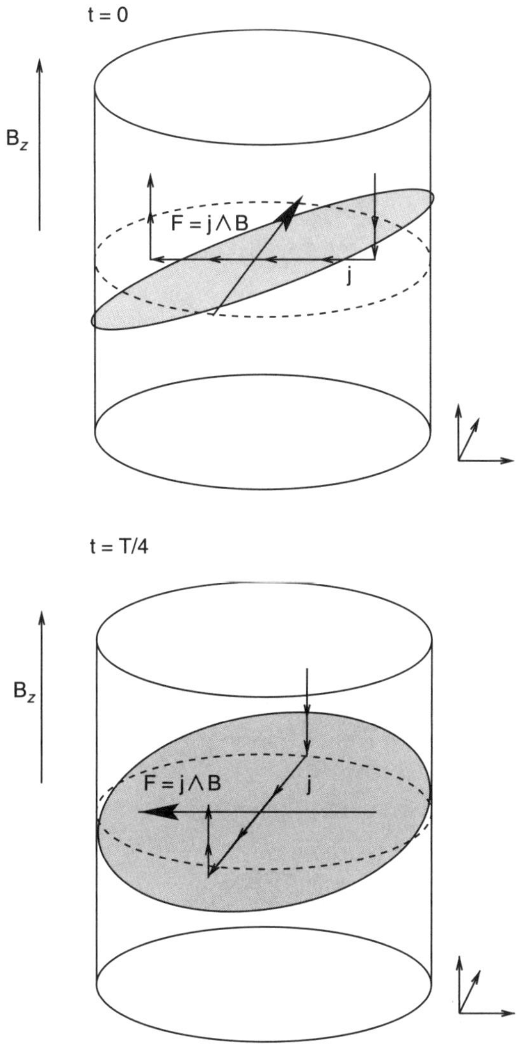

Fig. 6.9 Rolling phenomenon.

(see Section 6.2.2). A major interest of such a test is to check whether the result survives in the fully nonlinear setting.

The test case cell is a cylindrical cell of unit radius and height equal to 2. Initially, the fluids are at rest, the interface is located at the mean height and is flat and horizontal. On the wall, we impose during the whole simulation the magnetic boundary conditions corresponding to a uniform vertical electric current $-J_0\boldsymbol{e}_z$, where $J_0 > 0$. For $0 \leq t \leq 1$, the fluids are subjected to a gravity field, inclined at an angle of 5 degrees with respect to the vertical. This corresponds

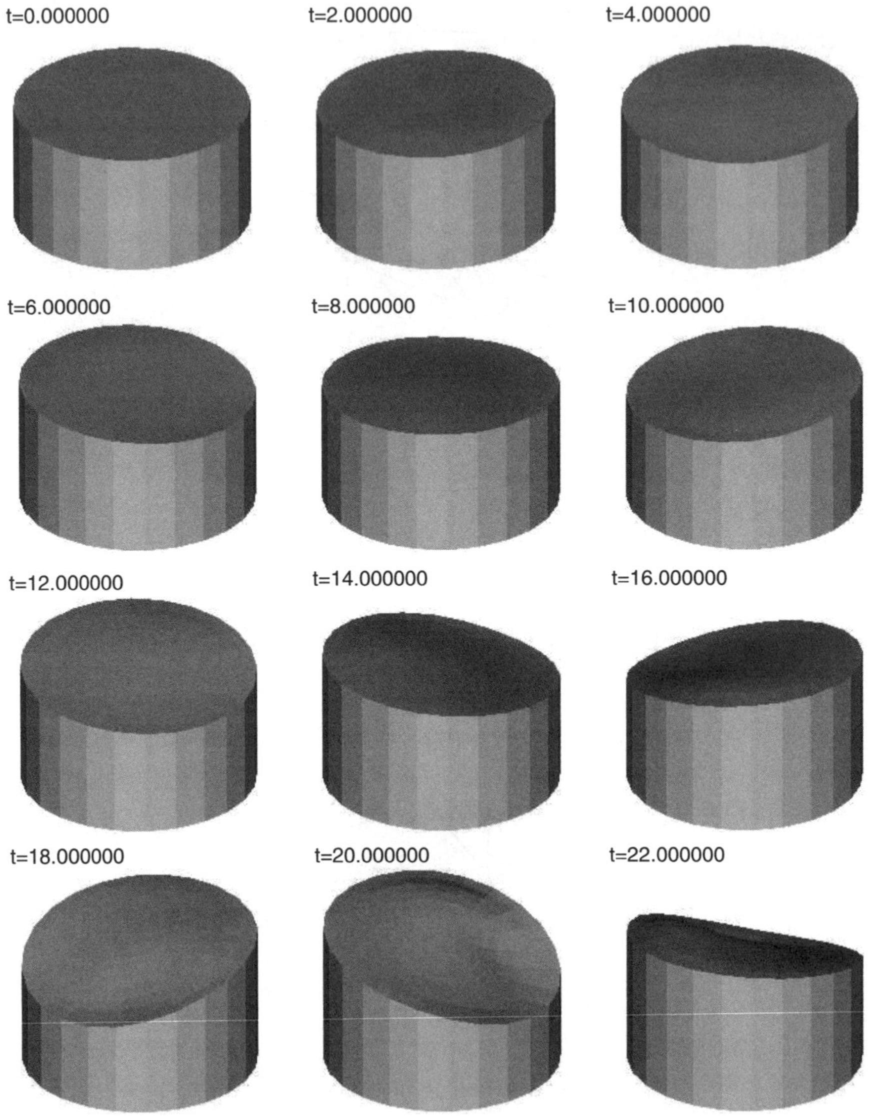

Fig. 6.10 The phenomenon of metal pad rolling in a cylindrical cell. Visualization of the interface and the lower fluid (the upper fluid is not represented for the sake of clarity). This is a case with $\boldsymbol{B}_z = 0.2$.

to slightly tilting the cell and creates an initial disturbance. For $1 < t \leq 25$, a vertical gravity is applied back, and a vertical magnetic field $\boldsymbol{B}_z$ is superimposed to the orthoradial magnetic field created by the electric current $-J_0\boldsymbol{e}_z$. This actually induces the metal rolling phenomenon. For $t > 25$, the vertical magnetic field is removed, and the system is allowed to recover its initial

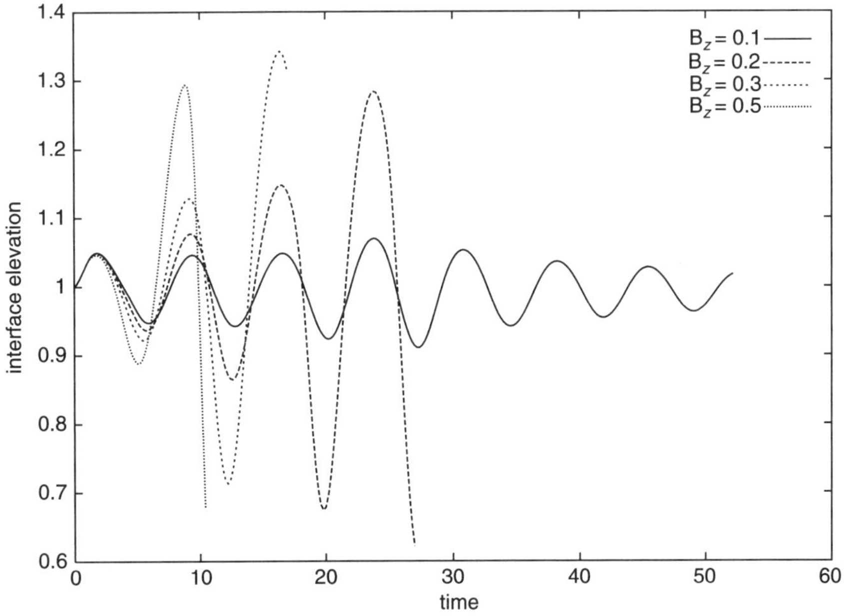

Fig. 6.11 Time evolution of the elevation of a point of the interface in the metal pad roll experiment for various values of $\boldsymbol{B}_z$. The stable simulation is obtained with $\boldsymbol{B}_z = 0.1$.

configuration. The non-dimensional parameters are the following: $Re_1 = Re_2 = 1000$, $M = 0.935$, $Fr = 0.1$, $Rm_1 = 10^{-4}$, $Rm_2 = 1$, $S = 1$, $We = 50$, $J_0 = 2$. We perform the simulations for various values of $\boldsymbol{B}_z$. We recall that J_0 and $\boldsymbol{B}_z$ are only used in the model to compute the boundary conditions for the magnetic field $\boldsymbol{B}$.

Figure 6.10 shows the lower fluid and the interface when $\boldsymbol{B}_z = 0.2$. Figure 6.11 shows the evolution in time of the elevation of the point of the interface initially situated at $(1, 0, 1)$ for $\boldsymbol{B}_z = 0.1, 0.2, 0.3$ and 0.5. The last three values lead to an "explosion" of the interface. The physical interpretation of the phenomenon proposed on Figure 6.9 is confirmed by computing the disturbed currents. In order to compute this disturbance, the initial current $\mathbf{curl}\,\boldsymbol{B}_0$ is subtracted from $\mathbf{curl}\,\boldsymbol{B}$. A few streamlines of the perturbed current are represented on Figure 6.12: there is indeed a "loop of current" as predicted in the theoretical explanation of the phenomenon (see Figure 6.9).

In other simulations, we have also observed that small disturbances of the initial state do not lead to instability, at least on a reasonable time scale. Likewise, a small positive $\boldsymbol{B}_z$ does not induce instability. This is in apparent contradiction with the results of the linear approach which claim the instability of the cell (see Section 6.2.4.1). At least our results show that, should the instability occur, it will occur only in the large-time limit, and therefore may not be relevant from the practical viewpoint.

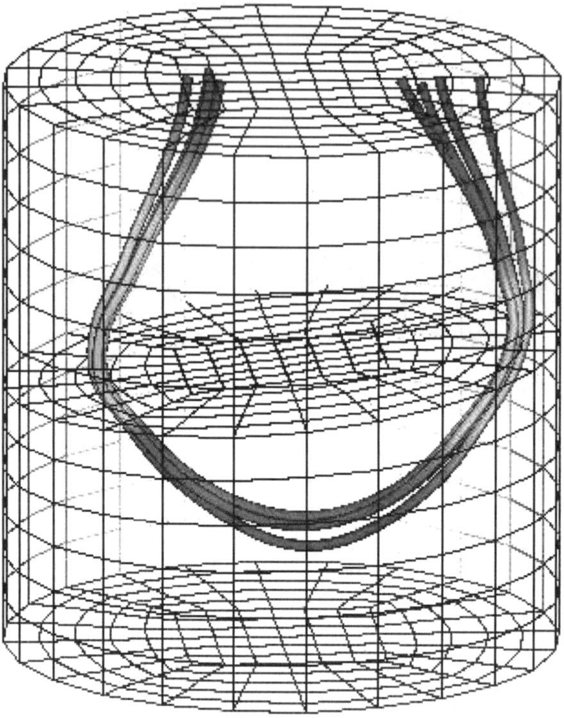

Fig. 6.12 Loops of current during metal pad roll. Some streamlines of the perturbed current $\boldsymbol{j}$ (*i.e.* the difference between the current and the current for a flat horizontal interface) are shown. The perturbed current flows here from the right-hand side of the figure (where the elevation of the interface is the highest) to the left-hand side.

This test case demonstrates the capability of the nonlinear approach to simulate complex MHD phenomena. These phenomena have been analyzed with models based on many simplifications of the original equations providing interesting qualitative results. But the influences of such simplifications need in any case to be tested and may become important when accurate results are desired.

The nonlinear approach is able to reproduce qualitatively the results predicted by simplified models, but also to give *quantitative* information on the transient evolution of the system. This is what we now show.

6.3.4 Spectral analysis

In this section, we would like to describe a way to compare quantitatively the results provided by the nonlinear approach and the spectral analysis on simplified linear systems, as described in Section 6.2. The idea is to use the evolution in time of the altitude of a point on the interface as the output signal. A spectral analysis of this signal (Fourier transform in time) allows us to identify the

	$g/15$	$g/10$	$g/5$	g
Analytical frequency (Hz)	0.028	0.034	0.049	0.109
Numerical frequency (Hz)	0.033	0.040	0.057	0.130

Table 6.3 Comparison of the numerical and analytical first gravitational frequency for various values of the gravity.

"eigenvalues" of the system, and to compare them with those provided by simpler approaches. More details on the results of this section may be found in [234].

6.3.4.1 Geometry and gravitational modes In this section, we consider a parallelepipedic cell. The height of aluminum and bath are 1 m. The horizontal dimensions are 2 m × 2 m. In the absence of an adequate turbulence model in our context, the viscosity is magnified by a factor of 1000.

We first compare the results provided by the nonlinear approach with a purely hydrodynamics experiment, without magnetic fields. To perturb the cell, we tilt it for a few seconds. We observe one peak in frequency, and compare it with the first gravitational frequency $\omega_{\pi/L_x,\pi/L_y}$, where $\omega_{k,l}$ is defined by (see [237]):

$$\omega_{k,l} = \sqrt{g \frac{(\rho_1 - \rho_2)\sqrt{k^2 + l^2}}{\rho_1 \coth(\sqrt{k^2 + l^2}\, h_1) + \rho_2 \coth(\sqrt{k^2 + l^2}\, h_2)}}. \tag{6.43}$$

This analytical formula is obtained by a linearization procedure similar to that given in Section 6.2.2, assuming in addition that the flow is potential. Notice that the classical formula (6.17) for shallow water models is recovered in the limit h_1 and h_2 go to zero. We observe a very good agreement between the gravitational frequency given by this analytical formula, and that obtained numerically by a Fourier analysis of the time evolution of the altitude of a point on the interface (see Table 6.3).

6.3.4.2 Spectral analysis of the metal pad rolling In this section, we now consider the classical framework to obtain metal pad rolling (see Section 6.2.4 and 6.3.3): an electric current runs through the cell and we add a vertical component to the magnetic field. In the nonlinear model, this is imposed through the boundary conditions on the magnetic field (see [213] for similar boundary conditions on a parallelepipedic cell):

$$\boldsymbol{B}_x = -\frac{\mu_0 j_0 y}{2}, \qquad \boldsymbol{B}_y = \frac{\mu_0 j_0 x}{2}, \qquad \boldsymbol{B}_z = \boldsymbol{B}_{0,z}. \tag{6.44}$$

We use a current density j_0 of about $10\,kA/m^2$. We decompose the time evolution of the altitude of a point on the interface $z(t)$ as:

$$z(t) = z_0 + \sum_{i=1}^{N} \alpha_i \exp(\tau_i t) \cos(2\pi t/T_i - \phi_i),$$

$\boldsymbol{B}_{0,z}$ (G)	0	30	50	65	80	95	130	160	180
$\tau_1(s^{-1})$	−0.0335	−0.0191	−0.0137	−0.0080	−0.0032	0.0000	0.0049	0.0092	0.0097
$T_1(s)$	29.4	29.3	29.2	29.1	29.3	29.3	30.0	29.8	30.0

Table 6.4 Growth factor (τ_1) and period (T_1) of the principal eigenmode for various vertical magnetic fields ($\boldsymbol{B}_{0,z}$).

$\boldsymbol{B}_{0,z}$ (G)	τ_1	T_1	τ_2	T_2	τ_3	T_3
250 G	−0.004	35	0.0051	20	0.0085	15
280 G	−0.01	35	0.0045	20	0.0092	15

Table 6.5 Growth factor (τ_i) and period (T_i) of the three principal eigenmodes for two values of the vertical magnetic field.

where N denotes the number of modes, τ_i the growth factor of each mode, and T_i the period of each mode. In Figure 6.13, we present a typical signal we obtain, for $\boldsymbol{B}_{0,z} = 280\,G$.

For small values of $\boldsymbol{B}_{0,z}$ ($\boldsymbol{B}_{0,z} < 200\,G$), one single Fourier mode ($N = 1$) is sufficient to describe the signal. Table 6.4 shows the growth factor τ_1 and period T_1 of this mode, for various values of the vertical component of the magnetic field. We observe that the period of the eigenmode only slightly depends on $\boldsymbol{B}_{0,z}$. More importantly, there exists a critical stability threshold on $\boldsymbol{B}_{0,z}$. For $\boldsymbol{B}_{0,z} < 95\,G$, τ_1 is negative and the cell is thus stable. For $\boldsymbol{B}_{0,z} > 95\,G$, τ_1 is positive and the cell is thus unstable. This contradicts the spectral analysis on simplified linear systems which predicts that a cell with $L_x = L_y$ is unstable irrespective of the value of $\boldsymbol{B}_{0,z}$ (see Section 6.2.4). Numerical experiments (not reproduced here) have shown that the above threshold is not significantly sensitive to variations of the artificial amplification factor used for the viscosity.

On the other hand, for larger values of $\boldsymbol{B}_{0,z}$, three Fourier modes ($N = 3$) are necessary to describe the signal (see Figure 6.13 for the case $\boldsymbol{B}_{0,z} = 280\,G$). We observe (Table 6.5) that as $\boldsymbol{B}_{0,z}$ increases, the first mode (namely that with the largest period) becomes stable again (τ_1 becomes negative again) and the instability of the cell is due to the second and third modes. Interestingly, this is also observed on the simplified systems we have considered above to perform the linear stability analysis (see Figure 6.4).

These numerical experiments show that linear stability analysis on simplified systems and numerical experiments on more complete nonlinear models can be compared, and advantageously complement one another.

6.4 Other nonlinear approaches and conclusions

As already mentioned, a complete linear stability analysis may be difficult to apply on the very complicated system of equations modeling an aluminum electrolysis cells. Therefore, some authors have tried to develop time-dependent nonlinear models to study the stability of the cells. The idea is to study the stability

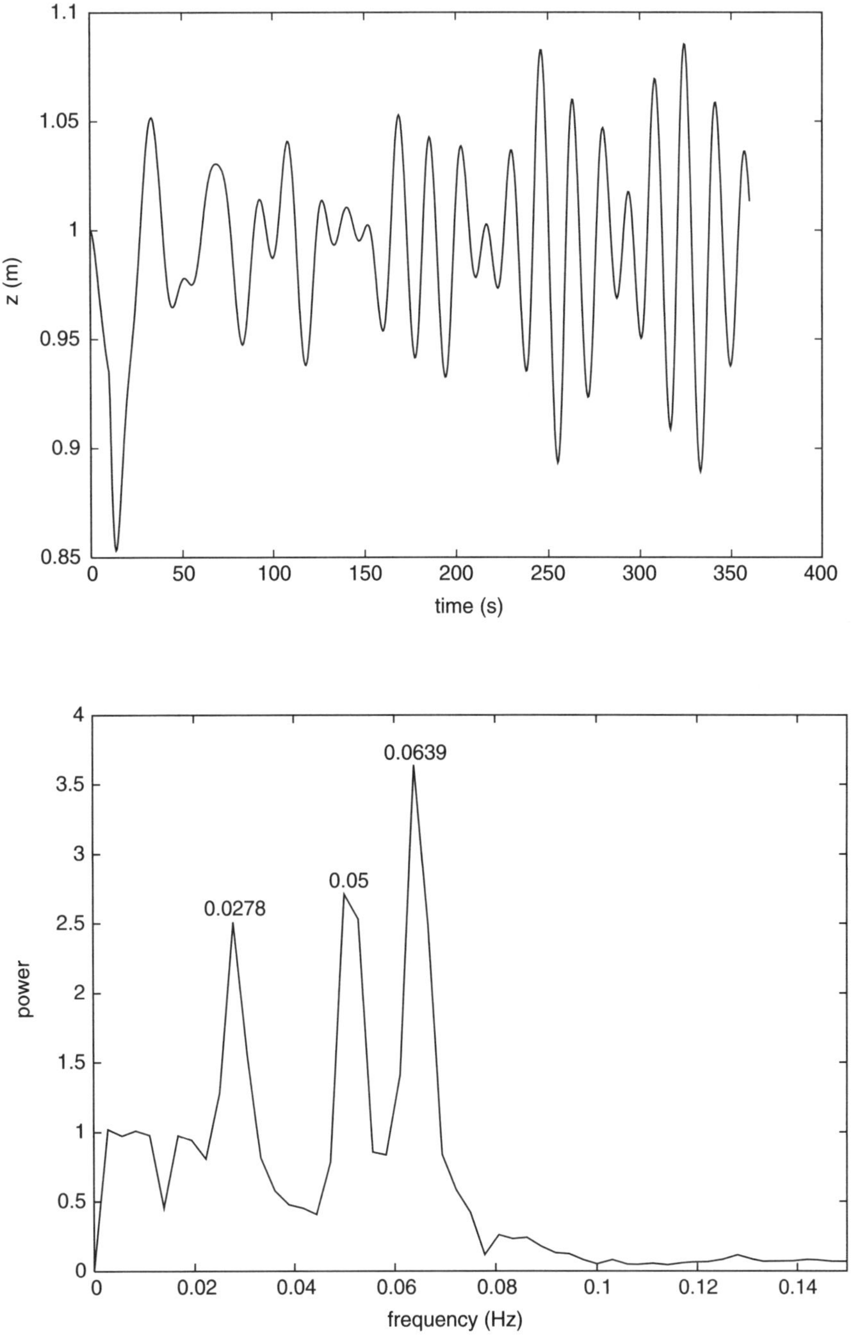

Fig. 6.13 Altitude of a node at the interface as a function of time and associated Fourier transform, for $\boldsymbol{B}_{0,z} = 280\,G$.

of the cell through some long-time computations on a genuinely time-dependent system.

We have already presented in Section 6.3 the results obtained with the numerical method presented in this book. Let us finally mention other attempts to study the stability of the cell by time-dependent numerical models.

- In [179], D. Munger uses a MHD model of the aluminum electrolysis cells similar to ours (1.52). In particular, this is the only other work we are aware of which does not use the magnetostatic approximation. To discretize the free-surface problem, he makes use of a level set method rather than a ALE approach. He particularly studies metal pad rolling instabilities, and also identifies a new potential instability mechanism (a so-called vortex instability).
- *Nonlinear time-dependent shallow water models*: in a first paper [243], O. Zikanov et al. introduce a shallow water model, which is further improved in a second paper [222] by H. Sun et al., using the Saint-Venant approach. The authors derive a two-dimensional model, by averaging the hydrodynamic equations along the vertical direction and using a magnetostatic approximation. The assumptions used by the authors are those used in linear approaches (except assumption (E4)), but they do not perform a linearization around a stationary state and rather discretize the nonlinear system they obtain. The difference with system (6.25) is thus in the hydrodynamic part (equations for the velocity field and the position of the interface) but the magnetic part is again treated using the magnetostatic approximation (E1)–(E2) and assuming a vertical current in the bath (E3) so that the magnetic part of the system of equations is reduced to a Poisson equation for the electric potential in aluminum, as in system (6.25). The authors have in particular used their models to study the influence of a non-zero background flow and background interface deformation on the stability of the cell, and to analyze competition between Kelvin–Helmholtz instabilities and purely MHD instabilities [241, 136].

REFERENCES

[1] Abramowitz, M. and Stegun, I. A. (ed.) (1992). *Handbook of mathematical functions with formulas, graphs, and mathematical tables.* Dover Publications Inc.

[2] Allain, G. (1987). Small-time existence for the Navier–Stokes equations with a free surface. *Appl. Math. Optim.*, **16**, 37–50.

[3] Ambrosio, L. and Soner, H. M. (1996). Level set approach to mean curvature flow in arbitrary codimension. *J. Differential Geom.* **43**(4), 693–737.

[4] Amrouche, C., Bernardi, C, Dauge, M., and Girault, V. (1998). Vector potentials in three-dimensional non-smooth domains. *Math. Methods Appl. Sci.*, **9**, 823–864.

[5] Antille, J. P., Descloux, J., Flueck, M., and Romerio, M. V. (1999). Eigenmodes and interface description in a Hall–Heroult cell. *Light Metals*, 333–338.

[6] Antille, J., Flueck, M., and Romerio, M. V. (1994). Steady velocity field in aluminium reduction cells derived from measurements of the anodic current fluctuations. *Light Metals*, 305–312.

[7] Antontsev, S. N., Kazhikov, A. V., and Monakhov, V. N. (1993). *Boundary values problems in mechanics of nonhomogeneous fluids.* North-Holland.

[8] Armero, F. and Simo, J. C. (1996). Long-time dissipativity of time-stepping algorithms for an abstract evolution equation with applications to the incompressible MHD and Navier–Stokes equations. *Comp. Methods Appl. Mech. Engrg.*, **131**, 41–90.

[9] Assous, F., Degond, P., Heintze, E., Raviart, P. A., and J., Segre (1993). On a finite-element method for solving the three dimensional Maxwell equations. *Jour. Comp. Phys.*, **109**, 222–237.

[10] Audusse, E. (2005). A multilayer Saint-Venant model: derivation and numerical validation. *Discrete Contin. Dyn. Syst. Ser. B* **5**(2), 189–214.

[11] Babuška, I. (1973). The finite element method with Lagrangian multipliers. *Num. Math.*, **20**, 179–192.

[12] Baer, T. A., Cairncross, R. A., Schunk, P. R., Sackinger, P. A., and Rao, R. R. (2000). A finite element method for free surface flows of incompressible fluids in three dimensions. Part II: dynamic wetting line. *Int. Jour. Num. Meth. Fluids*, **33**, 405–427.

[13] Baines, M. J. (1994). *Moving Finite Elements.* Oxford University Press.

[14] Beale, J. T. (1981). The initial value problem for the Navier–Stokes equations with a free surface. *Comm. Pure Appl. Math.*, **34**, 359–392.

[15] Beale, J. T. (1984). Large time regularity of viscous surface waves. *Arch. Rat. Mech. Anal.*, **84**, 307–352.

[16] Beale, J. T., Hou, T. Y., and Lowengrub, J. S. (1993). Growth rates for the linearized motion of fluid interfaces away from equilibrium. *Comm. Pure Appl. Math.*, **46**, 1269–1301.

[17] Beale, J. T. and Nishida, T. (1985). Large-time behaviour of viscous surface waves. In *Recent topics in nonlinear PDE II*, Volume 8 of *Lect. Notes Numer. Appl. Anal.*, 1–14. North-Holland.

[18] Beirao da Veiga, H. (1996). Long time behaviour of the solutions to the Navier–Stokes equations with diffusion. *Nonlinear Anal., Theory, Meth. Appl.* **27**(11), 1229–1239.

[19] Ben Salah, N., Soulaimani, A., and Habashi, W. G. (1999*a*). A fully coupled finite element method for the 3D MHD equations with a GMRES-based algorithm. Technical report, AIAA 99-3322.

[20] Ben Salah, N., Soulaimani, A., Habashi, W. G., and Fortin, M. (1999*b*). A conservative stabilized finite element method for the magnetohydrodynamic equations. *Int. Jour. Num. Meth. Fluids*, **29**, 535–554.

[21] Berton, R. (1991). *Magnétohydrodynamique.* Masson.

[22] Blanc, J. M. and Entner, P. (1980). Application of computer calculations to improve electromagnetic behaviour of pots. *AIME*, 285–295.

[23] Boffi, D. and Gastaldi, L. (2004). Stability and geometric conservation laws for ALE formulations. *Comp. Meth. Appl. Mech. Eng.*, **193**, 4717–4739.

[24] Bojarevics, V. (1998). Nonlinear waves with electromagnetic interaction in aluminium electrolysis cells. In *Progress in Fluid Flow Research: Turbulence and Applied MHD* (ed. H. Branover and Y. Unger), Volume 182 of *AIAA series: Progress in Astronautics and Aeronautics*, 833–848.

[25] Bojarevics, V. and Romerio, M. V. (1994). Long waves instability of liquid metal–electrolyte interface in aluminium electrolysis cells: a generalization of Sele's criterion. *Eur. J. Mech. B*, **13**, 33–56.

[26] Borchers, W. and Miyakawa, T. (1992). L^2-decay for Navier–Stokes flows in unbounded domains with application to exterior stationary flows. *Arch. Rat. Mech. Anal*, **118**, 273–295.

[27] Brackbill, J. U. (1976). Numerical magnetohydrodynamics for high-beta plasmas. *Methods in Computational Physics*, **16**, 1–41.

[28] Brackbill, J. U., Kothe, D. B., and Zemach, C. (1992). A continuum method for modeling surface tension. *J. Comput. Phys.*, **100**, 335–354.

[29] Brackbill, J. U. and Pracht, W. E. (1973). An implicit, almost-Lagrangian algorithm for magnetohydrodynamics. *J. Comp. Phys.*, **13**, 455–482.

[30] Brezis, Haïm (1983). *Analyse fonctionnelle.* Collection Mathématiques Appliquées pour la Maîtrise. Masson.

[31] Brezzi, F. (1974). On the existence uniqueness and approximation of saddle-point problems arising from Lagrange multipliers. *RAIRO*, **R. 2**, 129–151.

[32] Brezzi, F. and Fortin, M. (1991). *Mixed and hybrid finite element method.* Springer.

[33] Cairncross, R. A., Schunk, P. R., Baer, T. A., Rao, R., and Sackinger, P. A. (2000). A finite element method for free surface flows of incompressible fluids in three dimensions. Part I: Boundary fitted mesh motion. *Int. Jour. Num. Meth. Fluids*, **33**, 375–403.

[34] Cannone, M. (1995). *Ondelettes, Paraproduits et Navier–Stokes.* Diderot Editeur.

[35] Cannone, M. (1999). Rôle des oscillations et des espaces de Besov dans la résolution des équations de Navier–Stokes. Technical report, Université de Paris 7.

[36] Castro, Manuel J., García-Rodríguez, José A., González-Vida, José M., Macías, Jorge, Parés, Carlos, and Vázquez-Cendón, M. Elena (2004). Numerical simulation of two-layer shallow water flows through channels with irregular geometry. *J. Comput. Phys.* **195**(1), 202–235.

[37] Causin, P., Gerbeau, J. -F., and Nobile, F. (2005). Added-mass effect in the design of partitioned algorithms for fluid–structure problems. *Comp. Meth. Appl. Mech. Engng.* **194**(42–44), 4506–4527.

[38] Chen, S., Johnson, D. B., and Raad, P. E. (1991). *Computational Modelling of Free and Moving Boundary Problems*, Volume 1, Chapter The Surface Marker Method, 223–234. de Gruyter.

[39] Ciarlet, P. G. and Lions, J. -L. (ed.) (1991). *Handbook of Numerical Analysis, Vol. 2, Finite Element Methods (Part 1).* North-Holland.

[40] Constantin, P., Foias, C., Nicolaenko, B., and Temam, R. (1989). *Integral manifolds and inertial manifolds for dissipative partial differential equations.* Springer.

[41] Costabel, M. (1991). A coercive bilinear form for Maxwell's equations. *J. Math. Anal. and Appl.*, **157**, 527–541.

[42] Costabel, M. and Dauge, M. (2000). Singularities of electromagnetic fields in polyhedral domains. *Arch. Rat. Mech. Anal.* **151**(3), 221–276.

[43] Costabel, M. and Dauge, M. (2002). Weighted regularization of Maxwell equations in polyhedral domains. *Numer. Math.*, **93**, 239–277.

[44] Cottet, Georges-Henri and Maitre, Emmanuel (2004). A level-set formulation of immersed boundary methods for fluid–structure interaction problems. *C. R. Math. Acad. Sci. Paris* **338**(7), 581–586.

[45] Dautray, R. and Lions, J. -L. (1987). *Analyse mathématique et calcul numérique pour les sciences et les techniques*, Volume 2. Masson.

[46] Davidson, P. A. (1994). An energy analysis of unstable aluminium reduction cells. *Eur. J. Mech. B*, **13**, 15–32.

[47] Davidson, P. A. (2001). *An introduction to magnetohydrodynamics.* Cambridge Texts in Applied Mathematics. Cambridge University Press.

[48] Davidson, P. A. (2004). *Turbulence. An introduction for scientists and engineers.* Oxford University Press.

[49] Davidson, P. A. and Boivin, R. F. (1992). Hydrodynamics of aluminum reduction cells. *Light Metals*, 1199–1204.

[50] Davidson, P. A and Lindsay, R. I. (1998). Stability of interfacial waves in aluminium reduction cells. *J. Fluid Mech.*, **362**, 273–295.

[51] Davidson, Peter A. (ed.) and Thess, Andre (ed.) (2002). *Magnetohydrodynamics. Lectures given during the 10th IUTAM international summer school on magnetohydrodynamics, Udine, Italy, June 21–25, 1999.* CISM Courses and Lectures. 418. Springer.

[52] Delfour, M. C. and Zolésio, J. -P. (2001). *Shapes and Geometries.* Advances in design and control. SIAM.

[53] Denisova, I. V. (1994). Problem of the motion of two viscous incompressible fluids separated by a closed free interface. *Acta Applicandae Mathematicae*, **37**, 31–40.

[54] Denisova, I. V. and Solonnikov, V. A. (1996). Classical solvability of the problem on the motion of two viscous incompressible fluids. *St. Petersbg. Math. J.* **7**(5), 755–786.

[55] Descloux, J., Flueck, M., and Romerio, M. V. (1991*a*). Linear stability of aluminium electrolysis cells, part I. Technical Report 09. 91, Ecole Polytechnique Fédérale de Lausanne.

[56] Descloux, J., Flueck, M., and Romerio, M. V. (1991*b*). Modelling for instabilities in Hall–Heroult cells: mathematical and numerical aspects. *Magnetohydrodynamics in Process Metallurgy*, 107–110.

[57] Descloux, J., Flueck, M., and Romerio, M. V. (1992). Linear stability of aluminium electrolysis cells, part II. Technical Report 07. 92, Ecole Polytechnique Fédérale de Lausanne.

[58] Descloux, J., Flueck, M., and Romerio, M. V. (1994). Stability in aluminium reduction cells: a spectral problem solved by an iterative procedure. *Light Metals*, 275–281.

[59] Descloux, J., Flueck, M., and Romerio, M. V. (1998*a*). A modelling of the stability of aluminium electrolysis cells. In *Nonlinear partial differential equations and their applications. Collège de France Seminar, Vol. XIII (Paris, 1994/1996)*, Volume 391 of *Pitman Res. Notes Math. Ser.*, 117–133. Longman, Harlow.

[60] Descloux, J., Flueck, M., and Romerio, M. V. (1998*b*). A problem of magnetostatics related to thin plates. *RAIRO, Modélisation Math. Anal. Numér.* **32**(7), 859–876.

[61] Descloux, J., Flueck, M., and Romerio, M. V. (1995). Spectral aspects of an industrial problem. In *Spectral analysis of complex structures (Paris, 1993)*, Volume 49 of *Travaux en Cours*, 17–33. Hermann.

[62] Descloux, J., Frosio, R., and Flück, M. (1989). A two fluids stationary free boundary problem. *Comput. Methods Appl. Mech. Engrg.* **77**(3), 215–226.

[63] Descloux, J., Jaccard, Y., and Romerio, M. V. (1991). A bidimensional stability result for aluminium electrolytic cells. *Journal of Computational and Applied Mathematics*, **38**, 77–85.

[64] Descloux, J. and Romerio, M. V. (1989). On the analysis by perturbation methods of the anodic current fluctuations in an electrolytic cell for aluminium. *Light Metals*, 237–243.

[65] Desjardins, B. (1997). Regularity results for the two dimensional multiphase viscous flows. *Arch. Rat. Mech. Anal.*, **137**, 135–158.

[66] Desjardins, B. and Le Bris, C. (1998). Remarks on a nonhomogeneous model of magnetohydrodynamics. *Differential Integral Equations* **11**(3), 377–394.

[67] Dettmer, W., Saksono, P. H., and Peri, D. (2003). On a finite element formulation for incompressible newtonian fluid flows on moving domains in the presence of surface tension. *Commun. Numer. Meth. Engng.*, **19**, 659–668.

[68] Di Perna, R. J. and Lions, P. L. (1989). Ordinary differential equations, transport theory and Sobolev spaces. *Invent. Math.* **98**(3), 511–547.

[69] Dominguez de la Rasilla, J. -M. (1982). *Etude des équations de la magnétohydrodynamique stationnaires et de leur approximation par éléments finis.* Thèse, Université de Paris VI.

[70] Douglas, J. and Wang, J. (1989). An absolutely stabilized finite element method for the Stokes problem. *Math. of Comp.* **52**(186), 495–508.

[71] Droste, Ch. (2000). PHOENICS applications in the aluminium smelting industry. Technical report, VAW Aluminium Technology.

[72] Dupuis, M., Bojarevics, V., and Freibergs, J. (2003). Demonstration thermo-electric and MHD mathematical models of a 500 kA Al electrolysis cell. In *COM 2003* (42nd Conference of Metallurgists), *Vancouver, Canada.*

[73] Duvaut, G. and Lions, J. -L. (1972). *Les inéquations en mécanique et en physique.* Dunod.

[74] Engelman, M. S., Sani, R. L., and Gresho, P. M. (1982). The implementation of normal and/or tangential velocity boundary conditions in finite element codes for incompressible fluid flow. *Int. J. Num. Meth. Fluids* **2**(3), 225–238.

[75] Ern, A. and Guermond, J. -L. (2004). *Theory and practice of finite elements.* Springer.

[76] Errate, D., Esteban, M. J., and Maday, Y. (1994). Couplage fluide–structure. Un modèle simplifié en dimension 1. *C. R. Acad. Sci. Paris, Série I*, **318**, 275–281.

[77] Farhat, C., Lesoinne, M., and Maman, N. (1995). Mixed explicit/implicit time integration of coupled aeroelastic problems: three-field formulation,

geometry conservation and distributed solution. *Int. J. Num. Meth. Fluids* **21**(10), 807–835.

[78] Foias, C. and Saut, J. C. (1984). Asymptotic behaviour, as $t \longrightarrow +\infty$ of solutions of Navier–Stokes equations and nonlinear spectral manifolds. *Ind. Univ. Math. J.* **33**(3), 459–471.

[79] Formaggia, L., Gerbeau, J. -F., Nobile, F., and Quarteroni, A. (2002). Numerical treatment of defective boundary conditions for the Navier–Stokes equations. *SIAM J. Numerical Analysis* **40**(1), 376–401.

[80] Formaggia, L. and Nobile, F. (1999). A stability analysis for the arbitrary Lagrangian Eulerian formulation with finite elements. *East–West J. Numer. Math.* **7**(2), 105–131.

[81] Formaggia, L. and Nobile, F. (2004). Stability analysis of second order time accurate schemes for ALE-FEM. *Comput. Methods Appl. Mech. Engrg.*, **193**, 4097–4116.

[82] Franca, L. P. and Frey, S. L. (1992). Stabilized finite element methods: II. The incompressible Navier–Stokes equations. *Comp. Meth. Appl. Mech. Eng.*, **99**, 209–233.

[83] Fujita, H. and Kato, T. (1962). On the non-stationary Navier–Stokes equations system. *Rend. Sem. Mat. Univ. Padova*, **32**, 243–260.

[84] Galdi, G. P., Heywood, J. G., and Shibata, Y. (1997). On the global existence and convergence to steady state of Navier–Stokes flow past an obstacle that is started from rest. *Arch. Rat. Mech. Anal.*, **138**, 307–318.

[85] Gastaldi, L. (2001). A priori error estimates for the arbitrary Lagrangian Eulerian formulation with finite elements. *East–West J. Numer. Math.* **9**(2), 123–156.

[86] Georgescu, V. (1979). Some boundary value problems for differential forms on compact Riemannian manifolds. *Annali di Matematica Pura ed. Applicata* **4**(122), 159–198.

[87] Gerbeau, J. -F. (1998*a*). Comparison of numerical methods for solving a magnetostatic problem. application to the magnetohydrodynamic equations. In *Proceedings of the Fourth European Computational Fluid Dynamics, ECCOMAS 98*, 821–825. Wiley.

[88] Gerbeau, J. -F. (1998*b*). *Problèmes mathématiques et numériques posés par la modélisation de l'électrolyse de l'aluminium.* Thèse, Ecole Nationale des Ponts et Chaussées.

[89] Gerbeau, J. -F. (2000). A stabilized finite element method for the incompressible magnetohydrodynamic equations. *Numerische Mathematik* **87**(1), 83–111.

[90] Gerbeau, J. -F. and Le Bris, C. (1997). Existence of solution for a density-dependent magnetohydrodynamic equation. *Advances in Differential Equations* **2**(3), 427–452.

[91] Gerbeau, J. -F. and Le Bris, C. (1999*a*). On a coupled system arising in magnetohydrodynamics. *Appl. Math. Lett.*, **12**, 53–57.

[92] Gerbeau, J. -F. and Le Bris, C. (1999*b*). On the long time behaviour of the solution to the two-fluids incompressible Navier–Stokes equations. *Differential and Integral Equations* **12**(5), 691–740.

[93] Gerbeau, J. -F. and Le Bris, C. (2000*a*). A basic remark on some Navier–Stokes equations with body forces. *Appl. Math. Lett.* **13**(3), 107–112.

[94] Gerbeau, J. -F. and Le Bris, C. (2000*b*). Comparison between two numerical methods for a magnetostatic problem. *Calcolo* **37**(1), 1–20.

[95] Gerbeau, J. -F. and Le Bris, C. (2000*c*). *Mathematical study of a coupled system arising in Magnetohydrodynamics*, Volume 215 of *Lecture Notes in Pure and Applied Mathematics*, 355–367. Marcel Dekker Inc.

[96] Gerbeau, J. -F., Le Bris, C., and Bercovier, M. (1997). Spurious velocities in the steady flow of an incompressible fluid subjected to external forces. *Int. Jour. Num. Meth. Fluids*, **25**, 679–695.

[97] Gerbeau, J. -F. and Lelièvre, T. (2006). Variational formulation of the generalized Navier boundary condition in an ALE formulation. In preparation.

[98] Gerbeau, J. -F., Lelièvre, T., and Le Bris, C. (2002). Numerical simulations of two-fluids MHD flows. *Fundamental and Applied MHD. Proceedings of the Fifth international Pamir Conference.*, **1**, 101–105.

[99] Gerbeau, J. -F., Lelièvre, T., and Le Bris, C. (2003). Simulations of MHD flows with moving interfaces. *Jour. Comp. Phys.* **184**(1), 163–191.

[100] Gerbeau, J. -F., Lelièvre, T., and Le Bris, C. (2004). Modeling and simulation of the industrial production of aluminium: The nonlinear approach. *Comput. Fluids* **33**(5–6), 801–814.

[101] Gerbeau, J. -F., Lelièvre, T., Le Bris, C., and Ligonesche, N. (2002). Metal pad roll instabilities. In *2002 TMS Annual Meeting and Exhibition, Light Metals*, 483–487.

[102] Gerbeau, J. -F. and Perthame, B. (2001). Derivation of viscous Saint–Venant system for laminar shallow water; numerical validation. *Discrete and Continuous Dynamical Systems: Series B* **1**(1), 89–102.

[103] Geuzaine, P., Grandmont, C., and Farhat, C. (2003). Design and analysis of ALE schemes with provable second-order time-accuracy for inviscid and viscous flow simulations. *Journal of Computational Physics* **191**(1), 206–227.

[104] Ghidaglia, J. -M. (1984). *Etude d'écoulements de fluides visqueux incompressibles: comportement pour les grands temps et applications aux attracteurs.* Thèse, Université de Paris-Sud, Orsay.

[105] Gingold, R. A. and Monaghan, J. J. (1977). Smoothed particle hydrodynamics: theory and application to non-spherical stars. *Mon. Not. R. astr. Soc.*, **181**, 375–389.

[106] Girault, V. and Raviart, P. -A. (1986). *Finite element methods for Navier–Stokes equations.* Springer.

[107] Givry, J. P. (1967). Computer calculation of magnetic effects in the bath of aluminium cells. *Transactions of the Metallurgical Society of AIME*, **239**, 1161–1166.

[108] Glimm, J., Grove, J. W., Li, X. L., Shyue, K. M., Zeng, Y., and Zhang, Q. (1998). Three dimensional front tracking. *SIAM Journal on Scientific Computing* **19**(3), 703–727.

[109] Glowinski, R. (2003). Finite elements methods for incompressible flows. In *Handbook of numerical analysis* (ed. P. Ciarlet and J. Lions), Volume IX. North Holland.

[110] Gresho, P. M. and Sani, R. L. (2000*a*). *Incompressible Flow and the Finite Element Method Volume 1: Advection–Diffusion.* John Wiley & Sons.

[111] Gresho, P. M. and Sani, R. L. (2000*b*). *Incompressible Flow and the Finite Element Method Volume 2: Isothermal Laminar Flow.* John Wiley & Sons.

[112] Grigoriev, R. O. and Handel, A. (2002). Spectral theory for the failure of linear control in a nonlinear stochastic system. *Phys. Rev. E*, **66**, 065301.

[113] Grjotheim, K. and Kvande, H. (1993). *Introduction to aluminium electrolysis.* Aluminium-Verlag.

[114] Guermond, J. L. and Minev, P. D. (2003). Mixed finite element approximation of an MHD problem involving conducting and insulating regions: the 3D case. *Numer. Meth. PDE*, **19**, 709–731.

[115] Gueyffier, D., Li, J., Nadim, A., Scardovelli, S., and Zaleski, S. (1999). Volume of fluid interface tracking with smoothed surface stress methods for three-dimensional flows. *J. Comput Phys.*, **152**, 423–456.

[116] Guillard, H. and Farhat, C. (2000). On the significance of the geometric conservation law for flow computations on moving meshes. *Comput. Methods Appl. Mech. Engrg.* **190**(11–12), 1467–1482.

[117] Guillope, C. (1982). Comportement à l'infini des solutions des équations de Navier–Stokes et propriétés des ensembles fonctionnels invariants (ou attracteurs). *Ann. Inst. Fourier Grenoble* **32**(3), 1–37.

[118] Gunzburger, M. D., Meir, A. J., and Peterson, J. S. (1991, April). On the existence, uniqueness, and finite element approximation of solutions of the equations of stationary, incompressible magnetohydrodynamics. *Mathematics of Computation* **56**(194), 523–563.

[119] Harlow, F. H. and Welch, J. E. (1965). Numerical calculation of time-dependent viscous incompressible flow. *Phys. Fluids*, **8**, 2182–2189.

[120] Hasler, U., Schneebeli, A., and Schötzau, D. (2004). Mixed finite element approximation of incompressible MHD problems based on weighted regularization. *Appl. Numer. Math.* **51**(1), 19–45.

[121] Heywood, J. G. (1970). On stationary solutions of the Navier–Stokes equations as limits of nonstationary solutions. *Arch. Rat. Mech. Anal.*, **37**, 48–60.

[122] Heywood, J. G. (1980). The Navier–Stokes equations: on the existence, regularity and decay of solutions. *Indiana Univ. Math. J.* **29**(5), 639–681.

[123] Hirt, C. W., Amsden, A. A., and Cook, J. L. (1974). An Arbitrary Lagrangian–Eulerian computing method for all flow speeds. *J. Comput. Phys.* **14**(3), 227–253.

[124] Hirt, C. W. and Nichols, B. D. (1981). Volume of fluids VOF method for the dynamics of free boundaries. *J. Comput. Phys.*, **39**, 201–225.

[125] Ho, L. W. (1989). *A legendre spectral element method for simulation of incompressible unsteady viscous free-surface flows.* PhD thesis, Massachussets Institute of Technology.

[126] Hou, T. Y, Teng, Z. -H., and Zhang, P. (1996). Well-posedness of linearized motion for 3-D water waves far from equilibrium. *Comm. Part. Diff. Eq.* **21**(9–10), 1551–1585.

[127] Huang, W., Ren, Y., and Russell, R. D. (1994). Moving mesh partial differential equations (MMPDEs) based on the equidistribution principle. *SIAM J. Numer. Anal.*, **31**, 709–730.

[128] Huerta, A. and Liu, W. K. (1988). Viscous flow with large surface motion. *Comp. Meth. Appl. Mech. Engng.*, **69**, 277–324.

[129] Hughes, T. J. R., Franca, L. P., and Balestra, M. (1986). A new finite element formulation for computational fluid dynamics: V. Circumventing the Babuška–Brezzi condition: a stable Petrov–Galerkin formulation of the Stokes problem accommodating equal-order interpolations. *Comp. Meth. App. Mech. Eng.*, **59**, 85–99.

[130] Hughes, W. F. and Young, F. J. (1966). *The electromagnetodynamics of fluids.* Wiley.

[131] Kelley, C. T. (1995). *Iterative methods for linear and nonlinear equations.* SIAM.

[132] Kherief, K. (1984). *Quelques propriétés des équations de la magnétohydrodynamique stationnaires et d'évolution.* Thèse, Université de Paris VII.

[133] Kikuchi, F. (1987). Mixed and penalty formulations for finite element analysis of an eigenvalue problem in electromagnetism. In *Proceedings of the first world congress on computational mechanics (Austin, Tex., 1986)*, Volume 64, 509–521.

[134] Kikuchi, F. (1993, June). Numerical analysis electrostatic and magnetostatic problem. *Sugaku expositions* **6**(1), 332–345.

[135] Klouček, P. and Romerio, M. V. (2002). The detachment of bubbles under a porous rigid surface during aluminum electrolysis. *Math. Models Methods Appl. Sci.* **12**(11), 1617–1652.

[136] Kurenkov, A., Thess, A., Zikanov, O., Segatz, M., Droste, C., and Vogelsang, D. (2004). Stability of aluminium reduction cells with mean flow. *Magnetohydrodynamics* **40**(2), 3–13.

[137] LaCamera, A. F., Ziegler, D. P., and Kozarek, R. L. (1992). Magnetohydrodynamics in the Hall–Héroult process, an overview. *Light Metals*, 1179–1186.

[138] Lafaurie, B., Nardone, C., Scardovelli, R., Zaleski, S., and Zanetti, G. (1994). Modelling merging and fragmentation in multiphase flows with SURFER. *J. Comput. Phys.*, **113**, 134–147.

[139] Landau, L. and Lifchitz, E. (1984). *Course of Theoretical Physics*, volume 6. Pergamon Press.

[140] Le Bris, C. (1998). Some theoretical and numerical issues related to the multifluid magnetohydrodynamics equations. In *Proceedings of the 4th European Computational Fluid Dynamics Conference ECCOMAS 98*, 800–802. Wiler.

[141] Le Bris, C. (2005). *Encyclopedia of Mathematical Physics*, Chapter on Magnetohydrodynamics. Elsevier.

[142] Le Méhauté, B. (1976). *An introduction to hydrodynamics and water waves.* Springer.

[143] Le Tallec, P. and Mouro, J. (2001). Fluid structure interaction with large structural displacements. *Comput. Meth. Appl. Mech. Engrg.*, **190**, 3039–3067.

[144] Leboucher, L., Pericleous, K., Panaitescu, I., and Repetto, M. (1999). A finite-volume shallow layer method for the computation of MHD instabilities in an aluminium production cell. In *2nd Int. Conf. on CFD in the Minerals and Process Industries*, 335–338.

[145] Leray, J. (1933). Etude de diverses équations intégrales non linéaires et de quelques problèmes que pose l'hydrodynamique. *J. Math. Pures Appl.*, **12**, 1–82.

[146] Leray, J. (1934*a*). Essai sur le mouvement d'un liquide visqueux emplissant l'espace. *Acta Math.*, **63**, 193–248.

[147] Leray, J. (1934*b*). Essai sur les mouvements plans d'un liquide visqueux que limitent des parois. *J. Math. Pures Appl.*, **13**, 331–418.

[148] Lesieur, M. (1994). *La turbulence. (Turbulence).* Collection Grenoble Sciences. Grenoble: Presses Univ. de Grenoble.

[149] Lesieur, M. (1997). *Turbulence in fluids. 3rd rev. and enlarg. ed.* Fluid Mechanics and its Applications. 40. Dordrecht: Kluwer Academic Publishers.

[150] Lesoinne, M. and Farhat, C. (1996). Geometric conservation laws for flow problems with moving boundaries and deformable meshes and their impact on aeroelastic computations. *Computer Methods in Applied Mechanics and Engineering*, **134**, 71–90.

[151] Liao, G. J. and Anderson, D. (1992). A new approach to grid generation. *Appl. Anal.*, **44**, 285–298.

[152] Lieb, Elliott H. and Loss, Michael (2001). *Analysis.* 2nd edn. Graduate Studies in Mathematics. 14. American Mathematical Society (AMS).

[153] Lions, J. -L. (1969). *Quelques méthodes de résolution des problèmes aux limites non linéaires.* Etudes Mathématiques. Dunod.

[154] Lions, J. -L. (2002). *Quelques méthodes de résolution des problèmes aux limites non linéaires* (Reedition of the 1969 original edn). Les cours de références. Dunod.

[155] Lions, P. -L. (1996). *Mathematical Topics in Fluid Mechanics. Vol. 1: incompressible models.* Oxford University Press.

[156] Lukyanov, A., El, G., and Molokov, S. (2001). Instability of MHD-modified interfacial gravity waves revisited. *Phys. Lett. A*, **290**, 165–172.

[157] Lympany, S. D., Evans, J. W., and Moreau, R. (1982). Magnetohydrodynamic effects in aluminium reduction cells. In *Proc. IUTAM Symp. on Metallurgical Applications of Magnetohydrodynamics. Cambridge*, 15–23. London: The Metals Society.

[158] Maillard, P. (1992). *Sur la stabilité magnétohydrodynamique d'une cellule de Héroult de longueur infinie.* PhD thesis, Ecole Polytechnique Fédérale de Lausanne.

[159] Maillard, P. and Romerio, M. V. (1996). A stability criterion for an infinitely long Hall–Héroult cell. *Journal of Computational and Applied Mathematics*, **71**, 47–65.

[160] Maury, B. (1996). Characteristics ALE methods for the unsteady 3D Navier–Stokes equations with a free surface. *Comp. Fluid. Dyn.*, **6**, 175–188.

[161] Meir, A. J. and Schmidt, P. G. (1994). A velocity–current formulation for stationary MHD flow. *Appl. Math. Comp.*, **65**, 95–109.

[162] Meir, A. J. and Schmidt, P. G. (1996). Variational methods for stationary MHD flow under natural interface conditions. *Nonlinear Anal., Theor. Meth. Appl.* **26**(4), 659–689.

[163] Meir, A. J. and Schmidt, P. G. (1999). Analysis and numerical approximation of a stationnary MHD flow problem with nonideal boundary. *SIAM J. Numer. Anal.* **36**(4), 1304–1332.

[164] Meir, A. J. (1995). Thermally coupled, stationary, incompressible MHD flow; existence, uniqueness, and finite element approximation. *Num. Meth. Partial Diff. Eq.*, **11**, 311–337.

[165] Miller, K. (1981). Moving finite elements. II. *SIAM J. Numer. Anal.*, **18**, 1033–1057.

[166] Miller, K. and Miller, R. N. (1981). Moving finite elements. I. *SIAM J. Numer. Anal.*, **18**, 1019–1032.

[167] Miranville, A. and Wang, X. (1996). Upper bound on the dimension of the attractor for nonhomogeneous Navier–Stokes equations. *Disc. and Cont. Dyn. syst.* **2**(1), 95–110.

[168] Mohammadi, B. and Pironneau, O. (1994). *Analysis of the k-epsilon turbulence model.* Masson.

[169] Molokov, S., El, G., and Lukyanov, A. (2003). On the nature of interfacial instability in aluminium reduction cells. Technical report, Coventry University. Internal Report AM-01.

[170] Monaghan, J. J. (1994). Simulating free surface flows with SPH. *J. Comput. Phys.*, **110**, 399–406.

[171] Mooney, J. R. and Stokes, A. N. (1998). Time-varying MHD flows with free surfaces. *Applied Mathematical Modelling*, **22**, 949–962.

[172] Moreau, R. (1990). *Magnetohydrodynamics.* Kluwer Academic Publishers.

[173] Moreau, R. (1992). Ecoulement d'un métal liquide en présence d'un champ magnétique. In *Traité de Génie électrique*, Volume D4, D2950–3–D2950–30. Editions Techniques de l'ingénieur.

[174] Moreau, R. (2003). On turbulence in electromagnetic processing. In *Proceedings of the Electromagnetic Processing of Materials International Conference 2003.*

[175] Moreau, R. and Evans, J. W. (1984). An analysis of the hydrodynamics of aluminium in reduction cells. *J. Electrochem. Soc. : Electrochem. Sci. Tech.* **131**(10), 2251–2259.

[176] Moreau, R. and Ziegler, D. (1986). Stability of aluminum cells: a new approach. *Light Metals*, 359–364.

[177] Moreau, R. and Ziegler, D. (1988). The Moreau–Evans hydrodynamic model applied to actual Hall–Hroult cells. *Metal. Trans. B.*, **19B**, 737–744.

[178] Mori, K., Shiota, K., Urata, N., and Ikeuchi, H. (1976). The surface oscillation of liquid metal in aluminium reduction cells. *TMS, AIME*, **1**, 77–95.

[179] Munger, D. (2004). Simulation numérique des instabilités magnétohydrodynamiques dans les cuves de production de l'aluminium. Master's thesis, Université de Montréal.

[180] Nkonga, B. and Guillard, H. (1994). Godunov type method on nonstructured meshes for three-dimensional moving boundary problems. *Comput. Methods Appl. Mech. Engrg.* **113**(1–2), 183–204.

[181] Nobile, F. (2001). *Numerical approximation of fluid-structure interaction problems with application to haemodynamics.* PhD thesis, EPFL, Switzerland.

[182] Nouri, A. and Poupaud, F. (1995). An existence theorem for the multifluid Navier–Stokes problem. *J. Diff. Eq.* **122**(1), 71–88.

[183] Peskin, C. S. (2002). The immersed boundary method. *Acta Numerica*, **11**, 479–517.

[184] Peterson, J. S. (1988). On the finite element approximation of incompressible flows of an electrically conducting fluid. *Num. Meth. Partial Diff. Eq.*, **4**, 57–68.

[185] Pigny, S. and Moreau, R. (1992). Stability of fluid interfaces carrying an

electric current in the presence of a magnetic field. *Eur. J. Mech., B/, Fluids* **11**(1), 1–20.

[186] Pironneau, O. (1989). *Finite element methods for fluids.* Wiley.

[187] Pironneau, O., Liou, J., and Tezduyar, T. (1992). Characteristic Galerkin and Galerkin/least-squares space-time formulations for the advection-diffusion equation with time-dependent domains. *Computer Methods in Applied Mechanics and Engineering* **100**(1), 117–141.

[188] Potocnik, V. (1989). Modelling of metal–bath interface waves in Hall–Héroult cells using ESTER/PHOENICS. *Light Metals*, 227–235.

[189] Potocnik, V. and Laroche, F. (2001). Comparison of measured and calculated metal pad velocities for different prebake cell designs. *Light Metals*, 419–425.

[190] Qian, T. Z., Wang, X. -P., and Sheng, P. (2003). Molecular scale contact line hydrodynamics of immiscible flows. *Phys. Rev. E*, **68**, 016306.

[191] Qian, T. Z., Wang, X. -P., and Sheng, P. (2006). Molecular hydrodynamics of the moving contact line in two-phase immiscible flows. *Comm. Comp. Phys.* **1**(1), 1–52.

[192] Quarteroni, A. and Valli, A. (1997). *Numerical Approximation of Partial Differential equations.* Springer.

[193] Rappaz, J. and Touzani, R. (1991*a*). Modelling of a two-dimensional magnetohydrodynamic problem. *Eur. Jour. Mech., B, Fluids* **10**(5), 451–453.

[194] Rappaz, J. and Touzani, R. (1991*b*). On a two-dimensional magnetohydrodynamic problem. I) modelling and analysis. *Modélisation mathématique et Analyse numérique* **26**(2), 347–364.

[195] Rappaz, J. and Touzani, R. (1996). On a two-dimensional magnetohydrodynamic problem. II) numerical analysis. *Modélisation mathématique et Analyse numérique* **30**(2), 215–235.

[196] Rasmussen, H. K., Hassager, O., and Saasen, A. (1998). Viscous flow with large fluid–fluid interface displacement. *Int. Jour. Num. Meth. Fluids*, **28**, 859–881.

[197] Romerio, M. V. and Antille, J. (2000). The numerical approach to analyzing flow stability in the aluminium reduction cell. *Aluminium, International Journal for Industry, Research and Application* **76**(12) 1031–1037.

[198] Romerio, M. V. and Secrétan, M. -A. (1986). Magnetohydrodynamics equilibrium in aluminium electrolytic cells. *Computer Physics Reports*, **3**, 327–360.

[199] Rothman, D. H. and Zaleski, S. (1994). Lattice–gas models of phase separation: interfaces, phase transitions, and multiphase flow. *Rev. Mod. Phys.*, **66**, 1417–1479.

[200] Saad, Y. and Schultz, M. H. (1986). GMRES: a generalized minimal residual algorithm for solving nonsymmetric linear systems. *SIAM J. Sci. Statist. Comput.* **7**(3), 856–869.

[201] Samulyak, R., Glimm, J., Oh, W., Kirk, H., and McDonald, K. (2003). Numerical simulation of free surface MHD flows: Richtmyer–Meshkov instability and applications. In *Lecture Notes in Computer Science*, Volume 2667, 558–567. Springer.

[202] Sanchez-Palancia, E. (1968). Existence des solutions de certains problèmes aux limites en magnétohydrodynamique. *Journal de Mécanique* **7**(3), 405–426.

[203] Sanchez-Palancia, E. (1969). Quelques résultats d'existence et d'unicité pour des écoulements magnétohydrodynamiques non stationnaires. *Journal de Mécanique* **8**(4), 509–541.

[204] Saramito, B. (1994). *Stabilité d'un plasma : modélisation mathématique et simulation numérique.* Masson.

[205] Schötzau, D. (2004). Mixed finite element methods for stationary incompressible magneto-hydrodynamics. *Numer. Math.*, **96**, 771–800.

[206] Segatz, M. and Droste, C. (1994). Analysis of magnetohydrodynamic instabilities in aluminium reduction cells. *Light Metals*, 313–322.

[207] Sele, T. (1977). Instabilities of the metal surface in electrolytic cells. *Light Metals*, 7–24.

[208] Sermange, M. and Temam, R. (1983). Some mathematical questions related to the MHD equations. *Comm. Pure Appl. Math.*, **36**, 635–664.

[209] Sethian, J. A. and Smereka, P. (2003). Level set methods for fluid interfaces. *Annual Review of Fluid Mechanics*, **35**, 341–372.

[210] Simon, J. (1978). Ecoulement d'un fluide non homogène avec une densité initial s'annulant. *C. R. Acad. Sc.*, **A15**, 1009–1012.

[211] Simon, J. (1990). Non-homogeneous viscous incompressible fluids: existence of velocity, density and pressure. *SIAM J. Math. Anal.* **5**(21), 1093–117.

[212] Sneyd, A. D. (1985). Stability of fluid layers carrying a normal electric current. *J. Fluid Mech.*, **156**, 223–236.

[213] Sneyd, A. D. (1992). Interfacial instabilities in aluminium reduction cells. *J. Fluid Mech.*, **236**, 111–126.

[214] Sneyd, A. D. and Wang, A. (1994). Interfacial instability due to MHD mode coupling in aluminium reduction cells. *J. Fluid Mech.*, **263**, 343–359.

[215] Solonnikov, V. A. (1986). Solvability of the problem of evolution of an isolated volume of viscous, incompressible capillary fluid. *Jour. Sov. Math*, **32**, 223–228.

[216] Solonnikov, V. A. (1987). Evolution of an isolated volume of a viscous incompressible capillary fluid for large time values (in Russian). *Vestnik Leningrad Univ. Mat. Mekh. Astro.*, **3**, 49–55.

[217] Solonnikov, V. A. (1988*a*). On the transient motion of an isolated volume of viscous incompressible fluid. *Math. USSR Izvestiya* **31**(2), 381–405.

[218] Solonnikov, V. A. (1988*b*). Unsteady motion of a finite mass of fluid, bounded by a free surface. *Jour. Sov. Math.*, **40**, 672–686.

[219] Solonnikov, V. A. (1990). On nonstationary motion of a finite isolated mass of selfgravitating fluid. *Leningrad Math. J.* **1**(1), 227–276.

[220] Soulaïmani, A., Fortin, M., Dhatt, G., and Ouellet, Y. (1991). Finite element simulation of two- and three-dimensional free surface flows. *Comp. Meth. Appl. Mech. Engng*, **86**, 265–296.

[221] Soulaïmani, A. and Saad, Y. (1998). An arbitrary Lagrangian–Eulerian finite element method for solving three-dimensional free surface flows. *Comput. Meth. Appl. Mech. Engng.*, **162**, 79–106.

[222] Sun, H., Zikanov, O., and Ziegler, D. P. (2004). Non-linear two-dimensional model of melt flows and interface instability in aluminum reduction cells. *Fluid Dynamics Research* **35**(4), 255–274.

[223] Sussman, M., Smereka, P., and Osher, S. (1994). A level set approach for computing solutions to incompressible two-phase flow. *J. Comput. Phys.*, **114**, 146–159.

[224] Tanaka, N. (1993). Global existence of two-phase non-homogeneous viscous incompressible fluid flow. *Com. Par. Diff. Eq.* **18**(1-2), 41–81.

[225] Tani, A. (1991). Global existence of incompressible viscous capillary fluid flow in a field of external forces. *Lect. Notes in Num. Appl. Anal.*, **11**, 153–185.

[226] Tani, A. and Tanaka, N. (1995). Large-time existence of surface waves in incompressible viscous fluids with or without surface tension. *Arch. Rat. Mech. Anal.* **130**(4), 303–314.

[227] Temam, R. (1979). *Navier–Stokes Equations, Theory and Numerical Analysis.* North-Holland.

[228] Temam, R. (1988). *Infinite-dimensional dynamical systems in mechanics and physics.* Springer.

[229] Temam, R. (1995). *Navier–Stokes Equations and Nonlinear Functional Analysis* (2nd edn). CBMS-NSF Regional Conference Series in Applied Mathematics, SIAM.

[230] Tennekes, H. and Lumley, J. L. (1973). *A first course in turbulence. 2nd printing.* Cambridge, Mass. - London: The MIT Press. XII.

[231] Tezduyar, T. E., Behr, M., and Liou, J. (1992*a*). A new strategy for finite element computations involving moving boundaries and interfaces, the deforming-spatial-domain/space-time procedure: I. the concept and the preliminary numerical tests. *Computer Methods in Applied Mechanics and Engineering* **94**(3), 339–351.

[232] Tezduyar, T. E., Behr, M., and Liou, J. (1992*b*). A new strategy for finite element computations involving moving boundaries and interfaces, the deforming-spatial-domain/space-time procedure: II. Computation of free-surface flows, two-liquid flows, and flows with drifting cylinders. *Computer Methods in Applied Mechanics and Engineering* **94**(3), 353–371.

[233] Tobiska, L. and Verfürth, R. (1996). Analysis of a streamline diffusion finite element method for the Stokes and Navier–Stokes equations. *SIAM J. Numer. Anal.* **33**(1), 107–127.

[234] Tomasino, T., Le Hervet, M., Martin, O., and Lelièvre, T. (2006). Stability analysis of simplified electrolysis cells with MISTRAL. In *2006 TMS Annual Meeting and Exhibition, Light Metals.*

[235] Unverdi, S. O. and Tryggvason, G. (1992). Computations of multi-fluid flows. *Physica D*, **60**, 70–83.

[236] Urata, N. (1985). Magnetics and metal pad instability. *Light Metals*, 581–591.

[237] Urata, N., Mori, K., and Ikeuchi, H. (1976). Behavior of bath and molten metal in aluminum electrolytic cell. *J. Japan Inst. Light Metals* **26**(11), 573–583.

[238] Weatherburn, C. E. (1947). *Differential geometry of three dimensions*, Volume 1. Cambridge University Press.

[239] Zhu, J., Quartapelle, L., and Loula, A. F. D. (1996). Uncoupled variational formulation of a vector Poisson problem. *C. R. Acad. Sci. Paris, série 1*, **323**, 971–976.

[240] Ziegler, D. P. (1993). Stability of metal/electrolyte interface in Hall–Héroult cells: effect of the steady velocity. *Metal. Trans. B*, **24B**, 899–906.

[241] Zikanov, O., Sun, H., and Ziegler, D. P. (2004). Shallow water model of flows in Hall–Héroult cells. *Light Metals*, 445–451.

[242] Zikanov, O. and Thess, A. (2004). Direct numerical simulation as a tool for understanding MHD liquid metal turbulence. *Applied Mathematical Modelling* **28**(1), 1–13.

[243] Zikanov, O., Thess, A., Davidson, P. A., and Ziegler, D. P. (2000). A new approach to numerical simulation of melt flows and interface instability in Hall–Héroult cells. *Metal. Trans. B*, **B31**, 1541–1550.

INDEX

COPYRIGHTS

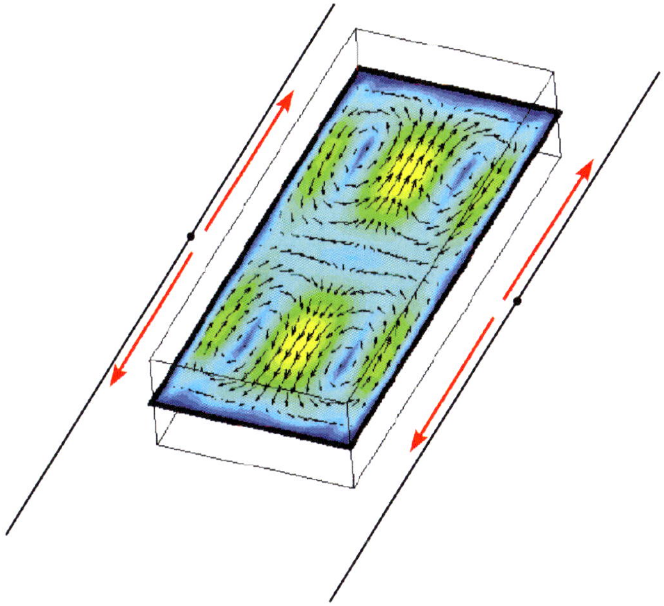

(i) Configuration with four external conductors (four symmetric vortices).

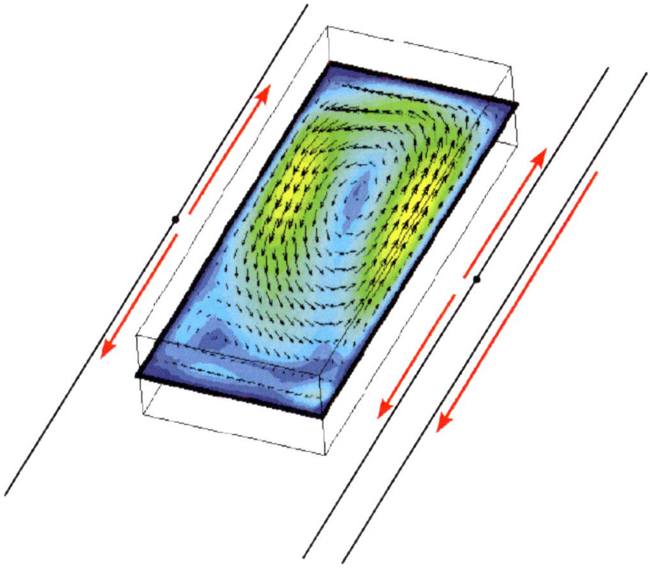

(ii) Configuration with five external conductors (asymmetric vortices).

PLATE 1 Simulations performed in the setting of Figure 3.9.

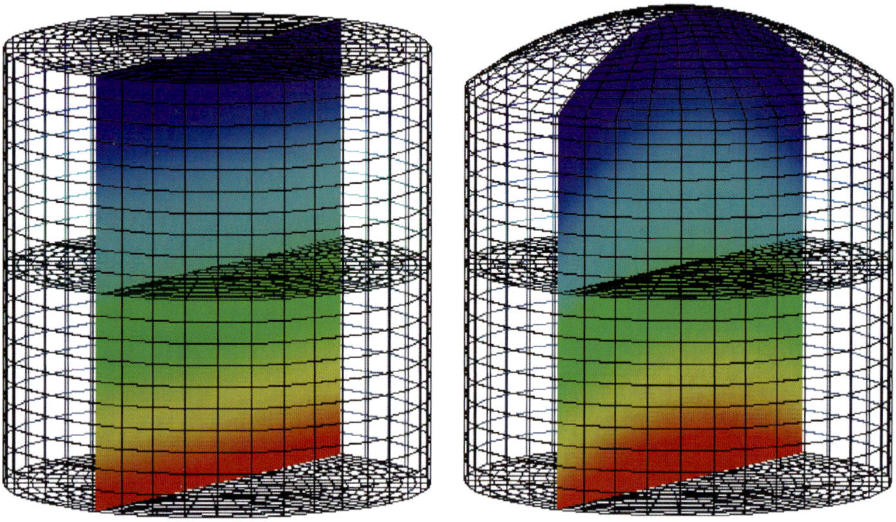

PLATE 2 Mesh and Isovalues of pressure on a portion of the mesh, in the presence only of gravity (left), and after the application of an electrical current (right).

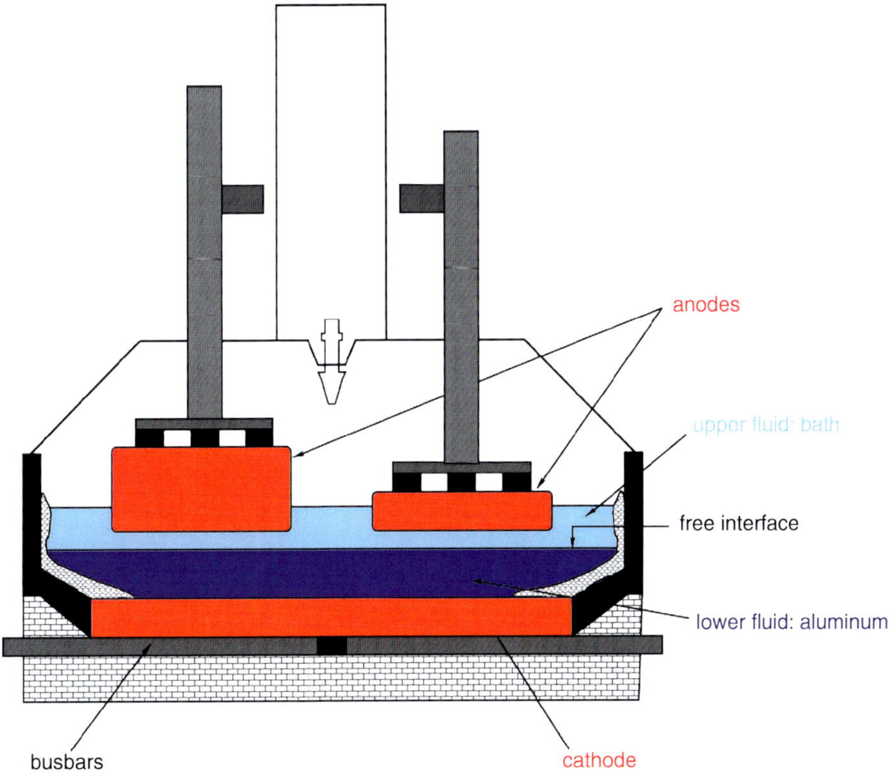

PLATE 3 Vertical cut of an aluminum cell.

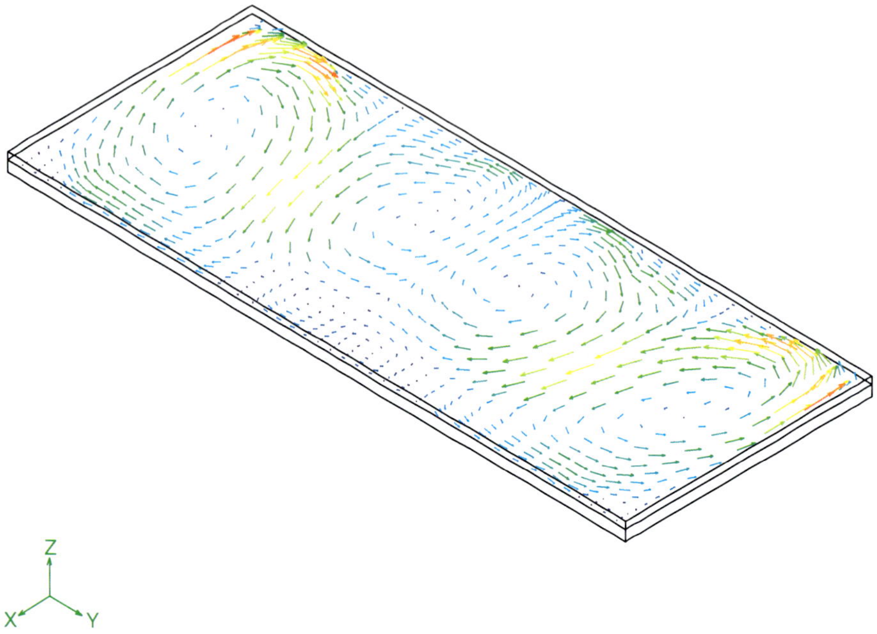

PLATE 4 Velocity profile in a plane situated 5 cm under the interface. Velocity magnitude is between 1 $\mathrm{cm\,s^{-1}}$ and 9 $\mathrm{cm\,s^{-1}}$.

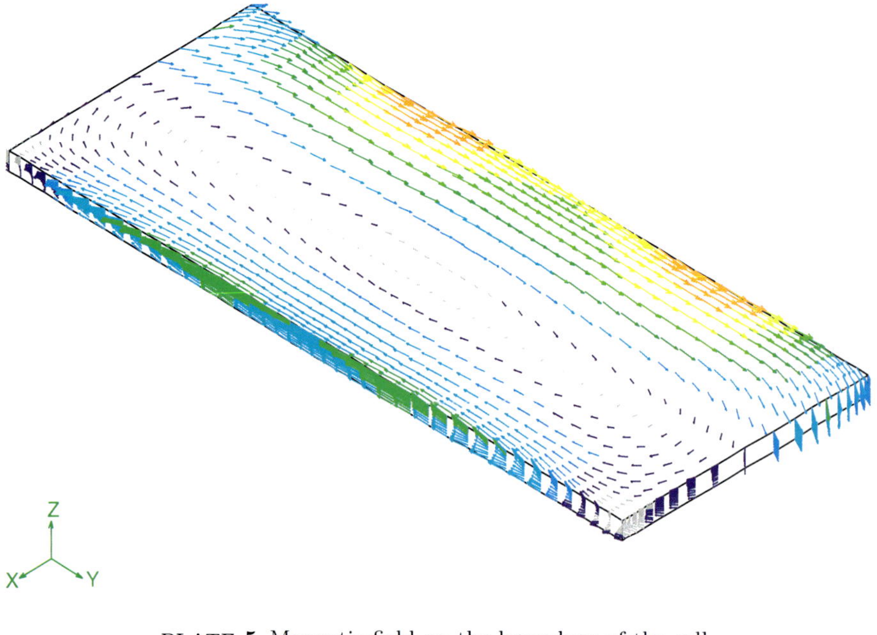

PLATE 5 Magnetic field on the boundary of the cell.